Origin of Life in Chemically Complex Messy Environments

Origin of Life in Chemically Complex Messy Environments

Editors

Ranajay Saha
Alberto Vázquez-Salazar

MDPI
Basel • Beijing • Wuhan • Barcelona • Belgrade • Novi Sad • Cluj • Manchester

Editors
Ranajay Saha
Chemical and Biomolecular Engineering
University of California Los Angeles
Los Angeles
United States

Alberto Vázquez-Salazar
Chemical and Biomolecular Engineering
University of California Los Angeles
Los Angeles
United States

Editorial Office
MDPI
St. Alban-Anlage 66
4052 Basel, Switzerland

This is a reprint of articles from the Special Issue published online in the open access journal *Life* (ISSN 2075-1729) (available at: www.mdpi.com/journal/life/special_issues/Origin_of_Life).

For citation purposes, cite each article independently as indicated on the article page online and as indicated below:

Lastname, A.A.; Lastname, B.B. Article Title. *Journal Name* **Year**, *Volume Number*, Page Range.

ISBN 978-3-7258-0178-7 (Hbk)
ISBN 978-3-7258-0177-0 (PDF)
doi.org/10.3390/books978-3-7258-0177-0

Contents

Preface

Unraveling the environmental conditions of early Earth, where life first emerged, is a challenge that has captivated scientists for decades. Traditionally, the vast majority of prebiotic chemistry research has consisted of experiments focused on the behavior of a restricted spectrum of organic molecules, usually studying the components of life separately, where the results are limited by the simulated scenario, also described as "clean and isolated" experiments. However, research on the origin of life needs to think big and outside the box in order to enable continuous progress in the field. Considering the prebiotic Earth four billion years ago (a messy atmosphere, in other words), a chaotic mélange of diverse starting materials appears realistic. As prebiotic chemists and origin-of-life researchers, we must modify our current approach and consider more chemical and geological scenarios in which both physical processes and driving forces towards primitive life formation are examined.

In this Special Issue, we present a collection of original research articles, reviews, communications, opinions, hypotheses, and concept papers that demonstrate and summarize advances related to the origin of life in various complex chemical and prebiotically feasible environments.

We are honored to have led this endeavor and are excited to invite you to join us in rethinking the origin of life, considering the prebiotically plausible and messy environmental conditions.

Ranajay Saha and Alberto Vázquez-Salazar
Editors

life

MDPI

Review

Origins of Life Research: The Conundrum between Laboratory and Field Simulations of Messy Environments

David Deamer

The Department of Biomolecular Engineering, University of California, Santa Cruz, CA 95064, USA; deamer@soe.ucsc.edu

Abstract: Most experimental results that guide research related to the origin of life are from laboratory simulations of the early Earth conditions. In the laboratory, emphasis is placed on the purity of reagents and carefully controlled conditions, so there is a natural tendency to reject impurities and lack of control. However, life did not originate in laboratory conditions; therefore, we should take into consideration multiple factors that are likely to have contributed to the environmental complexity of the early Earth. This essay describes eight physical and biophysical factors that spontaneously resolve aqueous dispersions of ionic and organic solutes mixed with mineral particles and thereby promote specific chemical reactions required for life to begin.

Keywords: origin of life; messy conditions; laboratory simulations

Citation: Deamer, D. Origins of Life Research: The Conundrum between Laboratory and Field Simulations of Messy Environments. *Life* **2022**, *12*, 1429. https://doi.org/10.3390/life12091429

Academic Editor: Paul Higgs

Received: 12 August 2022
Accepted: 6 September 2022
Published: 14 September 2022

1. Introduction

Virtually all experimental results that guide research related to the origin of life are from laboratory simulations of the early Earth conditions. How good are the simulations, and how confident can we be that the results are applicable to understanding the process by which life began 4 billion years ago in chemically complex and messy environments? One way to think about this question is to ask what we mean by "messy." In the laboratory, emphasis is placed on the purity of reagents and carefully controlled conditions, so there is a natural tendency to reject impurities and lack of control. The reason is obvious: scientists want their results to be repeatable by others.

Another meaning of messiness refers to environments that incorporate multiple biological and non-biological compounds [1–3]. An example is Bumpass Hell, a hydrothermal site on the Mount Lassen volcano in northern California (Figure 1). The image shows extensive deposits of clay, a common mineral in hydrothermal pools associated with volcanism. The pool is acidic, with pH 3, due to sulfur dioxide dissolved in the water, and the temperature is maintained at 91 °C by residual heat from the underlying magma that powered the eruption. It seems improbable that life could begin in such conditions, which, yet, are ubiquitous in volcanic regions of today's Earth and presumably on the prebiotic Earth 4 billion years ago.

That brings us to the question being addressed in this essay: How could orderly structures and functions emerge that led to the first forms of life? The answer must involve physical and chemical processes that occur even in messy conditions. We will refer to the result of such processes as resolution, meaning that relatively pure compounds are produced and then concentrated in such a way that they can react despite the complexity of their surroundings. However, it is helpful to realize that messiness can be divided into chemical and physical components, and that there are synergistic combinations in which certain physical conditions can promote specific chemical reactions. Each of the following sub-topics are examples of such interactions.

Figure 1. A messy, complex environment: Bumpass Hell on Mount Lassen, CA, USA. Credit: author.

2. Physical Self-Assembly Processes Tend to Resolve Complex Mixtures

A number of concerns arise when a chemist considers a natural environment in comparison to a laboratory one. Some of these are related to the composition of a mixture, while others come from physical variations such as temperature, pH, ionic solutes, adsorption to mineral surfaces and concentration, all of which are controlled in the laboratory. Let us begin with a few examples of such concerns:

1. Desired reactions cannot occur if a solution is composed of so many solutes that the reactants are diluted.
2. Side reactions interfere with yields if other solutes are present as potential reactants.
3. If the temperature is too low, there will be insufficient activation energy, while too high temperatures will result in what a chemist refers to as tar [4,5]. Rather than controlled condensation reactions that synthesize specific bonds such as esters and peptides, elevated temperatures and high activation energy drive the formation of multiple, random bonds that produce the intractable polymers called asphalt and tar.
4. Extreme pH ranges markedly affect the kinds of reactions that can occur and their yields.
5. High salt concentrations, particularly, of divalent cations such as calcium, precipitate anions that are essential for life. Calcium, for instance, reacts with phosphate to form the insoluble mineral apatite.

How can we address these concerns? The approach described here is to realize that certain physical properties of complex mixtures can add order to a disorderly system, thereby promoting specific chemical reactions. For example, hydrothermal vents have been proposed as a site conducive to the origin of life [6], and the hot seawater flowing through alkaline vents is a highly complex mixture with available free energy and reducing power [7,8]. The effluent is produced by a serpentinization reaction, and ionic solutes precipitate as mineral crystals when the hot fluid encounters the cooler seawater bathing the vents. The minerals are mostly calcium carbonate, which illustrates an important point. Ionic compounds in a solution are disordered and in a high entropy state, but when a hot, alkaline solution is mixed with cool seawater the solubility limits of calcium and carbonate are exceeded, and crystallization begins. Crystals are the most ordered state of ionic solutes; therefore, simply mixing a vent fluid with seawater leads to a vast reduction of the entropy of the system when relatively pure calcium carbonate crystals precipitate.

3. Three Conditions Are Key to Resolving Messy Reaction Mixtures

Here, we will focus on three conditions that are ubiquitous in the laboratory but seldom invoked as solutions to deal with messy conditions on the early Earth. The conditions are related to concentration, water activity and temperature. For instance, when organic chemists synthesize an ester, they will mix pure, highly concentrated solutions of organic acids and alcohols rather than dilute solutions in an aqueous phase. The obvious reason is that water is a product of the condensation reaction that forms ester bonds, so the highest yields require that the water activity is low. The chemist will also heat the reaction mixture to provide activation energy that increases the rate of the reaction.

An important question is how linking bonds could be synthesized on the prebiotic Earth to provide the more complex molecules required for life to begin. Examples that have been explored in the laboratory include polyesters [9], peptides [10] and depsipeptides [11]. Taking a lesson from organic chemistry, the potential reactants must be concentrated, dry and at an elevated temperature. There is only one natural environment that meets these requirements: hydrothermal pools on subaerial volcanic land masses [12]. Precipitation provides fresh water distilled from a salty ocean, and the water in hydrothermal pools undergoes continuous cycles of evaporation and rewetting. If organic solutes are in the water, they will become extremely concentrated films on mineral surfaces during the dry phase of a cycle and then redissolve in the wet phase. Finally, the temperature is elevated, but typically less than the 100 °C boiling point of water because of the lower atmospheric pressure at higher altitudes. An important point is that even though a variety of solutes might be present in a messy solution, only those solutes capable of self-assembly or forming linking bonds will serve as reactants for polymerization, while the other solutes will be inert. Given the significance of concentration, we can now consider processes that can increase the local concentration of potential reactants.

4. Polar vs. Non-Polar: Oils and Monolayers at the Air–Water Interface

We tend to focus on water-soluble polar and ionized solutes when we think about constituents of messy mixtures, but we should also consider non-polar compounds such as aliphatic and aromatic hydrocarbons and their derivatives. One of the simplest resolutions in a messy environment is the fact that oils are relatively insoluble in water and are also less dense. Hydrocarbon oils therefore tend to separate spontaneously from other solutes and float on water surfaces. Furthermore, some of the hydrocarbons present in unrefined oil are likely to be amphiphilic compounds that have both polar and non-polar moieties in their structure. Amphiphiles are classified as surfactants because they spontaneously accumulate at air–water interfaces as monomolecular films and reduce the surface tension as a result.

Fatty acids are common amphiphiles that take their name from the fact that biological fat is a triglyceride with three fatty acids attached to a glycerol molecule by ester bonds. When the fatty acids are released from fat by hydrolysis of the ester bonds in alkaline solutions, the result is a natural detergent called soap. Palmitic acid is a 16carbon non-polar hydrocarbon chain with the terminal carbon in the form of a hydrophilic carboxyl group:

$$CH_3(CH_2)_{14}\text{-}COOH \rightleftharpoons CH_3(CH_2)_{14}\text{-}COO^- + H^+ \quad (1)$$

It is an organic acid because the terminal carboxy group can either be protonated (-COOH) or become anionic (-COO$^-$) when it releases the proton in an aqueous solution.

If a crystal of palmitic acid is placed on the surface of water in a beaker, not much will happen, but if it is warmed to approximately 63 °C, the crystal will melt and suddenly spread to form a monomolecular layer that fills the available surface area. At the molecular level, the hydrophilic carboxyl head group is interacting with the aqueous phase, while the hydrophobic hydrocarbon chains are standing vertically, each occupying around 0.2 nm^2 of surface area. A measurement of surface tension will show an immediate decrease from 72 mN/m of water to approximately 57 mN/m. The difference between these two values—15 mN/m—is referred to as the surface pressure.

Now we can consider an example demonstrating how the surfactant effect can resolve a complex system. The Murchison meteorite is a mixture of silicate mineral grains with approximately 0.1% by weight of soluble organic compounds that include monocarboxylic acids, amino acids, alcohols, polycyclic aromatics such as pyrene and fluoranthene and even some purines such as adenine. Figure 2 shows a result reported by Mautner et al. [13] in which 20 mg of Murchison powder in 2 mL aqueous buffer was heated briefly to 100 °C at low and high pH ranges. The initial surface pressure was near zero mN/m at room temperature but increased dramatically to 8.1 mN/m when the powder was heated at pH 2, then increased further to 10.3 mN/m when heated at pH 11. The heat allowed small amounts of surface-active compounds to escape from the mineral powder and form a monolayer at the air–water interface. This represents a purification of the amphiphilic compounds by simply heating the original mixture in an aqueous solution. When the surface was cleaned, the surface pressure decreased but then rapidly increased as more surfactant molecules migrated to the interface and formed a monolayer.

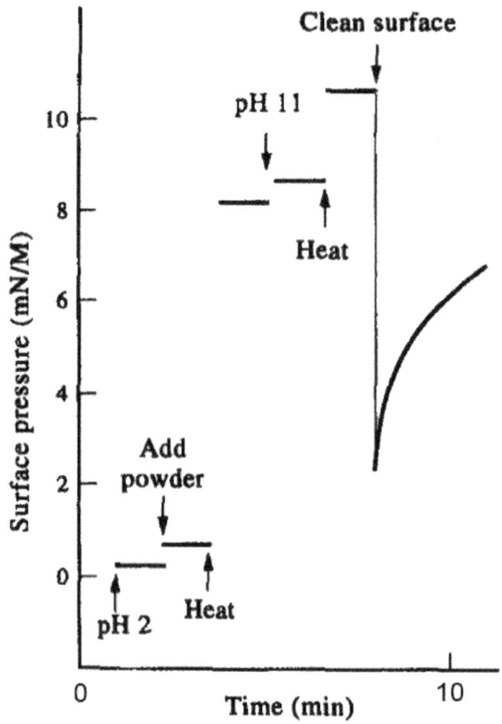

Figure 2. Monolayer assembly of amphiphilic compounds in the Murchison meteorite [13].

Why might the assembly of surfactant monolayers and oils at air–water interfaces be significant for the origin of life? The reason is that ultraviolet light was likely to be an important energy source driving photochemical reactions, and polycyclic aromatic hydrocarbons (PAH) are pigments that are elevated to excited states when they absorb UV photons. Oils and amphiphiles that accumulate on aqueous surfaces would be exposed to UV light, and a variety of photochemical reactions would then occur. It has been proposed that such reactions could have served as a primitive version of photosynthesis involving electron transfer from a PAH such as pyrene to an acceptor such as benzophenone [14].

5. Bilayer Membranes Form Compartments by the Self-Assembly of Amphiphilic Compounds

Figure 3 illustrates how an increasing concentration can drive the self-assembly of amphiphilic compounds into membranous compartments. The amphiphile is decanoic acid, a monocarboxylic acid with a 9-carbon hydrocarbon chain and a carboxyl head group. In a dilute solution, the acid is present as individual molecules dissolved in an aqueous phase (A). If the concentration increases, for instance by evaporation, at some point the critical micelle concentration (CMC) is exceeded, and micelles begin to form (B). When the critical vesicle concentration (CVC) is exceeded, the micelles fuse into vesicles bounded by bilayers (C). At the highest concentration near dryness, the vesicles fuse into a multilamellar matrix (D), but then reassemble into vesicles upon rehydration (E).

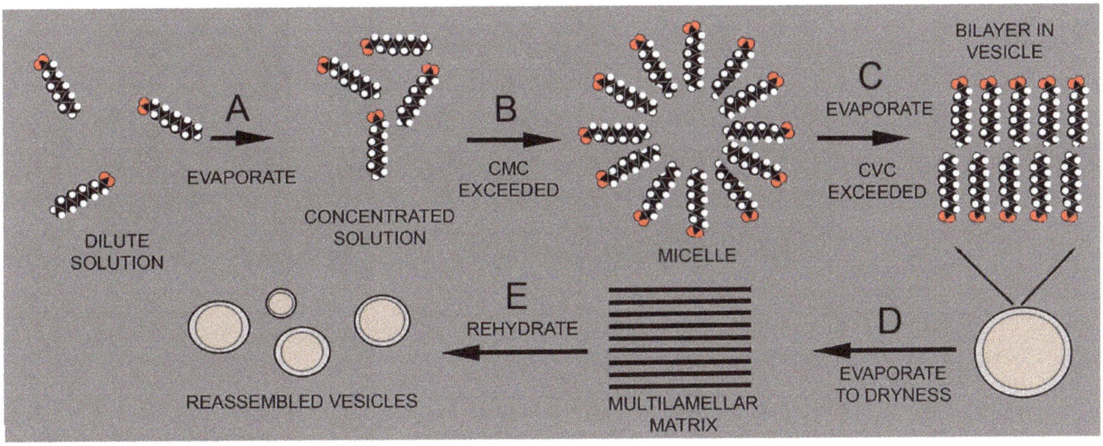

Figure 3. Self-assembly processes in solutions of amphiphilic molecules such as fatty acids.

This is another example of orderly structures emerging from disorderly, messy conditions. Membrane-bounded compartments were essential prerequisites for the origin of cellular life, and it seems likely that micelles and vesicles were abundant in the prebiotic environment [15]. Assuming that various polymers were also being synthesized, the microscopic membranous compartments would be able to encapsulate random peptides and oligonucleotides to generate protocells. Although protocells are not alive in the usual sense, they represent an essential step toward life. Each protocell differs in composition from all the rest and represents a microscopic experiment. Most are inert, and their components would be recycled, but a few might happen to have properties that allowed them to survive stresses imposed by the environment. Populations of protocells would therefore undergo the first stages of selection and evolution [12,16].

6. Adsorption as an Organizing Factor

Many physical and chemical processes cannot occur in the absence of a threshold concentration; so, a physical process that specifically increases the local concentration of a potential reactant can act as a catalyst. Clays are aluminum silicate minerals that have an enormous capacity to adsorb polar and ionic solutes. As shown in Figure 4, they are microscopic particles that are often present as layered structures called smectic phases.

Various clays have surface areas ranging from 100 to 400 square meters per gram to which solutes can bind. James Ferris and co-workers tested whether clay surfaces could promote the polymerization of activated nucleotide monomers. In their initial studies, they reported that when the nucleotides saturated the surface of montmorillonite clay, oligomers ranging up to 10 nucleotides could be detected [17]. Later research demonstrated chain lengths as long as 30–50 nucleotides [18].

Figure 4. Scanning electron micrograph of clay crystals. The bar corresponds to 10 μm, equivalent to the diameter of a human red blood cell. Photograph courtesy of Evelyne Delbos, the James Hutton Institute.

A chemist accustomed to reactions in aqueous solutions might be disappointed by the lack of polymerization in a solution of activated nucleotides, but then would be surprised to see polymers appear when the solution becomes more complex and "messier" with the addition of clay. The catalytic effect of clay minerals is an example of why messy conditions should be explored more extensively. There may be other surprises in store.

7. Eutectic Phases

The most common eutectic phase occurs when an aqueous solution freezes. As ice crystals form, they tend to exclude solutes, which become highly concentrated films between the crystals. The concentrating effect is similar to evaporation because reactions can occur that cannot proceed in a dilute, disordered solution. The power of eutectic phases to promote reactions was demonstrated by Monnard et al. [19] who froze complex mixtures of nucleotides at −18 °C. The nucleotides were chemically activated as imidazole esters and spontaneously formed ester bonds to polymerize into strands 5 to 14 nucleotides in length. Attwater et al. [20] also tested freezing temperatures as a way to promote the non-enzymatic polymerization of RNA. They used in vitro selection to evolve a cold-adapted ribozyme that was capable of catalyzing the template-directed synthesis of a 206-nucleotide RNA sequence. The reaction was markedly promoted by a cycle of freezing that concentrated the reactants in a eutectic phase.

Bada et al. [21] proposed that the early Earth may have passed through a "snowball" phase related to decreased solar luminosity at the time. They speculated that freeze–thaw cycles related to bolide impacts may have promoted reactions that synthesized organic compounds required for life's origin. Eutectic phases associated with freezing would have played a role by concentrating reactants such as HCN, NH_3 and aldehydes that would then undergo Strecker synthesis to form amino acids. Although this idea remains speculative, it does illustrate how another physical process, freezing and thawing in this case, has the potential to resolve specific reactions in complex conditions.

8. Duplex Structures of Nucleic Acids—Stacking and Base Pairing

Although three-dimensional crystals are common, and crystallization often serves to purify specific compounds from complex mixtures, the assembly of compounds into one-dimensional linear crystals is less familiar. We tend to take it for granted that the double helix of DNA is primarily stabilized by Watson–Crick base pairing between the two strands, but in fact stacking of the base pairs is even more important in terms of stabilization. Guanine monophosphate (GMP) quadruplexes are examples of spontaneous stacking that produces a linear crystal [22]. Figure 5 shows how four guanines form a quadruplex stabilized by eight hydrogen bonds indicated as dashed lines.

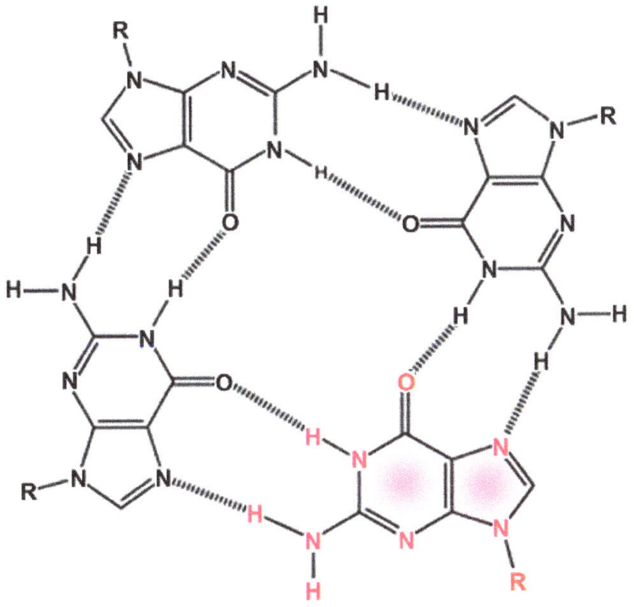

Figure 5. Guanine monophosphate assembles into a quadruplex structure stabilized by hydrogen bonding, indicated by dashed lines. The R groups represent the ribose phosphates linked to the guanine by N-glycoside bonds. One of the guanines is indicated by red fonts.

Atomic force microscopy can be used to visualize the linear crystals produced when a solution of GMP and cytidine monophosphate (CMP) is concentrated by evaporation at room temperature on a freshly cleaved mica sheet (Figure 6A). The GMP stacks into long, linear crystals that exclude the CMP. The linear crystals are adsorbed to three axes of the mica crystal, which accounts for their obvious alignment along three axes at 60°. Significantly, if the solution of GMP and CMP is evaporated on mica at 80 °C to provide activation energy, long polymeric strands emerge from the mixture [23], presumably composed of GMP and CMP linked by ester bond synthesis (Figure 6B).

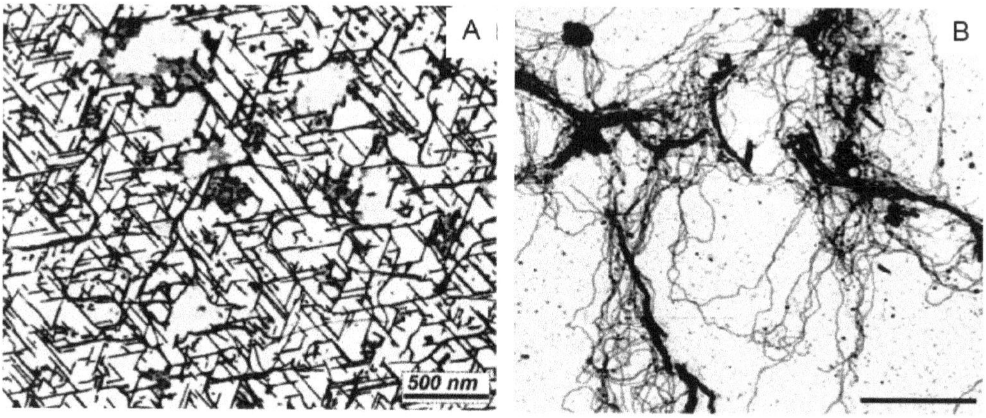

Figure 6. Linear strands of stacked guanosine monophosphate quadruplexes are shown in (**A**). The strands assemble when a mixture of GMP and CMP is dried on a freshly cleaved mica surface at room temperature, but CMP is excluded when the GMP forms quadruplexes aligned with the crystalline surface of the mica substrate. A different result occurs if the mixture is exposed to three wet–dry cycles at 80 °C (**B**). Instead of linear crystals, long strands of apparent polymers emerge, presumably composed of GMP and CMP linked by ester bonds [23].

This is an example of how two physical processes can drive a significant chemical reaction. The first process is evaporation that concentrates potential reactants on a mica surface so that one of the reactants forms quadruplexes and linear crystals. The second process is heating the mixture to provide activation energy that drives a condensation reaction leading to polymerization. These are conditions that are rarely used in the laboratory but common in hydrothermal sites associated with volcanism.

9. Thermodynamics, Kinetics and Wet–Dry Cycles

An important test of any chemical reaction proposed to be relevant to prebiotic chemistry is whether it is feasible in terms of thermodynamic principles [24]. A simplified description of the primary thermodynamic principle is that the free energy (ΔG) available to drive a reaction is a function of the change in enthalpy (ΔH) and entropy (ΔS) terms: $\Delta G = \Delta H - T\Delta S$. If heat can be released by a reaction, and disorder increases, the Gibbs free energy (ΔG) is negative, and the reaction is spontaneous.

If one visits volcanic regions on today's Earth, a striking feature is that precipitation constantly supplies water to hydrothermal sites. A well-known example is Yellowstone National Park, but similar sites can be found in Kamchatka, Russia, New Zealand and Iceland. We refer to these as prebiotic analogues because similar sites would have been abundant on the early Earth 4 billion years ago before life began. A characteristic of hydrothermal sites is that the pools and hot springs undergo continuous cycles of evaporation and rewetting. There are two important properties of the wet–dry cycles that go beyond traditional laboratory methods. First, as a solution of potential reactants is concentrated to dryness, it makes chemical free energy available through the favorable changes in enthalpy and entropy described earlier. The enthalpy change is due to the reduction in water activity which draws the equilibrium of a condensation reaction to the right. The entropy change is favorable because as the reactants in a dilute solution become concentrated, they have an increased probability of interacting and losing water to the environment.

Finally, in the laboratory a typical reaction undergoes a single cycle in which reactions proceed downhill to products. However, multiple cycles available in natural conditions can take advantage of kinetic traps in which a forward reaction can be fast, but a back reaction is slow. For instance, the synthesis of ester bonds during the dry phase of a cycle can be very fast, but the spontaneous hydrolysis of polymers is slow. As a result, polymers accumulate in a steady state away from the thermodynamic equilibrium [24,25].

10. Stepping Out of the Laboratory into the Wild

A given reaction can work under carefully controlled laboratory conditions using pure reagents, buffered solutions at a specific pH and ionic solutes such as Mg^{2+} that may be required, but how valid is the assumption that it would have also been able to proceed on the prebiotic Earth? There is one way to become more confident of this assumption, which is to test whether a significant experiment also works in a prebiotic analogue site on today's Earth. If the reaction is sufficiently robust, it will be reproduced even though such analogue sites are, by definition, messy and complex.

A few preliminary studies have been undertaken, and it is worth describing them here so that others might accept the challenge. Given that liquid water is essential for the origin of life, we should first note that on the Earth today there are two types of water that have been proposed as sites for the origin of life. Over 99% of the Earth's water is salty seawater with 580 mM NaCl, 53 mM $MgCl_2$ and 10 mM $CaCl_2$; therefore, a common assumption is that life must have begun in the ocean. The assumption was supported when hydrothermal vents called "black smokers" were discovered, in which abundant microbial life used the reducing power of hydrogen sulfide (H_2S) as a source of chemical energy [7]. The second type consists of the alkaline vents described earlier that emit a strongly alkaline solution containing up to 10 mM hydrogen gas in solution, which may serve as a source of reducing power [26].

The other 1% of the Earth's water is distilled from salty seawater and falls as precipitation on subaerial continental land masses and volcanic islands like Hawaii and Iceland. Today, most of the freshwater is in the form of ice covering Antarctica and Greenland, but 4 billion years ago such extensive ice would have been absent when the global temperatures were estimated to be 55–80 °C. However, active volcanoes were likely to have been emerging from the ocean, and distilled water falling as precipitation would become incorporated into hydrothermal hot spring sites resembling those we see today.

Freshwater hot spring sites are an alternative to salty seawater for several reasons:

1. The concentration of soluble cations is very low, in the range of a few millimolar [27].
2. Because of sulfur dioxide emissions that dissolve in the water, the solution is in the acidic range, with pH 2–4.
3. Cycles of evaporation and rewetting serve to concentrate potential reactants as films on mineral surfaces.

Hydrothermal fields such as those illustrated in Figure 1 are accessible prebiotic analogue sites which can be used to test the assumption that laboratory simulations represent processes and reactions expected to occur 4 billion years ago. A few such tests have been performed. For instance, Milshteyn et al. [28] asked whether membrane-bounded vesicles could assemble in hot spring conditions. Water samples were taken from Yellowstone hot spring pools at acidic and neutral pH ranges. Because contemporary membrane lipids such as phospholipid and cholesterol are products of metabolism, neither would have been available on the prebiotic Earth. Therefore a 1:1 mole ratio mixture of 12-carbon lauric acid and its monoglyceride was used to model the kinds of amphiphiles likely to be present. It was observed that vesicles readily formed in the hot spring water but not in seawater (Figure 7) because divalent cations of Ca^{2+} and Mg^{2+} caused the fatty acid to crystallize.

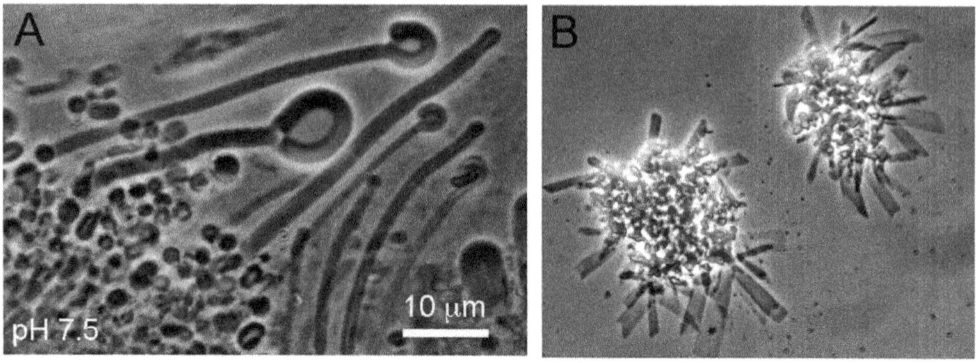

Figure 7. Membranous compartments assemble from a mixture of lauric acid and its monoglyceride exposed to hot spring water at pH 7.5 (**A**), but not in seawater at pH 8.1 (**B**). Image from Milshteyn et al. [28].

Similar experiments were performed by Joshi et al. [29] using a mixture of 40 mM decanoic acid and 20 mM monoglyceride exposed to hot spring water samples from the Puga region of Ladahk, India. Vesicles ranging from 5 to 10 μm in diameter were abundant, so it seems reasonable to assume that membranous compartments were present in freshwater hot springs on volcanic land masses emerging from the global ocean 4 billion years ago.

11. Conclusions

It is worth keeping in mind that a living cell is the messiest environment we can imagine, yet somehow, the biochemical reactions of metabolism proceed smoothly despite the incredible complexity of the cytoplasm. This is possible because enzymes catalyze specific reactions that allow those reactions to follow metabolic pathways established by several billion years of evolution. A second reason is that biophysical forces arrange structures within cells to optimize those functions. A prominent example is the self-assembly of biological membranes involved in electron transport and ATP synthesis. Although the kinds of lipids composing the membranes are under genetic control, the assembly process is not.

Another point is that the reaction of a pure, well-defined single reactant can generate what most researchers would call messy products. A classic example is the formose reaction in which hundreds of compounds are produced when formaldehyde (CHCO) reacts with itself in alkaline solutions. Another is the spontaneous polymerization of alpha hydroxy acids [9]. For instance, glycolic acid exposed to drying at 80 °C forms oligomers up to 12 subunits in length, with most between 3 to 6.

Although it is generally a good idea to use pure reagents and controlled conditions in the laboratory, it is also useful to realize that if we limit ourselves to the laboratory, we might miss an important ingredient or physical process. Which laboratory simulations of reactions related to the origin of life could be expected to proceed in prebiotic analogue conditions? This is a challenge to origins of life research because few investigators have been bold enough to perform such experiments in the field. The factors described here—concentration, crystallization, self-assembly of monolayers and bilayers, adsorption, eutectic phases, stacking and base pairing—can all occur in prebiotic environments and have the potential to promote desired reactions even in messy, complex conditions. Some of them, either singly or in combination, may provide significant clues to understand prebiotic chemistry that are not obvious in laboratory simulations.

Funding: This research received no external funding.

Acknowledgments: The author thanks Kathleen Campbell and her students and staff for organizing field work in New Zealand related to this essay and Bruce Damer for his collaborative efforts in the ongoing research. The author greatly appreciated the assistance of the UNSW graduate student Luke Steller in obtaining samples and of Bryan Drake for transportation to and from the Hell's Gate site.

Conflicts of Interest: The author declares no conflict of interest.

References

1. Schmitt-Kopplin, P.; Gabelica, Z.; Gougeon, R.D.; Fekete, A.; Kanawati, B.; Harir, M.; Gebefuegi, I.; Eckel, G.; Hertkorn, N. High molecular diversity of extraterrestrial organic matter in Murchison meteorite revealed 40 years after its fall. *Proc. Natl. Acad. Sci. USA* **2010**, *107*, 2763–2768. [CrossRef] [PubMed]
2. Guttenberg, N.; Virgo, N.; Chandru, K.; Scharf, C.; Mamajanov, I. Bulk measurements of messy chemistries are needed for a theory of the of life. *Philos. Trans. R. Soc. London. Ser. A Math. Phys. Eng. Sci.* **2017**, *375*, 20160347. [CrossRef] [PubMed]
3. Wołos, A.; Roszak, R.; Żądło-Dobrowolska, A.; Beker, W.; Mikulak-Klucznik, B.; Spólnik, G.; Dygas, M.; Szymkuć, S.; Grzybowski, B.A. Synthetic Connectivity, Emergence, and Self-Regeneration in the Network of Prebiotic Chemistry. *Science* **2020**, *369*, eaaw1955. [CrossRef] [PubMed]
4. Schwartz, A.W. Intractable Mixtures and the Origin of Life. *Chem. Biodivers.* **2007**, *4*, 656–664. [CrossRef]
5. Benner, S.A.; Kim, H.-J.; Carrigan, M.A. Asphalt, Water, and the Prebiotic Synthesis of Ribose, Ribonucleosides, and RNA. *Acc. Chem. Res.* **2012**, *45*, 2025–2034. [CrossRef]
6. Corliss, J.B.; Baross, J.A.; Hoffman, S.E. An hypothesis concerning the relationship between submarine hot springs and the origin of life on Earth. *Oceanol Acta* **1981**, *4*, 59–69.
7. Martin, W.; Russell, M.J. On the origins of cells: A hypothesis for the evolutionary transitions from abiotic geochemistry to chemoautotrophic prokaryotes, and from prokaryotes to nucleated cells. *Philos. Trans. R. Soc. B Biol. Sci.* **2003**, *358*, 59–85. [CrossRef]
8. Hudson, R.; de Graaf, R.; Rodin, M.S.; Ohno, A.; Lane, N.; McGlynn, S.E.; Yamada, Y.M.A.; Nakamura, R.; Barge, L.M.; Braun, D.; et al. CO_2 reduction driven by a pH gradient. *Proc. Natl. Acad. Sci. USA* **2020**, *117*, 22873–22879. [CrossRef]
9. Chandru, K.; Guttenberg, N.; Giri, C.; Hongo, Y.; Butch, C.; Mamajanov, I.; Cleaves, H.J. Simple prebiotic synthesis of high diversity dynamic combinatorial polyester libraries. *Commun. Chem.* **2018**, *1*, 30. [CrossRef]
10. Mayer, C. Spontaneous Formation of Functional Structures in Messy Environments. *Life* **2022**, *12*, 720. [CrossRef]
11. Fialho, D.M.; Karunakaran, S.C.; Greeson, K.W.; Martínez, I.; Schuster, G.B.; Krishnamurthy, R.; Hud, N.V. Depsipeptide Nucleic Acids: Prebiotic Formation, Oligomerization, and Self-Assembly of a New Proto-Nucleic Acid Candidate. *J. Am. Chem. Soc.* **2021**, *143*, 13525–13537. [CrossRef] [PubMed]
12. Damer, B.; Deamer, D. The Hot Spring Hypothesis for an Origin of Life. *Astrobiology* **2020**, *20*, 429–452. [CrossRef]
13. Mautner, M.N.; Leonard, R.L.; Deamer, D.W. Meteorite organics in planetary environments: Hydrothermal release, surface activity, and microbial utilization. *Planet. Space Sci.* **1995**, *43*, 139–147. [CrossRef]
14. Escabi-Perez, J.R.; Romero, A.; Lukac, S.; Fendler, J.H. Aspects of artificial photosynthesis. Photoionization and electron transfer in dihexadecyl phosphate vesicles. *J. Am. Chem. Soc.* **1979**, *101*, 2231–2233. [CrossRef]
15. Walde, P. Surfactant Assemblies and their Various Possible Roles for the Origin(S) of Life. *Orig. Life Evol. Biosph.* **2006**, *36*, 109–150. [CrossRef] [PubMed]
16. Sarkar, S.; Das, S.; Dagar, S.; Joshi, M.P.; Mungi, C.V.; Sawant, A.A.; Patki, G.M.; Rajamani, S. Prebiological Membranes and Their Role in the Emergence of Early Cellular Life. *J. Membr. Biol.* **2020**, *253*, 589–608. [CrossRef]
17. Ferris, J.P.; Ertem, G. Oligomerization of Ribonucleotides on Montmorillonite: Reaction of the 5′-Phosphorimidazolide of Adenosine. *Science* **1992**, *257*, 1387–1389. [CrossRef]
18. Ferris, J.P. Montmorillonite-catalysed formation of RNA oligomers: The possible role of catalysis in the origins of life. *Philos. Trans. R. Soc. B Biol. Sci.* **2006**, *361*, 1777–1786. [CrossRef]
19. Monnard, P.-A.; Kanavarioti, A.; Deamer, D.W. Eutectic Phase Polymerization of Activated Ribonucleotide Mixtures Yields Quasi-Equimolar Incorporation of Purine and Pyrimidine Nucleobases. *J. Am. Chem. Soc.* **2003**, *125*, 13734–13740. [CrossRef]
20. Attwater, J.; Wochner, A.; Holliger, P. In-ice evolution of RNA polymerase ribozyme activity. *Nat. Chem.* **2013**, *5*, 1011–1018. [CrossRef]
21. Bada, J.L.; Bigham, C.; Miller, S.L. Impact melting of frozen oceans on the early Earth: Implications for the origin of life. *Proc. Natl. Acad. Sci. USA* **1994**, *91*, 1248–1250. [CrossRef] [PubMed]
22. Davis, J.T. G-Quartets 40 years later: From 5′-GMP to molecular biology and supramolecular chemistry. *Angew. Chem.* **2004**, *43*, 668–698. [CrossRef] [PubMed]
23. Hassenkam, T.; Deamer, D. Visualizing RNA polymers produced by hot wet-dry cycling. *Sci. Rep.* **2022**, *12*, 1–11. [CrossRef] [PubMed]
24. Ross, D.S.; Deamer, D. Dry/Wet Cycling and the Thermodynamics and Kinetics of Prebiotic Polymer Synthesis. *Life* **2016**, *6*, 28. [CrossRef]

25. Higgs, P.G. The Effect of Limited Diffusion and Wet–Dry Cycling on Reversible Polymerization Reactions: Implications for Prebiotic Synthesis of Nucleic Acids. *Life* **2016**, *6*, 24. [CrossRef]
26. Sojo, V.; Herschy, B.; Whicher, A.; Camprubí, E.; Lane, N. The Origin of Life in Alkaline Hydrothermal Vents. *Astrobiology* **2016**, *16*, 181–197. [CrossRef]
27. Deamer, D.; Damer, B.; Kompanichenko, V. Hydrothermal Chemistry and the Origin of Cellular Life. *Astrobiology* **2019**, *19*, 1523–1537. [CrossRef]
28. Milshteyn, D.; Damer, B.; Havig, J.R.; Deamer, D. Amphiphilic compounds assemble into membranous vesicles in hydro-thermal hot spring water but not in seawater. *Life* **2018**, *8*, 11. [CrossRef]
29. Joshi, M.P.; Samanta, A.; Tripathy, G.R.; Rajamani, S. Formation and Stability of Prebiotically Relevant Vesicular Systems in Terrestrial Geothermal Environments. *Life* **2017**, *7*, 51. [CrossRef]

life

MDPI

Hypothesis

How Did Life Emerge in Chemically Complex Messy Environments?

Kenji Ikehara [1,2]

[1] G&L Kyosei Institute, The Keihanna Academy of Science and Culture (KASC), Keihanna Interaction Plaza, Lab. Wing 3F, 1-7 Hikaridai, Seika-cho, Souraku, Kyoto 619-0237, Japan; ikehara@cc.nara-wu.ac.jp; Tel.: +81-774-73-4478

[2] International Institute for Advanced Studies, Kizugawadai 9-3, Kizugawa, Kyoto 619-0225, Japan

Abstract: One of the problems that make it difficult to solve the mystery of the origin of life is determining how life emerged in chemically complex messy environments on primitive Earth. In this article, the "chemically complex messy environments" that are focused on are a mixed state of various organic compounds produced via prebiotic means and accumulated on primitive earth. The five factors described below are thought to have contributed to opening the way for the emergence of life: (1) A characteristic inherent in [GADV]-amino acids, which are easily produced via prebiotic means. [GADV] stands for four amino acids, Gly [G], Ala [A], Asp [D] and Val [V], which are indicated by a one-letter symbol. (2) The protein 0th-order structure or a [GADV]-amino acid composition generating water-soluble globular protein with some flexibility, which can be produced even by the random joining of [GADV]-amino acids. (3) The formation of versatile [GADV]-microspheres, which can grow, divide and proliferate even without a genetic system, was the emergence of proto-life. (4) The [GADV]-microspheres with a higher proliferation ability than others were able to be selected. Proto-Darwin evolution made it possible to proceed forward to the creation of a core life system composed of the $(GNC)_n$ gene, anticodon stem-loop tRNA or AntiC-SL tRNA (GNC genetic code), and [GADV]-protein. (5) Eventually, the first genuine life with a core life system emerged. Thus, the formation processes of [GADV]-protein and the $(GNC)_n$ gene in chemically complex messy environments were the steps to the emergence of genuine life.

Keywords: GADV hypothesis; origin of life; protein 0th-order structure; origin of protein; [GADV]-microsphere; origin of gene: the core life system

Citation: Ikehara, K. How Did Life Emerge in Chemically Complex Messy Environments? *Life* **2022**, *12*, 1319. https://doi.org/10.3390/life12091319

Academic Editor: Bruce Damer

Received: 27 June 2022
Accepted: 25 August 2022
Published: 26 August 2022

1. Introduction

Human beings have tried for many years to determine how life emerged on primitive Earth. In other words, human beings are interested in the origin of life. In addition, there is another significance to life-origin research. For example, answers to fundamental about genes and proteins, such as "why are there four types of nucleotides or bases in RNA and DNA?", and similarly, "why are modern proteins made up of 20 types of amino acids?", could be obtained if the mystery of the origin of life was solved. However, irrespective of the strenuous efforts of many researchers, the mystery of the origin of life remains unsolved. The main reasons for this are as follows.

A. Difficulties elucidating the establishment process of the fundamental life system

The most important point for solving the mystery of the origin of life is to clarify the fundamental life system establishment process, which involves six members (gene, genetic code, tRNA, metabolism, cell structure and protein) [1] (Chapter 2). However, it has been difficult to understand this establishment process for the following reasons.

1. The "chicken–egg relationship" between genes and proteins: The so-called "chicken–egg relationship" has made it difficult to solve the mystery of the origin of life for many

years. In such a situation, the RNA world hypothesis, which assumes that first life arose from the RNA world, which was formed by the self-replication of RNA, was proposed by Gilbert in 1986 [2]. However, it would be principally impossible to solve the mystery from the standpoint of this hypothesis because any gene encoding a mature protein could never be formed in the absence of a target protein or in the absence of protein, even if RNA could be produced on primitive Earth.

On the other hand, I proposed the [GADV]-protein world hypothesis (in short, the GADV hypothesis) approximately 20 years ago [1,3,4], assuming that the first life emerged from the [GADV]-protein world, which was formed by the pseudo-replication of [GADV]-proteins [5]. I believe now that the mystery could be solved based on the GADV hypothesis. [GADV] represents four amino acids, Gly [G], Ala [A], Asp [D] and Val [V]. The square brackets ([ ]) are used to discriminate the one-letter symbols of amino acids, particularly Gly [G] and Ala [A], from the one-letter symbols of nucleobases guanine G and adenine A.

2. The emergence of life under random processes: This theory suggests that any occurrences on primitive Earth proceeded as random processes. However, it is easily supposed that a mature gene and a mature protein with an ordered sequence could never be formed through a random process at one stroke, because the sequence diversities of gene encoding a protein composed of 100 amino acids and the protein itself are extraordinarily large at $(4^3)^{100}$ = $\sim 10^{180}$ and 20^{100} = $\sim 10^{130}$, respectively [1] (Chapter 3) [6]. Regarding this problem, I believe that the question of how a gene acquired the genetic information of a mature protein could be solved if the question were considered from the standpoint of the GADV hypothesis [1] (Chapter 8), as explained in detail later.

B. The problem of a research strategy based on experiments

1. The mystery of the origin of life could not be solved with experiments only: Needless to say, experiments are quite important or even crucial in studying the origin of life. However, it would be impossible to solve the mystery with experiments only, because the events that occurred on primitive Earth approximately 4 billion years ago could never be reproduced by experiments carried out in a present-day laboratory.

2. The establishment process of the core life system, composed of gene, genetic code (tRNA) and protein [1] (Chapter 2), could not be determined using bottom-up approaches only: In order to clarify the origin of life, it is undoubtedly important to answer the questions regarding what happened on primitive Earth to lead to the emergence of life. Therefore, many researchers have conducted studies for many years according to bottom-up approaches in a bid to discover where and what kind of organic compounds were produced on primitive Earth. However, it would probably be impossible to solve the mystery of the origin of life this way, because the establishment process of the core life system would never be made clear by simply using bottom-up approaches, unless newly born life or RNA/DNA containing the most primitive genetic information for protein synthesis could be found from rocks approximately 4 billion years ago.

3. The difference between an experimental condition in a laboratory and a primitive Earth situation: Experiments confirming nucleotide synthesis on primitive Earth have been carried out by many researchers, and results showing that nucleosides were actually produced with ribose and nucleobases are frequently reported in scientific journals [7–12]. However, it seems that it would be impossible to apply these results to the real occurrences on primitive Earth, as the experimental conditions, such as the concentrations and purity of reactants, are always very different from the situation on primitive Earth. Therefore, I consider that a parallel use of both bottom-up approaches, which have been carried out thus far, and top-down approaches, for example, using database analyses of modern genes and proteins, could hold one of the keys to solving the mystery of the origin of life [13].

C. The issue of how genes and proteins, which are composed of relatively small types of components, could be produced in chemically complex messy environments on primitive Earth

As described in the "Special Issue Information" of the Special Issue (Origin of Life in Chemically Complex Messy Environments), "Considering the prebiotic Earth four billion years ago (a messy atmosphere, in other words), a chaotic mélange of diverse starting materials appears realistic", there is another problem that makes it difficult to solve the mystery of the origin of life. Biopolymers as genes and proteins, composed of only four types of nucleotides and twenty types of amino acids, respectively, are used in extant organisms. However, such biopolymers must have been formed in chemically complex messy environments on primitive Earth. Therefore, how such biopolymers, using small types of components, could be formed in the messy environments of primitive Earth is an issue. When the mechanism producing genes and proteins, which uses a relatively small types of components, was acquired by something, that something would become the first life. Therefore, understanding the steps to the emergence of life should lead to the discovery of the correct answer to the question of how genes and proteins could be produced in chemically complex messy environments.

In this paper, I would like to discuss the processes of how biopolymers, composed of rather small types of relatively simple monomers, were formed in chemically messy environments from the standpoint of the GADV hypothesis [1,3,4]. For this purpose, it is necessary to understand the processes through which first life emerged on primitive Earth. Then, I will explain only the main points regarding how the GADV hypothesis could be proposed, as my ideas on the origin of life have already been described in detail in the book, *Towards revealing the origin of life: Presenting the GADV hypothesis*, which was published last year by Springer [1]. Thereafter, I will explain how genes and proteins were formed during repeated random reactions in chemically messy environments using small types of the respective components, and I will attempt to answer the third question (C) regarding how genes and proteins could have been produced using small types of, respectively, selected monomers from the messy environments.

Therefore, this article is described as providing an answer to the question that has been proposed in the Special Issue: "Origin of Life in Chemically Complex Messy Environments". In other words, I discuss in the article how a small types of amino acids or nucleotides were selected from the chemically complex messy environments of primitive Earth based on the GADV hypothesis, which I propose. The answer to the question described in the article is the novelty aspect of this paper.

2. The Key to Solving the Mystery of the Origin of Life (Protein 0th-Order Structure: [GADV]-Amino Acids)

The most important concept in the GADV hypothesis is one of protein 0th-order structures or [GADV]-amino acid composition [14]. First, I will explain the significance of protein 0th-order structure in solving the mystery of the origin of life (Figure 1).

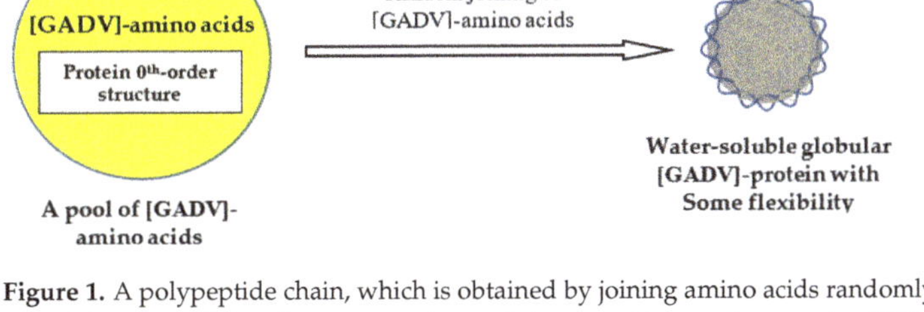

Figure 1. A polypeptide chain, which is obtained by joining amino acids randomly selected out from a pool (a protein 0th-order structure) containing roughly equal amounts of [GADV]-amino acids, should be folded into a water-soluble globular structure with some flexibility. The protein, actually aggregates of [GADV]-peptides, has a pluripotency that makes it possible to exhibit many catalytic activities owing to the structure flexibility. The wavy lines surrounding the gray circle and the thin yellow curved lines within the circle represent the flexible structure of an immature [GADV]-protein.

Consider here the problem based on the GADV hypothesis, how a gene, which encodes one amino acid sequence of a mature protein, was created. It is well known that it is impossible to produce one mature protein like a precise polymer machine through the random joining of even simple [GADV]-amino acids at one stroke, because the amino acid sequence diversity of a protein composed of only four types of one hundred [GADV]-amino acids is extraordinarily large ($4^{100} = \sim 10^{60}$) [1] (Chapter 3).

How were the genes that encoded mature proteins formed? For this purpose, one immature water-soluble globular protein with some flexibility, produced by the expression of one double-stranded (ds)-RNA encoding one essentially random [GADV]-amino acid sequence in the protein 0th-order structure, is indispensable for the formation of the gene (Figure 2). An RNA with a random $(GNC)_n$ codon sequence could be formed by the random joining of GNC anticodons carried by AntiC-SL RNAs [1] (Chapters 7 and 8).

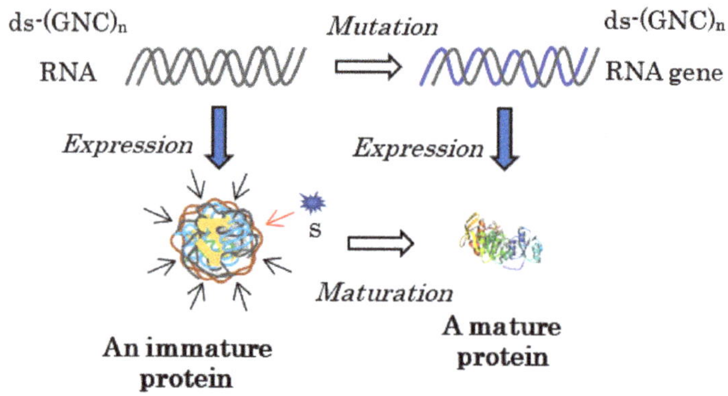

Figure 2. Every mature protein must be generated by the maturation of an immature protein, produced by the expression of an RNA strand encoding a random $(GNC)_n$ sequence for the immature protein. This is because a substrate binding site (a key hole) must be formed as accumulating appropriate mutations to adjust the site to fit closely to a substrate (a key). For this purpose, the protein 0th-order structure is indispensable for producing an immature protein with some flexibility. The brown wavy lines surrounding the immature protein and the blue curved lines represent the flexible structure of the immature [GADV]-protein. The curved gray lines and the blue line indicate random $(GNC)_n$ RNA strands and a $(GNC)_n$ RNA gene encoding a mature [GADV]-protein, respectively.

Specifically, every gene that encodes a mature protein has been formed as a result of maturation from an immature or incomplete water-soluble globular protein with some flexibility, which generates various weak catalytic activities or demonstrates pluripotency [1] (Chapter 3), to a mature protein with a rigid structure and high catalytic activity.

This means that the formation of a mature protein always requires an immature protein, because the formation of a mature protein must always be led by the elevation of the weak catalytic activity of the immature protein, which is easily understood as the relationship between a key and a key hole (Figure 2). This is one of the reasons why the mystery of the origin of life has not been solved using the RNA world hypothesis; that is, RNA never acquires genetic information for mature protein synthesis independently of an immature protein, even if RNA could first be produced by a random process or by the random joining of nucleotides in prebiotic environments.

As described above, both an immature [GADV]-protein and ds-RNA are required to form the first RNA gene (Figure 2). At present, an essentially random codon sequence similar to a random $(SNS)_n$ codon sequence or a non-stop frame on an antisense strand of a GC-rich gene is used as the field for creating an entirely new gene [15]. On primitive Earth, immature [GADV]-proteins were produced using a [GADV]-amino acid sequence encoded by a random $(GNC)_n$ codon sequence on one of two RNA strands, which were formed by the random joining of GNC anticodons carried by four AntiC-SL tRNAs (Figure 2) [1]

(Chapters 3 and 8) [4]. The reason it was possible is because one amino acid sequence randomly selected from a [GADV]-amino acid pool (protein 0th-order structure) is essentially the same as one amino acid sequence arranged by a random $(GNC)_n$ RNA sequence. Thus, the GADV hypothesis, which I have proposed, is an idea based on one of the protein 0th-order structures or [GADV]-amino acid composition.

3. Possible Steps from Chemical Evolution to the Emergence of First Life

Next, explicitly consider the steps to the emergence of first life on primitive Earth according to the GADV hypothesis (Figure 3) [1]. The reason why the steps to the emergence of life must be described here is that these steps are intimately related to the establishment process of the core life system composed of genes, genetic code (tRNA) and proteins, and to the formation process of proteins composed of four types of [GADV]-amino acids, which were encoded by the $(GNC)_n$ RNA gene composed of four types of nucleotides. In other words, the emergence of a genuine life was intimately related to the formation process of the $(GNC)_n$ RNA gene and [GADV]-protein using a relatively small type of components in chemically messy environments on primitive Earth.

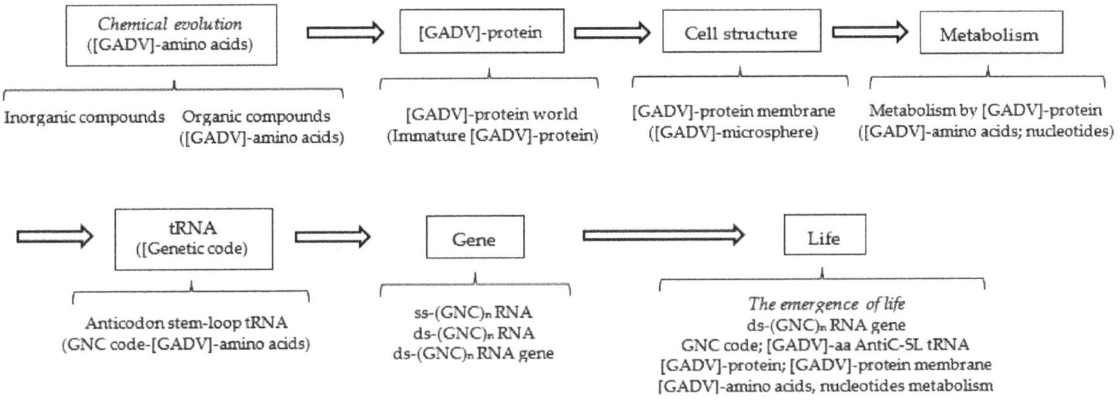

Figure 3. Possible steps from chemical evolution to the emergence of life, deduced from the [GADV]-protein world hypothesis (GADV hypothesis). All the steps are related to [GADV]-amino acids or [GADV]-protein, including GNC genetic code and the $(GNC)_n$ gene, both of which encode [GADV]-amino acids and [GADV]-protein, respectively. The steps to the emergence of life can be reasonably explained using the GADV hypothesis, assuming that life emerged as the piling up of the six members (protein, cell structure, metabolism, tRNA, genetic code and gene) in order from [GADV]-protein to $(GNC)_n$ gene encoding a mature [GADV]-protein.

Next, I explain the main object of this article regarding how biopolymers (genes and proteins) composed of small types of the respective monomers were formed in chemically messy environments on primitive Earth.

In this section, the steps from chemical evolution to the emergence of the first genuine life, equipped with the fundamental life system composed of six members, are discussed as divided into three (Table 1): the first part—from chemical evolution to the formation of the [GADV]-microsphere (proto-life); the second part—from the formation of the [GADV]-microsphere to the formation of AntiC-SL RNA; and the third part—from AntiC-SL RNA to the emergence of the first genuine life. The processes are discussed in three parts because the aspect of selection or of evolution changed before and after the formations of the [GADV]-microsphere. In addition, the second and third parts are separated because the use of nucleotides is restricted for the first time into four, A, G, U and C, by the formation of AntiC-SL RNA, which is folded with base pairs, A-U and G-C. On the other hand, the number of amino acids could not still be restricted to four at the time point as the establishment of the core system, composed of genes, tRNA (genetic code) and proteins, made it possible to use the four [GADV]-amino acids for the first time.

Table 1. Steps from chemical evolution to the emergence of life, which are discussed in this article as dividing into three parts, I, II and III.

Part I: From chemical evolution to formation of [GADV]-microsphere
1. [GADV]-amino acid synthesis with prebiotic means 2. Formation of [GADV]-microspheres ([GADV]-protein world) 3. Selection and evolution of [GADV]-microsphere
Part II: After formation of [GADV]-microsphere to formation of AntiC-SL tRNAs
1. Formation of proto-metabolism in [GADV]-microsphere [GADV]-amino acid and [GADV]-peptide syntheses, Nucleotide synthesis with immature [GADV]-proteins (actually aggregates of [GADV]-peptides) 2. Formation of AntiC-SL tRNAs
Part III: After formation of AntiC-SL tRNAs to the emergence of life
1. Formation of a ss-$(GNC)_n$ RNA 2. Formation of a ds-$(GNC)_n$ RNA 3. Formation of the first ds-$(GNC)_n$ RNA gene 4. The emergence of life

In addition, only random processes should have been naturally repeated before the first genuine life emerged or in the absence of any gene encoding an ordered amino acid sequence of a [GADV]-protein. Therefore, it is also essential to clarify how [GADV]-proteins with an ordered amino acid sequence were formed or how genes encoding the amino acid sequence were formed during the repetition of random processes in order to solve the mystery of the origin of life.

3.1. Part I. Steps from Chemical Evolution to Formation of [GADV]-Microsphere ([GADV]-Protein World)

Many studies on chemical evolution have been carried out to clarify what types of organic compounds can be synthesized from inorganic compounds under what conditions [16,17]. In these studies, the types of biomolecules that can be synthesized, such as amino acids, sugars, nucleobases, etc., which are necessary for first life to emerge, were mainly investigated.

As can be easily understood from previous studies, messy organic compounds, including various organic acids and amines in addition to various non-natural amino acids, should be produced via prebiotic means. Steps to the emergence of life proceeded through reaction processes, with the compounds selected from messy organic compounds. However, it would be quite difficult to select out only the necessary organic compounds, for example, only [GADV]-amino acids, from messy mixtures of organic compounds (Figure 4). In this Part I, I aim to explain a principle that can advance towards a solution to the problem from perspective of the GADV hypothesis.

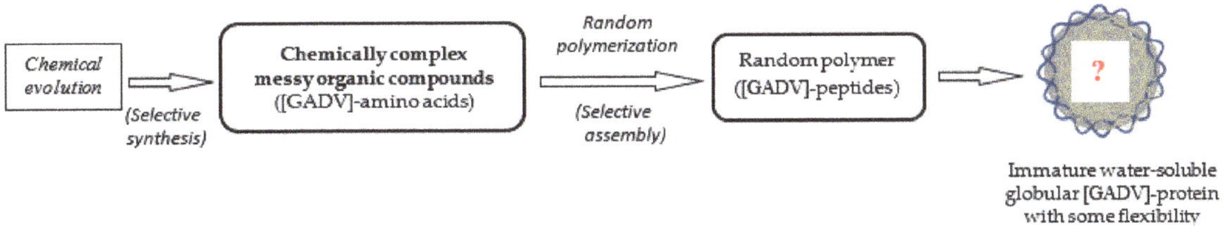

Figure 4. The organic compounds that were synthesized from inorganic compounds on primitive Earth via various prebiotic means were, naturally, chemically complex messy organic compounds. Therefore, a pool containing only [GADV]-amino acids did not exist on primitive Earth as expected by the GADV hypothesis. Therefore, one important question regarding how immature water-soluble globular [GADV]-proteins, which is a prerequisite in the GADV hypothesis, could be produced arises. Presenting an answer to the questions of how [GADV]-amino acids were selected and how immature [GADV]-protein could be formed is the main purpose of this article.

3.1.1. Preferential Synthesis of [GADV]-Amino Acids via Prebiotic Means

In previous studies, it was confirmed that simple organic compounds, such as amino acids and nucleobases with a small number of carbon atoms, were synthesized via various prebiotic means, as described below:

(1) Electric discharge into the primitive atmosphere (Miller-type experiments) [16,17];
(2) Catalytic reactions on pyrite, clay, hydrothermal vents, and so on [18–20];
(3) Organic synthesis via heavy bombardments of meteorites or asteroids [21,22];
(4) Introduction of organic compounds via meteorites, asteroids, space dust, and so on, from space [23,24].

Thus, various organic compounds including, especially, oxygen atom(s) were synthesized through physical and physico-chemical reactions, and messy compounds were introduced from space, accumulating on primitive Earth.

Amino acids were produced with organic compounds, which were selectively synthesized and accumulated in large amounts on primitive Earth via prebiotic means. Note that, at this point in time, a type of selection among organic compounds had been carried out during the synthetic processes of the organic compounds based on the nature of the chemical compounds themselves used as reactants, as described below.

Preferential Synthesis of α-Amino Acids

Amino acids were produced using prebiotic methods more easily than fatty acids and hydrocarbons, of which constituent atoms are connected chiefly with inactive carbon-carbon and carbon–hydrogen bonds. In fact, Miller describes in his book that there is no good method for fatty acid synthesis via prebiotic means [16]. On the contrary, glyoxylate and pyruvate must be rather easily produced via prebiotic means, as those keto-acids carrying a small number of carbon atoms have active carbon atoms with a localized electron, although it must be confirmed experimentally that glyoxylate and pyruvate could be synthesized via prebiotic means. Furthermore, it is well known that α-amino acids were obtained in larger amounts than β-amino acids in Miller's experiments [16]. These results indicate that α-amino acids were rather easily synthesized via prebiotic means and accumulated in large amounts.

3.1.2. Direct Random Joining of [GADV]-Amino Acids

Described next is the mechanism or principle of how only [GADV]-amino acids could eventually be selected out at a relatively high rate for primeval protein synthesis in the messy mixture of various organic compounds containing non-natural amino acids. The reasons why α-amino acids could be preferentially selected for the synthesis of protein are as follows.

[GADV]-proteins, actually aggregates of [GADV]-peptides, could be also produced via prebiotic means in messy organic compounds, which accumulated in large amounts on primitive Earth, although the proteins were incomplete in the sense that various amino acids other than [GADV]-amino acids were contained in the proteins.

Preferential Polymerization of [GADV]-Amino Acids

Amino acids should be selectively linked with each other during repeated wet-drying processes, as amino acids have both positive and negative charges in the molecules to facilitate easy association in water and to create a peptide bond between two amino acids [25,26]. Nevertheless, various organic compounds other than [GADV]-amino acids were naturally incorporated into [GADV]-proteins or [GADV]-peptide aggregates during the direct random joining of [GADV]-amino acids.

Preferential Association of [GADV]-Peptides Containing Val

All four [GADV]-amino acids with a rather simple structure were easily synthesized via prebiotic means. However, the synthetic amount of Val, which has a more complex molecular structure, should be much less than that of the other three amino acids, Gly, Ala and Asp. Nevertheless, the lesser amount of Val could be compensated for through the formation of [GADV]-peptide aggregates, owing to the high hydrophobicity of Val, because peptides containing a larger amount of Val could be preferentially associated with each other through large hydrophobic interactions. This also contributed to the formation of more active immature [GADV]-proteins.

3.1.3. Preferential Synthesis of [GADV]-Amino Acids by Immature [GADV]-Proteins

It should be considered that various organic compounds, especially [GADV]-amino acids, could be preferentially synthesized using simple organic compounds with functional groups as glyoxylate and pyruvate, even by immature [GADV]-proteins with weak catalytic activities, and [GADV]-proteins also could be produced by the random joining of [GADV]-amino acids under the protein 0th-order structure [1] (Chapters 3, 5) [14]. On the contrary, it must be difficult to synthesize hydrocarbons with immature [GADV]-proteins. It is considered that the selective synthesis of [GADV]-amino acids by immature [GADV]-proteins, which was performed before cell structure formation, further advanced the steps to the emergence of life at a faster rate than before.

On the other hand, many inactive and useless [GADV]-peptides could also be produced during the direct random joining of [GADV]-amino acids, because of the incorporation of various organic compounds into the peptides. In the reaction, [GADV]-peptides with sufficiently high catalytic activity could always be produced as a result of a wide distribution of [GADV]-peptides, which were synthesized through a random process, although the formation rate of active [GADV]-proteins might be low [26]. This made it possible to proceed towards the emergence of life at a faster rate than in the era of chemical evolution without immature [GADV]-peptide catalysts.

3.1.4. Incorporation of Non-Natural Amino Acids into Immature [GADV]-Proteins

Various α-amino acids and β-amino acids, other than [GADV]-amino acids, such as 2-aminobutylic acid (2-ABA), 2-aminopentanoic acid, β-alanine and so on, could be also produced via prebiotic means. Therefore, various amino acids other than [GADV]-amino acids would also be incorporated into immature [GADV]-proteins during the polymerization of [GADV]-amino acids, because non-natural amino acids could not be effectively excluded during simple polymerization among amino acids with both positive and negative charges in the molecule. Thus, it would be significant to form [GADV]-microspheres facilitating chemical evolution, as described below.

3.2. Part II. Steps from Formation of [GADV]-Microsphere to AntiC-SL RNA Formation

3.2.1. Significance of Cell Structure as a [GADV]-Microsphere for Facilitating Chemical Evolution

Incorporation of Various Organic Compounds into [GADV]-Microspheres

After sufficient amounts of [GADV]-amino acids accumulated on primitive Earth, [GADV]-microspheres were formed, for example, by repeated wet-drying processes in depressions of rocks on primitive Earth [1] (Chapter 4). The [GADV]-microspheres inevitably contained large amounts of [GADV]-peptides in the cell structure so that the [GADV]-peptides were synthesized by immature [GADV]-proteins in the microspheres. The supposition that the immature [GADV]-proteins could synthesize [GADV]-peptides is supported by the fact that even Gly-Gly and Gly-Gly-Gly have peptide synthetic activity [27]. The formation of [GADV]-microspheres made it possible to hold oligomeric [GADV]-peptides, which were synthesized in the microsphere, owing to the semi-permeable [GADV]-protein membrane. The accumulation of [GADV]-peptides in the [GADV]-microsphere generated higher osmotic pressure to induce the further incorporation of low molecular weight organic compounds as glyoxylate and pyruvate into the microsphere (Figure 5).

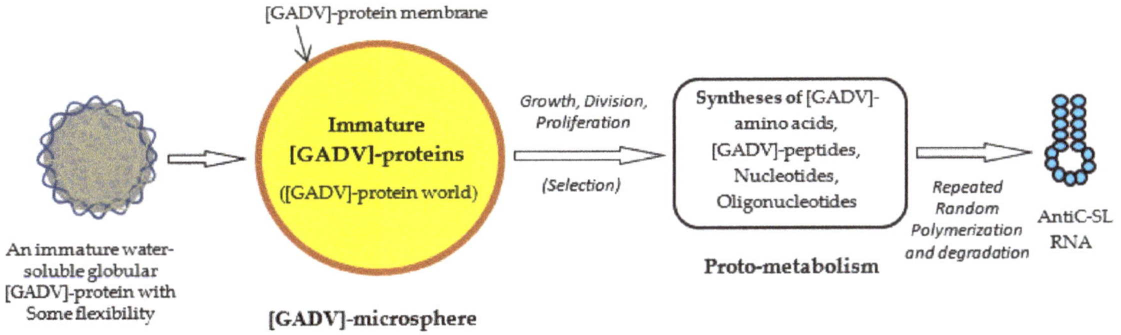

Figure 5. Possible steps from the formation of immature [GADV]-proteins to the formation of AntiC-SL RNA [1]. [GADV]-microspheres surrounded by [GADV]-protein membranes could be formed with [GADV]-peptides during repeated wet-drying processes. A [GADV]-protein world was formed in the [GADV]-microsphere. [GADV]-amino acids and nucleotides were synthesized by immature but pluripotent [GADV]-protein catalysts through proto-metabolism in the microsphere. Successively, AntiC-SL RNA was produced by the repeated random polymerization of nucleotides and their degradation. The core life system was established in [GADV]-microspheres, with a higher proliferation ability generated through the processes and, eventually, the first life arose on primitive Earth.

However, such proteins, into which amino acids other than [GADV]-amino acids were incorporated at a higher rate, would be gradually excluded, because [GADV]-microspheres using [GADV]-peptides composed of a higher rate of [GADV]-amino acids would be selected at a higher probability during proliferation followed by evolution. The selection became possible because [GADV]-amino acid composition is one of the protein 0th-order structures, and the incorporation of non-natural amino acids into [GADV]-proteins caused malfunction of the [GADV]-proteins. For the same reason, 2-ABA was excluded from natural amino acids as [GADV]-amino acids in order to avoid the duplicate use of Ala and 2-ABA, both of which are α-helix-forming amino acids [28]. The most important aspect of [GADV]-microsphere formation would be growth and proliferation, induced by high osmotic pressure accompanied by [GADV]-peptide synthesis (as shown in Figure 6).

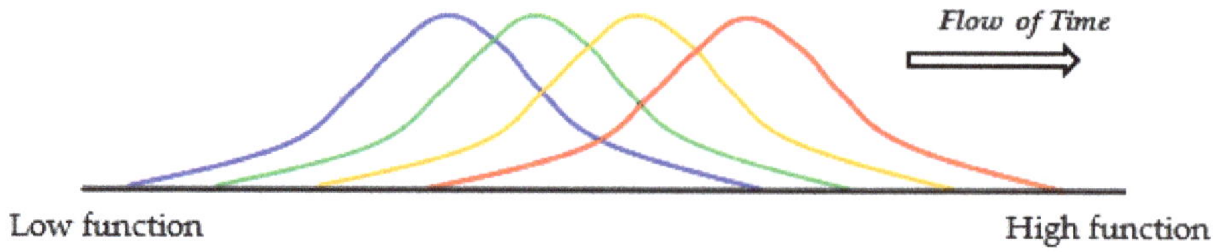

Figure 6. Evolution of [GADV]-microspheres without the genetic system. Random joining of amino acids carried out in the protein 0th-order structure inevitably generates [GADV]-peptides or [GADV]-proteins with a large distribution due to the random process. Therefore, at least part of [GADV]-proteins always had sufficiently high catalytic activities, which should have been effective for proceeding to the emergence of life. Thus, [GADV]-proteins with a higher catalytic activity than before could be generated in the microspheres. The steps to the emergence of life were the processes for the acquisition of [GADV]-proteins with higher catalytic activity than before, step-by-step. Changes in the distribution of [GADV]-microspheres from lower to higher functions are indicated in order from blue to red curves.

Selection of [GADV]-Microspheres with a High Proliferation Ability

In fact, at first, incomplete [GADV]-peptides containing non-natural amino acids would be produced at a high probability. However, [GADV]-microspheres containing non-natural amino acids at a lower rate could grow, proliferate and evolve faster than others to leave more descendants even before the establishment of the genetic system, because the [GADV]-microspheres containing at a higher rate of [GADV]-amino acids could acquire [GADV]-proteins with the higher function necessary to proceed to the emergence of life (Figure 6). Thus, [GADV]-microspheres containing lesser amounts of non-natural amino acids than others were consequently selected and proliferated. Furthermore, the formation of [GADV]-microspheres or [GADV]-protein membranes made it possible to protect against the dissipation of [GADV]-amino acids into environments, because immature [GADV]-proteins, produced with [GADV]-amino acids, could not ooze out through the [GADV]-protein membrane. This also contributed to a more efficient chemical evolution.

Growth, Division, Proliferation and Inactivation of [GADV]-Microspheres

[GADV]-microspheres with a higher ability for growth, division and proliferation could be consequently selected out from among many [GADV]-microspheres, as described above (Figure 6). The selected [GADV]-microspheres could leave more descendants and evolve further, even if the microspheres did not hold any genetic system. Contrasted with that, many other microspheres, which were defeated in the struggle for existence, disappeared, leaving many inactive bodies, for example, in the depressions of rocks on primitive Earth. However, those inactive bodies were reused for the growth and prosperity of the selected [GADV]-microspheres. This situation is similar to that observed on the present Earth where components of withered plants and dead bodies of animals are usually reused by presently living organisms after they have been degraded by various organisms, including bacteria.

Formation of Proto-Metabolic Pathways for [GADV]-Amino Acid Synthesis

As described above, [GADV]-amino acids with a sufficiently high stability were easily produced and accumulated in large amounts on primitive Earth. In addition, it was easy to refill [GADV]-amino acids, even if the amino acids were exhausted upon consumption for growth, because [GADV]-amino acids could be synthesized with simple organic compounds as glyoxylate and pyruvate, which accumulated on primitive Earth in large amounts and could be easily supplied from the environments into the microspheres (Figure 5) [1] (Chapter 5). Thus, [GADV]-amino acids were optimal compounds for advancing the steps to the emergence of life on primitive Earth.

Even such [GADV]-amino acids, which supported the proliferation of [GADV]-microspheres, would be depleted from inside of the microspheres and the proliferation of the [GADV]-microspheres would terminate soon after the deprivation due to the exponential growth of [GADV]-microspheres. The only way to avoid the situation was the construction of a proto-metabolic system for the synthesis of [GADV]-amino acids in the microspheres that were growing exponentially [1] (Chapter 5).

Proto-metabolic reactions using immature [GADV]-proteins started, in [GADV]-microspheres, to produce [GADV]-amino acids and [GADV]-peptides, just after the formation of [GADV]-microspheres. This would have been supported by the fact that [GADV]-amino acids can be synthesized using glyoxylate and pyruvate as the starting materials in a few reaction steps in modern metabolic pathways (Figure 7A). Inversely, the cycle of growth, division and proliferation of [GADV]-microspheres would terminate if the proto-metabolic pathways were not formed in the [GADV]-microspheres, and if sufficient osmotic pressure could not be maintained. Therefore, it can be considered that only [GADV]-microspheres, which had a high synthetic ability to produce [GADV]-peptides, were evolutionally selected and could leave more descendants than others. From these considerations, it can be concluded that the cell structures, which were indispensable to selection and evolution and in which [GADV]-peptides were synthesized to maintain high osmotic pressure, is the most essential function for life, not the genetic system.

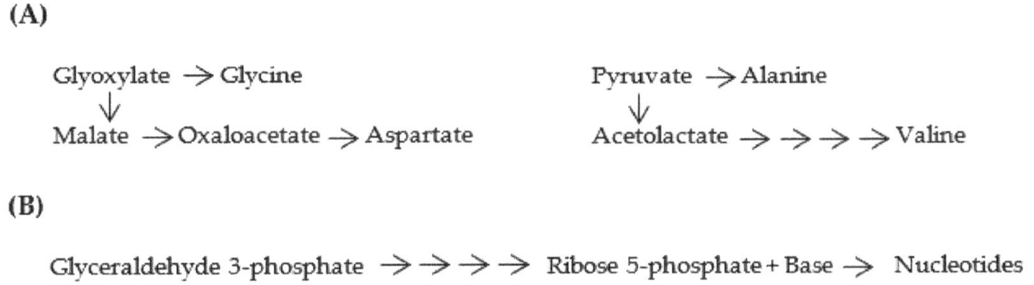

Figure 7. (**A**) Proto-metabolic pathways for [GADV]-amino acid synthesis. Four [GADV]-amino acids were produced from glyoxylate and pyruvate as starting materials through the proto-metabolic pathways. (**B**) Proto-metabolic pathways for nucleotide synthesis. Four nucleotides were produced with ribose 5-phosphate, which was synthesized from glyceraldehyde 3-phosphate as a starting material through the proto-metabolic pathways.

As is well known, modern cell membranes are composed of phospholipids and membrane proteins. Membrane proteins, but not phospholipids, exhibit various membrane functions. On the contrary, phospholipids are used for filling the gaps among membrane proteins and for expediting the migration of membrane proteins to express membrane functions more efficiently. Therefore, it would be valid to consider that phospholipids were inserted into [GADV]-protein membranes after the formation of phospholipid synthetic pathways via proto-metabolism in [GADV]-microspheres.

Formation of Proto-Metabolic Pathways for Nucleotide Synthesis

On the contrary, nucleotides, which are necessary to produce RNA, could not be synthesized in large amounts via prebiotic means because of the complex chemical structures of nucleotides. It can be seen from previous studies that nucleotides could not be produced by Miller's experiments [16,17] and that nucleotides have not been found in meteorites. It is still controversial whether or not nucleotides and nucleosides were actually synthesized using the reactants of simple inorganic compounds via prebiotic means, although some experimental results show that nucleotides and nucleosides could be produced with ribose-5-phspate or ribose and nucleobases, respectively, have been reported [9,29,30], because the concentrations and purity of the reactants used in the experiments are usually quite different from the conditions on primitive Earth.

Even papers showing that nucleosides and nucleotides were synthesized from formamide with meteorite catalysts under proton irradiation, were published [31,32]. Nevertheless, nucleotides must be depleted from the surroundings of [GADV]-microspheres, which proliferated exponentially. This means that proto-metabolism for nucleotide synthesis must be established in [GADV]-microspheres before deprivation of nucleotides. To establish the core life system, sufficient amounts of nucleotides necessary to synthesize RNA must be produced with previously existing immature [GADV]-proteins. For this purpose, proto-metabolic pathways, through which nucleotides could be synthesized, must be formed (Figure 7B). Needless to say, the formation of metabolic pathways in the absence of genes must rely on random processes.

Therefore, I consider that nucleotides could be produced through proto-metabolism using glyceraldehyde as a starting compound for ribose 5-phosphate synthesis [1] (Chapter 5). It is supposed that glyceraldehyde, having three electronically localized carbon atoms, could accumulate in large amounts, similar to glyoxylate and pyruvate, although it must be confirmed that glyceraldehyde could be synthesized via prebiotic means. Furthermore, nucleotide synthetic pathways could be formed with immature [GADV]-proteins because, in addition to the pluripotency of the immature [GADV]-proteins [1] (Chapter 3), [GADV]-microspheres, which acquired more favorable metabolic pathways for proliferation than others, even accidentally, could leave more descendants than others. I would like to name this phenomenon as "proto-Darwin evolution". In this way, during the evolutionary process, metabolic pathways, including nucleotide synthesis, which were favorable for the proliferation of [GADV]-microspheres, were formed in the microspheres.

However, many researchers may consider that nucleotides could never be synthesized from glyceraldehyde with immature [GADV]-proteins if they do not know the significance of the pluripotency of immature [GADV]-proteins, which could be synthesized by the direct random joining of [GADV]-amino acids under the protein 0th-order structure [1] (Chapter 3). Therefore, I would like to stress that the mystery of the origin of life will never be solved as long as they rely on nucleotide synthesis via prebiotic means only. Similarly, it would be reasonable to consider that the RNA world could never be formed on primitive Earth, as it would be impossible to produce a sufficient amount of RNA leading to the emergence of life without nucleotide metabolic pathways.

3.2.2. Formation of [GADV]-aa-AntiC-SL tRNA

Evolution of Activated [GADV]-Amino Acids

Initially, the synthesis of [GADV]-peptides in [GADV]-microspheres was carried out with [GADV]-amino acids by immature [GADV]-proteins, which were produced by the direct joining of [GADV]-amino acids, such as through wet-drying processes in the depressions of rocks on primitive Earth [25,26]. However, [GADV]-peptides could be produced at a much faster rate by using activated [GADV]-amino acids as [GADV]-adenosine monophosphate ([GADV]-AMP) than with the direct use of [GADV]-amino acids. Subsequently, activated [GADV]-amino acids were used for the more efficient synthesis of [GADV]-peptides in the order below, although there was no difference between the direct joining of [GADV]-amino acids and [GADV]-peptide synthesis with activated [GADV]-amino acids, except the difference in reaction rate.

1-1. Use of [GADV]-aminoacyl (aa)-AMP: [GADV]-aa-AMPs were synthesized with immature [GADV]-proteins (actually [GADV]-peptide aggregates) to accelerate peptide synthesis after adenosine triphosphate (ATP) was synthesized and accumulated in large amounts in [GADV]-microspheres. At this point in time, it is supposed that the activated [GADV]-amino acids were exclusively used for peptide synthesis with immature [GADV]-proteins owing to the accumulation of ATP in the microspheres

1-2. Use of [GADV]-aa-3′-ACC: [GADV]-aa-3′-ACCs were successively used for the synthesis of [GADV]-peptides. The stability of single-stranded (ss)-trinucleotide, CCA-3′, against the RNase activity of immature [GADV]-proteins made it possible for use in the synthesis. It is easy to understand that the ss-3′-ACC is stable against the RNase activity, because the 3′-ACC end of modern tRNA is also stable against RNase. Needless to say, the synthesis of [GADV]-peptides with activated [GADV]-amino acids was carried out non-specifically, because such activators, as AMP and 3′-ACC, cannot generate specificity to the respective [GADV]-amino acids. However, the use of [GADV]-aa-3′-ACCs contributed to a more efficient synthesis of [GADV]-peptides, as more sites of 3′-ACC than ATP, itself, could be used for binding with immature [GADV]-protein enzymes.

Of course, not only amino acids, but also other organic compounds might be activated with ATP and 3′-ACC. In the case of peptide synthesis with activated amino acids too, messy organic compounds could be incorporated into the peptides (Figure 4). However, it is supposed that activated organic compounds other than [GADV]-amino acids did not eventually contribute to the emergence of life, because the functions of [GADV]-peptides containing meaningless organic compounds would be lowered.

1-3. Use of 3′-ACC-AntiC-SL RNA: After the use of 3′-ACC, [GADV]-peptide synthesis was carried out using 3′-ACC-AntiC-SL RNA (Figure 8). One of the reasons why AntiC-SL RNA was used for the activation of [GADV]-amino acids is that AntiC-SL RNA, which was produced during cycles of oligonucleotide synthesis and degradation of the oligonucleotides, was the smallest but was a sufficiently stable RNA against hydrolysis by immature [GADV]-proteins [33]. In addition, the association of two AntiC-SL tRNAs side by side through base pairing between U and A in the two AntiC-loop RNAs facilitated the peptide bond formation between two amino acids bound to the 3′-ACC-end [1] (Chapter 7).

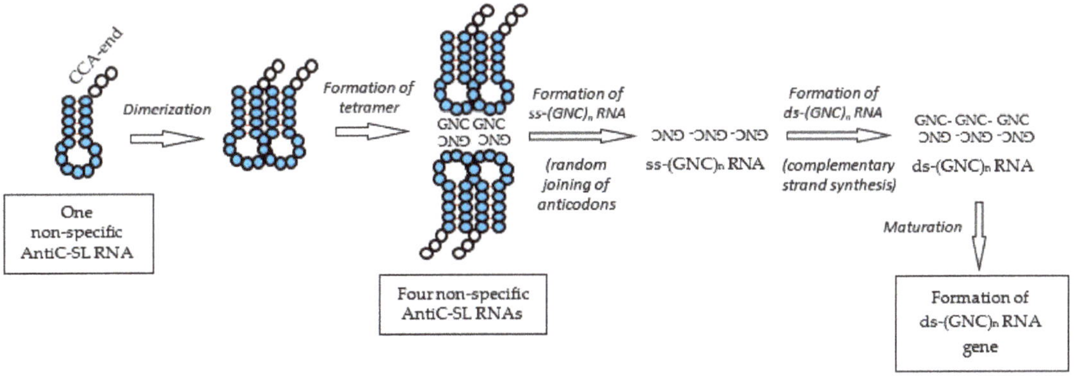

Figure 8. Formation process of the ds-RNA gene. The ds-RNA gene was formed through ss-$(GNC)_n$ RNA and ds-$(GNC)_n$ RNA. The ss-$(GNC)_n$ RNA was formed by the random joining of GNC anticodons carried by AntiC-SL RNAs. The use of amino acids was restricted to four [GADV]-amino acids accompanied by the establishment of GNC primeval genetic code and the ds-$(GNC)_n$ RNA gene for the first time.

The AntiC-SL RNA primitive tRNA hypothesis is supported by the fact that base pairs between two complementary GNCs are stable [34], as well as the fact that any base in an anticodon loop of three AntiC-SLs, except Asp-tRNA modified with queuosine and 2-methyladenosine, out of four modern *Escherichia coli* [GADV]-aa-tRNAs is not chemically modified [35].

3.3. Part III. Steps from Formation of AntiC-SL RNA to the Emergence of Life

3.3.1. Formation of ds-$(GNC)_n$ RNA Gene

Establishment of the Core Life System in [GADV]-Microspheres

Of course, the initial [GADV]-proteins, which were produced with immature [GADV]-proteins in a [GADV]-microsphere, were not literally [GADV]-proteins, meaning that

[GADV]-proteins contained non-natural amino acids other than [GADV]-amino acids. The synthesis of such impure polypeptides should always occur before the formation of the first gene or in the absence of genes on primitive Earth. However, even the impure and immature [GADV]-proteins could advance catalytic reactions in the microspheres, although the activities were low. Therefore, literal [GADV]-proteins could not be produced until the GNC primeval genetic code and the $(GNC)_n$ gene were formed. Inversely, the selection of [GADV]-microspheres with purer [GADV]-protein with higher catalytic activity made it possible to form the first gene, leading to the synthesis of pure [GADV]-proteins. Consequently, the genetic system or the core life system composed of the $(GNC)_n$ gene, AntiC-SL tRNA (GNC code) and [GADV]-protein was invented to improve [GADV]-protein functions through the complete exclusion of non-natural amino acids [1] (Chapter 6). Thus, the genetic system was formed to establish the most primitive, but pure, [GADV]-protein synthesizing system.

Needless to say, the first gene encoding the first mature [GADV]-protein with an ordered amino acid sequence must have been generated through an evolutionary process containing at least one random process, as described below.

(1) Synthesis of ss-$(GNC)_n$ RNA: In this case, the key point is to understand the process of how the first ss-$(GNC)_n$ RNA was formed, because an RNA with a random $(GNC)_n$ codon sequence could be formed by the random joining of GNC anticodons carried by the four AntiC-SL RNAs, although it must be confirmed that ss-$(GNC)_n$ RNA could be formed as I expected. [1] (Chapters 7 and 8). Note that the synthetic process of an immature [GADV]-protein with a random [GADV]-amino acid sequence through a random $(GNC)_n$ codon sequence is essentially the same as [GADV]-protein synthesis by the direct joining of [GADV]-amino acids. Then, how was the ss-$(GNC)_n$ RNA formed? I consider the formation process as follows [1] (Chapter 8).

1-1. Two pairs of two AntiC-SL tRNAs, which were bound in a column using two complementary GNC anticodons, were further aligned side-by-side to make a tetramer of four AntiC-SL RNAs (Figure 8).

1-2. Two anticodons of the two AntiC-SL tRNAs, which were aligned side-by-side, were connected by a phosphodiester bond. Thus, a random $(GNC)_n$ RNA sequence encoding a random [GADV]-amino acid sequence was created (Figure 8).

However, many researchers may doubt whether the ss-$(GNC)_n$ RNA could be formed as described above. My answer to the question is as follows.

1-3. The consecutive codons on mRNA have been actually read by two anticodons of two adjacent modern tRNAs.

1-4. The serial reading mechanism of the codon sequence on mRNA with the anticodon of tRNA clearly indicates that it is reasonable from a stereochemical viewpoint to consider that a ss-codon sequence (mRNA) can be produced by the joining of tRNA anticodons. That is the reason why a successive codon sequence can be read by the anticodons of two tRNAs tightly bound side-by-side. However, it would be difficult for two tRNAs to read a successive codon sequence if the codon sequence was formed independently of tRNA. The genetic sequence on mRNA was translated by AntiC-SL tRNA as the reverse process of the formation of the $(GNC)_n$ codon sequence in [GADV]-microspheres (Figure 8) [1] (Chapter 8). Thus, the fact that a comma-less codon sequence can be translated by tRNAs indicates that a genetic sequence was formed by the random joining of the anticodons carried by tRNAs.

Note that the random $(GNC)_n$ codon sequence is the simplest but most meaningful sequence, able to be produced through a random process and also able to be used for immature [GADV]-protein synthesis. This also indicates that the formation of a $(GNC)_n$ codon sequence via the random joining of the GNC anticodons of AntiC-SL tRNA is only one way under which an immature but meaningful [GADV]-protein could be produced through the RNA sequence, formed by an essentially random process.

(2) Formation of a ds-$(GNC)_n$ RNA: Successively, a ds-$(GNC)_n$ RNA was formed by the complementary strand synthesis of the ss-$(GNC)_n$ RNA.
(3) Formation of the first ds-$(GNC)_n$ RNA gene: Finally, the first $(GNC)_n$ RNA gene was formed. The formation of the first $(GNC)_n$ gene was triggered by the synthesis of an immature [GADV]-protein from one strand of the ds-RNA. The $(GNC)_n$ codon sequence encoding a random [GADV]-amino acid sequence evolved to one ds-RNA gene encoding a mature [GADV]-protein as led by the promotion of the activity on the immature [GADV]-protein (Figures 2 and 8).

Thus, ds-$(GNC)_n$ RNA genes encoding a mature [GADV]-protein were formed as the most effective means for producing mature [GADV]-proteins with high catalytic activity. It should be noted again that the steps towards the formation of the first gene were the processes for selecting only [GADV]-amino acids more efficiently from chemically complex messy environments. Thus, the key to the first ds-$(GNC)_n$ gene formation was the formation of ss-$(GNC)_n$ RNA with a random GNC codon sequence through a random process.

Thus, the synthesis of mature [GADV]-proteins composed of only [GADV]-amino acids became possible for the first time after the ds-$(GNC)_n$ RNA gene was acquired and the translation system using four types of [GADV]-aa-tRNAs was established. Note that all members needed to execute both mRNA synthesis and the translational process had already been created at the point in time when the first ds-RNA gene was created, as both the use of the transcription and translation systems had already become possible, just after not the ds-$(GNC)_n$ RNA gene, but ds-$(GNC)_n$ RNA was formed [1] (Chapter 6).

The Emergence of Life

(1) Genuine life arose after the acquisition of various $(GNC)_n$ genes through the creation of new homologous genes and entirely new $(GNC)_n$ genes, which were derived from sense strands [36] and antisense strands [15] of previously established $(GNC)_n$ genes, respectively [1] (Chapter 8).
(2) The first life emerged not at one moment, but during consecutive changes. In other words, any critical moment of the emergence of the first life did not exist. The emergence of life would have been such a gradual change as it can be confirmed that a typical life had arisen after a point in time. Finally, ds-$(GNC)_n$ RNA genes encoding a mature [GADV]-protein were formed as the most effective means for producing mature [GADV]-proteins with high catalytic activity. It should be noted again that the steps towards the formation of the first gene were the processes for selecting only [GADV]-amino acids more efficiently from chemically complex messy environments. Thus, the key to the first ds-$(GNC)_n$ gene formation was the formation of ss-$(GNC)_n$ RNA with a random GNC codon sequence through a random process.

4. Discussion

It is difficult to provide an answer to the question of how life emerged in "chemically complex messy environments". The problem is essentially the same as the question of how the most primitive core life system, which is composed of [GADV]-protein, AntiC-SL tRNA (GNC primeval genetic code) and the $(GNC)_n$ RNA gene using the respective small number of monomers, as four types of [GADV]-amino acids and four types of nucleotides, was established in those complex messy environments, because an answer to the mystery of the origin of life is, simultaneously, the answer to the question of how the small types of monomers were selected out from the chemically complex messy environments on primitive Earth. Therefore, it is necessary to consider focusing the discussion on how life arose using the core life system on primitive Earth in order to find the answer to the question of how life emerged in chemically complex messy environments.

The answer to the question is given based on the GADV hypothesis on the origin of life by considering several factors comprehensively, as described below.

1. Small types of, not so many, organic compounds including [GADV]-amino acids were able to be synthesized from inorganic compounds with prebiotic means on primitive Earth.
2. [GADV]-polypeptides composed of mainly four types of [GADV]-amino acids were able to be produced through a random process on primitive Earth. This is because [GADV]-amino acids, having both positive and negative charges in the molecule, pulled against each other, owing to the electrostatic attraction. Immature [GADV]-proteins, actually aggregates of [GADV]-peptides, which were produced by the joining of [GADV]-amino acids randomly selected from [GADV]-amino acid composition or one of the protein 0th-order structures, were able to be folded into water-soluble globular structures with some flexibility to exhibit various weak but effective functions [26].

The reason the four types of [GADV]-amino acids were chosen as components of the most primitive protein is not only because the [GADV]-amino acids were able to be easily produced via prebiotic means and accumulated in large amounts on primitive Earth, but also because [GADV]-amino acids satisfy the four conditions (hydropathy, α-helix, β-sheet, turn/coil formabilities) for forming water-soluble globular proteins, which were obtained based on amino acid compositions [3,4]. In fact, it was confirmed that [GADV]-proteins, actually aggregates of [GADV]-peptides, which were obtained by repeated wet-dry cycles of [GADV]-amino acids, have various catalytic functions [26].

3. [GADV]-microspheres were able to be formed with immature [GADV]-proteins, actually [GADV]-peptide aggregates. The formation of [GADV]-microspheres, which demonstrated individuality, made it possible for growth, proliferation and evolution. It is supposed that lipids were incorporated into [GADV]-protein membranes to enhance the membrane function through an increase in membrane fluidity during the evolution of [GADV]-microspheres.
4. Furthermore, metabolic pathways for [GADV]-amino acid synthesis using simple organic compounds as glyoxylate and pyruvate, which accumulated on primitive Earth, were able to be formed using immature [GADV]-proteins in the microspheres ([GADV]-protein world). The formation of the [GADV]-amino acid metabolic pathways assured the continuous growth, division and proliferation of [GADV]-microspheres through the synthesis of [GADV]-peptides.

It may be a matter of speculation whether or not [GADV]-amino acids were able to be synthesized with immature [GADV]-proteins using, as starting compounds, glyoxylate and pyruvate. The reason [GADV]-amino acid synthetic pathways were able to be formed is because such immature [GADV]-proteins, which have various catalytic activities or pluripotency as catalysts, could be used as biocatalysts [1] (Chapter 3) [26,27].

5. The metabolic pathway for the synthesis of four types of nucleotides was also able to be formed using immature [GADV]-proteins in [GADV]-microspheres, similar to the case of the synthetic pathways of [GADV]-amino acids.
6. Further, it is considered that AntiC-SL tRNAs were formed with four types of nucleotides synthesized through metabolic pathways. Thus, the use of nucleotides were restricted for the first time into four, adenine (A), guanine (G), uracil (U) and cytosine (C), because two base pairs, A-U and G-C, were indispensable to folding RNA strands into AntiC-SL through hydrogen bonds with high directionality.
7. Next, the GNC primeval genetic code, which determines the framework composed of four [GADV]-amino acids and four GNC codons, was established, although the corresponding relationships between [GADV]-amino acids and the GNC codons were accidentally determined and frozen, as assumed by the GNC code frozen-accident theory [1] (Chapter 7).
8. Eventually, the first genuine life emerged on primitive Earth after $(GNC)_n$ RNA genes were formed successively in order of ss-$(GNC)_n$ RNA, ds-$(GNC)_n$ RNA and the ds-$(GNC)_n$ RNA gene, and the core life system was established.

Thus, the first genuine life using biopolymers composed of small types of components emerged in the chemically complex messy environments on primitive Earth approximately 4 billion years ago. I believe that such steps to the emergence of life were inevitable, and there was no way to the emergence of life. In this way, it can be reasonably explained, based on the GADV hypothesis on the origin of life, how the core life system is composed of the $(GNC)_n$ gene, AntiC-SL tRNA and [GADV]-proteins, all of which are composed of small types of components, could be established in complex messy environments on primitive Earth. This indicates that the GADV hypothesis is a valid concept for explaining the steps to the emergence of life.

In addition, the GADV hypothesis is testable. Therefore, in order to further confirm the validity of the GADV hypothesis experimentally, I propose several experiments at the end of this article described below:

1. A structural analysis and measurement of the various catalytic activities of a pluripotent immature [GADV]-protein with a random amino acid sequence;
2. Confirmation of the growth, division and proliferation of [GADV]-microspheres, formed by the repeated wet-drying cycles of [GADV]-amino acids;
3. Syntheses of [GADV]-amino acids and nucleotides with immature pluripotent [GADV]-proteins.
4. Formation of AntiC-SL RNA during repeated random joining of nucleotides and its degradation.
5. Formation of ss-$(GNC)_n$ RNA by random joining of anticodons carried by four AntiC-SL RNAs.
6. Establishment of the core life system accompanied by ds-$(GNC)_n$ RNA gene, etc.

In the abstract of a recent review article about the Hot Spring Hypothesis on the origin of life of Damer and Deamer [37], it is described that "*a continuity is observed for biogenesis beginning with simple protocell aggregates, through the transitional form of the progenote, to robust microbial mats that leave the fossil imprints of stromatolites so representative in the rock record. A roadmap to future testing of the hypothesis is presented*". I want to pay attention to the future development of the Hot Spring Hypothesis.

Funding: This research received no external funding.

Institutional Review Board Statement: Not applicable.

Informed Consent Statement: Not applicable.

Data Availability Statement: Not applicable.

Acknowledgments: I am very grateful to Tadashi Oishi (G&L Kyosei Institute, Emeritus professor of Nara Women's University) for encouragement throughout my research on the origin and evolution of the fundamental life system.

Conflicts of Interest: The author declares no conflict of interest.

References

1. Ikehara, K. *Towards Revealing the Origin of life.—Presenting the GADV Hypothesis*; (Monograph) Springer Nature, Gewerbestrasse: Cham, Switzerland, 2021.
2. Gilbert, W. The RNA world. *Nature* **1986**, *319*, 618. [CrossRef]
3. Ikehara, K. Origins of gene, genetic code, protein and life: Comprehensive view of life system from a GNC-SNS primitive genetic code hypothesis. *J. Biosci.* **2002**, *27*, 165–186. [CrossRef] [PubMed]
4. Ikehara, K. Possible steps to the emergence of life: The [GADV]-protein world hypothesis. *Chem. Rec.* **2005**, *5*, 107–118. [CrossRef]
5. Ikehara, K. Pseudo-replication of [GADV]-proteins and origin of life. *Int. J. Mol. Sci.* **2009**, *10*, 1525–1537. [CrossRef] [PubMed]
6. Dill, K.A. Dominant forces in protein folding. *Biochemistry* **1990**, *297*, 133–7155. [CrossRef] [PubMed]
7. Kim, S.C.; O'Flaherty, D.K.; Giurgiu, C.; Zhou, L.; Szostak, J.W. The emergence of RNA from the heterogeneous products of prebiotic nucleotide synthesis. *J. Am. Chem. Soc.* **2021**, *143*, 3267–3279. [CrossRef]
8. Kim, S.C.; Zhou, L.; Zhang, W.; O'Flaherty, D.K.; Rondo-Brovetto, V.; Szostak, J.W. A Model for the Emergence of RNA from a Prebiotically Plausible Mixture of Ribonucleotides, Arabinonucleotides, and 2'-Deoxynucleotides. *J. Am. Chem. Soc.* **2020**, *142*, 2317–2326. [CrossRef]

9. Banfalvi, G. Prebiotic Pathway from Ribose to RNA Formation. *Int. J. Mol. Sci.* **2021**, *22*, 3857. [CrossRef]
10. Benner, S.A.; Kim, H.J.; Carrigan, M.A. Asphalt, water, and the prebiotic synthesis of ribose, ribonucleosides, and RNA. *Acc. Chem. Res.* **2012**, *45*, 2025–2034. [CrossRef]
11. Orgel, L.E. Prebiotic chemistry and the origin of the RNA world. *Crit. Rev. Biochem. Mol. Biol.* **2004**, *39*, 99–123. [CrossRef]
12. Unrau, P.J.; Bartel, D.P. RNA-catalysed nucleotide synthesis. *Nature* **1998**, *395*, 260–263. [CrossRef] [PubMed]
13. Ikehara, K. Evolutionary Steps in the Emergence of Life Deduced from the Bottom-Up Approach and GADV Hypothesis (Top-Down Approach). *Life* **2016**, *6*, 6. [CrossRef] [PubMed]
14. Ikehara, K. Protein ordered sequences are formed by random joining of amino acids in protein 0th-order structure, followed by evolutionary process. *Orig. Life Evol. Biosph.* **2014**, *44*, 279–281. [CrossRef] [PubMed]
15. Ikehara, K.; Amada, F.; Yoshida, S.; Mikata, Y.; Tanaka, A. A possible origin of newly-born bacterial genes: Significance of GC-rich nonstop frame on antisense strand. *Nucl. Acids Res.* **1996**, *24*, 4249–4425. [CrossRef]
16. Miller, S.L.; Orgel, L.E. *The Origins of Life on the Earth*; Prentice-Hall: Englewood Cliffs, NJ, USA, 1974.
17. Cleaves, H.J.; Chalmers, J.H.; Lazcano, A.; Miller, S.L.; Bada, J.L. A reassessment of prebiotic organic synthesis in neutral planetary atmosphere. *Orig. Life Evol. Biosph.* **2008**, *38*, 105–115. [CrossRef]
18. Clay, A.P.; Cooke, R.E.; Kumar, R.; Yadav, M.; Krishnamurthy, R.; Springsteen, G. A plausible prebiotic one-pot synthesis of orotate and pyruvate suggestive of common protometabolic pathways. *Angew. Chem. Int. Ed. Engl.* **2022**, *61*, e202112572. [CrossRef]
19. Zhou, R.; Basu, K.; Hartman, H.; Matocha, C.J.; Sears, S.K.; Vali, H.; Guzman, M.I. Catalyzed synthesis of zinc clays by prebiotic central metabolites. *Sci. Rep.* **2017**, *7*, 533. [CrossRef]
20. Schwartz, A.W.; Chittenden, G.J. Synthesis of uracil and thymine under simulated prebiotic conditions. *Biosystems* **1977**, *9*, 87–92. [CrossRef]
21. Pearce, B.K.D.; Tupper, A.S.; Pudritz, R.E.; Higgs, P.G. Constraining the Time Interval for the Origin of Life on Earth. *Astrobiology* **2018**, *18*, 343–364. [CrossRef]
22. Ehrenfreund, P.; Spaans, M.; Holm, N.G. The evolution of organic matter in space. *Philos Trans. A Math. Phys. Eng. Sci.* **2011**, *369*, 538–554. [CrossRef] [PubMed]
23. Pizzarello, S.; Weber, A.L. Prebiotic amino acids as asymmetric catalysis. *Science* **2004**, *303*, 1151. [CrossRef] [PubMed]
24. Furukawa, Y.; Chikaraishi, Y.; Ohkouchi, N.; Ogawa, N.O.; Glavin, D.P.; Dworkin, J.P.; Abe, C.; Nakamura, T. Extraterrestrial ribose and other sugars in primitive meteorites. *Proc. Natl. Acad. Sci. USA* **2019**, *116*, 24440–24445. [CrossRef] [PubMed]
25. Suwannachot, Y.; Rode, B.M. Catalysis of dialanine formation by glycine in the salt-induced peptide formation reaction. *Orig. Life Evol. Biosph.* **1988**, *28*, 79–90. [CrossRef] [PubMed]
26. Oba, T.; Fukushima, J.; Maruyama, M.; Iwamoto, R.; Ikehara, K. Catalytic activities of [GADV]-peptides: Formation and establishment of [GADV]-protein world for the emergence of life. *Orig. Life Evol. Biosph.* **2005**, *35*, 447–460. [CrossRef]
27. Luisi, P.L. *The Emergence of Life-from Chemical Origins to Synthetic Biology*, 2nd ed.; Cambridge University Press: Cambridge, UK, 2016.
28. Ikehara, K. GADV Hypothesis on the Origin of Life—Life Emerged in This Way!? LAP LAMBERT Academic Publishing: Saarbrucken, Germany, 2016.
29. Yadav, M.; Kumar, R.; Krishnamurthy, R. Chemistry of Abiotic Nucleotide Synthesis. *Chem. Rev.* **2020**, *120*, 4766–4805. [CrossRef]
30. Kruse, F.M.; Teichert, J.S.; Trapp, O. Prebiotic Nucleoside Synthesis: The Selectivity of Simplicity. *Chemistry* **2020**, *26*, 14776–14790. [CrossRef]
31. Costanzo, G.; Saladino, R.; Crestini, C.; Ciciriello, F.; Di Mauro, E. Nucleoside phosphorylation by phosphate minerals. *J. Biol. Chem.* **2007**, *282*, 16729–16735. [CrossRef]
32. Saladino, R.; Carota, E.; Botta, G.; Kapralov, M.; Timoshenko, G.N.; Rozanov, A.Y.; Krasavin, E.; Di Mauro, E. Meteorite-catalyzed syntheses of nucleosides and of other prebiotic compounds from formamide under proton irradiation. *Proc. Natl. Acad. Sci. USA* **2015**, *112*, E2746–E2755. [CrossRef]
33. Ikehara, K. The Origin of tRNA deduced from *Pseudomonas aeruginosa 5′* Anticodon-stem Sequence-Anticodon-stem loop hypothesi. *Orig. Life Evol. Biosph.* **2019**, *49*, 61–75. [CrossRef]
34. Taghavi, A.; van der Schoot, P.; Berryman, J.T. DNA partitions into triplets under tension in the presence of organic cations, with sequence evolutionary age predicting the stability of the triplet phase. *Q. Rev. Biophys.* **2017**, *50*, e15. [CrossRef]
35. tRNAdb. Transfer RNA Database (Universitat Leipzig). Available online: http//trnadb.bioinf.uni-leipzig.de (accessed on 26 June 2022).
36. Ohno, S. *Evolution by Gene Duplication*; Springer: Berlin/Heidelberg, Germany, 1970.
37. Damer, B.; Deamer, D. The Hot Spring Hypothesis for an Origin of Life. *Astrobiology* **2020**, *20*, 429–452. [CrossRef] [PubMed]

life

Concept Paper

Spontaneous Formation of Functional Structures in Messy Environments

Christian Mayer

Institute of Physical Chemistry, CENIDE, University of Duisburg-Essen, 45141 Essen, Germany; christian.mayer@uni-due.de; Tel.: +49-201-183-2570

Abstract: Even though prebiotic chemistry initially deals with simple molecules, its composition rapidly gains complexity with oligomerization. Starting with, e.g., 20 monomers (such as the 20 proteinogenic amino acids), we expect 400 different dimers, 3,200,000 pentamers, or more than 10^{13} decamers. Hence, the starting conditions are very messy but also form a very powerful pool of potentially functional oligomers. A selecting structure (a "selector" such as membrane multilayers or vesicles) may pick and accumulate those molecules from the pool that fulfill a simple function (such as the suitability to integrate into a bilayer membrane). If this "selector" is, in turn, subject to a superimposed selection in a periodic process, the accumulated oligomers may be further trimmed to fulfill more complex functions, which improve the survival rate of the selectors. Successful oligomers will be passed from generation to generation and further improved in subsequent steps. After thousands of generations, the selector, together with its integrated oligomers, can form a functional unit of considerable order and complexity. The actual power of this process of random formation and selection has already been shown in laboratory experiments. In this concept paper, earlier results are summarized and brought into a new context.

Keywords: messy environments; order; complexity; function; selection; origin of life

Citation: Mayer, C. Spontaneous Formation of Functional Structures in Messy Environments. *Life* **2022**, *12*, 720. https://doi.org/10.3390/life12050720

Academic Editors: Pasquale Stano, Ranajay Saha and Alberto Vázquez-Salazar

Received: 14 April 2022
Accepted: 10 May 2022
Published: 11 May 2022

1. The Creative Potential of Messy Prebiotic Chemistry

If we talk about a chemically messy environment, we usually have complex, chaotic systems in mind that contain a large variety of chemical compounds [1,2]. In principle, the full chemical space of small organic molecules, even with some restrictions in molecular size and contributing elements, contains millions of molecules [3,4]. Oligomers and polymers consisting of repetitive units with some limited variety form an interesting fraction of this chemical space [5]. This fraction is very likely to have special relevance for prebiotic chemistry for the simple reason that recent biological systems still make extensive use of it; proteins, nucleic acids, and carbohydrates are the most prominent examples. Other types of such polymers may have contributed in the past [6]. An interesting example is polyesters possibly formed by a variety of hydroxycarboxylic acids in analogy to proteins assembled from different amino acids [7,8]. In any case, we know that different types of polymers are actually the basis for the chemical variability of life.

Therefore, in order to find a potential way out of prebiotic "messiness", a good starting point is to think of a basic set of chain-forming molecules. If we choose the 20 proteinogenic amino acids for this purpose, we can expect $20 \times 20 = 400$ different dimers after the first condensation reaction. A further 4 condensation reactions lead to $20^5 = 3.2 \times 10^6$ pentamers, and 9 condensation reactions lead to more than 10^{13} different decamers. Hence, by oligomerization, the corresponding subset of the chemical space explodes into a huge variety. On the other hand, one has to consider the low concentrations of the oligomers in the thermal equilibrium. With knowledge of the average rate constants of condensation (k_c) and hydrolysis (k_h), these equilibrium concentrations c_n for oligomers with n repetitive units are determined by a set of differential equations [9]. For a situation with up to six units, the reactions that summarize all steps of formation and degradation for each oligomer A_n—with A_1 being the amino acid monomer—are listed in the following (water as a reaction partner is omitted in all cases for simplicity):

condensation reactions (with rate constant k_c)		hydrolysis reactions (with rate constant k_h)	
$A_1 + A_1 \rightarrow A_2$	x2 (1)	$A_2 \rightarrow A_1 + A_1$	
$A_1 + A_2 \rightarrow A_3$	x2 (1)	$A_3 \rightarrow A_1 + A_2$	x2 (2)
$A_1 + A_3 \rightarrow A_4$	x2 (1)	$A_4 \rightarrow A_1 + A_3$	x2 (2)
$A_1 + A_4 \rightarrow A_5$	x2 (1)	$A_4 \rightarrow A_2 + A_2$	
$A_1 + A_5 \rightarrow A_6$	x2 (1)	$A_5 \rightarrow A_1 + A_4$	x2 (2)
$A_2 + A_2 \rightarrow A_4$	x2 (1)	$A_5 \rightarrow A_2 + A_3$	x2 (2)
$A_2 + A_3 \rightarrow A_5$	x2 (1)	$A_6 \rightarrow A_1 + A_5$	x2 (2)
$A_2 + A_4 \rightarrow A_6$	x2 (1)	$A_6 \rightarrow A_2 + A_4$	x2 (2)
$A_3 + A_3 \rightarrow A_6$	x2 (1)	$A_6 \rightarrow A_3 + A_3$	

(1) These reactions have to be considered with a factor of two since each amino acid (or peptide) on the left side of the equation offers two possible reaction sites, -COOH and $-NH_2$.

(2) These reactions have to be considered with a factor of two since the hydrolysis can occur at two different positions, each leading to an equivalent pair of products.

This reaction network leads to a set of differential equations, each characterized by averaged (second-order) rate constant k_c for the condensation reactions and a corresponding (pseudo-first-order) rate constant k_h for the hydrolysis reactions [9]. These coupled differential equations can be numerically solved, leading to a time development of the concentrations c_n, as shown in Figure 1. If we start with $c_1(t) = c_0$ at $t = 0$, the concentration $c_1(t)$ of the monomers hardly changes over time; the reduction by approximately 20% with respect to the starting concentration c_0 is hardly visible on the logarithmic scale. The concentrations of the oligomers, on the other hand, vary dramatically. They start out at zero, then increase in the course of the condensation reactions, and finally approach the equilibrium values asymptotically over time. For the given settings of the rate constants, the equilibrium concentrations decrease by approximately an order of magnitude for each additional monomer unit. For the dimer, the equilibrium concentration c_2 may still be in the range of 0.1 c_0, which in practical experiments could be easily detected by NMR spectroscopy [10]. The equilibrium value for the hexamer, on the other hand, is expected near $c_6 = 3 \times 10^{-6}\ c_0$, where in practice, it can only be observed by sensitive analytics.

Based on this estimation, one may be tempted to disregard the presence of the hexamers that only form about one-millionth of the mass of the total organic constituents in the given model system. However, this fraction has something like a powerful creative potential. It contains a theoretical set of 20^6 = 64,000,000 different sequences. With ongoing condensation and hydrolysis processes, it forms a dynamic pool of potentially functional peptides. Even though their individual concentration may be astronomically small, they may be selected and accumulated out of the dynamic pool, while they will be constantly resupplied by the full set of equilibrium reactions summarized in Equations (1)–(6) in Ref. [9]. The situation could be seen as a representation of the infinite monkey theorem [11]; it is like a herd of monkeys repeatedly typing six consecutive letters representing the amino acid sequence of the hexamers. At this point, it just needs an efficient selector to pick out the meaningful or functional ones. Over time, the functional sequences will be typed again and again and just need to be collected.

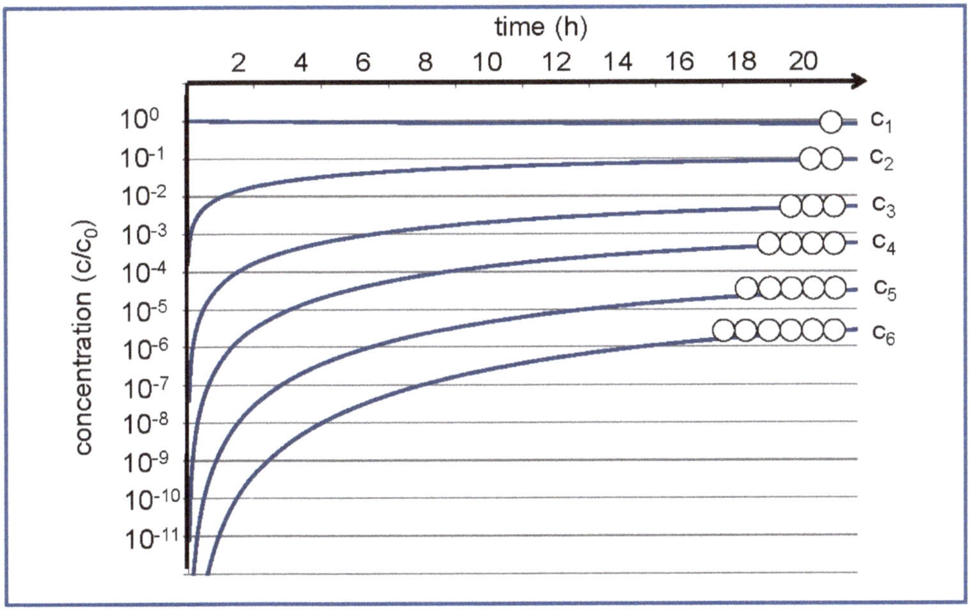

Figure 1. Time development of the relative concentrations of monomer (c_1) and oligomers (c_2–c_6) according to the given reaction network, starting with a monomer solution at t = 0 [9]. The approximated time scale is dferived from NMR experiments at 120 °C [10], where $k_c = 5.24 \times 10^{-3}$ L/(mol·min) and $k_h = 5.76 \times 10^{-2}$ L/min.

Creative pools of this kind are especially active in high-temperature environments, such as near primordial volcanic islands [12], in hot springs [2,13,14], or tectonic fault zones [15–17]. The formation of chain molecules by condensation is significantly accelerated by periodic variation of the water activity. Such a repetitive switching between high and low water concentration can be induced by wet–dry cycling (as it would naturally occur at the shores of a hot pond in a daily rhythm) or by pressure variations connected to a phase transition of carbon dioxide resulting in a corresponding change in water solubility [16]. Caused by the periodic change of the water activity, the system remains in a constant non-equilibrium state, with longer oligomers being favored under dry conditions.

2. The Potential Nature of a "Selector"

Which system could have the power to actively select useful sequences out of this creative pool? In principle, any structure capable of molecular recognition and somehow complementary to this sequence would be suitable [2]. Selectively binding to an oligomer with a particular sequence, such a "selector" could form a stable complex that prevents hydrolysis of the oligomer and, therefore, efficiently removes it from the cycle of condensation and hydrolysis [13,14]. Consequently, the oligomer with this particular sequence would accumulate and reach concentrations far above the equilibrium values depicted in Figure 1. Of course, the question remains: where should the complementary sequence come from? It necessarily must have at least the same degree of complexity as the collected oligomer. Consequently, it would have to be formed by a separate process in advance.

Alternatively, and with significantly reduced selectivity, the growing oligomer could be encapsulated or bind to the surface of a mesoscopic structure formed by amphiphiles. Membrane structures such as multilayers, micelles, or vesicles could have this capability and act as efficient selectors for specific sequences of biopolymers such as polypeptides or nucleic acids [18–25].

Another principle for selection could be simple amphiphilic interactions between the selector and the particular oligomer. Membrane multilayers, micelles, or vesicles tend to integrate those oligomers that reflect the amphiphilicity profile of their main constituents [9,16,26]. With a suitable geometry of the hydrophilic and hydrophobic part of the oligomer chain, intermolecular interactions build-up and a negative enthalpy change occurs; hence the integration process would be energy driven. The dissipation of this energy would compensate for the loss in entropy that is connected to i) the integration process and ii) the selection of a particular fraction of the oligomers (the latter would correspond to the entropy of mixing that would occur if this fraction returns into the original pool). Integrated oligomers reside in a protected environment, being efficiently shielded against hydrolysis or elution. Consequently, they are being selected and accumulated, while the others either remain part of the cycle of hydrolysis and condensation or cross-phase boundaries [9]. An example of such a selector, membrane vesicles in a two-phase environment of supercritical CO_2 and water in a high-pressure environment, is shown in Figure 2.

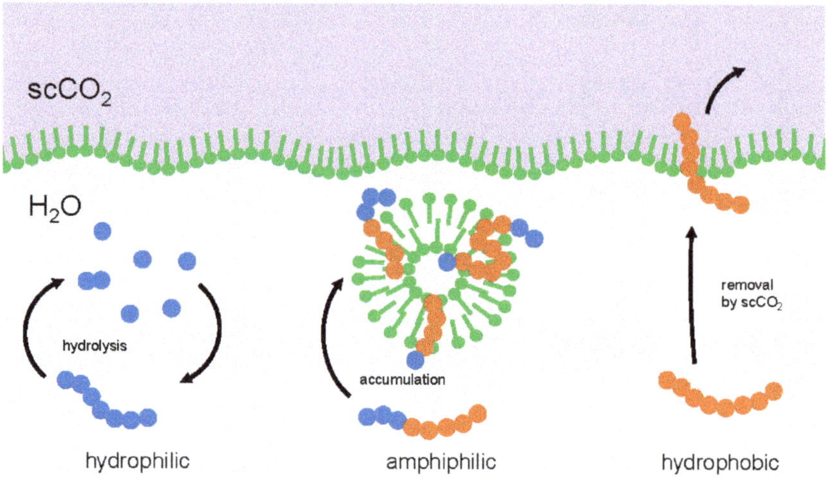

Figure 2. Example of a membrane structure acting as a selector for amphiphilic oligomers. Left: oligomers primarily made up of hydrophilic units (blue) form a cycle of condensation and hydrolysis in the aqueous phase. Right: oligomers made up primarily of hydrophobic units (red) tend to transfer into a hydrophobic environment, e.g., a separate phase of supercritical carbon dioxide. Center: amphiphilic oligomers integrate into membrane vesicles where they are protected against hydrolysis and phase transfer and accumulate over time.

Membrane multilayers, micelles, and membrane vesicles are mesoscopic structures that form spontaneously whenever amphiphilic molecules occur at a suitable concentration [27,28]. Under hydrothermal conditions, the occurrence of those molecules is very likely. The simplest pathway to their prebiotic generation would be the formation of aliphatic chains with an oxidized initial methyl group by Fischer–Tropsch chemistry [29]. Molecules of this kind were detected in meteorites [30] as well as in recent tectonic fault systems [31] and in remnants of very old hydrothermal environments [17]. Therefore, one can consider membrane multilayers, micelles, or membrane vesicles as especially likely first selectors for prebiotic oligomers. This is even more so since all these structures are highly dynamic [27]. They easily form and disintegrate under cyclic conditions such as wet–dry cycling or periodic phase transitions [26], and they may have even undergone steps of self-organization on their own [32,33]. In addition, and this may be the most important aspect, they themselves can be subject to a second stage of selection.

3. The Selection of the Selector

Thus far, the accumulation of the oligomers would just follow one predetermined selection principle: the amphiphilicity profile. In this case, the selected oligomer would be of the same complexity as the selector itself, so no progress would be achieved. However, the selector itself may be subject to a superimposed selection process [9,26]. In the simplest case, its thermal stability could depend on the interaction with the integrated oligomer. This interaction may follow three different stages of complexity in its development (Figure 3):

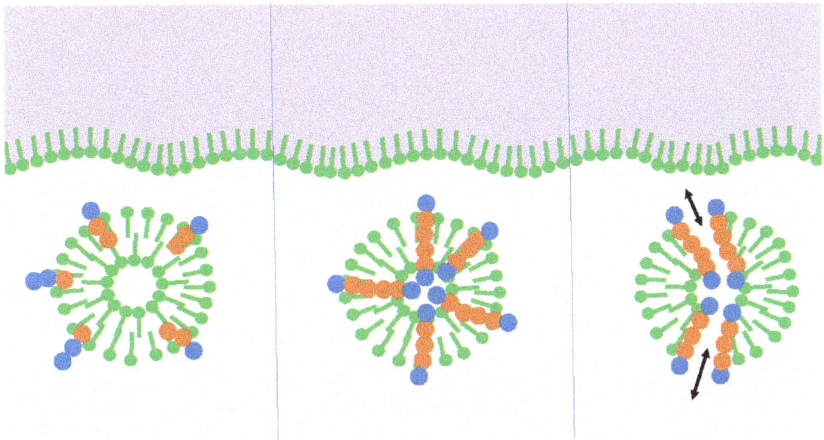

Figure 3. Three stages of the influence of the integrated oligomer on the survival of the membrane vesicle: (**a**) destabilizing effect, (**b**) stabilizing effect, (**c**) additional stabilizing function, e.g., the formation of pores leading to the relaxation of osmotic pressure [26].

(a) Oligomers with sequences of type (a) have a destabilizing effect on the membrane multilayer, the micelle, or the vesicle. In this case, the amphiphilic structure would come apart sooner than the competing ones, the amphiphiles would assemble at other surfaces, and the oligomers would be released. In the following, they are subject to hydrolysis just like the other oligomers in the pool and gain only a small temporary advantage.

(b) Oligomers with sequences of type (b) stabilize the membrane multilayer, the micelle, or the vesicle. The increase in thermal stability could derive from particular interactions between the oligomer and the adjacent amphiphiles. In this case, the lifetime of the structure would be extended, leading to extended protection of the oligomers with sequences of type (b) due to the reduced access of water molecules [26]. Hence, the sequences (b) would accumulate much more efficiently than the sequences (a).

(c) Oligomers with sequences of type (c) induce a more complex stabilizing function on the membrane multilayer, the micelle, or the vesicle. This effect goes beyond a simple thermodynamic stabilization by selective interactions. Instead, it compensates for destructive influences that shorten the lifetime of the amphiphilic structure. An example may be osmotic pressure, which regularly occurs during membrane formation. If an oligomer with a sequence (c) is capable of forming a transmembrane pore [26], this osmotic pressure can be released, leading to extended membrane longevity. Other specific functions may be the induction of a specific membrane curvature, the induction of a specific membrane mobility, or the accumulation of charges on the membrane surface. All these functions induced by sequences (c) could further extend the lifetime of the selector and hence give them an additional selective advantage.

The accumulation of the sequences (b) and (c) over time raises the chance that they will combine by condensation, leading to longer oligomers with superior survival strategies for the amphiphilic structures. After the stabilized selectors finally disintegrate, the oligomers of types (b) and (c) will be released into the pool, from where they will (at least partially) be collected by newly created selectors, creating a short-term molecular memory for the stabilized and functional structures. In all cases, the selection of the selectors leads to an increase in complexity far beyond the one that is observed for the basic selector itself and the oligomers that are initially accumulated.

The actual power of the overall development lies in the superposition of two selection processes (Figure 4). The original "creative pool" of oligomers (Figure 4, center) permanently delivers random sequences but also contains some chemical memory for successful stabilization and functionalization. The selector (in Figure 4, represented as membrane vesicles) picks out and integrates suitable oligomers and, by itself, is submitted to a selection process. In the course of this development, the selector develops more and more complex survival strategies.

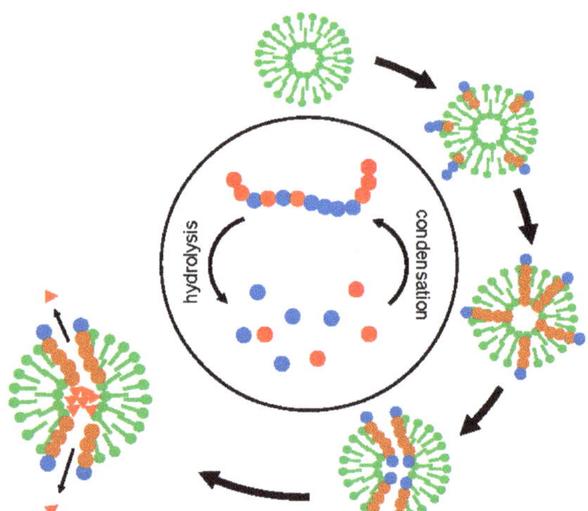

Figure 4. Superposition of two selection processes on the example of membrane vesicles. The vesicles ("selectors") collect oligomers from a central pool of condensation and hydrolysis. At the same time, they themselves are being selected for stability. "Successful" selectors accumulate oligomers with stabilizing and functional properties. In the end, a function like the relaxation of a given concentration gradient (triangles) may even become a source of free energy, representing a starting point for energy metabolism.

In laboratory experiments, these conditions led to three kinds of effects caused by a particular selected oligomer with the sequence KSPFPFAA [26]:

(i) Thermal stabilization of the vesicle membrane;
(ii) An increase in the permeability of the vesicle membrane;
(iii) A decrease in the vesicle size.

All these effects can be interpreted as possible survival strategies for the vesicles. This is obvious in the case of the effect (i). In the case of (ii), the selection advantage may be given by the relaxation of the destructive osmotic pressure. In the case of (iii), the vesicles may reduce the risk of being mechanically destroyed by bubble formation and shear. A molecular dynamic simulation of the resulting membrane structure with the selected peptide KSPFPFAA is shown in Figure 5 [34]. Within a period of less than 0.5 μs, the oligomers form intra-membrane clusters with a hydrophilic pore in the center. This pore facilitates the passage of water molecules and, possibly, ions through the membrane, thereby reducing the osmotic pressure load.

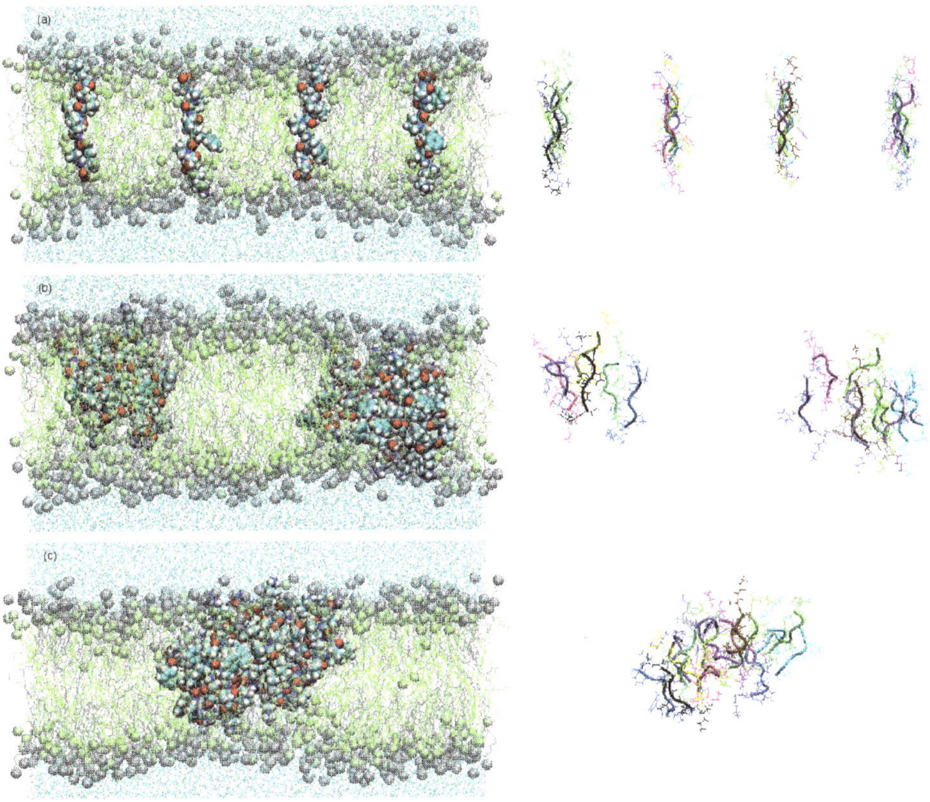

Figure 5. Membrane structure resulting from an experiment with superimposed selection processes in a molecular dynamics study [26,34]. Within 450 ns, the particularly successful oligomer KSPFPFAA adopts a stretched-out conformation across the membrane (**a**), starts to agglomerate in clusters (**b**), and forms pores with increased water permeability (**c**) [34].

4. Spontaneous Formation of Order and Complexity

The peptide cluster shown in Figure 5 is already of remarkable structural complexity. Nevertheless, it is obviously being formed in a spontaneous process from comparably simple molecules, all of them typical for a hydrothermal environment. It represents experimental evidence for the power of the described combination of random formation of oligomers with a two-step selection process.

The thermodynamic driving force is basically the increase of the overall entropy, its rate dS/dt given by a balance equation [35]:

$$\frac{dS}{dt} = \frac{d_iS}{dt} + \frac{d_eS}{dt} \tag{1}$$

where d_iS/dt stands for the entropy increase of internal irreversible processes connected to periodic phase transitions together with all the chemical reactions they induce while the reaction mixtures inside the system are permanently kept in a non-equilibrium state. The term d_eS/dt denotes the entropy flow rate due to the exchange of matter and energy with the environment, which is positive as well since processes like water evaporation or carbon dioxide flow are spontaneous on a large scale even if no internal processes occur.

In the proposed superimposed two-stage process, a tiny fraction of this overall entropy gain dS/dt is being sacrificed to generate local structural order by small-scale selection.

Focusing on this small local structure, the overall progress regarding complexity and order [36] may be illustrated, as shown in Figure 6. The molecules formed initially by hydrothermal chemistry (1) are low in complexity and occur in a diluted and mixed state, so they are low in structural order as well. As soon as those molecules form oligomers (2), the complexity rises dramatically (according to Kolmogorov's definition of complexity, more bytes are needed for their description [37]), whereas the structural order remains more or less unchanged. This is basically the messy environment we are starting with. At this point, a primary selection process by a "selector" (3) reduces complexity but at the same time increases order (represented by a suitable sequence), the increase corresponding to the negative mixing entropy. The secondary selection process of the "selector" itself (4) follows the same pattern, leading to an even more significant increase in order due to the formation of distinct membrane structures, as illustrated in Figure 5. Eventually, even the most stable vesicles will disintegrate, thereby losing a large part of the original progress in complexity and order (5). However, some of the successful oligomers will remain in the pool; therefore, the following sequence will not have to start from scratch. Instead, the next step will begin at position (5) in the diagram. If the superimposed selection process is repeated in numerous subsequent generations, the overall development will follow the path (6), indicated on the right of Figure 6. Therefore, the endpoint of the superimposed selection will gradually shift along the diagonal in the diagram. In essence, it is this development of rising order and complexity along this diagonal which can lead to specific survival mechanisms and, eventually, to something like a primitive functional unit [36]. Peptides grown by the condensation of randomly formed short oligopeptides [38] may further support the early functionality, especially since they may have developed catalytic functions [39].

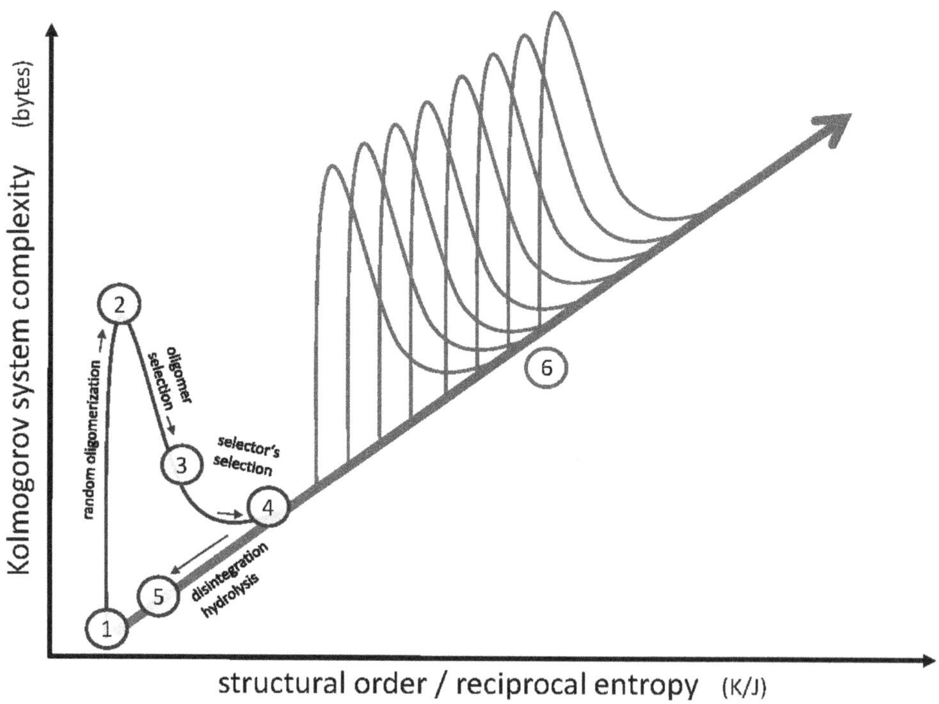

Figure 6. Representation of the effect of superimposed selection processes on a system's order and complexity. A single generation of selectors follows positions (**1**) to (**5**). The effect of subsequent generations is shown at position (**6**). For a detailed description of positions (**1**) to (**6**), see text.

5. Conclusions

The described interaction between two superimposed selection processes has a considerable power to turn a "messy" environment (in a chemical sense) into a system that contains complex functional structures. During this process of random formation and selection, the system's order and complexity increase gradually, eventually leading to structural functionality. The overall driving force is represented by the entropy increase of the vast majority of system components, which is only slightly reduced by the selected fractions of increased structural order. Overall, the superimposed selection process can lead to a significant degree of complexity and function of local structures.

Funding: This research received no external funding.

Institutional Review Board Statement: Not applicable.

Informed Consent Statement: Not applicable.

Data Availability Statement: Not applicable.

Conflicts of Interest: The author declares no conflict of interests.

References

1. Dyson, F.J. *Origins of Life*; Cambridge University Press: Cambridge, UK, 1999.
2. Deamer, D. *Assembling Life*; Oxford University Press: Oxford, UK, 2019.
3. Dobson, C.M. Chemical space and biology. *Nature* **2004**, *432*, 824–828. [CrossRef] [PubMed]
4. Reymond, J.-L. The chemical space project. *Acc. Chem. Res.* **2015**, *48*, 722–730. [CrossRef] [PubMed]
5. Matsuno, K. Polymer. In *Encyclopedia of Astrobiology*; Springer: Berlin, Germany, 2011; pp. 788–790.
6. Chandru, K.; Jia, T.Z.; Mamajanov, I.; Bapat, N.; Cleaves, H.J. Prebiotic oligomerization and self-assembly of structurally diverse xenobiological monomers. *Sci. Rep.* **2020**, *10*, 17560. [CrossRef] [PubMed]
7. Jia, T.Z.; Chandru, K.; Hongo, Y.; Afrin, R.; Usui, T.; Myojo, K.; Cleaves, H.J. Membraneless polyester microdroplets as primordial compartments at the origins of life. *Proc. Natl. Acad. Sci. USA* **2019**, *116*, 15830–15835. [CrossRef] [PubMed]
8. Chandru, K.; Mamajanov, I.; Cleaves, H.J.; Jia, T.Z. Polyesters as a model system for building primitive biologies from non-biological prebiotic chemistry. *Life* **2020**, *10*, 6. [CrossRef]
9. Mayer, C.; Schreiber, U.; Dávila, M.J. Selection of prebiotic molecules in amphiphilic environments. *Life* **2017**, *7*, 3. [CrossRef]
10. Waasmann, L.E.; Dávila, M.J.; Mayer, C.; University of Duisburg-Essen, Duisburg, Germany. 2018; Unpublished results.
11. Borel, E. Mécanique statistique et irréversibilité. 5e série. *J. Phys.* **1913**, *3*, 189–196.
12. Fox, S.; Strasdeit, H. Chemical evolution on primordial volcanic islands. *EPSC Abstr.* **2010**, *5*, 358.
13. Damer, B.; Deamer, D. The hot springs hypothesis of life. *Astrobiology* **2020**, *20*, 429–452. [CrossRef]
14. Damer, B.; Deamer, D. Coupled phases and combinatorial selection in fluctuating hydrothermal pools: A scenario to guide experimental approaches to the origin of cellular life. *Life* **2015**, *5*, 872–887. [CrossRef]
15. Schreiber, U.; Locker-Grütjen, O.; Mayer, C. Hypothesis: Origin of life in deep-reaching tectonic faults. *Orig. Life Evol. Biosph.* **2012**, *42*, 47–54. [CrossRef] [PubMed]
16. Mayer, C.; Schreiber, U.; Dávila, M.J. Periodic vesicle formation in tectonic fault zones—An ideal scenario for molecular evolution. *Orig. Life Evol. Biosph.* **2015**, *45*, 139–148. [CrossRef] [PubMed]
17. Schreiber, U.; Mayer, C.; Schmitz, O.J.; Rosendahl, P.; Bronja, A.; Greule, M.; Keppler, F.; Mulder, I.; Sattler, T.; Schöler, H.F. Organic compounds in fluid inclusions of Archean quartz—Analogues of prebiotic chemistry on early Earth. *PLoS ONE* **2017**, *12*, e0177570. [CrossRef] [PubMed]
18. Black, R.A.; Blosser, M.C.; Stottrup, B.L.; Tavakley, R.; Deamer, D.W.; Keller, S.L. Nucleobases bind to and stabilize aggregates of a prebiotic amphiphile, providing a viable mechanism for the emergence of protocells. *Proc. Natl. Acad. Sci. USA* **2013**, *110*, 13272–13276. [CrossRef]
19. Le Chevalier Isaad, A.; Carrara, P.; Stano, P.; Krishnakumar, K.S.; Lafont, D.; Zamboulis, A.; Buchet, R.; Bouchu, D.; Albrieux, F.; Strazewski, P. A hydrophobic disordered peptide spontaneously anchors a covalently bound RNA hairpin to giant lipidic vesicles. *Org. Biomol. Chem.* **2014**, *12*, 6363–6373. [CrossRef]
20. Strazewski, P. *Omne vivum ex vivo ... omne*? How to feed an inanimate evolvable chemical system so as to let it self-evolve into increased complexity and life-like behavior. *Isr. J. Chem.* **2015**, *55*, 851–864. [CrossRef]
21. Black, R.A.; Blosser, M.C. A self-assembled aggregate composed of a fatty acid membrane and the building blocks of biological polymers provides a first step in the emergence of protocells. *Life* **2016**, *6*, 33. [CrossRef]
22. Xue, M.J.; Black, R.A.; Cornell, C.E.; Drobny, G.P.; Keller, S.L. A step toward molecular evolution of RNA: Ribose binds to prebiotic fatty acid membranes, and nucleosides bind better than individual bases do. *ChemBioChem* **2020**, *21*, 2764–2767. [CrossRef]
23. Cohen, Z.R.; Nguyen, J.; Hazra, A.; Lalic, G.; Black, R.A.; Keller, S.L. Fatty acid membranes boost peptide yield and implications for the origins of cellular life. *Biophys. J.* **2020**, *118*, 228A. [CrossRef]

24. Xue, M.J.; Black, R.A.; Cohen, Z.R.; Roehrich, A.; Drobny, G.P.; Keller, S.L. Binding of dipeptides to fatty acid membranes explains their colocalization in protocells but does not select for them relative to unjoined amino acids. *J. Phys. Chem. B* **2021**, *125*, 7933–7939. [CrossRef]
25. Cohen, Z.R.; Cornell, C.E.; Catling, D.C.; Black, R.A.; Keller, S.L. Prebiotic protocell membranes retain encapsulated contents during flocculation, and phospholipids preserve encapsulation during dehydration. *Langmuir* **2022**, *38*, 1304–1310. [CrossRef] [PubMed]
26. Mayer, C.; Schreiber, U.; Dávila, M.J.; Schmitz, O.J.; Bronja, A.; Meyer, M.; Klein, J.; Meckelmann, S.W. Molecular evolution in a peptide vesicle system. *Life* **2018**, *8*, 16. [CrossRef] [PubMed]
27. Deamer, D. The role of lipid membranes in life's origin. *Life* **2017**, *7*, 5. [CrossRef] [PubMed]
28. Sakuma, Y.; Imai, M. From vesicles to protocells: The roles of amphiphilic molecules. *Life* **2015**, *5*, 651–675. [CrossRef]
29. Simoneit, B.R.T. Prebiotic organic synthesis under hydrothermal conditions: An overview. *Adv. Space Res.* **2004**, *33*, 88–94. [CrossRef]
30. Aponte, J.C.; Whitaker, D.; Powner, M.W.; Elsila, J.E.; Dworkin, J.P. Analyses of aliphatic aldehydes and ketones in carbonaceous chondrites. *ACS Earth Space Chem.* **2019**, *3*, 463–472. [CrossRef]
31. Großmann, Y.; Schreiber, U.; Mayer, C.; Schmitz, O.J. Origin of life: Aliphatic aldehydes in the Earth's crust—Remains of prebiotic chemistry? *Res. Sq.* **2022**. [CrossRef]
32. Segré, D.; Ben-Eli, D.; Deamer, D.W.; Lancet, D. The lipid world. *Orig. Life Evol. Biosph.* **2001**, *31*, 119–145. [CrossRef]
33. Kahana, A.; Maslov, S.; Lancet, D. Dynamic lipid aptamers: Non-polymeric chemical path to early life. *Chem. Soc. Rev.* **2021**, *50*, 11741–11746. [CrossRef]
34. Dávila, M.J.; Mayer, C. Membrane structure obtained in an experimental evolution process. *Life* **2022**, *12*, 145. [CrossRef]
35. Kondepudi, D.; Prigogine, I. *Modern Thermodynamics: From Heat Engines to Dissipative Structures*, 2nd ed.; John Wiley & Sons: New York, NY, USA, 2015.
36. Mayer, C. Life in the context of order and complexity. *Life* **2020**, *10*, 5. [CrossRef] [PubMed]
37. Li, M.; Vitányi, P. Preliminaries. In *An Introduction to Kolmogorov Complexity and Its Applications*; Texts in Computer Science; Springer: New York, NY, USA, 2008.
38. Chessari, S.; Thomas, R.; Polticelli, F.; Luisi, P.L. The prosuction of *de novo* folded proteins by a stepwise chain elongation: A model for prebiotic chemical evolution of macromolecular sequences. *Chem. Biodivers.* **2006**, *3*, 1202–1210. [CrossRef] [PubMed]
39. Wieczorek, R.; Adamala, K.; Gasperi, T.; Polticelli, F.; Stano, P. Small and random peptides: An unexplored reservoir of potentially functional primitive organocatalysts. The case of seryl-histidine. *Life* **2017**, *7*, 19. [CrossRef] [PubMed]

life

MDPI

Review

A Physicochemical Consideration of Prebiotic Microenvironments for Self-Assembly and Prebiotic Chemistry

Arpita Saha [1,2], Ruiqin Yi [3], Albert C. Fahrenbach [4,5,6], Anna Wang [4,5,6,*] and Tony Z. Jia [1,3,*]

1 Blue Marble Space Institute of Science, 600 1st Ave, Floor 1, Seattle, WA 98104, USA
2 Amity Institute of Applied Sciences, Amity University, Kolkata 700135, India
3 Earth-Life Science Institute, Tokyo Institute of Technology, 2-12-1-IE-1 Ookayama, Meguro-ku, Tokyo 152-8550, Japan
4 School of Chemistry, UNSW Sydney, Sydney, NSW 2052, Australia
5 Australian Centre for Astrobiology, UNSW Sydney, Sydney, NSW 2052, Australia
6 UNSW RNA Institute, UNSW Sydney, Sydney, NSW 2052, Australia
* Correspondence: anna.wang@unsw.edu.au (A.W.); tzjia@elsi.jp (T.Z.J.)

Abstract: The origin of life on Earth required myriads of chemical and physical processes. These include the formation of the planet and its geological structures, the formation of the first primitive chemicals, reaction, and assembly of these primitive chemicals to form more complex or functional products and assemblies, and finally the formation of the first cells (or protocells) on early Earth, which eventually evolved into modern cells. Each of these processes presumably occurred within specific prebiotic reaction environments, which could have been diverse in physical and chemical properties. While there are resources that describe prebiotically plausible environments or nutrient availability, here, we attempt to aggregate the literature for the various physicochemical properties of different prebiotic reaction microenvironments on early Earth. We introduce a handful of properties that can be quantified through physical or chemical techniques. The values for these physicochemical properties, if they are known, are then presented for each reaction environment, giving the reader a sense of the environmental variability of such properties. Such a resource may be useful for prebiotic chemists to understand the range of conditions in each reaction environment, or to select the medium most applicable for their targeted reaction of interest for exploratory studies.

Keywords: origin of life; chemical evolution; reaction microenvironments; physical chemistry; geochemistry

Citation: Saha, A.; Yi, R.; Fahrenbach, A.C.; Wang, A.; Jia, T.Z. A Physicochemical Consideration of Prebiotic Microenvironments for Self-Assembly and Prebiotic Chemistry. *Life* **2022**, *12*, 1595. https://doi.org/10.3390/life12101595

Academic Editors: Alberto Vázquez-Salazar and Ranajay Saha

Received: 17 September 2022
Accepted: 8 October 2022
Published: 13 October 2022

Publisher's Note: MDPI stays neutral with regard to jurisdictional claims in published maps and institutional affiliations.

1. Introduction

The early Earth was like a laboratory but without an intervening chemist. That is to say, early Earth possessed a variety of chemicals, reaction vessels/compartments, and conditions [1], generating complex chemical systems without a target, but which managed to self-organize into life. For many of these scenarios, interdisciplinary studies have been carried out to qualify and quantify their prebiotic plausibility. For example, the origins of life community have considered different geological settings (i.e., hot springs or oceans) as plausible "reaction vessels" on early Earth [2–5]. However, these geological settings, while informative about constraining the potential reaction environments, conditions, and chemicals, are mostly still at large length scales of centimeters, meters, or kilometers. Less attention has been paid to early Earth environments at the microscopic level, the length scales that have the potential to directly affect the dynamics of chemical reactions, self-assembly, and cellular/protocellular processes.

Indeed, within larger geological settings, a variety of physical and chemical environments exist at smaller length scales. These smaller 'microenvironments', which are as small as the microliter or micron scale, could vary significantly in physical and chemical properties. Examples include the widely varying temperature and pH conditions in hot springs [6] or the simultaneous existence of both aqueous environments as well as supercritical liquid carbon dioxide (CO_2) in the deep ocean [7].

Here, we consider a number of prebiotic microenvironments on Earth and discuss their physical and chemical characteristics and subsequent impact on primitive reactions and/or self-assembly. The intention is to create a resource that researchers can use to guide their laboratory reactions toward more realistic geological conditions or to use specific reaction microenvironments to constrain what prebiotic processes can take place. This review focuses on condensed phase reactions including but not limited to aqueous phase chemistry, interfacial chemistry, and hydrothermal/geochemical synthesis (Figure 1) [1]. We refer the reader to reviews of gas-phase chemistries and in particular, gas-phase photochemistry [8–10].

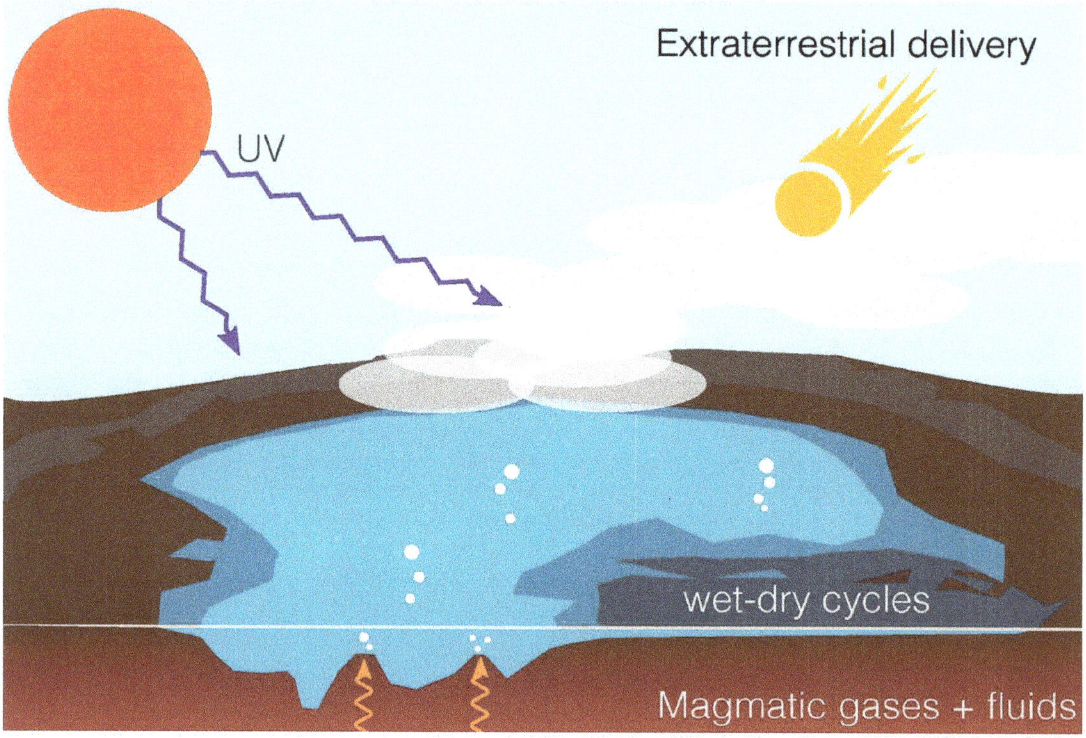

Figure 1. A variety of chemical processes could have occurred on early Earth. Prebiotic reactions could have occurred in the gas phase (atmospheric synthesis), the aqueous phase, or on material interfaces, just to name a few. Other reactions could have occurred extraterrestrially, followed by delivery to Earth; in this review, we particularly focus on condensed phase reactions on Earth. Figure adapted and reprinted with permission from [9] under a Creative Commons license.

The scope of this paper is primarily on environments generated purely from and/or residing in geological formations, although a number of potential environments generated from self-assembled primitive molecules (such as in the interior of vesicle bilayers [11] or within phase-separated polymer droplets [12,13]) are briefly mentioned. Although a number of reaction environments exist extraterrestrially such as methane surface lakes or ammonium-rich sub-surface lakes on Titan [14], we restrict the environments covered here to only those thought to have occurred on Earth. In the future, we hope to expand this analysis to more extraterrestrial environments as more concrete information on the chemical and physical properties of such environments becomes known, following additional planetary science studies and missions such as the Titan Dragonfly mission [15]. We refer the interested reader to papers discussing prebiotic chemistry in extraterrestrial environments such as meteorites [16], interstellar ice [17] or gas [18], or other planetary bodies [2,19–21].

2. Prebiotic Microenvironments and Where to Find Them

The goal in this section is to take a physicochemical perspective and categorize specific reaction microenvironments available on early Earth, which may be found within a number of (macroscopic) geological environments and scenarios. The overview of the physicochemical properties is expected to help point to opportunities that different environments present. For completeness, we encourage the reader to consider how the geological context is inextricable from prebiotic chemistry by reading the excellent works on the topic (e.g., [1,22–25]).

We start our overview with bulk polar and non-polar environments. Within these broad categories, different colloidal structures are discussed. Finally, we end with a brief discussion of molten and solid rocks.

2.1. *Aqueous Environments*

Living processes depend on water. One reason is that liquid water is important for diffusion, thus allowing reactants to meet. Water also hosts acid and base chemistries, and is a polar protic solvent that can readily solubilize ionic compounds and hydrogen bond with solutes. Bulk water's highly polar environment is responsible for the hydrophobic effect [26], liquid–liquid phase separation [27], and promotes amphiphile self-assembly into micelles, membranes, or other structures [28].

On early Earth, just as in the present day, there would have been sources of fresh water including streams, ponds, lakes, hot springs as well as salt water such as oceans and hydrothermal systems containing brines. The range of different salinities is covered in Section 3.1.

2.1.1. Bulk Aqueous Solution

In bulk systems there are still microenvironments that should be taken into account. One important factor to consider is the temporal stability of the aqueous environment. Surficial systems can be subjected to wet/dry or freeze/thaw cycles, and while this is not applicable to larger bodies of water, turbulent mixing still creates temporal instability in any aqueous system where the length scales are larger than the Reynolds number. That said, ocean stratification can broadly lead to different zones of stable salt/density and temperature, with other variables such as nutrient content potentially varying during mixing [29]. If the relevant length scales are small and the velocities are slow such as in rock pores or narrow channels [30], the flow could easily be laminar, and other types of transport processes such as thermophoresis [30,31] or diffusiophoresis [32] could begin to dominate.

2.1.2. Sea Spray (Aqueous Aerosols)

Another form of liquid aqueous environments is in aerosol droplets [33], which could have formed on early Earth due to turbulent waves or wind acting on different bodies of water. While the volumes of water in aerosols are small, the high surface area to volume ratio and ample exposure to sunlight means that surface-based processes and photochemistry can generate new molecules potentially not synthesizable in other environments [33,34] affording, for example, cross-linked lipids [35].

2.1.3. Gels and Other Hygroscopic Environments

Aqueous environments can also exist in forms that are intermediates to the solid and liquid phases. Siliceous hot-spring deposits, which likely would have been present on early Earth, contain hydrophilic amorphous silica that can retain water [36]. Mixtures of organic molecules likely present on early Earth have also been shown to form gels [37,38]. Hygroscopic salts also sequester water [39]. These environments enable unique aqueous-phase chemistries such as the synthesis of polymer-supported zinc sulfide nanocrystals [40] or photochemical phosphorylation [38], and can also reduce evaporation rates and prevent total desiccation.

2.1.4. Ice

The final form of an aqueous-derived environment considered here is ice, which could have been derived from the freezing of liquid aqueous solutions in different bodies of water on early Earth. While carbon-cycle modeling reveals that early Earth is thought to have been temperate [4], fluctuations in conditions could have potentially created sub-zero temperatures and thus ice. Eutectic phases in ice provide aqueous environments that are concentrated in solutes, leading to reactions that are unfavorable in a dilute aqueous environment; we cover eutectic phases in Section 2.2.1.

2.2. Alternative Liquid Environments

2.2.1. Non-Aqueous Solvents

Aside from water, other liquid environments on early Earth could have been in the form of non-aqueous solvents. Many reactions in prebiotic chemistry are formally known as condensations, which are reactions that covalently join together two compounds while eliminating a molecule of water in the process. The polymerization of amino acids and ribonucleotides into peptides and RNA, respectively, are examples of condensation reactions. In water, these condensations tend not to be spontaneous, partly as a consequence of Le Chatelier's principle, since water as a solvent is present in large excess and pushes the equilibrium towards the reactants. Hence, non-aqueous solvents (i.e., those based on organic compounds) have the potential to make condensation reactions more favorable.

While the modern synthetic organic laboratory has a large variety of non-aqueous (organic) solvents at its disposal, the majority would almost certainly not have been abundantly available on early Earth. One reason is simply the lack of prebiotically plausible synthetic pathways to achieve reservoir amounts of these organic compounds necessary to act as solvents, many of the chemical structures of which can be relatively complex. Another reason is that the temperature and pressure conditions of early Earth limit what potential solvents could have accumulated, even if endogenous prebiotic synthetic pathways were producing them in large quantities. The boiling points could be too low to exist as liquids at room temperature or be significantly lower than that of water so that their concentration from aqueous solutions is not realistic. Only organic liquids that have a higher boiling point than water could have accumulated to excess amounts, a circumstance required for a compound to act as a solvent. With these constraints in mind, it is conceivable, however, that some organic liquids could have accumulated in relatively large excess.

For example, an organic solvent that may have accumulated in certain early Earth geological scenarios is formamide. A formal hydration product of hydrogen cyanide (molecular formula: $HCONH_2$), formamide has a boiling point of 210 °C under standard pressure and has limited azeotropic associations with water [41]. Formamide could have been produced through multiple pathways [42] including mechanisms that involve atmospheric spark-discharge [43], ionizing radiation such as proton irradiation [44], UV irradiation [45], pyrolysis [46], or thermal reactions promoted by catalysis [47]. While it is unclear whether very large pools of formamide could have existed on early Earth (at least in comparison to aqueous pools), even transient accumulation of small volumes of formamide in different microenvironments on early Earth such as those that could occur in rock pores or on mineral surfaces following radiolytic synthesis and dehydration could have produced segregated organic formamide microenvironments that housed chemical reactions at the microscale [42].

2.2.2. Deep Eutectic Solvents

Deep eutectic solvents (DESs) have also been recently considered as alternative non-aqueous liquids in prebiotic chemistry and could have formed readily, for example, within different ice–water systems on early Earth [48]. A eutectic solvent is a liquid made from a specific mixture of two or more substances that taken individually exist as solids, but as a mixture forms a liquid that has a single melting point lower than either of its individual components. The eutectic is the specific ratio of component compounds that exhibits the lowest melting point. A *deep* eutectic solvent is a mixture of solids whose melting point becomes so depressed that it exists as a liquid at room temperature. For example, a 1:2 ratio of choline chloride (melting point = 302 °C) to urea (melting point = 133 °C) has a melting point of 12 °C, and thus is a liquid at room temperature [49]. The mechanism of melting point depression is thought to involve hydrogen bonds, and so the majority of known DES mixtures involve hydrogen bond donors and acceptors [50]. DESs have the characteristics of high viscosity, low volatility, and are typically polar enough to dissolve high concentrations of ionic compounds. These non-aqueous solvents have been shown to promote various prebiotic condensation reactions including phosphorylation [51] and peptide bond formation [51,52]. Some of the components of typical DES mixtures [53,54] such as urea [55], glycerol [56], or acetamide [57] are also organic molecules thought to be generally available in prebiotic chemical systems.

2.2.3. High Pressure Supercritical Fluids (CO_2, H_2O)

One alternative non-polar environment is supercritical fluids, found in the high pressure environments of deep ocean ridges and hydrothermal systems that likely also existed on early Earth [7]. Whilst supercritical CO_2 has traditionally been thought of as rather non-polar by some in the scientific community, this, in fact, is incorrect; the polarity of supercritical CO_2 can be tuned, and CO_2 in this phase can also act either as a Lewis acid or a Lewis base depending on the specific conditions [58]. Conversely, water when supercritical becomes as non-polar as 1-dodecanol [59].

2.2.4. Tars

Non-polar environments can also be found in tars, which can be made as a product of polymerization reactions containing prebiotically available organics, and result in thick, sticky substances that exhibit extremely slow diffusion times that, even in the presence of rainfall or aqueous solutions, is practically impossible to dilute [60,61]. From a microscopic perspective, it is a reaction environment that is 'hard to leave', but offers high concentrations and extremely complex reaction environments. Some molecules, however, are able to exit through the surface by sublimation or slowly leach out into surrounding fluids.

2.2.5. Inside Lipid Bilayers and Related Interfacial Assemblies

The polar nature of water can induce amphiphilic molecules to self-assemble into a variety of phases including micelles, cubic phases, lamellar phases, and liposome or lipid bilayer vesicles. Vesicles and micelles, in particular, have been proposed to be primitive compartments that were precursors to modern cells (i.e., protocells) that could have assembled on early Earth, and the presence of such amphiphilic molecular assemblies on early Earth means that an aqueous environment can host non-polar compounds and thus non-aqueous chemistries. For a review of how these environments could affect different reactions, please see [62].

Amphiphilic molecules can also reduce surface energies [63] by adsorbing onto surfaces such as that of mineral particles [64] or exist at liquid–gas interfaces [35]. The importance of creating such layers is apparent in chemistry. Lipid monolayers have been shown to nucleate mineral growth [65], and surfaces can assist in creating desirable lipid membrane structures [66,67]. The intermembrane spaces could also be a potential site for a range of reactions, including RNA polymerization [68], where confinement to two dimensions and a non-aqueous environment is beneficial.

2.2.6. Condensed Droplet Microenvironments

Non-amphiphilic molecules can also self-aggregate via non-covalent interactions into condensed phases. These condensed phases could form due to a process known as liquid–liquid phase separation, a common phenomenon in cells that forms membraneless organelles [13]. Such phase separation could have also occurred on early Earth and would have yielded membraneless droplets that can form associatively such as coacervation between nucleic acids and cationic peptides [12,69–71], or dissociatively such as aqueous two-phase systems [72–74] or polyester microdroplets [75–77]. These droplets can host and thus concentrate molecules via similar forces as the forces that lead to condensation [78,79]. The interior of such droplet microenvironments can also vary from apolar (mainly) polymer-based environments such as in polyester microdroplets [75] to polar aqueous (but polymer-rich) environments such as in coacervates [70].

2.3. Minerals/Rocks

2.3.1. Solid Mineral Surfaces

Solid minerals are found all over the Earth's crust in rock or suspended colloidal forms, and would have been present in abundance very early on in Earth's history. They are capable of increasing the local concentration of molecules via adsorption due to electrostatics or by reducing interfacial energies [80–83]. Mineral surfaces can also preorganize molecules while precluding water to increase reaction rates [84]. As a physical environment, minerals can contain large surface area to volume ratios, with much of the area being internal 2D interlayers such as in clays, or narrow networks of rock pores that are shielded from light as well as turbulent flow. For an overview of the importance of minerals for prebiotic chemistry, we refer readers to [85–90].

Minerals are also an important source of elements essential to prebiotic chemistry such as phosphorous [91,92]. Consequently, minerals create local microenvironments that can not only enrich molecules by adsorption, but also leach out materials to their surroundings. In particular, we note that most chemical reactions involving minerals will occur at the mineral surface, i.e., a mineral-air or mineral-water interface such as within mineral pores or cracks [87,93,94]. However, there are some mineral-based chemical processes that could occur exclusively in the solid phase, such as metamorphic changes in rocks at high temperature and pressure, which could affect the availability of certain minerals.

2.3.2. Mantle

Earth's solid mantle, which would also have formed very early on in Earth's history, is a source of minerals and gases that can partake in other chemistries once ejected/erupted onto the Earth's surface [95], and undergoes solid-state convection, a key to plate tectonics [96]. The oxidation state of the mantle is possibly driven by the disproportionation of Fe^{2+} [97], with metallic iron sinking and Fe^{3+} persisting in the mantle, rendering it oxidizing [98]. It is the high viscosity and physical inaccessibility of the mantle that enables it to be transiently out of equilibrium with the ocean, atmosphere, and crust. As a result, the oxidation state, not to mention the temperatures and pressures, can differ vastly from other regions of Earth, and enable novel (inorganic) chemistries within the mantle [99,100].

3. Physicochemical Properties

In this section, we broadly introduce the relevant physicochemical characteristics that have wide variability amongst the reaction environments introduced above. The aggregated data showing the values of each physicochemical characteristic serve to guide the design of future prebiotic chemistry studies as a way for researchers to better understand the relevance of each reaction environment to different chemical processes.

3.1. Ionic Strength

Ionic strength is, simplistically, the total concentration of charge (both positive and negative) contributed by all dissolved ions in a given solution [101]. Ionic strength contributions are proportional to the square of the charge on the ion, and are thus greater for divalent ions compared to monovalent ions.

Ionic strength affects the solubility of electrolytes, inter- and intramolecular supramolecular interactions, the dissociation constant of acids (which can result in more dissolved protons in solution and lower pH [102]), and the strength of electrostatic interactions [103]. It can also impact the osmotic pressure of semipermeable systems. It may be significant that no living cell today has an intracellular concentration of 0.6 M NaCl, the sodium chloride concentration of the ocean. Instead, most cells use active transport to maintain the internal concentration of NaCl at 0.015 M, while KCl is maintained at approximately 0.15 M within the cell [104].

High ionic strengths could result in the dissociation of molecular complexes bound through charge-charge interactions such as peptide-nucleotide complexes that form primitive phase separated coacervates upon binding [69]. Salt can also inhibit the self-assembly of phospholipids into vesicles.

Here, we report the range of ionic strengths found in each of the prebiotic reaction environments introduced above (Table 1).

Table 1. The ionic strength of different reaction environments.

Environment	Ionic Strength Range (M)	References
Aqueous solution	0.1–0.8 (oceans) 0.002–6 (lakes) 0.1–17 (lagoons) 0.1–7 (seas) 0.7–6 (hydrothermal brines)	[105]
Sea spray	Up to 6 (marine aerosol)	[106]
Gels	Variable, depending on components. Salt can dramatically alter gel properties.	[107,108]
Deep eutectic solvents	Ranges from 0 to >1, but ionic strength may not be the relevant principle.	[109,110]
Pure formamide	0.024 (commercially available pure formamide contains a significant amount of ionic impurities)	[111]
Lipid bilayer vesicle lumens (interior)	0–0.6, depends on the solution in which the amphiphilic molecules self-assemble.	[112,113]
Condensed droplet microenvironments	Up to 15 (within coacervate droplets)	[79]
Solid mineral surfaces	No ionic strength for solid mineral surfaces, surface charge density may be the more relevant parameter.	

3.2. *Surface Effects*

When considering microenvironments, surface effects must be taken into account because of the large surface area to volume ratios of such environments compared to the bulk. Whether that interface is solid-gas, liquid-gas, liquid-solid, or liquid-liquid, the interface could be a non-negligible site that concentrates materials and increases chemical reactivity [63]. This concentration mechanism could be relevant to prebiotic chemical reactions where the reactants are highly diluted in a mixed reactant pool and would otherwise not react to any appreciable degree [114]. The reactions that occur at the air–water interface of an aerosol or droplet could therefore be more important than reactions in the bulk of the aerosol/droplet [33,115] (Figure 2).

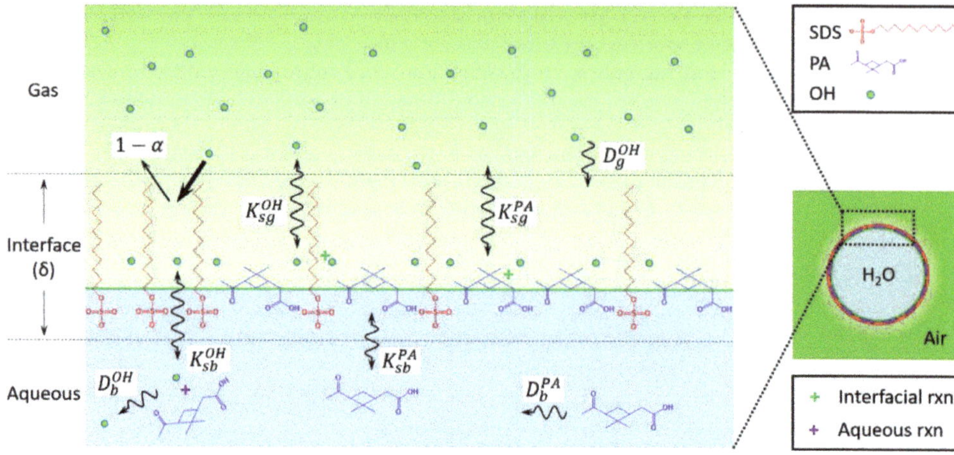

Figure 2. In a droplet system (blue) containing pinonic acid (PA), sodium dodecyl sulfate (SDS), and hydroxyl radicals (OH) in both aqueous and gas phases (green), various reactions can occur in the droplet (volume-dominated process), the gas phase (volume-dominated process), or the gas-droplet interface (surface-dominated process). For example, SDS participates only in reactions with OH (oxidation) at surface-dominated processes due to its high surface activity (as an amphiphile). However, PA can react with OH (oxidation) both at the air-droplet interface (surface-dominated process) as well as inside the bulk droplet (volume-dominated process) due to its lower surface activity than SDS. OH can also participate in reactions in the gas phase, the liquid phase, or at the interface. Reprinted with permission from Huang, Y. et al. "Probing the OH Oxidation of Pinonic Acid at the Air-Water Interface Using Field-Induced Droplet Ionization Mass Spectrometry (FIDI-MS)". *J. Phys. Chem A.* **122**(31), 6445–6456 (2018). [115] Copyright 2018 American Chemical Society.

There are several mechanisms by which materials can accumulate at liquid interfaces. A hydrodynamic mechanism is the 'coffee ring effect'. For example, gas bubbles within heated rock pores have been shown to concentrate catalytic nucleic acids at the bubble interface and increase catalytic activity [116]. This effect can also be driven by surface tensions. Because any surface or interface has a non-zero interfacial tension at the boundary (e.g., the air-water interface being ~72.8 mN/m at room temperature), materials have a propensity to adsorb to the boundary layer, which results in an overall lower free energy. Take, for example, amphiphilic molecules, which form monolayers at the liquid-liquid interfaces that can decrease the interfacial energy by an order of magnitude [63]. Finally, surfaces can adsorb molecules directly. Those studying chemical reactions in/around rock pores need to consider the effects of the mineral surface (e.g., roughness, chemical properties) on each species participating in the reaction [93,117].

Surface features and effects are specific to each system and are not explicitly presented here. Large surface areas are also implicit in microenvironments. For more information, we refer the interested reader to the interfacial catalysis literature [118–120].

3.3. Viscosity

Chemical reactions in solution are either diffusion-limited (where reactants will react instantaneously upon contact with each other, and the reaction is thus controlled by the speed at which the reactants diffuse toward each other in solution) or reaction-limited (e.g., due to some energetic barrier) [121].

For diffusion-limited reactions, the viscosity of materials will control the speed of the reaction with higher viscosities typically slowing down reaction rates. In some cases, high viscosity may aid reactions by limiting how far molecules can diffuse from each other.

Highly viscous media have been shown to support the replication and catalysis of primitive nucleic acids [48,122].

Here, we report the typical viscosities found in each of the prebiotic environments introduced above (Table 2). It should be noted that the rheology of materials (how materials deform and flow) depends on the applied stresses and strains, the temperature, and the length scales considered. Gels, for instance, can appear solid at larger length scales but still support flow inside their pores. Some materials exhibit viscous propreties at long-time scales, and elastic behavior at short-time scales.

Table 2. The viscosity of different reaction environments.

Environment	Typical Viscosity (mPas)	References
Aqueous solution	0.89–1.00 (freshwater at room temperature) Up to 1.3 (seawater at room temperature, depending on salinity)	[123–125]
Sea spray	Ranges from 1 (sea water) to 10–10,000 during evaporation or in presence of organics	[126]
Gels	Ranges from 1 up to 2×10^6 (colloidal silica gel)	[127]
Ice	10^{15}	[128]
Deep eutectic solvents	Variable; >100 and as high as 1700 possible	[129,130]
Formamide	3.23	[131]
High pressure supercritical fluids	0.02–0.16 (CO_2, depending on pressure) 2.98 (water)	[132–134]
Tars	10–over 10^{10}	[135]
Inside lipid bilayers	2D diffusion ~100–1000 1–1500 (heterogeneous)	[136]
Condensed droplet microenvironments	100 (coacervate)	[137]
Solid mineral surfaces	$<1.0 \times 10^{28}$ (crust)	[138]
Mantle	2.8×10^{25}	[138]

3.4. *Specific Heat Capacity*

Specific heat capacity is the amount of energy needed to increase the temperature of one kg of a material by one degree K. In other words, it can be used as a measure of the energy that it takes for a volume of a material to heat or cool down to a given temperature or the insulation or conduction ability of the material [139], and depends on a material's temperature and phase.

The specific heat capacity is important where temperature stability (or fluctuations) is critical. Given that the heat capacity of water (a good insulator) [140] and solids such as rocks (which could be good heat conductors) [141] are quite different, their close proximity in the form of water-rock interfaces (such as in hydrothermal vent environments or hot springs [142]) leads to significant heat transfer and could potentially affect processes such as self-assembly [87], geoelectrochemistry [143], transport in thermal gradients [30], evaporation, or even mineral composition [144].

Here, we report on the range of specific heat found in each of the prebiotic environments introduced above (Table 3).

Table 3. The specific heat of different reaction environments.

Environment	Specific Heat (kJ/Kg K)	References
Aqueous solution	4.18 (freshwater) 3.6–4.18 (saltwater), at room temperature.	[124,125,145,146]
Sea spray	Aerosols readily evaporate; specific heat is not very relevant.	
Gels	0.8–1.10 (silica gel) Specific heat for hydrogels depends on water level and temperature, for example, up to 30.	[147,148]
Ice	0.4873–0.3496 (from 0 to −80 °C, respectively)	[149]
Deep eutectic solvent	1.5–1.8 (example of salt eutectic)	[150]
Formamide	2.39	[151]
High pressure supercritical fluids	3–30 (CO_2, depending on pressure) 27–690 (water, depending on pressure)	[152,153]
Tars	1.25–2	[154]
Inside bilayers	0.3–0.9; higher near melting temperature	[155,156]
Condensed droplet microenvironments	1.483	[157]
Solid mineral surfaces	0.180 (bromyrite) to 1.510 (epsomite); however, most are between 0.3 and 0.9	[158]
Mantle	1.250	[159]

3.5. *pH*

Because pH is defined as a solution property, only liquids can exhibit a pH. pH is usually defined in terms of the autoionization of water, but the concept of pH can also be extended to neat nonaqueous solvents as long as they have some ability to donate a proton [160]. The solution pH affects chemical properties such as the protonation state of molecules in the solution and hence their potential to participate in chemical reactions or assembly into supramolecular structures. For example, pH fluctuations could give rise to the cyclical assembly and disassembly of coacervate droplets due to changes in the charge states of the constituent polymers [161], while pH changes also modulate RNA base-pairing, resulting in the ability to affect strand separation [162] and vesicle self-assembly [163]. While the pH of a solution is generally uniform, there are some cases where the pH within an environment exhibits changes and is not uniform such as in certain terrestrial lakes [164] or water mixing zones [165].

Here, we report on the range of pH found in each of the prebiotic environments introduced above (Table 4).

Table 4. The pH range of different reaction environments.

Environment	pH Range	References
Aqueous solution	6.3–7.2 (4.0 Ga ocean) 6.5–7.7 (2.5 Ga ocean) 8.2 (modern ocean) 6–8 (freshwater) Pure water is 7.0 Hot spring environments have more variability, and can range from very acidic (less than pH 3) to somewhat alkaline (as high as pH 10).	[4,166–168]

Table 4. *Cont.*

Environment	pH Range	References
Sea spray	Around 8.0	[169]
Gels	Variable, depending on components.	[75,170–173]
Deep eutectic solvents	1.2–13.5 (eutectic at room temperature; pH varies greatly between eutectics, and also changes with temperature, down to pH 0)	[174,175]
High pressure supercritical fluids	2.80–2.95 (of water around $scCO_2$)	[176]
Inside lipid bilayers	pH can be of a variety of ranges such as low as pH 2 or lower [177] or as high as pH 12 [178].	
Condensed droplet microenvironments	Highly dependent on the components, and especially their charge states at different pH (i.e., pKa).	
Solid mineral surfaces	Aqueous solutions containing solid mineral surfaces are mostly acidic. However, some have been found that were alkaline (pH 8.7–9.6).	[179]
Mantle	Mantle-derived igneous rocks can be alkaline, while mantle-derived minerals on the seafloor (around hydrothermal systems) can be around pH 9–11	[180,181]

3.6. Density

The density of reaction environments is important to consider, because differences in density could lead to the physical separation of different components, a process that occurs during ocean stratification or hydrodynamic sorting.

While solids are generally more dense than liquids, which are more dense than gases, there are cases where this is inverted such as tungsten hexafluoride gas [182] being at least 10 times denser than graphene aerogel solid [183] or solid pumice being able to float on water. The density of a material will increase upon increasing pressure (decreasing the volume due to pressure-driven compression (Section 3.10)), while increasing temperatures will usually, but not always [184], cause a density decrease. Some materials are non-uniform (e.g., rocks or minerals [185]), leading to different microenvironments even within the same material.

Here, we report on the density of each of the prebiotic environments introduced above (Table 5); however, as there are different environments on early Earth with variable temperatures (such as hot springs [6]) and pressured (such as near hydrothermal vent systems in the deep ocean [186]), the densities reported here may change accordingly.

Table 5. The density of the different reaction environments.

Environment	Density (g/mL)	References
Aqueous solution	0.9999749 (freshwater at 4 °C); 0.9970470 (freshwater at 25 °C) 1.025 (seawater, average; can be up to 1.09 depending on salinity)	[124,125,187]
Sea spray	1.12–2.16 (at room temperature)	[188,189]
Gels	Lower bound is that of the solvent for dilute gels.	
Ice	0.84–0.91 (sea ice)	[190]
Deep eutectic solvent	0.8–1.8 (example of a eutectic between 5 and 100 °C)	[174,191]

Table 5. *Cont.*

Environment	Density (g/mL)	References
Formamide	1.129 (at 25 °C)	[131]
High pressure supercritical fluids	0.1–1 (CO_2, depending on temperature and pressure) ~0.1–0.326 (water, depending on temperature and pressure)	[134,192,193]
Tars	1.1–1.23	[194,195]
Lipid bilayers	~0.9 for the lipid bilayer itself (e.g., decanoic acid density is 0.893 g/cm^3) In the aqueous lumen, values as per 'aqueous solution'.	[196]
Condensed droplet microenvironments	1.18–1.92	[197]
Solid mineral surfaces	1.2 (kerogen) to 10.969 (uraninite); however, most are typically between 2 and 7	[158,198]
Mantle	3.4 (mantle surface, and gets larger deeper)	[198]

3.7. *Dielectric Constant*

A general adage in chemistry is that "like dissolves like"; polar solvents are more likely to dissolve charged solutes or solutes with high dipole moments (i.e., polar compounds) [199]. This is because the ability of a solvent to disrupt solute-solute interactions depends on the specific intermolecular forces involved.

One parameter used to estimate solvent polarity is the zero-frequency component of the dielectric constant (ε). While other measures such as hydrogen bonding capacity, dipole moment, and acidity/basicity are also important, the dielectric constant remains a good rule of thumb for estimating the polarity of the solvent as well as how miscible solvents are with each other [200]. Formally, ε is the relative permittivity of a material compared to vacuum and is defined as the amount of polarization that a material will experience (i.e., the magnitude of dipole moments) when an electric field is applied to it [199,201]. This means that ε is a measure of the polarizability of a solvent, with solvents having $\varepsilon \gtrsim 10$–20 defined as polar. As points of reference, apolar organic solvents have a relatively low ε~2, while polar water has an ε~80 [202]. Solvents with similar dielectric constants are generally miscible.

Given the wide variety of chemistries thought necessary for the origin of life [203], it could have been possible for a variety of aqueous and nonaqueous media to contribute to the potential prebiotic reaction space. Polar solvents can be divided further into protic solvents (able to hydrogen bond or donate hydrogen) and aprotic solvents, which can be determined by looking at the solvent molecule's structure.

Here, we report the ε of each of the prebiotic environments introduced above (Table 6).

Table 6. The dielectric constant (ε) of different reaction environments.

Environment	ε (unitless)	References
Aqueous solution	~70–80 (decreases with increasing temperature and salinity; seawater may be slightly lower than freshwater)	[204,205]
Sea spray	2.5–50	[206]
Gels	1.008–1.9 (silica gel, depending on density)	[207]

Table 6. *Cont.*

Environment	ε (unitless)	References
Ice	30–130 (ice)	[208]
Deep eutectic solvent	22.8 (one example)	[109]
Formamide	105–113 (room temperature)	[209,210]
High pressure supercritical fluids	1.07–1.46 (CO_2, depending on temperature and pressure)	[211]
Tars	Up to 8 (coal tar)	[195,212]
Inside bilayers	2–3, can be higher for membranes that are more permeable than phospholipids	[213,214]
Condensed droplet microenvironments	40–50	[215]
Solid mineral surfaces	4.9–7.5	[216]
Mantle	~38 (water in the upper mantle at 300 km and 1000 K)	[217]

3.8. Boiling, Melting/Freezing Temperatures

At higher pressures such as in the deep ocean [7], different phase transitions can occur such as the direct sublimation of ice to water vapor upon increasing temperature [218]. Additionally, hysteresis, such as in rock pores, has also been observed (i.e., the freezing temperature is not identical to the melting temperature [219]).

Knowing the phase transition temperatures of materials is important for several reasons. Phase transitions accessible to the temperatures and pressures on early Earth impact the abundance of solvents. Additionally, materials in different phases have very different properties. Carbon dioxide gas and supercritical liquid carbon dioxide will have different affinities for various prebiotically plausible chemicals [203,220,221], resulting in differences in the reactivities or plausible chemistries residing within such environments. Furthermore, it has been shown that freeze-thaw cycles in water could have contributed to primitive genetic biopolymer (i.e., RNA) replication and assembly [222,223].

Here, we report on the boiling (liquid to gas transition) and melting (solid to liquid transition) temperatures of each of the prebiotic environments introduced above at atmospheric pressure as a point of reference (Table 7). However, as there were different environments on early Earth with variable pressure such as near hydrothermal vent systems in the deep ocean [186], the temperatures reported here will change accordingly (and at pressures below the triple point, there may only be one phase-transition temperature, i.e., sublimation, physically possible). Additionally, for some systems such as condensed droplet microenvironments, "melting" may refer to the transition from the condensed phase to the uniform phase, as increasing temperatures will inhibit the non-covalent bonds required for the structure to form, and depends on the composition of the system [69,224,225].

Table 7. The boiling and melting/freezing temperatures of different reaction environments.

Environment	Boiling Temperature	Melting/Freezing Temperature	References
Aqueous solution	Freshwater (100 °C); As high as 102 °C (seawater, depending on salinity)	Freshwater (0 °C); As low as −2 °C (seawater, depending on salinity)	[124,125,226–228]
Sea spray	70–100 °C	Close to 0 °C	[229,230]
Gels	2230 °C (silica gel)	1710 °C (silica gel)	[231]

Table 7. *Cont.*

Environment	Boiling Temperature	Melting/Freezing Temperature	References
Ice (eutectic)	In solid form, same as water (depending on salinity).		
Formamide	210 °C	2–3 °C	[232,233]
High pressure supercritical fluids		See footnote *	
Tars	190–400 °C		[234]
Inside lipid bilayers		See footnote ^	
Solid mineral surfaces	N/A	700–900 °C	[235]
Mantle	N/A	~3600 °C near the core–mantle boundary	[236]

* Typical "boiling" and "melting/freezing" transitions may not be applicable. Rather, the supercritical fluid to liquid, solid, and/or gas transition temperatures will depend on the pressure and is unique to each system based on the phase diagram. For example, $scCO_2$ will transition to the liquid state below 304 K at 100 bar, but will transition to the solid state below 304 K at 10,000 bar [220]. Supercritical water will transition to the liquid state below 647 K at any pressure above 22.1 MPa; supercritical water cannot directly transition to the solid form under any circumstances [237]. Neither $scCO_2$ nor supercritical water can transition to the gas phase based on temperature changes and can only transition to the gas phase upon decreasing pressure. ^ The lipid bilayer itself may not boil (as boiling requires the bilayer to vaporize, effectively resulting in the loss of the bilayer structure). However, the boiling point of the lipids that compose the bilayer vary depending on lipid composition. Typically, the boiling point increases with an increasing chain length; for example, caproic acid (C6 saturated) has a boiling point of 205.8 °C, while stearic acid (C18 saturated) has a boiling point of 376.1 °C [238]. The "melting" of a bilayer refers to the solid (gel) to liquid transition, and not the melting of the lipid components themselves. This also depends on the lipid composition; longer chain lipids typically have a higher phase transition temperature [239,240].

3.9. Vapor Pressure

Vapor pressure is related to the volatility of a material (i.e., the amount of gas that is released from a material at any given point) with the boiling point being defined as when the vapor pressure of the liquid material is equivalent to the ambient pressure (Section 3.8). Extreme cases include the highly volatile ammonia [241] and non-volatile mineral oil [242]. Higher temperatures will result in higher vapor pressure as per the Antoine relation [243]. Vapor pressure is also applicable to solids that sublimate (e.g., dry ice) [244]. A related concept is Henry's law for mixtures of gases, which relates the amount of a dissolved gas to the partial pressure of that gas.

The volatility of a prebiotic material impacts whether that reaction environment is stable at a given temperature, or whether it will spontaneously (and quickly) change form into a gas, even below the boiling point. For example, it has been reported that the vapor pressure of fatty acids [245] and fatty acid esters [246] decreases with increasing chain-length. This suggests that fatty acids on early Earth, which could undergo liquid-phase reactions at high temperatures such as in hot spring environments [6], may have been more biased toward longer-chain fatty acids, as shorter chain fatty acids would likely have been volatilized into the gas phase.

Here, we report on the vapor pressure of each of the prebiotic environments introduced above (Table 8).

Table 8. The vapor pressure of different reaction environments.

Environment	Vapor Pressure (kPa)	References
Aqueous solution	2.3–4.2 (freshwater, room temperature) 2.1–3.9 (seawater, room temperature, depending on salinity)	[124,125]
Gels	~0.13–2.3, depending on the gel formulation and conditions.	[247]

Table 8. *Cont.*

Environment	Vapor Pressure (kPa)	References
Ice	6.1 (ice at 0 °C), but decreases with decreasing temperature (for example, 0.1 at −20 °C and 0.0014 at −100 °C).	[248]
Deep eutectic solvent	1.48 ($CaCl_2$ eutectic in water at 20 °C). However, vapor pressure of other eutectics may vary depending on composition and temperature.	[249,250]
Formamide	0.008	[233]
Inside bilayers	Vapor pressure will be related to the vapor pressure of the bilayer components; vapor pressure typically decreases with increasing chain length (at constant temperature).	[251,252]
Condensed droplet microenvironments	Very low to negligible vapor pressure (ionic liquids)	[253]
Solid mineral surfaces	Around 0.05–0.25 (melted minerals >1900 K) Vapor pressure of solid mineral surfaces is negligible	[254]

3.10. Compressibility and Stiffness

For solids, one measure of deformability is the Young's Modulus (E) (Figure 3a). Materials with lower E are more compressible (less stiff), and vice versa. The E of coal is about 10 times less than limestone [255,256]. The Young's modulus is only applicable to solids, as fluids (such as liquids and gases) require zero force to change in size lengthwise.

Bulk modulus (K) is defined as the amount of pressure (equally from all sides) required to effect a resulting volume change on a material (Figure 3b) and is defined for both fluids and solids. A typical gas has a very low K of <0.1 GPa [255] whereas quartz or clay has a K of 20–40 GPa [255].

While not immediately obvious, the compressibility and stiffness of materials become relevant on early Earth either in high-pressure environments, or when environments encounter high pressures resulting from a large external force. Examples include pressure from water deep in the ocean [186] or the late heavy bombardment during impacts [257]. The temporary deformation or compression of surface minerals with low E, for example, could have affected primitive mineral-driven processes such as mechanochemical sugar [258] or peptide [94,258] synthesis, possibly within mica sheets [259].

Here, we report the K and E of each of the prebiotic environments introduced above (Table 9).

Table 9. The Young's (E) and bulk (K) modulus of different reaction environments.

Environment	E (GPa) *	K (GPa)	References
Aqueous solution	-	2.1	[260]
Gels	0.05–10 (of a silica aerogel, depending on gel density)	4–20 (of an alkaline-calcium silica hydrogel, depending on pressure)	[261,262]
Ice	8.6–12 (depends on the plane)	8.5–11.5 (depends on temperature)	[263,264]
High pressure supercritical fluids	-	1 (water at room temperature and pressure) 0.1–0.7 (CO_2, depending on temperature and pressure)	[192,265,266]
Inside bilayers	0.02–0.03	0.6–0.9 (depending on temperature and location)	[267–269]

Table 9. *Cont.*

Environment	*E* (GPa) *	*K* (GPa)	References
Condensed droplet microenvironments	These values will all depend on the droplet composition; "aging" is also an issue in these droplets.		
Solid mineral surfaces	6.38–288 (depending on the mineral and pressure)	40–120 (depending on mineral and pressure)	[270–272]
Mantle	150–720 (depending on depth)	100–600 (depending on depth)	[272]

* The Young's modulus is not defined for liquids and gases.

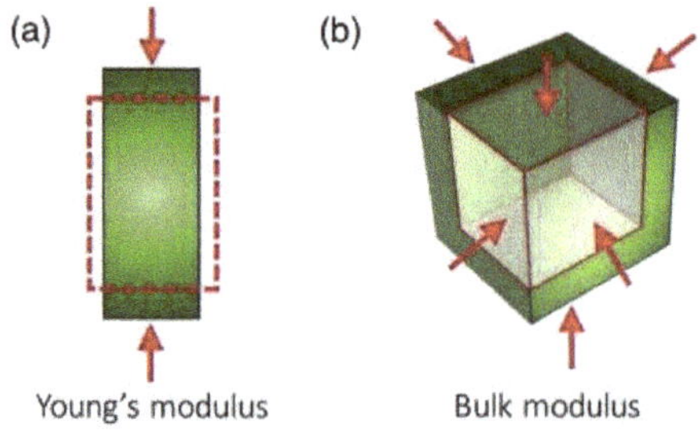

Figure 3. A physical description of the direction of forces used to calculate (**a**) the Young's modulus (*E*) and (**b**) bulk modulus (*K*). Reprinted with permission from Burtch, NC, et al. "Mechanical Properties in Metal-Organic Frameworks: Emerging Opportunities and Challenges for Device Functionality and Technological Applications". *Adv. Mater.* **30**(37), 1704124 (2018). [273] Copyright 2018 Wiley.

3.11. Exposure to Radiation

Radiation is the process of energy transmission in the form of photons or massive particles and includes electromagnetic radiation (radio waves, microwaves, infrared, visible light, ultraviolet, X-rays, and gamma radiation) and particle radiation (e.g., alpha and beta particles, neutrons). Electromagnetic waves carry radiant energy as photons wherein the wavelength determines the energy, whereas particle radiation is the result of fast-moving subatomic particles whose energies can vary depending on their mass and velocities. Radiation is further classified as non-ionizing versus ionizing depending on the energy of the photon or particle. The energy of ionizing radiation is broadly defined to be between 10 and 33 eV, which is typically enough to ionize molecules and break chemical bonds (the energy of a C–C bond, for example, is about 3.6 eV). The region of electromagnetic radiation including higher energy ultraviolet, X-rays, and gamma radiation as well as typical particle radiation (alpha radiation, beta radiation and neutron radiation) are all considered types of ionizing radiation.

On early Earth, UV radiation would have been a strong driver of chemical synthesis and evolution. While the presence of ozone on modern Earth can absorb all ionizing and 98% of non-ionizing UV light, the surface of early Earth was exposed to much higher fluences of UV light (in particular, wavelengths longer than ~200 nm) prior to the build-up of atmospheric oxygen and thus ozone. In addition, while the young Sun was about 25% less luminous than today [274], its output in the UV region was likely larger. UV radiation has been demonstrated to play a key role in the synthesis of prebiotic molecules [275]. For example, the Sutherland group reported a UV radiation-driven photosynthesis of simple sugars from HCN [276], while UV radiation has also been shown to produce amino acids in the atmosphere [277]. However, UV radiation is also a double-edged sword, and has been shown to induce the degradation or structural/configurational alteration of organic molecules including DNA [278], proteins, and lipids [279].

Radiolysis from ionizing radiation is also employed in prebiotic chemistry as a synthetic mechanism [57]. Ionizing radiation could have been present through multiple sources such as cosmic rays and radioactive minerals. Cosmic rays consist of high-energy protons and atomic nuclei originating from the Sun or outside of the Solar System [280]; the energy spectrum of primary cosmic rays is between 10^{16} eV (2.31 × 10^{17} kcal/mol or 3.89 × 10^{17} kT, at 298 K) and 10^{18} eV (2.31 × 10^{19} kcal/mol or 3.89 × 10^{19} kT, at 298 K) [281]. Cosmic rays impact Earth's upper atmosphere to produce showers of secondary photons and particles. A higher output of solar energetic particles from the young Sun via more frequent solar flares and coronal mass ejections has been proposed to have exposed the atmosphere of the early Earth to significantly higher fluxes of radiation in comparison to today [282]. Ancient radioactive mineral deposits containing, for example, monazite and uraninite [283], could also have provided locally high fluxes of alpha, beta, or gamma radiation on their surfaces [284]. Under the primordial conditions of the Hadean eon when the ^{235}U isotope was much more abundant, so-called natural nuclear reactors that can promote self-sustaining nuclear fission chain reactions may have been relatively commonplace and produced highly radioactive local environments [285]. Over a dozen individual fission zones are known from the Oklo locality in Gabon [283]. A typical fission zone comparable in size to those found at Oklo are thought to produce ~10 kilowatts of power output (radiolysis and heat) wherein ~13% of this power is composed of γ or β rays that can penetrate substantial distances beyond their host minerals [283,284].

The availability of radiation in prebiotic reaction environments is determined by how far the radiation is able to penetrate a given material and the characteristics of the radiation source. Gamma radiation, arising from the radioactive decay of atomic nuclei, is characterized by short-wavelength electromagnetic waves (~10^{-11} m) with the highest photon energies above 100 keV. Gamma radiation is capable of penetrating materials to significant depths, requiring thick layers of high-density materials to block it. In comparison, alpha radiation with a fast-moving helium-4 nucleus is halted by a sheet of paper, and beta radiation consisting of high-velocity electrons can be stopped by an aluminum plate. The spectrum of the young Sun and attenuation of UV light by gases and aqueous media is covered thoroughly by Ranjan and Sasselov [275].

4. Conclusions

Here, we introduced some important physicochemical properties of various prebiotic reaction environments, with some brief examples of relevant prebiotic processes that could have been modulated by those physicochemical properties. Correlating physicochemical properties with reaction environments, while considering such properties during the design of prebiotically plausible reactions is highly relevant to the origins of life field. However, values of physicochemical characteristics in some environments still remain to be elucidated. This work serves as a request for the community to contribute to "filling in the blanks" in future works, so that researchers in the field can have a more holistic understanding of the prebiotic geochemical reaction environment. We also acknowledge the fact that experiments that fill in these blanks could be rather tedious and may not lead to high-profile publications; however, such studies are essential to pursue.

From this overview, it is evident that an extremely wide range of physicochemical conditions can be accessed through the early Earth environment, which further supports the widely accepted notion that a large repertoire of chemical reactions were taking place on early Earth. Likewise, the prebiotic chemical repertoire is also highly diverse, with reactions or processes that may have had a range of tolerances to a wide variety of conditions and others that did not. The fact that modern biology, especially in the form of extremophiles, is also tolerant to a wide variety of conditions could be an artifact of the chemical reactions or processes that led to the life's origins.

Generally speaking, the robustness/tolerance of the entire prebiotic chemical reaction repertoire to changes in conditions is at present not well-understood due to the large parameter space. In future, probing the "limits" of a wide range of prebiotic processes is necessary to gain a better understanding of which prebiotic environments could have plausibly hosted certain prebiotic chemical reaction networks. By aggregating this type of data, it will also be possible to conjecture which groups of prebiotic chemical processes/reactions could have been co-localized with each other as well as those which likely could not have occurred simultaneously under the same conditions in the same location.

The environments, properties, and examples of prebiotic processes provided here are but a brief and general overview of the entire prebiotic chemical milieu, and are not meant to be an exhaustive or comprehensive resource. A number of parameters for many primitive environments are not known (and as such, cannot be presented here), which behooves the field to continue characterizing the unknown physicochemical properties of all reaction environments. We further look forward to the field's continuing exploration of extraterrestrial reaction environments, which could also provide insights into the possible reaction conditions on early Earth.

Author Contributions: All authors contributed to the preparation of this manuscript. All authors have read and agreed to the published version of the manuscript.

Funding: R.Y. is supported by the Japan Society for the Promotion of Science (JSPS) Grant-in-aid 21K14029. A.C.F acknowledges support from the University of New South Wales Strategic Hires and Retention Pathways (SHARP) program and the Australian Research Council Discovery Project Grant DP210102133. A.W. is supported by an Australian Research Council Discovery Early Career Award DE210100291. T.Z.J. is supported by a JSPS Grant-in-Aid 21K14746, a Start-up Grant from the Earth-Life Science Institute (ELSI), and by the Assistant Staffing Program by the Gender Equality Section, Diversity Promotion Office, Tokyo Institute of Technology.

Institutional Review Board Statement: Not applicable.

Informed Consent Statement: Not applicable.

Data Availability Statement: No new data was generated by this study.

Acknowledgments: A.S. and T.Z.J. acknowledge the Blue Marble Space Institute of Science (BMSIS) Young Scientist Program (YSP); A.S., currently a BMSIS visiting scholar, was a YSP fellow and connected with T.Z.J., a BMSIS YSP mentor, through the program.

Conflicts of Interest: The authors declare no conflict of interest.

References

1. Cleaves, H.J. Prebiotic Chemistry: Geochemical Context and Reaction Screening. *Life* **2013**, *3*, 331–345. [CrossRef]
2. Barge, L.M. Considering Planetary Environments in Origin of Life Studies. *Nat. Commun.* **2018**, *9*, 5170. [CrossRef] [PubMed]
3. Lunine, J.I. Physical Conditions on the Early Earth. *Philos. Trans. R. Soc. Lond. B Biol. Sci.* **2006**, *361*, 1721–1731. [CrossRef]
4. Krissansen-Totton, J.; Arney, G.N.; Catling, D.C. Constraining the Climate and Ocean pH of the Early Earth with a Geological Carbon Cycle Model. *Proc. Natl. Acad. Sci. USA* **2018**, *115*, 4105–4110. [CrossRef]
5. Van Kranendonk, M.J. Earth's Early Atmosphere and Surface Environments: A Review. In *Earth's Early Atmosphere and Surface Environment*; Geological Society of America Special Papers; Geological Society of America: Boulder, CO, USA, 2014; pp. 105–130. ISBN 9780813725048.
6. Damer, B.; Deamer, D. The Hot Spring Hypothesis for an Origin of Life. *Astrobiology* **2020**, *20*, 429–452. [CrossRef] [PubMed]
7. Zhang, X.; Li, L.-F.; Du, Z.-F.; Hao, X.-L.; Cao, L.; Luan, Z.-D.; Wang, B.; Xi, S.-C.; Lian, C.; Yan, J.; et al. Discovery of Supercritical Carbon Dioxide in a Hydrothermal System. *Sci. Bull.* **2020**, *65*, 958–964. [CrossRef]
8. Balucani, N. Elementary Reactions and Their Role in Gas-Phase Prebiotic Chemistry. *Int. J. Mol. Sci.* **2009**, *10*, 2304–2335. [CrossRef]
9. Rimmer, P.B.; Shorttle, O. Origin of Life's Building Blocks in Carbon- and Nitrogen-Rich Surface Hydrothermal Vents. *Life* **2019**, *9*, 12. [CrossRef]
10. Green, N.J.; Xu, J.; Sutherland, J.D. Illuminating Life's Origins: UV Photochemistry in Abiotic Synthesis of Biomolecules. *J. Am. Chem. Soc.* **2021**, *143*, 7219–7236. [CrossRef]
11. Cooper, G.M. Cell Membranes. In *The Cell: A Molecular Approach*, 2nd ed.; Sinauer Associates: Sunderland, MA, USA, 2000.
12. Ghosh, B.; Bose, R.; Tang, T.-Y.D. Can Coacervation Unify Disparate Hypotheses in the Origin of Cellular Life? *Curr. Opin. Colloid Interface Sci.* **2021**, *52*, 101415. [CrossRef]
13. Yoshizawa, T.; Nozawa, R.-S.; Jia, T.Z.; Saio, T.; Mori, E. Biological Phase Separation: Cell Biology Meets Biophysics. *Biophys. Rev.* **2020**, *12*, 519–539. [CrossRef]
14. Hayes, A.G. The Lakes and Seas of Titan. *Annu. Rev. Earth Planet. Sci.* **2016**, *44*, 57–83. [CrossRef]
15. Barnes, J.W.; Turtle, E.P.; Trainer, M.G.; Lorenz, R.D.; MacKenzie, S.M.; Brinckerhoff, W.B.; Cable, M.L.; Ernst, C.M.; Freissinet, C.; Hand, K.P.; et al. Science Goals and Objectives for the Dragonfly Titan Rotorcraft Relocatable Lander. *Planet. Sci. J.* **2021**, *2*, 130. [CrossRef]

16. Pizzarello, S.; Shock, E. The Organic Composition of Carbonaceous Meteorites: The Evolutionary Story ahead of Biochemistry. *Cold Spring Harb. Perspect. Biol.* **2010**, *2*, a002105. [CrossRef]
17. Arumainayagam, C.R.; Garrod, R.T.; Boyer, M.C.; Hay, A.K.; Bao, S.T.; Campbell, J.S.; Wang, J.; Nowak, C.M.; Arumainayagam, M.R.; Hodge, P.J. Extraterrestrial Prebiotic Molecules: Photochemistry vs. Radiation Chemistry of Interstellar Ices. *Chem. Soc. Rev.* **2019**, *48*, 2293–2314. [CrossRef]
18. Van Dishoeck, E.F.; Herbst, E.; Neufeld, D.A. Interstellar Water Chemistry: From Laboratory to Observations. *Chem. Rev.* **2013**, *113*, 9043–9085. [CrossRef]
19. Lasne, J.; Noblet, A.; Szopa, C.; Navarro-González, R.; Cabane, M.; Poch, O.; Stalport, F.; François, P.; Atreya, S.K.; Coll, P. Oxidants at the Surface of Mars: A Review in Light of Recent Exploration Results. *Astrobiology* **2016**, *16*, 977–996. [CrossRef]
20. Deamer, D.; Damer, B. Can Life Begin on Enceladus? A Perspective from Hydrothermal Chemistry. *Astrobiology* **2017**, *17*, 834–839. [CrossRef]
21. Cable, M.L.; Hörst, S.M.; Hodyss, R.; Beauchamp, P.M.; Smith, M.A.; Willis, P.A. Titan Tholins: Simulating Titan Organic Chemistry in the Cassini-Huygens Era. *Chem. Rev.* **2012**, *112*, 1882–1909. [CrossRef]
22. Kitadai, N.; Maruyama, S. Origins of Building Blocks of Life: A Review. *Geosci. Front.* **2018**, *9*, 1117–1153. [CrossRef]
23. Cleaves, H.J. Prebiotic Chemistry: What We Know, What We Don't. *Evolution* **2012**, *5*, 342–360. [CrossRef]
24. Ruiz-Mirazo, K.; Briones, C.; de la Escosura, A. Prebiotic Systems Chemistry: New Perspectives for the Origins of Life. *Chem. Rev.* **2014**, *114*, 285–366. [CrossRef]
25. Deamer, D.; Singaram, S.; Rajamani, S.; Kompanichenko, V.; Guggenheim, S. Self-Assembly Processes in the Prebiotic Environment. *Philos. Trans. R. Soc. Lond. B Biol. Sci.* **2006**, *361*, 1809–1818. [CrossRef]
26. Tanford, C. The Hydrophobic Effect and the Organization of Living Matter. *Science* **1978**, *200*, 1012–1018. [CrossRef]
27. Das, S.; Lin, Y.-H.; Vernon, R.M.; Forman-Kay, J.D.; Chan, H.S. Comparative Roles of Charge, and Hydrophobic Interactions in Sequence-Dependent Phase Separation of Intrinsically Disordered Proteins. *Proc. Natl. Acad. Sci. USA* **2020**, *117*, 28795–28805. [CrossRef]
28. Maibaum, L.; Dinner, A.R.; Chandler, D. Micelle Formation and the Hydrophobic Effect. *J. Phys. Chem. B* **2004**, *108*, 6778–6781. [CrossRef]
29. Farmer, J.R.; Sigman, D.M.; Granger, J.; Underwood, O.M.; Fripiat, F.; Cronin, T.M.; Martínez-García, A.; Haug, G.H. Arctic Ocean Stratification Set by Sea Level and Freshwater Inputs since the Last Ice Age. *Nat. Geosci.* **2021**, *14*, 684–689. [CrossRef]
30. Mast, C.B.; Schink, S.; Gerland, U.; Braun, D. Escalation of Polymerization in a Thermal Gradient. *Proc. Natl. Acad. Sci. USA* **2013**, *110*, 8030–8035. [CrossRef]
31. Herschy, B.; Whicher, A.; Camprubi, E.; Watson, C.; Dartnell, L.; Ward, J.; Evans, J.R.G.; Lane, N. An Origin-of-Life Reactor to Simulate Alkaline Hydrothermal Vents. *J. Mol. Evol.* **2014**, *79*, 213–227. [CrossRef]
32. Kreysing, M.; Keil, L.; Lanzmich, S.; Braun, D. Heat Flux across an Open Pore Enables the Continuous Replication and Selection of Oligonucleotides towards Increasing Length. *Nat. Chem.* **2015**, *7*, 203–208. [CrossRef]
33. Dobson, C.M.; Ellison, G.B.; Tuck, A.F.; Vaida, V. Atmospheric Aerosols as Prebiotic Chemical Reactors. *Proc. Natl. Acad. Sci. USA* **2000**, *97*, 11864–11868. [CrossRef]
34. Donaldson, D.J.; Tervahattu, H.; Tuck, A.F.; Vaida, V. Organic Aerosols and the Origin of Life: An Hypothesis. *Orig. Life Evol. Biosph.* **2004**, *34*, 57–67. [CrossRef]
35. Rapf, R.J.; Perkins, R.J.; Dooley, M.R.; Kroll, J.A.; Carpenter, B.K.; Vaida, V. Environmental Processing of Lipids Driven by Aqueous Photochemistry of α-Keto Acids. *ACS Cent. Sci.* **2018**, *4*, 624–630. [CrossRef] [PubMed]
36. Fournier, R.O.; Rowe, J.J. The Deposition of Silica in Hot Springs. *Bull. Volcanol.* **1966**, *29*, 585–587. [CrossRef]
37. Mamajanov, I. Wet-Dry Cycling Delays the Gelation of Hyperbranched Polyesters: Implications to the Origin of Life. *Life* **2019**, *9*, 56. [CrossRef]
38. Dass, A.V.; Jaber, M.; Brack, A.; Foucher, F.; Kee, T.P.; Georgelin, T.; Westall, F. Potential Role of Inorganic Confined Environments in Prebiotic Phosphorylation. *Life* **2018**, *8*, 7. [CrossRef]
39. Campbell, T.D.; Febrian, R.; McCarthy, J.T.; Kleinschmidt, H.E.; Forsythe, J.G.; Bracher, P.J. Prebiotic Condensation through Wet–dry Cycling Regulated by Deliquescence. *Nat. Commun.* **2019**, *10*, 1–7. [CrossRef]
40. Mamajanov, I.; Caudan, M.; Jia, T.Z. Protoenzymes: The Case of Hyperbranched Polymer-Scaffolded ZnS Nanocrystals. *Life* **2020**, *10*, 150. [CrossRef]
41. Nguyen, V.S.; Orlando, T.M.; Leszczynski, J.; Nguyen, M.T. Theoretical Study of the Decomposition of Formamide in the Presence of Water Molecules. *J. Phys. Chem. A* **2013**, *117*, 2543–2555. [CrossRef]
42. Bizzarri, B.M.; Saladino, R.; Delfino, I.; García-Ruiz, J.M.; Di Mauro, E. Prebiotic Organic Chemistry of Formamide and the Origin of Life in Planetary Conditions: What We Know and What Is the Future. *Int. J. Mol. Sci.* **2021**, *22*, 917. [CrossRef] [PubMed]
43. Saitta, A.M.; Saija, F. Miller Experiments in Atomistic Computer Simulations. *Proc. Natl. Acad. Sci. USA* **2014**, *111*, 13768–13773. [CrossRef]
44. Koike, T.; Kaneko, T.; Kobayashi, K.; Miyakawa, S.; Takano, Y. Formation of Organic Compounds from Simulated Titan Atmosphere: Perspectives of the Cassini Mission. *Biol. Sci. Space* **2003**, *17*, 188–189.
45. Gerakines, P.A.; Moore, M.H.; Hudson, R.L. Ultraviolet Photolysis and Proton Irradiation of Astrophysical Ice Analogs Containing Hydrogen Cyanide. *Icarus* **2004**, *170*, 202–213. [CrossRef]

46. Takano, Y.; Tsuboi, T.; Kaneko, T.; Kobayashi, K.; Marumo, K. Pyrolysis of High-Molecular-Weight Complex Organics Synthesized from a Simulated Interstellar Gas Mixture Irradiated with 3 MeV Proton Beam. *Bull. Chem. Soc. Jpn.* **2004**, *77*, 779–783. [CrossRef]
47. Saladino, R.; Crestini, C.; Pino, S.; Costanzo, G.; Di Mauro, E. Formamide and the Origin of Life. *Phys. Life Rev.* **2012**, *9*, 84–104. [CrossRef] [PubMed]
48. He, C.; Gállego, I.; Laughlin, B.; Grover, M.A.; Hud, N.V. A Viscous Solvent Enables Information Transfer from Gene-Length Nucleic Acids in a Model Prebiotic Replication Cycle. *Nat. Chem.* **2017**, *9*, 318–324. [CrossRef] [PubMed]
49. Abbott, A.P.; Capper, G.; Davies, D.L.; Rasheed, R.K.; Tambyrajah, V. Novel Solvent Properties of Choline Chloride/urea Mixtures. *Chem. Commun.* **2003**, 70–71. [CrossRef]
50. Hansen, B.B.; Spittle, S.; Chen, B.; Poe, D.; Zhang, Y.; Klein, J.M.; Horton, A.; Adhikari, L.; Zelovich, T.; Doherty, B.W.; et al. Deep Eutectic Solvents: A Review of Fundamentals and Applications. *Chem. Rev.* **2021**, *121*, 1232–1285. [CrossRef]
51. Gull, M.; Zhou, M.; Fernández, F.M.; Pasek, M.A. Prebiotic Phosphate Ester Syntheses in a Deep Eutectic Solvent. *J. Mol. Evol.* **2014**, *78*, 109–117. [CrossRef]
52. Chien, C.-Y.; Yu, S.-S. Ester-Mediated Peptide Formation Promoted by Deep Eutectic Solvents: A Facile Pathway to Proto-Peptides. *Chem. Commun.* **2020**, *56*, 11949–11952. [CrossRef]
53. Smith, E.L.; Abbott, A.P.; Ryder, K.S. Deep Eutectic Solvents (DESs) and Their Applications. *Chem. Rev.* **2014**, *114*, 11060–11082. [CrossRef]
54. Smith, P.J.; Arroyo, C.B.; Lopez Hernandez, F.; Goeltz, J.C. Ternary Deep Eutectic Solvent Behavior of Water and Urea? Choline Chloride Mixtures. *J. Phys. Chem. B* **2019**, *123*, 5302–5306. [CrossRef] [PubMed]
55. Jiménez-Serra, I.; Martín-Pintado, J.; Rivilla, V.M.; Rodríguez-Almeida, L.; Alonso Alonso, E.R.; Zeng, S.; Cocinero, E.J.; Martín, S.; Requena-Torres, M.; Martín-Domenech, R.; et al. Toward the RNA-World in the Interstellar Medium-Detection of Urea and Search of 2-Amino-Oxazole and Simple Sugars. *Astrobiology* **2020**, *20*, 1048–1066. [CrossRef]
56. Kaiser, R.I.; Maity, S.; Jones, B.M. Synthesis of Prebiotic Glycerol in Interstellar Ices. *Angew. Chem. Int. Ed. Engl.* **2015**, *54*, 195–200. [CrossRef]
57. Yi, R.; Tran, Q.P.; Ali, S.; Yoda, I.; Adam, Z.R.; Cleaves, H.J., 2nd; Fahrenbach, A.C. A Continuous Reaction Network That Produces RNA Precursors. *Proc. Natl. Acad. Sci. USA* **2020**, *117*, 13267–13274. [CrossRef]
58. Raveendran, P.; Ikushima, Y.; Wallen, S.L. Polar Attributes of Supercritical Carbon Dioxide. *Acc. Chem. Res.* **2005**, *38*, 478–485. [CrossRef] [PubMed]
59. Deguchi, S.; Tsujii, K. Supercritical Water: A Fascinating Medium for Soft Matter. *Soft Matter* **2007**, *3*, 797–803. [CrossRef]
60. Guttenberg, N.; Virgo, N.; Chandru, K.; Scharf, C.; Mamajanov, I. Bulk Measurements of Messy Chemistries Are Needed for a Theory of the Origins of Life. *Philos. Trans. R. Soc. A* **2017**, *375*, 20160347. [CrossRef]
61. Barranco, F.T.; Dawson, H.E. Influence of Aqueous pH on the Interfacial Properties of Coal Tar. *Environ. Sci. Technol.* **1999**, *33*, 1598–1603. [CrossRef]
62. Serrano-Luginbühl, S.; Ruiz-Mirazo, K.; Ostaszewski, R.; Gallou, F.; Walde, P. Soft and Dispersed Interface-Rich Aqueous Systems That Promote and Guide Chemical Reactions. *Nat. Rev. Chem.* **2018**, *2*, 306–327. [CrossRef]
63. Israelachvili, J.N. *Intermolecular and Surface Forces*, 3rd ed.; Academic Press: San Diego, CA, USA, 2010; ISBN 9780123751829.
64. Hanczyc, M.M.; Mansy, S.S.; Szostak, J.W. Mineral Surface Directed Membrane Assembly. *Orig. Life Evol. Biosph.* **2007**, *37*, 67–82. [CrossRef] [PubMed]
65. Dey, A.; Bomans, P.H.H.; Müller, F.A.; Will, J.; Frederik, P.M.; de With, G.; Sommerdijk, N.A.J.M. The Role of Prenucleation Clusters in Surface-Induced Calcium Phosphate Crystallization. *Nat. Mater.* **2010**, *9*, 1010–1014. [CrossRef] [PubMed]
66. Köksal, E.S.; Liese, S.; Kantarci, I.; Olsson, R.; Carlson, A.; Gözen, I. Nanotube-Mediated Path to Protocell Formation. *ACS Nano* **2019**, *13*, 6867–6878. [CrossRef] [PubMed]
67. Steller, L.H.; Van Kranendonk, M.J.; Wang, A. Dehydration Enhances Prebiotic Lipid Remodeling and Vesicle Formation in Acidic Environments. *ACS Cent. Sci.* **2022**, *8*, 132–139. [CrossRef]
68. Rajamani, S.; Vlassov, A.; Benner, S.; Coombs, A.; Olasagasti, F.; Deamer, D. Lipid-Assisted Synthesis of RNA-like Polymers from Mononucleotides. *Orig. Life Evol. Biosph.* **2008**, *38*, 57–74. [CrossRef]
69. Fraccia, T.P.; Jia, T.Z. Liquid Crystal Coacervates Composed of Short Double-Stranded DNA and Cationic Peptides. *ACS Nano* **2020**, *14*, 15071–15082. [CrossRef]
70. Poudyal, R.R.; Pir Cakmak, F.; Keating, C.D.; Bevilacqua, P.C. Physical Principles and Extant Biology Reveal Roles for RNA-Containing Membraneless Compartments in Origins of Life Chemistry. *Biochemistry* **2018**, *57*, 2509–2519. [CrossRef]
71. Koga, S.; Williams, D.S.; Perriman, A.W.; Mann, S. Peptide-Nucleotide Microdroplets as a Step towards a Membrane-Free Protocell Model. *Nat. Chem.* **2011**, *3*, 720–724. [CrossRef]
72. Cakmak, F.P.; Keating, C.D. Combining Catalytic Microparticles with Droplets Formed by Phase Coexistence: Adsorption and Activity of Natural Clays at the Aqueous/Aqueous Interface. *Sci. Rep.* **2017**, *7*, 3215. [CrossRef]
73. Guo, W.; Kinghorn, A.B.; Zhang, Y.; Li, Q.; Poonam, A.D.; Tanner, J.A.; Shum, H.C. Non-Associative Phase Separation in an Evaporating Droplet as a Model for Prebiotic Compartmentalization. *Nat. Commun.* **2021**, *12*, 3194. [CrossRef]
74. Jia, T.Z.; Hentrich, C.; Szostak, J.W. Rapid RNA Exchange in Aqueous Two-Phase System and Coacervate Droplets. *Orig. Life Evol. Biosph.* **2014**, *44*, 1–12. [CrossRef] [PubMed]
75. Jia, T.Z.; Chandru, K.; Hongo, Y.; Afrin, R.; Usui, T.; Myojo, K.; Cleaves, H.J. Membraneless Polyester Microdroplets as Primordial Compartments at the Origins of Life. *Proc. Natl. Acad. Sci. USA* **2019**, *116*, 15830–15835. [CrossRef] [PubMed]

76. Jia, T.Z.; Bapat, N.V.; Verma, A.; Mamajanov, I.; Cleaves, H.J.; Chandru, K. Incorporation of Basic α-Hydroxy Acid Residues into Primitive Polyester Microdroplets for RNA Segregation. *Biomacromolecules* **2021**, *22*, 1484–1493. [CrossRef] [PubMed]
77. Afrin, R.; Chen, C.; Sarpa, D.; Sithamparam, M.; Yi, R.; Giri, C.; Mamajanov, I.; Cleaves, H.J., II; Chandru, K.; Jia, T.Z. The Effects of Dehydration Temperature and Monomer Chirality on Primitive Polyester Synthesis and Microdroplet Assembly. *Macromol. Chem. Phys.* **2022**, 2200235. [CrossRef]
78. Mountain, G.A.; Keating, C.D. Formation of Multiphase Complex Coacervates and Partitioning of Biomolecules within Them. *Biomacromolecules* **2020**, *21*, 630–640. [CrossRef]
79. Frankel, E.A.; Bevilacqua, P.C.; Keating, C.D. Polyamine/Nucleotide Coacervates Provide Strong Compartmentalization of Mg^{2+}, Nucleotides, and RNA. *Langmuir* **2016**, *32*, 2041–2049. [CrossRef]
80. Zaia, D.A.M. A Review of Adsorption of Amino Acids on Minerals: Was It Important for Origin of Life? *Amino Acids* **2004**, *27*, 113–118. [CrossRef]
81. Pedreira-Segade, U.; Hao, J.; Razafitianamaharavo, A.; Pelletier, M.; Marry, V.; Le Crom, S.; Michot, L.J.; Daniel, I. How Do Nucleotides Adsorb Onto Clays? *Life* **2018**, *8*, 59. [CrossRef]
82. Hashizume, H. Adsorption of Nucleic Acid Bases, Ribose, and Phosphate by Some Clay Minerals. *Life* **2015**, *5*, 637–650. [CrossRef]
83. Lambert, J.-F. Adsorption and Polymerization of Amino Acids on Mineral Surfaces: A Review. *Orig. Life Evol. Biosph.* **2008**, *38*, 211–242. [CrossRef]
84. Erastova, V.; Degiacomi, M.T.; G Fraser, D.; Greenwell, H.C. Mineral Surface Chemistry Control for Origin of Prebiotic Peptides. *Nat. Commun.* **2017**, *8*, 2033. [CrossRef] [PubMed]
85. Rimola, A.; Sodupe, M.; Ugliengo, P. Role of Mineral Surfaces in Prebiotic Chemical Evolution. In Silico Quantum Mechanical Studies. *Life* **2019**, *9*, 10. [CrossRef] [PubMed]
86. Pérez-Aguilar, C.D.; Cuéllar-Cruz, M. The Formation of Crystalline Minerals and Their Role in the Origin of Life on Earth. *Prog. Cryst. Growth Charact. Mater.* **2022**, *68*, 100558. [CrossRef]
87. Gillams, R.J.; Jia, T.Z. Mineral Surface-Templated Self-Assembling Systems: Case Studies from Nanoscience and Surface Science towards Origins of Life Research. *Life* **2018**, *8*, 10. [CrossRef]
88. Cleaves, H.J.; Michalkova Scott, A.; Hill, F.C.; Leszczynski, J.; Sahai, N.; Hazen, R. Mineral-Organic Interfacial Processes: Potential Roles in the Origins of Life. *Chem. Soc. Rev.* **2012**, *41*, 5502–5525. [CrossRef]
89. Hazen, R.M. Chance, Necessity and the Origins of Life: A Physical Sciences Perspective. *Philos. Trans. A Math. Phys. Eng. Sci.* **2017**, *375*, 20160353. [CrossRef]
90. Li, Y. Minerals as Prebiotic Catalysts for Chemical Evolution towards the Origin of Life. In *Mineralogy*; IntechOpen: Rijeka, Croatia, 2022.
91. Pasek, M.; Herschy, B.; Kee, T.P. Phosphorus: A Case for Mineral-Organic Reactions in Prebiotic Chemistry. *Orig. Life Evol. Biosph.* **2015**, *45*, 207–218. [CrossRef]
92. Walton, C.R.; Shorttle, O.; Jenner, F.E.; Williams, H.M.; Golden, J.; Morrison, S.M.; Downs, R.T.; Zerkle, A.; Hazen, R.M.; Pasek, M. Phosphorus Mineral Evolution and Prebiotic Chemistry: From Minerals to Microbes. *Earth Sci. Rev.* **2021**, *221*, 103806. [CrossRef]
93. Gözen, İ. Did Solid Surfaces Enable the Origin of Life? *Life* **2021**, *11*, 795. [CrossRef]
94. Stolar, T.; Grubešić, S.; Cindro, N.; Meštrović, E.; Užarević, K.; Hernández, J.G. Mechanochemical Prebiotic Peptide Bond Formation*. *Angew. Chem. Int. Ed. Engl.* **2021**, *60*, 12727–12731. [CrossRef]
95. Cleaves, H.J.; Chalmers, J.H.; Lazcano, A.; Miller, S.L.; Bada, J.L. A Reassessment of Prebiotic Organic Synthesis in Neutral Planetary Atmospheres. *Orig. Life Evol. Biosph.* **2008**, *38*, 105–115. [CrossRef] [PubMed]
96. Ogawa, M. Mantle Convection: A Review. *Fluid Dyn. Res.* **2008**, *40*, 379–398. [CrossRef]
97. Frost, D.J.; Liebske, C.; Langenhorst, F.; McCammon, C.A.; Trønnes, R.G.; Rubie, D.C. Experimental Evidence for the Existence of Iron-Rich Metal in the Earth's Lower Mantle. *Nature* **2004**, *428*, 409–412. [CrossRef] [PubMed]
98. Benner, S.A.; Kim, H.-J.; Biondi, E. Prebiotic Chemistry That Could Not Not Have Happened. *Life* **2019**, *9*, 84. [CrossRef]
99. Labrosse, S.; Hernlund, J.W.; Coltice, N. A Crystallizing Dense Magma Ocean at the Base of the Earth's Mantle. *Nature* **2007**, *450*, 866–869. [CrossRef]
100. Hu, Q.; Kim, D.Y.; Liu, J.; Meng, Y.; Yang, L.; Zhang, D.; Mao, W.L.; Mao, H.-K. Dehydrogenation of Goethite in Earth's Deep Lower Mantle. *Proc. Natl. Acad. Sci. USA* **2017**, *114*, 1498–1501. [CrossRef]
101. Sastre de Vicente, M.E. The Concept of Ionic Strength Eighty Years after Its Introduction in Chemistry. *J. Chem. Educ.* **2004**, *81*, 750. [CrossRef]
102. Khouri, S.J. Titrimetric Study of the Solubility and Dissociation of Benzoic Acid in Water: Effect of Ionic Strength and Temperature. *Am. J. Analyt. Chem.* **2015**, *06*, 429–436. [CrossRef]
103. Hu, Y.; Li, K.; Li, Y.; Liu, H.; Guo, M.; Ye, X.; Wu, Z.; Lee, K. Dyes Adsorption onto Fe_3O_4 -bis(trimethoxysilylpropyl)amine Composite Particles: Effects of pH and Ionic Strength on Electrostatic Interactions. *ChemistrySelect* **2019**, *4*, 617–622. [CrossRef]
104. Melkikh, A.V.; Sutormina, M.I. Model of Active Transport of Ions in Cardiac Cell. *J. Theor. Biol.* **2008**, *252*, 247–254. [CrossRef]
105. Millero, F.J. The Physical Chemistry of Natural Waters. *Pure Appl. Chem.* **1985**, *57*, 1015–1024. [CrossRef]
106. Mekic, M.; Gligorovski, S. Ionic Strength Effects on Heterogeneous and Multiphase Chemistry: Clouds versus Aerosol Particles. *Atmos. Environ.* **2021**, *244*, 117911. [CrossRef]
107. Feng, Y.; Taraban, M.; Yu, Y.B. The Effect of Ionic Strength on the Mechanical, Structural and Transport Properties of Peptide Hydrogels. *Soft Matter* **2012**, *8*, 11723–11731. [CrossRef] [PubMed]

108. Anderson, M.A.; Tomić, M.; Gieselmann, M.J.; Villegas, M.A. The Critical Gelling Point in Silica Gels Containing Lithium, Sodium and Potassium. *J. Non-Cryst. Solids* **1989**, *110*, 17–25. [CrossRef]
109. Sanchez-Fernandez, A.; Jackson, A.J.; Prévost, S.F.; Doutch, J.J.; Edler, K.J. Long-Range Electrostatic Colloidal Interactions and Specific Ion Effects in Deep Eutectic Solvents. *J. Am. Chem. Soc.* **2021**, *143*, 14158–14168. [CrossRef]
110. Abbott, A.P.; Edler, K.J.; Page, A.J. Deep Eutectic Solvents-The Vital Link between Ionic Liquids and Ionic Solutions. *J. Chem. Phys.* **2021**, *155*, 150401. [CrossRef]
111. Porras, S.P.; Kenndler, E. Formamide as Solvent for Capillary Zone Electrophoresis. *Electrophoresis* **2004**, *25*, 2946–2958. [CrossRef]
112. Redondo-Morata, L.; Oncins, G.; Sanz, F. Force Spectroscopy Reveals the Effect of Different Ions in the Nanomechanical Behavior of Phospholipid Model Membranes: The Case of Potassium Cation. *Biophys. J.* **2012**, *102*, 66–74. [CrossRef]
113. Deamer, D. The Role of Lipid Membranes in Life's Origin. *Life* **2017**, *7*, 5. [CrossRef]
114. Walton, C.R.; Shorttle, O. Scum of the Earth: A Hypothesis for Prebiotic Multi-Compartmentalised Environments. *Life* **2021**, *11*, 976. [CrossRef]
115. Huang, Y.; Barraza, K.M.; Kenseth, C.M.; Zhao, R.; Wang, C.; Beauchamp, J.L.; Seinfeld, J.H. Probing the OH Oxidation of Pinonic Acid at the Air-Water Interface Using Field-Induced Droplet Ionization Mass Spectrometry (FIDI-MS). *J. Phys. Chem. A* **2018**, *122*, 6445–6456. [CrossRef] [PubMed]
116. Morasch, M.; Liu, J.; Dirscherl, C.F.; Ianeselli, A.; Kühnlein, A.; Le Vay, K.; Schwintek, P.; Islam, S.; Corpinot, M.K.; Scheu, B.; et al. Heated Gas Bubbles Enrich, Crystallize, Dry, Phosphorylate and Encapsulate Prebiotic Molecules. *Nat. Chem.* **2019**, *11*, 779–788. [CrossRef] [PubMed]
117. Hazen, R.M.; Sverjensky, D.A. Mineral Surfaces, Geochemical Complexities, and the Origins of Life. *Cold Spring Harb. Perspect. Biol.* **2010**, *2*, a002162. [CrossRef] [PubMed]
118. Gong, J.; Bao, X. Fundamental Insights into Interfacial Catalysis. *Chem. Soc. Rev.* **2017**, *46*, 1770–1771. [CrossRef] [PubMed]
119. Scott, D.L.; White, S.P.; Otwinowski, Z.; Yuan, W.; Gelb, M.H.; Sigler, P.B. Interfacial Catalysis: The Mechanism of Phospholipase A2. *Science* **1990**, *250*, 1541–1546. [CrossRef]
120. Pera-Titus, M.; Leclercq, L.; Clacens, J.-M.; De Campo, F.; Nardello-Rataj, V. Pickering Interfacial Catalysis for Biphasic Systems: From Emulsion Design to Green Reactions. *Angew. Chem. Int. Ed. Engl.* **2015**, *54*, 2006–2021. [CrossRef]
121. Carroll, K.M.; Knoll, A.W.; Wolf, H.; Duerig, U. Explaining the Transition from Diffusion Limited to Reaction Limited Surface Assembly of Molecular Species through Spatial Variations. *Langmuir* **2018**, *34*, 73–80. [CrossRef]
122. He, C.; Lozoya-Colinas, A.; Gállego, I.; Grover, M.A.; Hud, N.V. Solvent Viscosity Facilitates Replication and Ribozyme Catalysis from an RNA Duplex in a Model Prebiotic Process. *Nucleic Acids Res.* **2019**, *47*, 6569–6577. [CrossRef]
123. Qasem, N.A.A.; Generous, M.M.; Qureshi, B.A.; Zubair, S.M. A Comprehensive Review of Saline Water Correlations and Data: Part II—Thermophysical Properties. *Arab. J. Sci. Eng.* **2021**, *46*, 1941–1979. [CrossRef]
124. Nayar, K.G.; Sharqawy, M.H.; Banchik, L.D.; Lienhard, J.H., V. Thermophysical Properties of Seawater: A Review and New Correlations That Include Pressure Dependence. *Desalination* **2016**, *390*, 1–24. [CrossRef]
125. Sharqawy, M.H.; Lienhard, J.H., V; Zubair, S.M. Thermophysical Properties of Seawater: A Review of Existing Correlations and Data. *Desalination Water Treat.* **2010**, *16*, 354–380. [CrossRef]
126. Tumminello, P.R.; James, R.C.; Kruse, S.; Kawasaki, A.; Cooper, A.; Guadalupe-Diaz, I.; Zepeda, K.L.; Crocker, D.R.; Mayer, K.J.; Sauer, J.S.; et al. Evolution of Sea Spray Aerosol Particle Phase State across a Phytoplankton Bloom. *ACS Earth Space Chem.* **2021**, *5*, 2995–3007. [CrossRef]
127. Ghaffari, Z.; Rezvani, H.; Khalilnezhad, A.; Cortes, F.B.; Riazi, M. Experimental Characterization of Colloidal Silica Gel for Water Conformance Control in Oil Reservoirs. *Sci. Rep.* **2022**, *12*, 9628. [CrossRef] [PubMed]
128. Fowler, A.C. Glaciers and Ice Sheets. In *The Mathematics of Models for Climatology and Environment*; Springer: Berlin/Heidelberg, Germany, 1997; pp. 301–336. ISBN 9783642644726.
129. Wang, Y.; Ma, C.; Liu, C.; Lu, X.; Feng, X.; Ji, X. Thermodynamic Study of Choline Chloride-Based Deep Eutectic Solvents with Water and Methanol. *J. Chem. Eng. Data* **2020**, *65*, 2446–2457. [CrossRef]
130. Lemaoui, T.; Darwish, A.S.; Attoui, A.; Abu Hatab, F.; Hammoudi, N.E.H.; Benguerba, Y.; Vega, L.F.; Alnashef, I.M. Predicting the Density and Viscosity of Hydrophobic Eutectic Solvents: Towards the Development of Sustainable Solvents. *Green Chem.* **2020**, *22*, 8511–8530. [CrossRef]
131. Cases, A.M.; Gómez Marigliano, A.C.; Bonatti, C.M.; Sólimo, H.N. Density, Viscosity, and Refractive Index of Formamide, Three Carboxylic Acids, and Formamide + Carboxylic Acid Binary Mixtures. *J. Chem. Eng. Data* **2001**, *46*, 712–715. [CrossRef]
132. Heidaryan, E.; Hatami, T.; Rahimi, M.; Moghadasi, J. Viscosity of Pure Carbon Dioxide at Supercritical Region: Measurement and Correlation Approach. *J. Supercrit. Fluids* **2011**, *56*, 144–151. [CrossRef]
133. Deng, J.; Zhao, G.; Zhang, L.; Ma, H.; Song, F.; Cao, Q.; Zhang, X. Simple and Accurate Calculation Model of Viscosity for Supercritical CO2. *J. Phys. Conf. Ser.* **2021**, *2076*, 012030. [CrossRef]
134. Zheng, H.; Yu, T.; Qu, C.; Li, W.; Wang, Y. Basic Characteristics and Application Progress of Supercritical Water. *IOP Conf. Ser. Earth Environ. Sci.* **2020**, *555*, 012036. [CrossRef]
135. Wood, L.J.; Downer, M. Viscosity/temperature Equations for Coal Tar Pitches and Refined Tars. *J. Appl. Chem.* **2007**, *15*, 431–438. [CrossRef]
136. Nojima, Y.; Iwata, K. Viscosity Heterogeneity inside Lipid Bilayers of Single-Component Phosphatidylcholine Liposomes Observed with Picosecond Time-Resolved Fluorescence Spectroscopy. *J. Phys. Chem. B* **2014**, *118*, 8631–8641. [CrossRef]

137. Lu, T.; Spruijt, E. Multiphase Complex Coacervate Droplets. *J. Am. Chem. Soc.* **2020**, *142*, 2905–2914. [CrossRef]
138. She, Y.; Fu, G. Viscosities of the Crust and Upper Mantle Constrained by Three-Dimensional GPS Rates in the Sichuan–Yunnan Fragment of China. *Earth Planets Space* **2019**, *71*, 33. [CrossRef]
139. Yousefi, Y.; Tariku, F. Thermal Conductivity and Specific Heat Capacity of Insulation Materials at Different Mean Temperatures. *J. Phys. Conf. Ser.* **2021**, *2069*, 012090. [CrossRef]
140. Mallamace, F.; Corsaro, C.; Mallamace, D.; Fazio, E.; Chen, S.-H.; Cupane, A. Specific Heat and Transport Functions of Water. *Int. J. Mol. Sci.* **2020**, *21*, 622. [CrossRef] [PubMed]
141. Waples, D.W.; Waples, J.S. A Review and Evaluation of Specific Heat Capacities of Rocks, Minerals, and Subsurface Fluids. Part 2: Fluids and Porous Rocks. *Nat. Resour. Res.* **2004**, *13*, 123–130. [CrossRef]
142. Pirajno, F. Subaerial Hot Springs and near-Surface Hydrothermal Mineral Systems Past and Present, and Possible Extraterrestrial Analogues. *Geosci. Front.* **2020**, *11*, 1549–1569. [CrossRef]
143. Kitadai, N.; Nakamura, R.; Yamamoto, M.; Okada, S.; Takahagi, W.; Nakano, Y.; Takahashi, Y.; Takai, K.; Oono, Y. Thioester Synthesis through Geoelectrochemical CO_2 Fixation on Ni Sulfides. *Commun. Chem.* **2021**, *4*, 37. [CrossRef]
144. Yang, F.; Wang, G.; Hu, D.; Zhou, H.; Tan, X. Influence of Water-Rock Interaction on Permeability and Heat Conductivity of Granite under High Temperature and Pressure Conditions. *Geothermics* **2022**, *100*, 102347. [CrossRef]
145. Feistel, R. A New Extended Gibbs Thermodynamic Potential of Seawater. *Prog. Oceanogr.* **2003**, *58*, 43–114. [CrossRef]
146. Millero, F.J.; Perron, G.; Desnoyers, J.E. Heat Capacity of Seawater Solutions from 5° to 35 °C and 0.5 to 22‰ Chlorinity. *J. Geophys. Res.* **1973**, *78*, 4499–4507. [CrossRef]
147. Islam, M.A.; Pal, A.; Saha, B.B. Experimental Study on Thermophysical and Porous Properties of Silica Gels. *Int. J. Refrig.* **2020**, *110*, 277–285. [CrossRef]
148. Di Maggio, R.; Dirè, S.; Callone, E.; Bergamonti, L.; Lottici, P.P.; Albatici, R.; Rigon, R.; Ataollahi, N. Super-Adsorbent Polyacrylate under Swelling in Water for Passive Solar Control of Building Envelope. *SN Appl. Sci.* **2020**, *2*, 45. [CrossRef]
149. Barnes, W.H.; Maass, O. Specific Heats and Latent Heat of Fusion of Ice. *Can. J. Res.* **1930**, *3*, 205–213. [CrossRef]
150. Seo, J.; Shin, D. Enhancement of Specific Heat of Ternary Nitrate ($LiNO_3$-$NaNO_3$-KNO_3) Salt by Doping with SiO_2 Nanoparticles for Solar Thermal Energy Storage. *Micro Nano Lett.* **2014**, *9*, 817–820. [CrossRef]
151. De Wit, H.G.M.; De Kruif, C.G.; Van Miltenburg, J.C. Thermodynamic Properties of Molecular Organic Crystals Containing Nitrogen, Oxygen, and Sulfur II. Molar Heat Capacities of Eight Compounds by Adiabatic Calorimetry. *J. Chem. Thermodyn.* **1983**, *15*, 891–902. [CrossRef]
152. Li, W.; Yu, Z. Heat Exchangers for Cooling Supercritical Carbon Dioxide and Heat Transfer Enhancement: A Review and Assessment. *Energy Rep.* **2021**, *7*, 4085–4105. [CrossRef]
153. Pioro, I.; Mokry, S. Thermophysical Properties at Critical and Supercritical Pressures. In *Heat Transfer-Theoretical Analysis, Experimental Investigations and Industrial Systems*; InTechOpen: Rijeka, Croatia, 2011; ISBN 9789533072265.
154. Hyman, D.; Kay, W.B. Heat Capacity and Content of Tars and Pitches. *Ind. Eng. Chem.* **1949**, *41*, 1764–1768. [CrossRef]
155. Marsh, D. Thermodynamics of Phospholipid Self-Assembly. *Biophys. J.* **2012**, *102*, 1079–1087. [CrossRef]
156. Heimburg, T. Mechanical Aspects of Membrane Thermodynamics. Estimation of the Mechanical Properties of Lipid Membranes close to the Chain Melting Transition from Calorimetry. *Biochim. Biophys. Acta Biomembr.* **1998**, *1415*, 147–162. [CrossRef]
157. Yang, P.H.; Rupley, J.A. Protein–Water Interactions. Heat Capacity of the Lysozyme–Water System. *Biochemistry* **1979**, *18*, 2654–2661. [CrossRef] [PubMed]
158. Waples, D.W.; Waples, J.S. A Review and Evaluation of Specific Heat Capacities of Rocks, Minerals, and Subsurface Fluids. Part 1: Minerals and Nonporous Rocks. *Nat. Resour. Res.* **2004**, *13*, 97–122. [CrossRef]
159. Jaupart, C.; Labrosse, S.; Lucazeau, F.; Mareschal, J.-C. Temperatures, Heat, and Energy in the Mantle of the Earth. In *Treatise on Geophysics*; Elsevier: Amsterdam, The Netherlands, 2015; pp. 223–270. ISBN 9780444538031.
160. Kahlert, H.; Leito, I. Generalization of Acid-Base Diagrams Based on the Unified pH-Scale. *Chemphyschem* **2019**, *20*, 1779–1785. [CrossRef] [PubMed]
161. Love, C.; Steinkühler, J.; Gonzales, D.T.; Yandrapalli, N.; Robinson, T.; Dimova, R.; Tang, T.-Y.D. Reversible pH-Responsive Coacervate Formation in Lipid Vesicles Activates Dormant Enzymatic Reactions. *Angew. Chem. Int. Ed. Engl.* **2020**, *59*, 5950–5957. [CrossRef] [PubMed]
162. Mariani, A.; Bonfio, C.; Johnson, C.M.; Sutherland, J.D. pH-Driven RNA Strand Separation under Prebiotically Plausible Conditions. *Biochemistry* **2018**, *57*, 6382–6386. [CrossRef]
163. Rubio-Sánchez, R.; O'Flaherty, D.K.; Wang, A.; Coscia, F.; Petris, G.; Di Michele, L.; Cicuta, P.; Bonfio, C. Thermally Driven Membrane Phase Transitions Enable Content Reshuffling in Primitive Cells. *J. Am. Chem. Soc.* **2021**, *143*, 16589–16598. [CrossRef]
164. Komatsu, G.; Senthil Kumar, P.; Goto, K.; Sekine, Y.; Giri, C.; Matsui, T. Drainage Systems of Lonar Crater, India: Contributions to Lonar Lake Hydrology and Crater Degradation. *Planet. Space Sci.* **2014**, *95*, 45–55. [CrossRef]
165. Lee, J.; Kim, G. Dependence of pH in Coastal Waters on the Adsorption of Protons onto Sediment Minerals. *Limnol. Oceanogr.* **2015**, *60*, 831–839. [CrossRef]
166. Dutta, S.; Sarma, D.; Nath, P. Ground and River Water Quality Monitoring Using a Smartphone-Based pH Sensor. *AIP Adv.* **2015**, *5*, 057151. [CrossRef]
167. Sasaki, M. Classification of Water Types of Acid Hot-Spring Waters in Japan. *J. Geotherm. Res. Soc. Jpn.* **2018**, *40*, 235–243.

168. Poddar, A.; Das, S.K. Microbiological Studies of Hot Springs in India: A Review. *Arch. Microbiol.* **2018**, *200*, 1–18. [CrossRef] [PubMed]
169. Angle, K.J.; Crocker, D.R.; Simpson, R.M.C.; Mayer, K.J.; Garofalo, L.A.; Moore, A.N.; Mora Garcia, S.L.; Or, V.W.; Srinivasan, S.; Farhan, M.; et al. Acidity across the Interface from the Ocean Surface to Sea Spray Aerosol. *Proc. Natl. Acad. Sci. USA* **2021**, *118*, e2018397118. [CrossRef] [PubMed]
170. Sing, K.S.W.; Madeley, J.D. The Surface Properties of Silica Gels. I. Importance of pH in the Preparation from Sodium Silicate and Sulphuric Acid. *J. Appl. Chem.* **2007**, *3*, 549–556. [CrossRef]
171. Balköse, D. Effect of Preparation pH on Properties of Silica Gel. *J. Chem. Technol. Biotechnol.* **2007**, *49*, 165–171. [CrossRef]
172. Ülkü, S.; Balköse, D.; Baltacboğlu, H. Effect of Preparation pH on Pore Structure of Silica Gels. *Colloid Polym. Sci.* **1993**, *271*, 709–713. [CrossRef]
173. Toyama, Y.; Sahara, R.; Iino, Y.; Kubota, K. PH Dependence of Rheological Properties of Gelatin Gel Mixed with Agar or Agarose. *Trans. Mater. Res. Soc. Jpn.* **2011**, *36*, 383–386. [CrossRef]
174. Naser, J.; Mjalli, F.; Jibril, B.; Al-Hatmi, S.; Gano, Z. Potassium Carbonate as a Salt for Deep Eutectic Solvents. *Int. J. Chem. Eng. Appl.* **2013**, *4*, 114–118. [CrossRef]
175. Skulcova, A.; Russ, A.; Jablonsky, M.; Sima, J. The pH Behavior of Seventeen Deep Eutectic Solvents. *BioResources* **2018**, *13*, 5042–5051. [CrossRef]
176. Toews, K.L.; Shroll, R.M.; Wai, C.M.; Smart, N.G. PH-Defining Equilibrium between Water and Supercritical CO2. Influence on SFE of Organics and Metal Chelates. *Anal. Chem.* **1995**, *67*, 4040–4043. [CrossRef]
177. Goertz, M.P.; Goyal, N.; Montano, G.A.; Bunker, B.C. Lipid Bilayer Reorganization under Extreme pH Conditions. *Langmuir* **2011**, *27*, 5481–5491. [CrossRef]
178. Petelska, A.D.; Figaszewski, Z.A. Effect of pH on the Interfacial Tension of Lipid Bilayer Membrane. *Biophys. J.* **2000**, *78*, 812–817. [CrossRef]
179. Ryan, D.F.; Kahler, D.M. Geochemical and Mineralogical Indications of pH in Lakes and Soils in Central New Hampshire in the Early Holocene1. *Limnol. Oceanogr.* **1987**, *32*, 751–757. [CrossRef]
180. Hutchison, W.; Finch, A.A.; Borst, A.M.; Marks, M.A.W.; Upton, B.G.J.; Zerkle, A.L.; Stüeken, E.E.; Boyce, A.J. Mantle Sources and Magma Evolution in Europe's Largest Rare Earth Element Belt (Gardar Province, SW Greenland): New Insights from Sulfur Isotopes. *Earth Planet. Sci. Lett.* **2021**, *568*, 117034. [CrossRef]
181. Sleep, N.H.; Bird, D.K.; Pope, E.C. Serpentinite and the Dawn of Life. *Philos. Trans. R. Soc. Lond. B Biol. Sci.* **2011**, *366*, 2857–2869. [CrossRef] [PubMed]
182. Rohrback, G.H.; Cady, G.H. The Liquid—Vapor Equilibrium of the System Tungsten Hexafluoride-Perfluorocycyclopentane1. *J. Am. Chem. Soc.* **1951**, *73*, 4250–4251. [CrossRef]
183. Gorgolis, G.; Galiotis, C. Graphene Aerogels: A Review. *2D Mater.* **2017**, *4*, 032001. [CrossRef]
184. Mallamace, F.; Branca, C.; Broccio, M.; Corsaro, C.; Mou, C.-Y.; Chen, S.-H. The Anomalous Behavior of the Density of Water in the Range 30 K < T < 373K. *Proc. Natl. Acad. Sci. USA* **2007**, *104*, 18387–18391. [CrossRef]
185. Schön, J.H. Density. In *Developments in Petroleum Science*; Developments in Petroleum Science; Elsevier: Amsterdam, The Netherlands, 2015; pp. 109–118. ISBN 9780081004043.
186. Minic, Z.; Thongbam, P.D. The Biological Deep Sea Hydrothermal Vent as a Model to Study Carbon Dioxide Capturing Enzymes. *Mar. Drugs* **2011**, *9*, 719–738. [CrossRef]
187. Tanaka, M.; Girard, G.; Davis, R.; Peuto, A.; Bignell, N. Recommended Table for the Density of Water between 0 C and 40 C Based on Recent Experimental Reports. *Metrologia* **2001**, *38*, 301–309. [CrossRef]
188. Sarangi, B.; Aggarwal, S.G.; Sinha, D.; Gupta, P.K. Aerosol Effective Density Measurement Using Scanning Mobility Particle Sizer and Quartz Crystal Microbalance with the Estimation of Involved Uncertainty. *Atmos. Meas. Tech.* **2016**, *9*, 859–875. [CrossRef]
189. Sofieva, S.; Asmi, E.; Atanasova, N.S.; Heikkinen, A.E.; Vidal, E.; Duplissy, J.; Romantschuk, M.; Kouznetsov, R.; Kukkonen, J.; Bamford, D.H.; et al. Effects of Temperature and Salinity on Sea-Spray-Aerosol Formation Simulated with a Bubble-Generating Chamber. *Atmos. Meas. Tech. Discuss.* **2022**, 1–40. [CrossRef]
190. Timco, G.W.; Frederking, R.M.W. A Review of Sea Ice Density. *Cold Reg. Sci. Technol.* **1996**, *24*, 1–6.
191. Hou, X.-J.; Yu, L.-Y.; Wang, Y.-X.; Wu, K.-J.; He, C.-H. Comprehensive Prediction of Densities for Deep Eutectic Solvents: A New Bonding-Group Interaction Contribution Scheme. *Ind. Eng. Chem. Res.* **2021**, *60*, 13127–13139. [CrossRef]
192. Yam, H.; Schmitt, D.R. CO Rock Physics: A Laboratory Study. In Proceedings of the Recovery–Joint CSPG CSEG CWLS Annual Convention, Pittsburgh, PA, USA, 19–22 May 2011; pp. 1–7.
193. Guo, Z.; Rüpke, L.; Tao, C. *HydrothermalFoam* v1.0: A 3-D Hydrothermal Transport Model for Natural Submarine Hydrothermal Systems. *Geosci. Model Dev.* **2020**, *13*, 6547–6565.
194. Lewis, R.A. *Hawley's Condensed Chemical Dictionary*; John Wiley & Sons: Hoboken, NJ, USA, 2016; ISBN 9781118135150.
195. Iravani, M.A.; Deparis, J.; Davarzani, H.; Colombano, S.; Guérin, R.; Maineult, A. The Influence of Temperature on the Dielectric Permittivity and Complex Electrical Resistivity of Porous Media Saturated with DNAPLs: A Laboratory Study. *J. Appl. Geophys.* **2020**, *172*, 103921. [CrossRef]
196. Rajput, M.K.; Konwar, M.; Sarma, D. Hydrophobic Natural Deep Eutectic Solvent THY-DA as Sole Extracting Agent for Arsenic (III) Removal from Aqueous Solutions. *Environ. Technol. Innov.* **2021**, *24*, 102017. [CrossRef]

197. Kim, S.; Huang, J.; Lee, Y.; Dutta, S.; Yoo, H.Y.; Jung, Y.M.; Jho, Y.; Zeng, H.; Hwang, D.S. Complexation and Coacervation of like-Charged Polyelectrolytes Inspired by Mussels. *Proc. Natl. Acad. Sci. USA* **2016**, *113*, E847–E853.
198. Earle, S. *Physical Geology*; BCcampus: Victoria, BC, Canada, 2019; ISBN 9781774200285.
199. Zhuang, B.; Ramanauskaite, G.; Koa, Z.Y.; Wang, Z.-G. Like Dissolves like: A First-Principles Theory for Predicting Liquid Miscibility and Mixture Dielectric Constant. *Sci. Adv.* **2021**, *7*, eabe7275.
200. Griffiths, T.R.; Pugh, D.C. Correlations among Solvent Polarity Scales, Dielectric Constant and Dipole Moment, and a Means to Reliable Predictions of Polarity Scale Values from Cu. *Coord. Chem. Rev.* **1979**, *29*, 129–211.
201. Ahmad, Z. Polymer Dielectric Materials. In *Dielectric Material*; InTech: Vienna, Austria, 2012; ISBN 9789535107644.
202. Kato, C.; Nishihara, S.; Tsunashima, R.; Tatewaki, Y.; Okada, S.; Ren, X.-M.; Inoue, K.; Long, D.-L.; Cronin, L. Quick and Selective Synthesis of Li6[α-P2W18O62]·28H2O Soluble in Various Organic Solvents. *Dalton Trans.* **2013**, *42*, 11363–11366.
203. Mayer, C.; Schreiber, U.; Dávila, M.J. Selection of Prebiotic Molecules in Amphiphilic Environments. *Life* **2017**, *7*, 3. [CrossRef]
204. Klein, L.; Swift, C. An Improved Model for the Dielectric Constant of Sea Water at Microwave Frequencies. *IEEE J. Ocean. Eng.* **1977**, *2*, 104–111. [CrossRef]
205. Malmberg, C.G.; Maryott, A.A. Dielectric Constant of Water from 0 C to 100 C. *J. Res. Natl. Bur. Stand.* **1956**, *56*, 2641. [CrossRef]
206. Davison, S.W.; Gentry, J.W. Differences in Diffusion Charging of Dielectric and Conducting Ultrafine Aerosols. *Aerosol Sci. Technol.* **1985**, *4*, 157–163. [CrossRef]
207. Hrubesh, L.W.; Pekala, R.W. Dielectric Properties and Electronic Applications of Aerogels. In *Sol-Gel Processing and Applications*; Springer US: Boston, MA, USA, 1994; pp. 363–367. ISBN 9781461360988.
208. Aragones, J.L.; MacDowell, L.G.; Vega, C. Dielectric Constant of Ices and Water: A Lesson about Water Interactions. *J. Phys. Chem. A* **2011**, *115*, 5745–5758. [CrossRef]
209. Essex, J.W.; Jorgensen, W.L. Dielectric Constants of Formamide and Dimethylformamide via Computer Simulation. *J. Phys. Chem.* **1995**, *99*, 17956–17962. [CrossRef]
210. Bass, S.J.; Nathan, W.I.; Meighan, R.M.; Cole, R.H. Dielectric Properties of Alkyl Amides. II. Liquid Dielectric Constant and Loss. *J. Phys. Chem.* **1964**, *68*, 509–515. [CrossRef]
211. Leeke, G.; Santos, R.; Al-Duri, B.; Seville, J.; Smith, C.; Holmes, A.B. Solubilities of 4-Phenyltoluene, Phenylboric Acid, Biphenyl, and Iodobenzene in Carbon Dioxide from Measurements of the Relative Permittivity. *J. Chem. Eng. Data* **2005**, *50*, 1370–1374. [CrossRef]
212. Li, Z.X.; Zhou, J.; Guo, X.S.; Ji, B.B.; Zhou, W.; Li, D.H. Terahertz Spectral Properties of Coal Tar. *J. Appl. Spectrosc.* **2018**, *85*, 840–844. [CrossRef]
213. Gramse, G.; Dols-Perez, A.; Edwards, M.A.; Fumagalli, L.; Gomila, G. Nanoscale Measurement of the Dielectric Constant of Supported Lipid Bilayers in Aqueous Solutions with Electrostatic Force Microscopy. *Biophys. J.* **2013**, *104*, 1257–1262. [CrossRef] [PubMed]
214. Dilger, J.P.; McLaughlin, S.G.; McIntosh, T.J.; Simon, S.A. The Dielectric Constant of Phospholipid Bilayers and the Permeability of Membranes to Ions. *Science* **1979**, *206*, 1196–1198. [CrossRef]
215. Yewdall, N.A.; André, A.A.M.; Lu, T.; Spruijt, E. Coacervates as Models of Membraneless Organelles. *Curr. Opin. Colloid Interface Sci.* **2021**, *52*, 101416. [CrossRef]
216. Takubo, J.; Ukai, Y.; Kuo, C.C. On the dielectric constants of rocks. *Mineral. J.* **1953**, *1*, 25–35. [CrossRef]
217. Pan, D.; Spanu, L.; Harrison, B.; Sverjensky, D.A.; Galli, G. Dielectric Properties of Water under Extreme Conditions and Transport of Carbonates in the Deep Earth. *Proc. Natl. Acad. Sci. USA* **2013**, *110*, 6646–6650. [CrossRef]
218. Attinger, D.; Frankiewicz, C.; Betz, A.R.; Schutzius, T.M.; Ganguly, R.; Das, A.; Kim, C.-J.; Megaridis, C.M. Surface Engineering for Phase Change Heat Transfer: A Review. *MRS Energy Sustain.* **2014**, *1*, 4. [CrossRef]
219. Petrov, O.; Furó, I. A Study of Freezing–melting Hysteresis of Water in Different Porous Materials. Part I: Porous Silica Glasses. *Microporous Mesoporous Mater.* **2011**, *138*, 221–227. [CrossRef]
220. Budisa, N.; Schulze-Makuch, D. Supercritical Carbon Dioxide and Its Potential as a Life-Sustaining Solvent in a Planetary Environment. *Life* **2014**, *4*, 331–340. [CrossRef] [PubMed]
221. Schreiber, U.; Locker-Grütjen, O.; Mayer, C. Hypothesis: Origin of Life in Deep-Reaching Tectonic Faults. *Orig. Life Evol. Biosph.* **2012**, *42*, 47–54. [CrossRef]
222. Zhang, S.J.; Duzdevich, D.; Ding, D.; Szostak, J.W. Freeze-Thaw Cycles Enable a Prebiotically Plausible and Continuous Pathway from Nucleotide Activation to Nonenzymatic RNA Copying. *Proc. Natl. Acad. Sci. USA* **2022**, *119*, e2116429119. [CrossRef]
223. Mutschler, H.; Wochner, A.; Holliger, P. Freeze-Thaw Cycles as Drivers of Complex Ribozyme Assembly. *Nat. Chem.* **2015**, *7*, 502–508. [CrossRef] [PubMed]
224. Jia, T.Z.; Fraccia, T.P. Liquid Crystal Peptide/DNA Coacervates in the Context of Prebiotic Molecular Evolution. *Crystals* **2020**, *10*, 964. [CrossRef]
225. Lu, T.; Nakashima, K.K.; Spruijt, E. Temperature-Responsive Peptide-Nucleotide Coacervates. *J. Phys. Chem. B* **2021**, *125*, 3080–3091. [CrossRef] [PubMed]
226. Chang, H. The Myth of the Boiling Point. *Sci. Prog.* **2008**, *91*, 219–240. [CrossRef]
227. Ming, F.; Chen, L.; Li, D.; Du, C. Investigation into Freezing Point Depression in Soil Caused by NaCl Solution. *Water* **2020**, *12*, 2232. [CrossRef]

228. Miyake, Y. Chemical Studies of the Western Pacific Ocean. III. Freezing Point, Osmotic Pressure, Boiling Point, and Vapour Pressure of Sea Water. *Bull. Chem. Soc. Jpn.* **1939**, *14*, 58–62. [CrossRef]
229. Rosen, J.M. The Boiling Point of Stratospheric Aerosols. *J. Appl. Meteorol.* **1971**, *10*, 1044–1046.
230. DeMott, P.J.; Hill, T.C.J.; McCluskey, C.S.; Prather, K.A.; Collins, D.B.; Sullivan, R.C.; Ruppel, M.J.; Mason, R.H.; Irish, V.E.; Lee, T.; et al. Sea Spray Aerosol as a Unique Source of Ice Nucleating Particles. *Proc. Natl. Acad. Sci. USA* **2016**, *113*, 5797–5803. [CrossRef]
231. Wypych, G. *Handbook of Fillers*; Elsevier: Amsterdam, The Netherlands, 2016; ISBN 9781927885109.
232. Saladino, R.; Crestini, C.; Ciciriello, F.; Pino, S.; Costanzo, G.; Di Mauro, E. From Formamide to RNA: The Roles of Formamide and Water in the Evolution of Chemical Information. *Res. Microbiol.* **2009**, *160*, 441–448. [PubMed]
233. Michael, T. Formamide [MAK Value Documentation, 2013]. In *The MAK-Collection for Occupational Health and Safety*; J. Wiley & Sons: Hoboken, NJ, USA, 2013; pp. 1–26.
234. Clayton, G.D. *Patt's Industrial Hygiene and Toxicology: Toxicology*; J. Wiley & Sons: Hoboken, NJ, USA, 1981; Volume 2, ISBN 9780471160427.
235. Chapman, J.B.; Runyon, S.E.; Shields, J.E.; Lawler, B.L.; Pridmore, C.J.; Scoggin, S.H.; Swaim, N.T.; Trzinski, A.E.; Wiley, H.N.; Barth, A.P.; et al. The North American Cordilleran Anatectic Belt. *Earth Sci. Rev.* **2021**, *215*, 103576. [CrossRef]
236. Kennedy, G.C.; Higgins, G.H. Melting Temperatures in the Earth's Mantle. In *Developments in Geotectonics*; Developments in geotectonics; Elsevier: Amsterdam, The Netherlands, 1972; pp. 221–232. ISBN 9780444410153.
237. Canıaz, R.O.; Erkey, C. Process Intensification for Heavy Oil Upgrading Using Supercritical Water. *Chem. Eng. Res. Des.* **2014**, *92*, 1845–1863. [CrossRef]
238. Marinos-Kouris, D.; Krokida, M.; Oreopoulou, V. Frying of Foods. In *Handbook of Industrial Drying*, 3rd ed.; CRC Press: Boca Raton, FL, USA, 2006; ISBN 9781574446685.
239. Martinotti, C.; Ruiz-Perez, L.; Deplazes, E.; Mancera, R.L. Molecular Dynamics Simulation of Small Molecules Interacting with Biological Membranes. *Chemphyschem* **2020**, *21*, 1486–1514. [CrossRef] [PubMed]
240. Kalepu, S.; Sunilkumar, K.T.; Betha, S.; Mohanvarma, M. Liposomal Drug Delivery System—A Comprehensive Review. *Int. J. Drug Dev. Res.* **2013**, *5*, 62–75.
241. Lowry, C.A.; Kay, L.M. Chemical Factors Determine Olfactory System Beta Oscillations in Waking Rats. *J. Neurophysiol.* **2007**, *98*, 394–404. [CrossRef]
242. Chamberlin, J.C. Heavy Mineral Oil as a Permanent Non-Volatile Preservative for Valuable Biological Material. *Science* **1925**, *61*, 634–635. [CrossRef]
243. Gutmann, F.; Simmons, L.M. A Theoretical Basis for the Antoine Vapor Pressure Equation. *J. Chem. Phys.* **1950**, *18*, 696–697. [CrossRef]
244. Meyers, C.H. *The Vapor Pressure of Liquid and Solid Carbon Dioxide (Classic Reprint)*; Forgotten Books: London, UK, 2018; ISBN 9781390303124.
245. Orthoefer, F.T.; List, G.R. Dynamics of Frying. In *Deep Frying*; Elsevier: Amsterdam, The Netherlands, 2007; pp. 253–275. ISBN 9781893997929.
246. Mishra, V.K.; Temelli, F.; Ooraikul, B. Vapor Pressure of Fatty Acid Esters: Correlation and Estimation. *J. Food Eng.* **1994**, *23*, 467–480. [CrossRef]
247. Bonnell, D.G.R. Studies in Gels III. Vapour Pressure of Silica Gels. *Trans. Faraday Soc.* **1932**, *28*, 463. [CrossRef]
248. Wexler, A. Vapor Pressure Formulation for Ice. *J. Res. Natl. Bur. Stand. A Phys. Chem.* **1977**, *81*, 5. [CrossRef]
249. Girnik, I.; Aristov, Y. An Aqueous $CaCl_2$ Solution in the Condenser/evaporator instead of Pure Water: Application for the New Adsorptive Cycle "heat from Cold". *Energies* **2020**, *13*, 2904. [CrossRef]
250. Xin, K.; Roghair, I.; Gallucci, F.; van Sint Annaland, M. Total Vapor Pressure of Hydrophobic Deep Eutectic Solvents: Experiments and Modelling. *J. Mol. Liq.* **2021**, *325*, 115227. [CrossRef]
251. Bockish, M. Composition, Structure, Physical Data, and Chemical Reactions of Fats and Oils, Their Derivatives, and Their Associates. In *Fats and Oils Handbook*; Elsevier: Amsterdam, The Netherlands, 1998; pp. 53–120. ISBN 9780981893600.
252. Matricarde Falleiro, R.M.; Akisawa Silva, L.Y.; Meirelles, A.J.A.; Krähenbühl, M.A. Vapor Pressure Data for Fatty Acids Obtained Using an Adaptation of the DSC Technique. *Thermochim. Acta* **2012**, *547*, 6–12. [CrossRef]
253. Van Lente, J.; Pazos Urrea, M.; Brouwer, T.; Schuur, B.; Lindhoud, S. Complex Coacervates as Extraction Media. *Green Chem.* **2021**, *23*, 5812–5824. [CrossRef] [PubMed]
254. Fegley, B., Jr.; Schaefer, L.; Kargel, J.S. Vapor Pressure, Vapor Composition and Fractional Vaporization of High Temperature Lavas on Io. *LPI Contrib.* **2003**, 1686.
255. Guo, Z.; Qin, X.; Zhang, Y.; Niu, C.; Wang, D.; Ling, Y. Numerical Investigation of the Effect of Heterogeneous Pore Structures on Elastic Properties of Tight Gas Sandstones. *Front. Earth Sci.* **2021**, *9*, 641637. [CrossRef]
256. Molina, O.; Vilarrasa, V.; Zeidouni, M. Geologic Carbon Storage for Shale Gas Recovery. *Energy Procedia* **2017**, *114*, 5748–5760. [CrossRef]
257. Lowe, D.R.; Byerly, G.R. The Terrestrial Record of Late Heavy Bombardment. *New Astron. Rev.* **2018**, *81*, 39–61. [CrossRef]
258. Lamour, S.; Pallmann, S.; Haas, M.; Trapp, O. Prebiotic Sugar Formation Under Nonaqueous Conditions and Mechanochemical Acceleration. *Life* **2019**, *9*, 52. [CrossRef]
259. Hansma, H.G. Potassium at the Origins of Life: Did Biology Emerge from Biotite in Micaceous Clay? *Life* **2022**, *12*, 301. [CrossRef]

260. Shashi Menon, E. *Pipeline Planning and Construction Field Manual*; Elsevier: Amsterdam, The Netherlands, 2011.
261. Lei, J.; Liu, Z.; Yeo, J.; Ng, T.Y. Determination of the Young's Modulus of Silica Aerogels—An Analytical–numerical Approach. *Soft Matter* **2013**, *9*, 11367. [CrossRef]
262. Phair, J.W.; Tkachev, S.N.; Manghnani, M.H.; Livingston, R.A. Elastic and Structural Properties of Alkaline-Calcium Silica Hydrogels. *J. Mater. Res.* **2005**, *20*, 344–349. [CrossRef]
263. Neumeier, J.J. Elastic Constants, Bulk Modulus, and Compressibility of H_2O Ice Ihfor the Temperature Range 50 K–273 K. *J. Phys. Chem. Ref. Data* **2018**, *47*, 033101. [CrossRef]
264. Schulson, E.M. The Structure and Mechanical Behavior of Ice. *JOM* **1999**, *51*, 21–27. [CrossRef]
265. Arriaga, M.-C.S. Supercritical Thermodynamics of the Rock/Fluid Geothermal System. In Proceedings of the World Geothermal Congress 2020+1, Reykjavik, Iceland, 24–27 October 2021; pp. 1–15.
266. Lumley, D.; Sherlock, D.; Daley, T.; Huang, L.; Lawton, D.; Masters, R.; Verliac, M.; White, D. Highlights of the 2009 SEG Summer Research Workshop on CO_2 Sequestration. *Lead. Edge* **2010**, *29*, 138–145. [CrossRef]
267. Terzi, M.M.; Deserno, M.; Nagle, J.F. Mechanical Properties of Lipid Bilayers: A Note on the Poisson Ratio. *Soft Matter* **2019**, *15*, 9085–9092. [CrossRef] [PubMed]
268. Picas, L.; Rico, F.; Scheuring, S. Direct Measurement of the Mechanical Properties of Lipid Phases in Supported Bilayers. *Biophys. J.* **2012**, *102*, L01–L03. [CrossRef] [PubMed]
269. Momeni Bashusqeh, S.; Rastgoo, A. Elastic Modulus of Free-Standing Lipid Bilayer. *Soft Mater.* **2016**, *14*, 210–216. [CrossRef]
270. Marsden, L.H.; Neuberg, J.W.; Thomas, M.E.; Mothes, P.A.; Ruiz, M.C. Combining Magma Flow and Deformation Modeling to Explain Observed Changes in Tilt. *Front. Earth Sci.* **2019**, *7*, 219. [CrossRef]
271. Heap, M.J.; Faulkner, D.R.; Meredith, P.G.; Vinciguerra, S. Elastic Moduli Evolution and Accompanying Stress Changes with Increasing Crack Damage: Implications for Stress Changes around Fault Zones and Volcanoes during Deformation. *Geophys. J. Int.* **2010**, *183*, 225–236. [CrossRef]
272. Gutenberg, B. Elastic Constants, and Elastic Processes in the Earth. In *International Geophysics*; International geophysics series; Elsevier: Amsterdam, The Netherlands, 1959; pp. 165–184. ISBN 9780123106506.
273. Burtch, N.C.; Heinen, J.; Bennett, T.D.; Dubbeldam, D.; Allendorf, M.D. Mechanical Properties in Metal–Organic Frameworks: Emerging Opportunities and Challenges for Device Functionality and Technological Applications. *Adv. Mater.* **2018**, *30*, 1704124. [CrossRef]
274. Gough, D.O. Solar Interior Structure and Luminosity Variations. In *Physics of Solar Variations*; Springer: Dordrecht, The Netherlands, 1981; pp. 21–34. ISBN 9789401096355.
275. Ranjan, S.; Sasselov, D.D. Influence of the UV Environment on the Synthesis of Prebiotic Molecules. *Astrobiology* **2016**, *16*, 68–88. [CrossRef]
276. Ritson, D.; Sutherland, J.D. Prebiotic Synthesis of Simple Sugars by Photoredox Systems Chemistry. *Nat. Chem.* **2012**, *4*, 895–899. [CrossRef] [PubMed]
277. Zang, X.; Ueno, Y.; Kitadai, N. Photochemical Synthesis of Ammonia and Amino Acids from Nitrous Oxide. *Astrobiology* **2022**, *22*, 387–398. [CrossRef] [PubMed]
278. Rastogi, R.P.; Richa; Kumar, A.; Tyagi, M.B.; Sinha, R.P. Molecular Mechanisms of Ultraviolet Radiation-Induced DNA Damage and Repair. *J. Nucleic Acids* **2010**, *2010*, 592980. [CrossRef] [PubMed]
279. Santos, A.L.; Moreirinha, C.; Lopes, D.; Esteves, A.C.; Henriques, I.; Almeida, A.; Domingues, M.R.M.; Delgadillo, I.; Correia, A.; Cunha, A. Effects of UV Radiation on the Lipids and Proteins of Bacteria Studied by Mid-Infrared Spectroscopy. *Environ. Sci. Technol.* **2013**, *47*, 6306–6315. [CrossRef]
280. Indriolo, N.; McCall, B.J. Cosmic-Ray Astrochemistry. *Chem. Soc. Rev.* **2013**, *42*, 7763–7773. [CrossRef]
281. Bertaina, M.; Apel, W.D.; Arteaga-Velázquez, J.C.; Bekk, K.; Blümer, J.; Bozdog, H.; Brancus, I.M.; Buchholz, P.; Cantoni, E.; Chiavassa, A.; et al. The Cosmic Ray Energy Spectrum in the Range 10^{16}–10^{18} eV Measured by KASCADE-Grande. *Astrophys. Space Sci. Trans.* **2011**, *7*, 229–234. [CrossRef]
282. Airapetian, V.S.; Glocer, A.; Gronoff, G.; Hébrard, E.; Danchi, W. Prebiotic Chemistry and Atmospheric Warming of Early Earth by an Active Young Sun. *Nat. Geosci.* **2016**, *9*, 452–455. [CrossRef]
283. Ebisuzaki, T.; Maruyama, S. Nuclear Geyser Model of the Origin of Life: Driving Force to Promote the Synthesis of Building Blocks of Life. *Geosci. Front.* **2017**, *8*, 275–298. [CrossRef]
284. Adam, Z.R.; Hongo, Y.; Cleaves, H.J.; Yi, R.; Fahrenbach, A.C.; Yoda, I.; Aono, M. Estimating the Capacity for Production of Formamide by Radioactive Minerals on the Prebiotic Earth. *Sci. Rep.* **2018**, *8*, 265. [CrossRef]
285. Coogan, L. Did Natural Reactors Form as a Consequence of the Emergence of Oxygenic Photosynthesis during the Archean? *GSA Today* **2009**, *19*, 4–10. [CrossRef]

life

MDPI

Article

Prebiotic Chemistry of Phosphite: Mild Thermal Routes to Form Condensed-P Energy Currency Molecules Leading Up to the Formation of Organophosphorus Compounds

Maheen Gull [1,*], Tian Feng [1], Harold A. Cruz [2], Ramanarayanan Krishnamurthy [2] and Matthew A. Pasek [1]

[1] School of Geosciences, University of South Florida, Tampa, FL 33584, USA
[2] Department of Chemistry, The Scripps Research Institute, La Jolla, CA 92037, USA
* Correspondence: ambermaheen@yahoo.com

Abstract: The in-fall of meteorites and interstellar dust particles during the Hadean–Archean heavy bombardment may have provided the early Earth with various reduced oxidation state phosphorus compounds and minerals, including phosphite (HPO_3^{2-})([Pi(III)]). The ion phosphite ([Pi(III)])has been postulated to be ubiquitous on the early Earth and consequently could have played a role in the emergence of organophosphorus compounds and other prebiotically relevant P species such as condensed P compounds, e.g., pyrophosphite ([PPi(III)]) and isohypophosphate ([PPi(III–V)]). In the present study, we show that phosphite ([Pi(III)]) oxidizes under mild heating conditions (e.g., wet–dry cycles and a prebiotic scenario mimicking a mildly hot-evaporating/drying pool on the early Earth at 78–83 °C) in the presence of urea and other additives, resulting in changes to orthophosphate ([Pi(V)]) alongside the formation of reactive condensed P compounds (e.g., pyrophosphite ([PPi(III)]) and isohypophosphate ([PPi(III–V)])) through a one-pot mechanism. Additionally, we also show that phosphite ([Pi(III)]) and the condensed P compounds readily react with organics (nucleosides and organic alcohol) to form organophosphorus compounds.

Keywords: phosphite; phosphorus; organophosphorus compounds; origin of life; condensed phosphates; condensation; phosphorylation; wet–dry cycles

Citation: Gull, M.; Feng, T.; Cruz, H.A.; Krishnamurthy, R.; Pasek, M.A. Prebiotic Chemistry of Phosphite: Mild Thermal Routes to Form Condensed-P Energy Currency Molecules Leading Up to the Formation of Organophosphorus Compounds. *Life* **2023**, *13*, 920. https://doi.org/10.3390/life13040920

Academic Editors: Ranajay Saha and Alberto Vázquez-Salazar

Received: 3 March 2023
Revised: 25 March 2023
Accepted: 28 March 2023
Published: 31 March 2023

1. Introduction

Phosphorus (P, hereafter) is a key biologic element that is ubiquitous in biochemistry because phosphorylated biomolecules play central roles in many life-sustaining processes such as replication and information (as an essential component in RNA and DNA), in metabolism (as ATP and NADPH, etc.), and cellular structure (as phospholipids) [1]. It exists in various inorganic chemical forms including orthophosphate ([Pi(V)]), pyrophosphate ([PPi(V)]), triphosphate ([PPPi(V)]), phosphite ([Pi(III)]), phosphine ([Pi(III)]), and hypophosphite ([Pi(I)]). Living organisms use these various forms of inorganic P for the formation of organophosphorus compounds with C-O-P and C-P type linkages by utilizing various enzymes [1]. Orthophosphates ([Pi(V)]) (mainly in the form of minerals) are considered to be the major carriers of P on the surface of the Earth [1,2].

The geochemistry of P on the Hadean Earth may have been significantly altered by the meteoritic mineral schreibersite $(Fe,Ni)_3P$ ([Pi(0)]), which is believed to have been supplied by meteorites during the heavy bombardment period on the early Earth [3–5]. This mineral is found in many types of meteorites and in interplanetary dust particles [6] and readily reacts and corrodes into water to give inorganic P species such as phosphate ([Pi(V)]), condensed phosphates, and reduced oxidation state P compounds (simply called reduced P, hereafter), including phosphite ([Pi(III)]) and even hypophosphite ([Pi(I)]) [7,8]. This extraterrestrial mineral would have supplied a significant amount of P to the early Earth [9,10]. The total mass of reduced P that the late accretion period could possibly have delivered to Earth from 4.50 Ga to 3.50 Ga is estimated to be around 1.32×10^{19} kg [9].

In addition, for high-velocity impacts of large extraterrestrial objects (>100 m in diameter), the projectile partially evaporates and is distributed to the surroundings as fine-grained particles [1,11]. During the heavy bombardment period, the whole Earth could have been covered by mafic and metallic particles, both extremely reducing in nature relative to the surface of the Earth [12]. Furthermore, the vapor plume of the material resulting from an impact is also postulated to be reducing in nature and could potentially reduce phosphates present in the target material to their reduced form as phosphides [1]. The evidence to support this phenomenon is the detection of vast amounts of P as schreibersite in Lunar melts [13]. This indicates that a substantial amount of P delivered during meteoritic impacts was in a reduced form and also that the impact process itself could also have reduced phosphates to phosphides [14].

The possibility of such reduced P compounds being relevant to early Earth is further supported by the occurrence of phosphonic acids in the Murchison meteorite [15] and phosphite in ancient Archean marine carbonates [16], in natural glasses called fulgurites [17], in hydrothermal systems [18], in natural waters [19], and by the geochemical reduction of phosphates into phosphite [20].

Addison Gulick was the first to propose that reduced P compounds such as hypophosphite ([Pi(I)]) and phosphite ([Pi(III)]) could plausibly have been more relevant to the origin of life on the early Earth than phosphates [21]. These reduced P compounds tend to be around 10^3 to 10^6 times more soluble in water as compared to orthophosphate in the presence of certain common divalent cations [22,23]. The reduced P compounds are released from the aqueous corrosion of schreibersite and can react with organic compounds to form C-O-P and C-P type compounds, thus establishing this mineral as highly relevant to the prebiotic chemistry and the origin of life [6–8,16,24,25].

Recent studies have shown that despite its reactivity, phosphite ([Pi(III)]) can be stable enough to be detected in various natural environments [26]. It is therefore highly likely that this reduced P compound would have played an important role in the prebiotic P chemistry. Kee and colleagues suggested that phosphite ([Pi(III)]) would have played a significant role in the formation of reactive condensed inorganic P compounds including pyrophosphite ([PPi(III)]), isohypophosphate ([PPi(III–V)]), and pyrophosphate ([PPi(V)]), with the more energetically accessible pyrophosphite ([PPi(III)]) enabling the formation of pyrophosphate ([PPi(V)]) via an isohypophosphate ([PPi(III–V)]) intermediary [27]. Condensed P compounds, including pyrophosphate ([PPi(V)]), play an important role in biochemistry.

In the present study, starting with phosphite species, we report the formation of pyrophosphite ([PPi(III)]) in the presence of urea, various salts, and other additives such as minerals/clays, as potentially plausible condensation agents under mild conditions (78–83 °C, 1 atm) and through wet–dry cycles. In some reactions where phosphite ([Pi(III)]) condensed, we also observed the formation of inorganic phosphate possibly from the auto-oxidation of phosphite ([Pi(III)]) during heating through the wet–dry cycles or heating leading to dryness, which was primarily facilitated by the presence of urea. Besides urea, we also studied various other additives to test their potential roles in the formation of various prebiotically relevant P compounds. The additives included prebiotically relevant cations, e.g., Ca^{2+}, Mg^{2+} [28–30], Na^+ [31], NH_4^+ [32,33], CO_3^{2-} [34], clays and other minerals [35,36], and urea [37,38]. We also show that the reaction mixture containing phosphite and the condensed P species readily reacts with nucleosides and organic alcohols to form organophosphites.

2. Materials and Methods

Sodium hypophosphite hydrate ($NaH_2PO_2 \cdot H_2O$, 98%), phosphorous acid (H_3PO_3, 98%), sodium phosphite dibasic pentahydrate ($Na_2HPO_3 \cdot 5H_2O$), adenosine ($C_{10}H_{13}N_5O_4$, 98%), and deuterium oxide (D_2O, 99.8% atom % D) were from Acros Organic; Uridine ($C_9H_{12}N_2O_6$, 98%), standard compounds, e.g., uridine-5-monophosphate (5′-UMP) and adenosine-5-monophosphate (5′-AMP) were from Sigma Aldrich, urea, thiourea, kaolinite clay, calcium sulphate dihydrate ($CaSO_4 \cdot 2H_2O$), magnesium chloride, sodium chloride, ammonium carbonate, ammonium chloride from TCI, calcium chloride ($CaCl_2$, 98%), white sand (SiO_2) and ferrous chloride tetrahydrate ($FeCl_2 \cdot 4H_2O$, 98%), and instant ocean were from Alfa Aesar. Deionized water was obtained in-house using a Barnstead (Dubuque, IA, USA) NANO pure® Diamond Analytical combined reverse osmosis-deionization system [24–26].

2.1. *Synthesis of Inorganic Condensed P Compounds through Wet–Dry Cycles*

0.100–0.150 g of P source (Table 1) was added to a clean glass vial (20 mL capacity) containing 7 mL DDI water (doubly deionized or ultrapure water). In some reaction samples, various additives were also added (Table 1) to test their plausible role as condensation agents for the formation of condensed P compounds. The contents were mixed and the initial pH was noted using pH paper. A small magnetic stirrer was added to the solution. The sample was allowed to heat on a hot plate, uncapped at 78–83 °C. After 24 h, the heat-dried mixture was rehydrated with 7 mL DDI water. The rehydrated sample was heated and after the completion of 48 h, it was rehydrated once again with 7 mL DDI water and heated, leading to a complete dryness. The reaction was stopped at exactly 72 h.

Table 1. Reaction conditions of various reaction samples for the synthesis of various inorganic condensed P compounds.

Sample	Description
P3	0.1 g Na_2HPO_3 $5H_2O$, 7 mL DDI water, 0.5 g urea, pH = 8.5
P3-gyp	0.1 g Na_2HPO_3 $5H_2O$, 7 mL DDI water, 0.5 g urea, 0.2 g $CaSO_4 \cdot 2H_2O$, pH = 8.5
P3-$MgCl_2$	0.1 g Na_2HPO_3 $5H_2O$, 7 mL DDI water, 0.5 g urea, 0.2 g $MgCl_2$, pH = 8.5
P3-NaCl	0.1 g Na_2HPO_3 $5H_2O$, 7 mL DDI water, 0.5 g urea, 0.2 g NaCl, pH = 8.0
P3-Am.Carb.1	0.1 g Na_2HPO_3 $5H_2O$, 7 mL DDI water, 0.5 g urea, 0.4 g $(NH_4)_2CO_3$, pH = 8.0
P3-Am.Carb.2	0.1 g Na_2HPO_3 $5H_2O$, 7 mL DDI water, 0.5 g urea, 0.8 g $(NH_4)_2CO_3$, pH = 8.0
P3-No.Ad.	0.1 g Na_2HPO_3 $5H_2O$, 7 mL DDI water, no additive, pH = 7.5
P3-Kao.	0.1 g Na_2HPO_3 $5H_2O$, 7 mL DDI water, 0.5 g urea, 0.3 g kaolinite, pH = 8.5
P3-SiO_2	0.1 g Na_2HPO_3 $5H_2O$, 7 mL DDI water, 0.5 g urea, 0.25 g white sand, pH = 8.5
P3-IO	0.1 g Na_2HPO_3 $5H_2O$, 7 mL DDI water, 0.5 g urea, 0.25 g instant ocean, pH = 8.0
P3-NH_4Cl	0.1 g Na_2HPO_3 $5H_2O$, 7 mL DDI water, 0.5 g urea, 0.4 g NH_4Cl, pH = 8.0
HP3-No.Ad.	0.1 g H_3PO_3, 7 mL DDI water, no additive, pH = 2
HP3-U	0.1 g H_3PO_3, 7 mL DDI water, 0.5 g urea, pH = 2
P3-NWD	0.1 g Na_2HPO_3 $5H_2O$, 7 mL DDI water, 0.5 g urea, pH = 8.5
P1-U	0.1 g $NaH_2PO_2 \cdot H_2O$, 7 mL DDI water, 0.5 g urea, pH = 6.0, NWD
P3-thio.	0.1 g Na_2HPO_3 $5H_2O$, 7 mL DDI water, 0.5 g thiourea, pH = 8.5

Prebiotic synthesis of condensed-P compounds. Various conditions tried in the study. Each of the samples were heated uncovered at 78–83 °C for 3 days and were given wet–dry cycle treatment for 3 days e.g., every 24 h. Each sample received three wet and three dry cycles. The meanings of the abbreviations used are; P3 (phosphite), gyp (gypsum), Am.Carb.1 (ammonium carbonate), P3-No.Ad. (phosphite with no additive), Kao. (kaolinite), IO (instant ocean), HP3-No.Ad. (phosphorous acid with no additive), HP3-U (phosphorous acid with urea), P3-NWD (phosphite with no wet–dry cycles), P1 (hypophosphite), and thio means (thiourea). Samples P3-NWD and P1-U represent 'Warm-Pool Model' Theme suggested in Section 2.2, whereas all the other samples represent 'wet–dry cycles' scenario suggested in Section 2.1, respectively.

One reaction was also performed to specifically compare the possible role of urea in promoting the condensation reactions of phosphite ([Pi(III)]). In this reaction sample, instead of urea, thiourea was added to determine if it also promotes the heat driven oxidation of phosphite to phosphate. The reaction conditions were similar and the only difference was that instead of using urea, thiourea was added as an additive (Table 1, also see SI).

2.2. *Synthesis of Inorganic Condensed P under 'Warm-Pool Model' Theme*

This study was carried out to investigate the formation of condensed P compounds in a prebiotic scenario mimicking a mildly hot, evaporating/drying pool on the early Earth, as previously described [39]. 0.1 g of P source; hypophosphite ($NaH_2PO_2 \cdot H_2O$) (Sample P1-U) or phosphite (Na_2HPO_3 $5H_2O$) and 0.5 g urea (Sample P3-NWD) (Table 1) were added to a clean glass vial (20 mL capacity) containing 7 mL DDI water. The contents were mixed and the initial pH (around 8.5) was noted using pH paper. A small magnetic stirrer was added to the solution. The sample was allowed to heat on a hot plate, uncapped at 78–83 °C for 2 days. After 2 days, the heat-dried mixture was removed from heating and was prepared to be analyzed by ^{31}P-NMR.

2.3. Synthesis of Organophosphites from the Reactive Condensed P through Wet–Dry Cycles

0.1 g P source (sodium phosphite), 0.6 g–0.8 g organic compound (either a nucleoside: uridine or adenosine or an organic alcohol: 0.8 g glycerol), and 0.5 g urea were added to a clean glass vial of 20 mL capacity containing 7 mL of DDI water. The pH of the reaction mixture solution was 8.5. This solution was stirred using a small magnetic stirrer and was heated at 70–78 °C, uncapped for 24 h, to complete dryness. After 24 h, the dried reaction mixture was rehydrated with 7 mL DDI water. This reaction mixture was heated (uncapped) for another 24 h, after which the heat-dried reaction was stopped. In another set of experiments, identical reactions were carried out omitting urea (Table 2).

Table 2. Reaction conditions of various reaction samples to study phosphonylation of organic molecules.

Sample	Description
A	0.1 g Na_2HPO_3 $5H_2O$, 7 mL DDI water, 0.6 g uridine, pH = 8, 70–72 °C
B	0.1 g Na_2HPO_3 $5H_2O$, 7 mL DDI water, 0.5 g urea, 0.6 g uridine, pH = 8.5–9, 70–72 °C
C	0.1 g Na_2HPO_3 $5H_2O$, 7 mL DDI water, 0.8 g glycerol, pH = 7.5, 73–75 °C
D	0.1 g Na_2HPO_3 $5H_2O$, 7 mL DDI water, 0.5 g urea, 0.8 g glycerol, pH = 8.5–9, 73–75 °C
E	0.1 g Na_2HPO_3 $5H_2O$, 7 mL DDI water, 0.65 g adenosine, pH = 7.5, 78–80 °C
F	0.1 g Na_2HPO_3 $5H_2O$, 7 mL DDI water, 0.65 g adenosine, 0.5 g urea, pH = 8.5–9, 78–80 °C

Prebiotic synthesis of organophosphites. Various conditions used in the study. Each of the samples were heated uncovered at from 70–78 °C for 2 days and were given wet–dry cycle treatment for 2 days, e.g., every 24 h. Each sample, therefore, received two wet and two dry cycle.

2.4. Analyses and Characterization of Inorganic and Organic P Compounds

For ^{31}P-NMR analyses, the samples were analyzed on a 400-MHz Varian Unity Inova NMR operating at 161.9 MHz in both H-coupled and H-decoupled modes. The width of the spectrum was 200 ppm, and the running temperature was 22 °C. Various P products e.g., both inorganic and organic P compounds were quantified by peak integration method as previously reported [16,24–26,39–41].

The specific details of the ^{31}P-NMR instrument and its related parameters have already been reported in our previous studies [16,23–26,39–41]. Each sample, completely dried out from heating, was cooled down to room temperature and was rehydrated with 5 mL DDI water. The reaction sample was mixed and stirred until a suspension was formed, which was filtered and centrifuged. The contents (2 mL) were then transferred to a clean watch glass followed by air-drying at room temperature. The air-dried room temperature sample was then rehydrated with a 2 mL D_2O (90%) and DDI water (10%) solution (or only DDI water in the case of analysis for MS) and was centrifuged once again. The total volume of the solution was 2 mL. About 400 μL of the sample solution was transferred to a clean NMR tube and was analyzed by ^{31}P-NMR.

Mass spectrometry (MS) analyses were formed in negative ion mode on a 6130 Single Quadrupole Mass Spectrometer (Agilent, Santa Clara, CA, USA) attached to an Agilent 1200 HPLC by direct injection, and deionized water was used as a solvent as reported previously [16,40,41].

Organophosphorus compounds including 5′-AMP and 5′-UMP were confirmed by spiking with the standard compounds as previously [33] (see also SI). Remaining organophosphorus compounds including organic phosphites were identified and characterized by studying their characteristic peak splitting in the H-coupled ^{31}P-NMR, measuring their *J* coupling constants, and finding the target peaks in the mass spectrometer.

3. Results

Heating inorganic P compounds through wet–dry cycles at 78–83 °C resulted in the formation of condensed P species. The reaction samples produced high-energy condensed P compounds that reacted with organic substrates (Table 3, Figure 1, see also SI, Figures S1 and S2). When sodium phosphite ([Pi(III)]) was heated in the presence of urea, pyrophosphite [PPi(III)] was generated. In some samples of the above-mentioned reactions, to our surprise, we also detected orthophosphate [Pi(V)]. To further confirm the presence and source of orthophosphate [Pi(V)], ^{31}P-NMR of blank sodium phosphite [Pi(III)] did not show any presence of orthophosphate [Pi(V)] (SI, Figure S2). This confirmed that phosphite under mild heating conditions and through the wet–dry cycling has a tendency to autoxidize to orthophosphate [Pi(V)]. Subsequently, this orthophosphate [Pi(V)] reacts with pyrophosphite [PPi(III)] to form a mixed valence condensed P species isohypophosphate [PPi(III–V)] [27].

Table 3. The relative abundances (%) of various inorganic P products produced in various reactions.

Sample	Unreacted P	Phosphate	Isohypophosphate	Pyrophosphate	Pyrophosphite	P_T
P3	37	11	9	25	18	52
P3-gyp	72	3	10	BDL	15	25
P3-$MgCl_2$	100	BDL	BDL	BDL	BDL	BDL
P3-NaCl	69	13	10	BDL	8	18
P3-Am.Carb.1	39	1	6	1	53	60
P3-Am.Carb.2	75	10	7	6	2	15
P3-No.Ad.	95	5	BDL	BDL	BDL	BDL
P3-Kao.	98	2	BDL	BDL	BDL	BDL
P3-SiO_2	83	1	5	3	8	16
P3-IO	98	2	BDL	BDL	BDL	BDL
P3-NH_4Cl	95	5	BDL	BDL	BDL	BDL
HP3-No.Ad.	90	10	BDL	BDL	BDL	BDL
HP3-U	95	5	BDL	BDL	BDL	BDL
P3-NWD	5	1	8	4	82	94
P3-thio.	85.5	0.1	BDL	BDL	14.4	14.4

The relative abundances (%) of the inorganic P products were calculated on the basis of the total P dissolved and by the peak integration method as previously reported [24,37–39]. Various P sources used in the samples include; $Na_2HPO_3{\cdot}5H_2O$, H_3PO_3, or $NaH_2PO_2{\cdot}H_2O$ (Table 1). Furthermore, the amount (%) of orthophosphate detected was produced by the oxidation of phosphite. Some of the abbreviations meanings are as follows: BDL (below detection limit) and P_T (total inorganic condensed P compounds generated). The meanings of the abbreviations used are; P3 (phosphite), gyp (gypsum), Am.Carb.1 (ammonium carbonate), P3-No.Ad. (phosphite with no additive), Kao. (kaolinite), IO (instant ocean), HP3-No.Ad. (phosphorous acid with no additive), HP3-U (phosphorous acid with urea), P3-NWD (phosphite with no wet–dry cycles), P1-U (hypophosphite), and thio means (thiourea).

H-coupled ^{31}P-NMR analysis confirmed isohypophosphate [PPi(III–V)] in the form of three doublets in the −3.0 to −6.0 ppm region, as reported previously [27]. The peaks in our results were slightly shifted from the previously reported values: −2.5 to −7.0 ppm [27], to −3.0 to −6.0 ppm. This slight shift in the location of the peak and chemical shift values was attributed to the pH changes [42]. Peak c (isohypophosphate) [PPi(III–V)] was identified by the following coupling constant values [27]: δ–4.44 [dd, c_1J_{PH} = 645 Hz, c_1J_{PP} = 17.36 Hz, Pi(III)]; δ–5.50 [d, c_2J_{PP} = 17.22 Hz, Pi(V)]. These values are within the range as reported previously [27]. Pyrophosphite [Ppi(III)] was identified as two triplets in the H-coupled mode of ^{31}P-NMR; one triplet around −3.6ppm and the other around −6.5 ppm, as also reported previously [27]. Peaks e (pyrophosphite) at δ–5.05 [dd, c_3 J_{PH} = 660 Hz, PPi (III)] correspond to phosphite triplet splitting (Figure 2a,b).

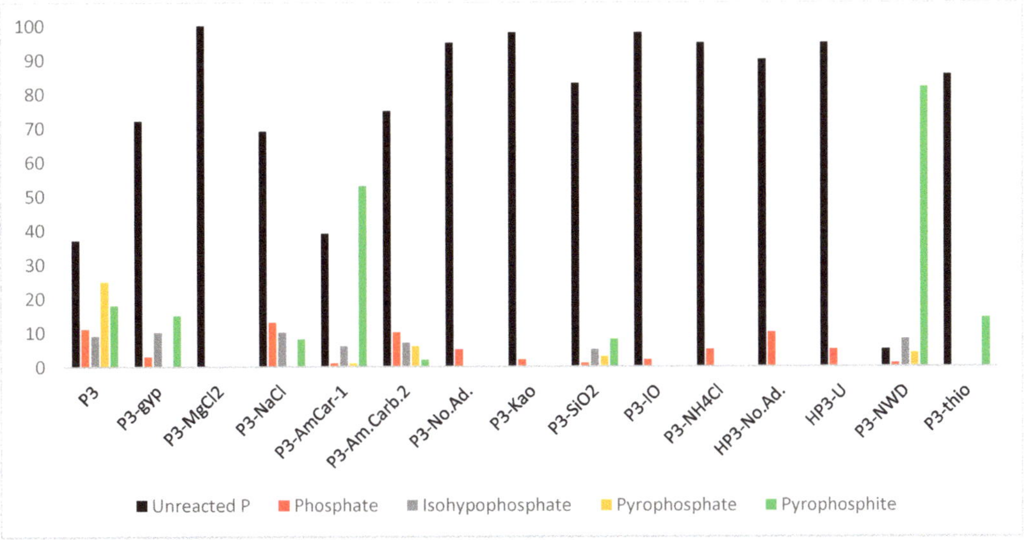

Figure 1. Yields (abundances (%)) of various condensed P compounds in various samples. Where unreacted P means the starting P source, which in this case is either phosphite or phosphorous acid. The *Y*-axis represents the abundance (%), whereas the *X*-axis represents the various samples.

Pyrophosphate [PPi(V)] was identified as a singlet in H-coupled mode of ^{31}P-NMR around the −7 ppm region. Figure 2 shows the ^{31}P-NMR spectrum of the reaction sample containing phosphite, urea, and heating leading to dryness at 78–83 °C for 3–4 days (without wet–dry cycles).

In order to confirm that heating leading to dryness and wet–dry cycles accompanied with heating can result in the auto-oxidation of phosphite [Pi(III)], we also studied the similar reactions of sodium hypophosphite [Pi(I)], representing another source of reduced P. When the latter was heated leading to dryness for 3–4 days at 78–83 °C, we observed phosphite [Pi(III)] and even phosphate [Pi(V)], implying that the orthophosphate [Pi(V)] was not an impurity but was actually a product of auto-oxidation of a reduced P compound (SI, Figures S3 and S4). We also analyzed (^{31}P-NMR), the solutional blanks containing 1 pure hypophosphite [Pi(I)] (SI, Figures S2 and S4) phosphite [Pi(III)] (SI, Figure S2). The ^{31}P-NMR analysis did not reveal any peaks containing phosphate [Pi(V)], suggesting the reduced P compounds did undergo auto-oxidation (SI) on heating through wet–dry cycles or heating leading to dryness.

Heating phosphite [Pi(III)] via wet–dry cycles at 78–83 °C in the presence of urea produced condensed P compounds up to 60% (% abundance) (Table 1, Sample P3-Am.Carb.1) and 52% in the case of Sample P3 (Table 1, Sample P3). Sodium carbonate (along with urea) seemed to promote the condensation of phosphite (Table 1, Sample P3-Am.Carb.1; see also Supplementary information (SI)). However, when the concentration of ammonium carbonate was doubled, it significantly declined the rate of condensation of phosphite [Pi(III)]. The other additives including salts, instant ocean (IO), minerals, and clays did not seem to positively enhance the condensation process of phosphite. Urea seemed to be an excellent additive (and condensation agent) in all the different reactions attempted. The highest production of condensed P products (pyrophosphite [PPi(III)], pyrophosphate [PPi(V)], and isohypophosphate [PPi(III–V)]) was obtained by the simple heating of the solution mixtures of phosphite [Pi(III)] and urea to dryness without wet–dry cyclic treatment with yields (% abundances) of up to 94% (Figure 2, Sample P3-NWD, Table 3). Overall, the condensation reactions required urea and heating the solutions at 78–83 °C for

3–4 days and proceeded smoothly following either route i.e., wet–dry cycles or heating leading to dryness.

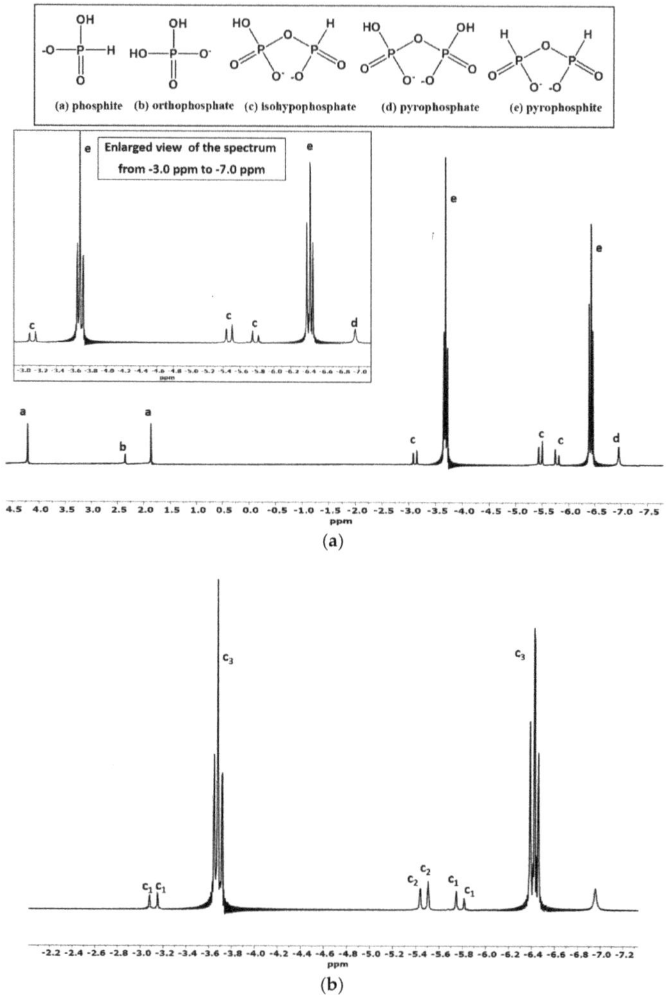

Figure 2. (**a**). H-coupled ^{31}P-NMR of condensation (and auto-oxidation) reaction of phosphite [Pi(III)] in the presence of urea (Sample P3-NWD). The labeled peaks represent the following compounds: (a) phosphite [PPi(III)], (b) orthophosphate [Pi(V)], (c) isohypophosphate [PPi(III–V)], (d) pyrophosphate [PPi(V)], and (e) pyrophosphite [PPi(III)]; (**b**) enlarged spectrum from −2.0 ppm to −7.2 ppm showing various coupling constants: C1 and C2 (coupling constants (δ values) show the presence of isohypophosphate, and C3 represents the coupling constants values for pyrophosphite [PPi(III)].

Both starting reduced P compounds (hypophosphite [Pi(I)] and phosphite [Pi(III)]) seemed to undergo auto-oxidation (see also SI, Figures S2–S4). This auto-oxidation of phosphite [Pi(III)] produced phosphate, whereas the auto-oxidation of hypophosphite [Pi(I)] produced phosphite [Pi(III)] and phosphate [Pi(V)]. These reduced P (hypophosphite [Pi(I)] and phosphite [Pi(III)]) compounds at the same time also condensed to form various condensed P compounds, including mixed-valence state P compounds such as isohypophosphate [PPi(III–V)].

Furthermore, the reactions seemed to be pH sensitive. The best yields were obtained around pH = 8, whereas at lower pH (in case of H_3PO_3, pH = 2), we did not observe any condensed P compounds, though heating and the wet–dry cycle promoted the auto-oxidation of H_3PO_3 to form orthophosphate [Pi(V)] (Table 1 reaction Samples HP3-No.Ad. and HP3-U). As mentioned above, when reaction samples (solutions) were heated to a complete dryness at 78–83°C for two days without wet–dry cycles, the relative abundances (% yields) of the condensed compounds were around 94%, with pyrophosphite [PPi(III)] yield being around 82%. This particular reaction gave the best possible yields of the condensed P compounds but the observed yield of orthophosphate [Pi(V)] (oxidation product of phosphite [Pi(III)]) was only 1% (Table 3, reaction Sample P3-NWD, Figure 2). The other samples with better yields (% abundances) of inorganic condensed P species including pyrophosphite [PPi(III)], pyrophosphate [PPi(V)], and isohypophosphate were seen when phosphite [Pi(III)] was heated in the presence of urea and heated through wet–dry cycles (Table 3). Heating leading to dryness, therefore, favored the formation of pyrophosphite [PPi(III)] over other condensed P compounds. The rationale behind would be that the formation of pyrophosphite [PPi(III)] occurs quicker than the multistep conversion of isohyphopshate [PPi(III–V)], for (1) oxidation of phosphite [Pi(III)], to orthophosphate [Pi(V)], (2) condensation of phosphite [Pi(III)], with phosphate [Pi(V)], to form isohypophopshate [PPi(III–V)], (3) condensation of orthophosphate [Pi(V)], to form pyrophosphate [PPi(V)].

Urea seemed to promote the condensation as well as oxidation reactions of the reduced P compounds. Although, the auto-oxidation of the reduced P compounds could be promoted without the urea (Tables 2 and 3, Sample P3-No.Ad.), higher (relative abundances) yields of phosphate [Pi(V)] were observed when urea was present. The supportive role of urea was further explored when thiourea was used in place of urea. In case of thiourea as an additive, no phosphate [Pi(V)] was observed (it was below detection limits) and pyrophosphite [PPi(III)] was the only product observed, suggesting that thiourea did promote the condensation but not the oxidation of phosphite [Pi(III)] (SI, Figure S7).

In order to further investigate the reactivity of these high-energy condensed P compounds, we also studied the phosphonylation reaction of nucleosides (adenosine and uridine) and organic alcohol (glycerol) with the reaction mixtures containing phosphite [Pi(III)] and urea (Table 2). The reaction mixture readily reacted with an organic compound in the presence of urea at 70–80 °C through wet–dry cycles and produced organic phosphites as expected. However, some organophosphates were also observed. The organophosphates were observed possibly due to the reaction between the orthophosphate [Pi(V)] formed as a consequence of the auto-oxidation of phosphite [Pi(III)] and a nucleoside. In the case of organic alcohol (glycerol), we did not observe any organophosphates, which implied that the rate of phosphonylation of glycerol was faster than the auto-oxidation of phosphite [Pi(III)] (Figure 3).

The presence of various organophosphorus compounds was confirmed by ^{31}P-NMR peak characterization as well as by MS (the direct injection method), as reported previously [33,39–41]. The direct injection MS of reaction sample containing glycerol showed the following major peaks: [$C_3H_9O_5P$-H] at m/z 155.02 corresponding to glycerol phosphite and [$C_3H_{10}O_7P_2$-H] at m/z: 218.99 corresponding to glycerol diphosphite. In the reaction samples with uridine, we observed: [$C_9N_2O_6H_{11}$-H] at m/z 243 corresponding to uridine nucleoside, [$C_9N_2O_9PH_{13}$-H] at m/z 323.04 corresponding to uridine-monophosphate (2′, 3′ and 5′-UMP species), and [$C_9N_2O_8PH_{12}$-H] at m/z 307 corresponding to uridine-monophosphite. Similarly, the major peaks in MS were identified for the adenosine reaction in the reaction samples. For the solutions with adenosine, we observed: [$C_{10}H_{13}N_5O_4$-H] at m/z 266 corresponding to adenosine nucleoside, [$C_{10}H_{13}N_5O_7P$-H] at m/z 346 corresponding to monophosphate (2′, 3′ and 5′-AMP species), and, finally, [$C_{10}H_{14}N_5O_6P$-H] at m/z 330 corresponding to adenosine-monophosphite.

The formation of the organophosphorus compounds was improved by the presence of urea (Table 4). Heating uridine with an aqueous solution of phosphite [Pi(III)] and urea produced various uridine phosphites. Figure 3 (Sample B) shows ^{31}P-NMR analysis of

Sample B in H-coupled mode. Various uridine phosphite species were identified and characterized by observing their chemical shift values; C-O-P (carbon, oxygen and phosphorus) and P-H interactions [43]. Figure 3 shows the various P species, including both organic and inorganic P compounds, and is without any spiking with the standard compounds. In the reaction sample B, we did not observe any organophosphates. However, various species of uridine phosphites were observed. The organophosphite 5′-uridine phosphite (peak g) was identified in the form of two triplets around 5.2 and 7.8 ppm; 2′- and 3′-uridine phosphites (peaks i) were identified as two doublets. Various other doublets and triplets labeled as peaks k represent uridine diphosphite species. These uridine diphosphite species (peaks k) represent uridine-P species having one phosphite group attached to 5′- position and another either to 2′- or to 3′-positions but not through a pyrophospite [PPi(III)] (P-O-P) type linkage. Each triplet and a doublet, in the case of phosphite, [Pi(III)] splits further into another triplet and a doublet, representing uridine-diphosphite species. The total yield (% abundance) of organophosphite in the case of uridine was 99%. When phosphite [Pi(III)] was heated without urea (Sample A), the yields (% abundances) of the uridine phosphates and phosphites were lower. These yields represent the relative abundances (%) of the phosphonylated/phosphorylated products and were calculated on the basis of the total P dissolved and by the peak integration method as reported before [24–26].

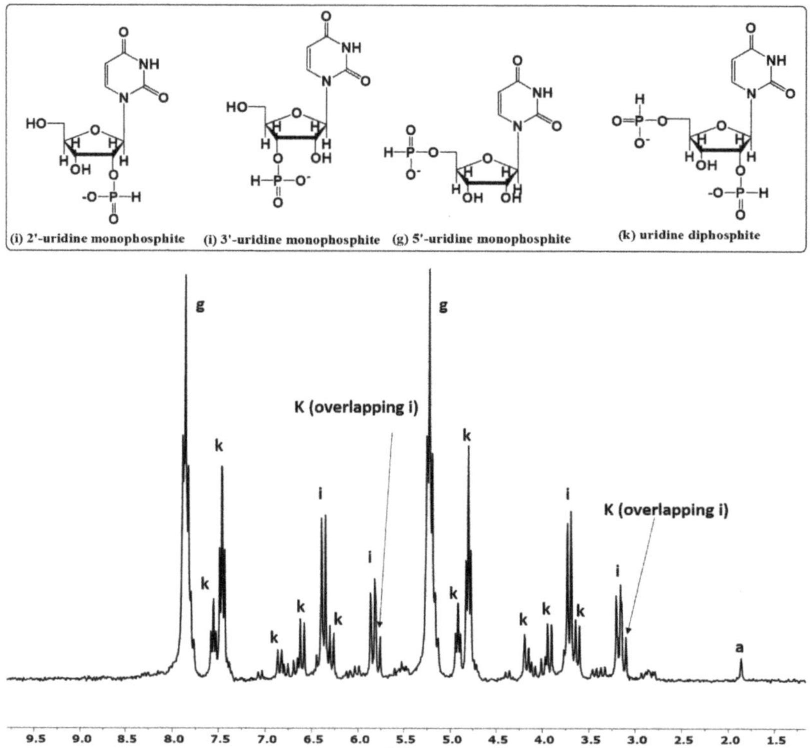

Figure 3. H-coupled ^{31}P-NMR of phosphorylation and phosphonylation reactions of uridine (sample B). The labeled peaks correspond to the following compounds: (a) phosphite, (g) 5′-uridine phosphite, (i) 2′and 3′-uridine phosphites, and (k) various diphosphite (not pyrophosphite) species of uridine. Various isomers are possible in this case, e.g., uridine 2,3-diphosphite, uridine 2,5-diphosphite, etc. Each isomer in the H-coupled mode of ^{31}P-NMR splits into two triplets and two doublets (peaks k). Such diphosphite species were identified by distinctive peak splitting patterns i.e., doublet of doublets and doublet of triplets indicated the presence of various organic diphosphite species.

Table 4. ^{31}P-NMR relative abundances [1] (%) of the organic P compounds detected in various reaction samples.

Sample Name	Phosphite (Unreacted)	Orthophosphate	5′-mono-PO_3	5′-mono-PO_4	2′-or 3′-mono-PO_3	2′-or 3′-mono-PO_4	Nucleoside Diphosphite Species	Glycerol-1-PO_3	Glycerol-1-PO_4	Glycerol-2-PO_3	Glycerol-2-PO_4	Total Org. PO_4	Total Org. PO_3	^{T}C-O-P
	a	b	g	h	i	j	k	l	m	n	o			
A	93	3	3	BDL	1	BDL	BDL	—	—	—	—	BDL	4	4
B	1	BDL	44	BDL	20	BDL	35	—	—	—	—	BDL	99	99
C	88	6.5	—	—	—	—	—	3.5	BDL	2	BDL	BDL	5.5	5.5
D	78	2	—		—	—	—	10	6	3	1	7	13	20
E	98	1	1	BDL	BDL	BDL	—	—	—	—	—	BDL	1	1
F	44	0.5	36	1	18	0.5	—	—	—	—	—	1.5	54	55.5

[1] The relative abundances (%) of the phosphonylated/phosphorylated products were calculated on the basis of the total P dissolved and by the peak integration method, as reported previously [24–26]; ^{T}C-O-P means total C-O-P (carbon-oxygen-phosphorus) type organophosphorus compounds e.g., total sum of organic phosphates and phosphites for that particular reaction (and may not represent the exact sum due to rounding). The blank lines in the table show that these compounds are not present in the sample. BDL signifies below detection limit.

In the glycerol and phosphite sample without urea (Sample C), only 5% glycerol phosphites were detected by ^{31}P-NMR, whereas for the other sample containing glycerol, phosphite [Pi(III)], and urea, the yield (relative abundances) of the glycerol phosphites reached around 20% (reaction Sample D, Figure 4). Figure 4 shows ^{31}P-NMR analysis of Sample D in H-coupled mode. Again, various glycerol phosphite species were identified and characterized by observing their chemical shift values; C-O-P (carbon, oxygen and phosphorus) and P-H interactions [42–44]. Glycerol-1-phosphite was identified by two triplets (peaks l) and glycerol-2-phosphite was confirmed by two doublets (peaks n). Similarly, triplet peak m represents glycerol-1-phosphate, and peak o (doublet) represents glycerol-2-phosphate (Figure 4).

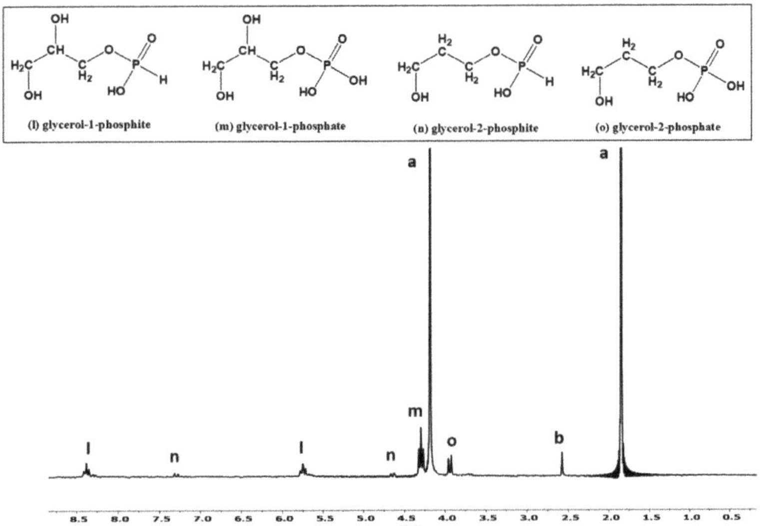

Figure 4. H-coupled ^{31}P-NMR of phosphorylation and phosphonylation reactions of glycerol (sample D). The labeled peaks correspond to the following P compounds: (a) phosphite, (b) phosphate, (l) glycerol-1-phosphite, (m) glycerol-1-phosphate, (n) glycerol-2-phosphite, and (o) glycerol-2-phosphate. The figure also shows the enlarged peaks from 2.8 to 8.4 ppm.

Similar reaction trends were seen in the case of adenosine nucleoside. On heating (and through wet–dry cycles) adenosine with the phosphite solution in the presence of urea, the yields of adenosine-P (both phosphates and phosphites) reached around 55.5%; this declined to only 1% when urea was not included. Urea played a significant role in the C-O-P bond formation. Various reaction products of adenosine-P are shown in Figure 5. Peak labeling and identification is consistent with that of Figure 4. Nucleotides including 5′-AMP and 5′-UMP, if any, present in each one of the respective reaction samples were also spiked with standard 5′-AMP and 5′-UMP solutions, as mentioned in Section 2.4 and as previously described [33] (Figure 6, see also SI, Figure S6a,b).

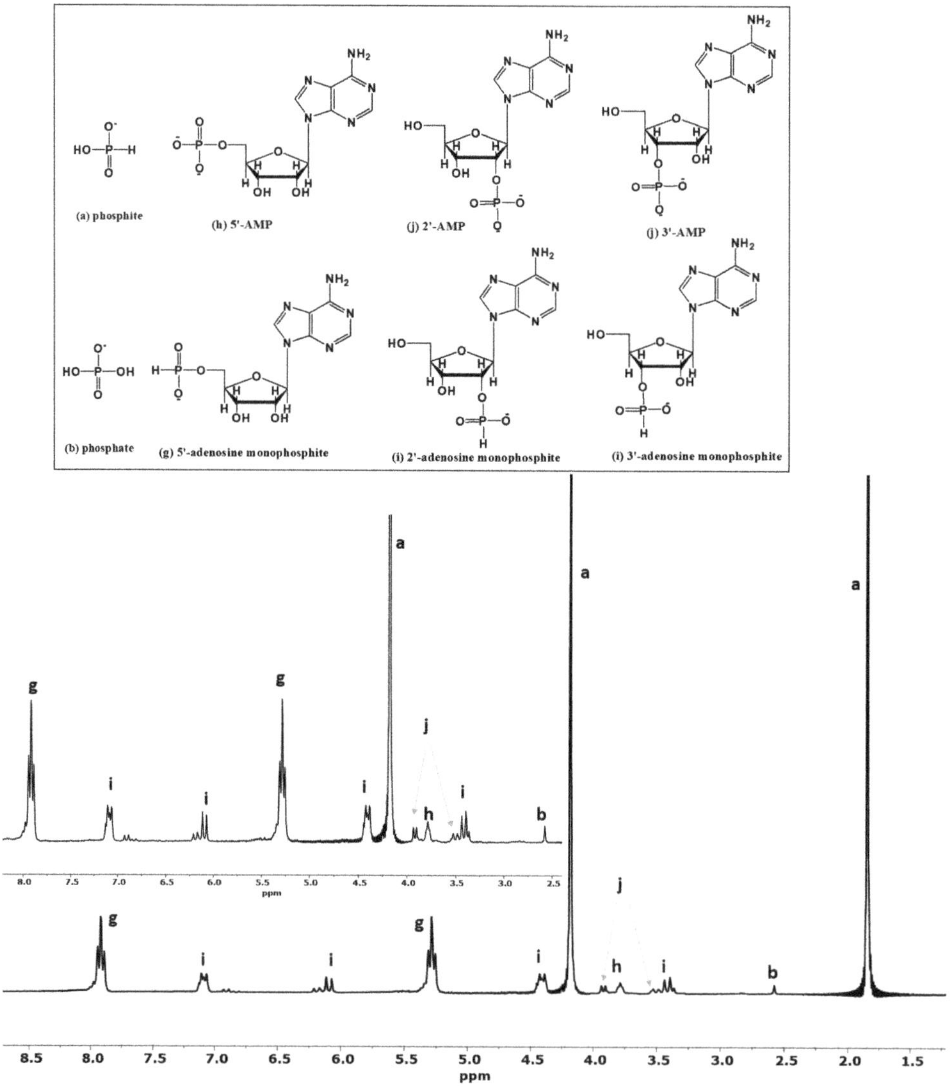

Figure 5. H-coupled ^{31}P-NMR of phosphorylation and phosphonylation reactions of adenosine (Sample F). The labeled peaks represent the following compounds: (a) phosphite, (b) phosphate, (g) 5′-adenosine phosphite, (h) 5′-AMP, (i) 2′and 3′-adenosine phosphites, and (j) 2′and 3′-AMP.

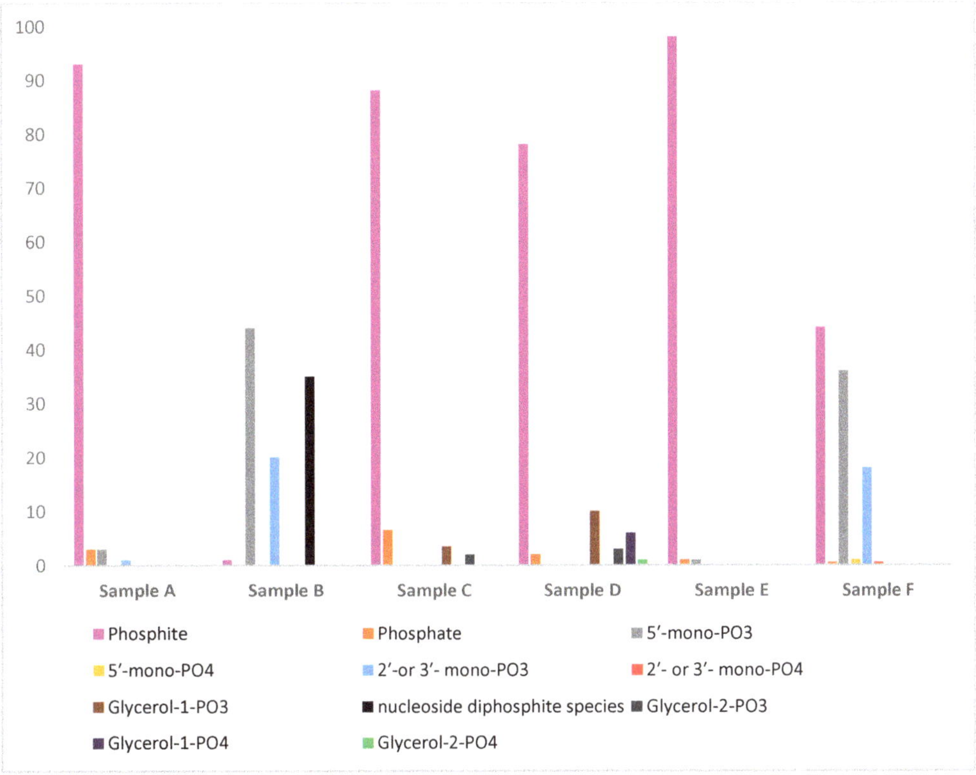

Figure 6. The relative abundances (%) of the phosphonylated and phosphorylated as well as amount of inorganic phosphate (oxidation product of phosphite) observed in each sample. Where Samples B, D, and F where with urea as an additive, and Samples A, C, and E were without urea. The relative abundances (%) were calculated using the peak integration method of ^{31}P-NMR [24–26].

Organophosphites (phosphites of glycerol, uridine, and adenosine) were also identified by observing their coupling constants; for example, the coupling constant (d (delta) value for the two organophosphite doublets or triplets in the H-coupled ^{31}P-NMR were around 640–655 Hz, a clear indication of the presence of phosphites derivatives of organics [43].

4. Discussion

Heating phosphite with urea at 78–83 °C with wet–dry cycles (or heating leading to dryness e.g., without wet–dry cycles 'warm alkaline pool scenario') lead to the oxidation of some of the phosphite [Pi(III)] to orthophosphate [Pi(V)] (1–13%), along with the formation of various condensed inorganic P compounds including pyrophosphate[Pi(V)], pyrophosphite [PPi(III)], and isohypophosphate [PPi(III–V)]. An interesting finding in the reaction system was the auto-oxidation of phosphite into phosphate [Pi(III)] and hypophosphite [Pi(I)] into phosphite [Pi(III)] and even phosphate. At present, we do not know the exact mechanism. Since phosphate [Pi(V)] was also detected in all samples, even the one without urea (Table 1, Sample Na-12), no additive seemed to be required to promote the auto-oxidation under the air of phosphite [PPi(III)] into phosphate [Pi(V)] and hypophosphite [Pi(I)] into phosphate [Pi(V)] and phosphite [Pi(III)]. A plausible explanation would be that evaporation and heating leading to dryness somehow causes this autooxidation of phosphite into phosphate [Pi(V)].

In our studies, we made the following observations (Figure 7): (1) some amount of phosphite [Pi(III)] was oxidized to orthophosphate [Pi(V)] (1–13%) during the prolonged heating (3–4 days) at 78–83 °C through wet–dry cycles; (2) phosphite [Pi(III)] condensed in the presence of urea into pyrophosphite [PPi(III)]; (3) this pyrophosphite [PPi(III)] subsequently hydrolyzed into phosphite [Pi(III)] to react with the orthophosphate [Pi(V)] generated via self-oxidation of phosphite to form isohypophosphate [PPi(III–V)]; and (4) some of the orthophosphate [Pi(V)] also condensed in the presence of urea to form pyrophosphate [PPi(V)]. Although, these reaction steps are *one-pot*, it is not clear how isohypophosphate [PPi(III–V)] is being produced (e.g., either by the hydrolysis of pyrophosphite [PPi(III)], as reported previously [27], or the reactant (phosphite) [Pi(III)] reacting with the phosphate [Pi(V)] present in the solution to form this mixed valence state P compound [PPi(III–V)].

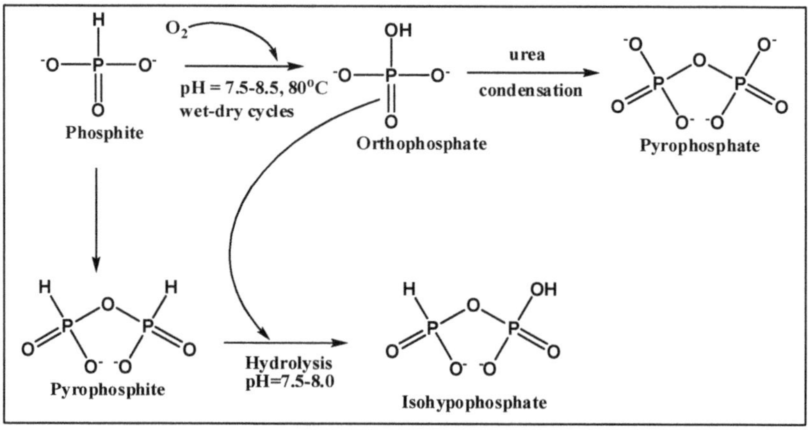

Figure 7. Description of various reaction steps suggested in the text.

To better understand the mechanism, we also studied the heating (leading to complete dryness) reactions of hypophosphite [Pi(I)] at 78–83 °C in the presence of urea for 2 days. We observed the oxidation and condensation reactions of hypophosphite [Pi(I)]. In the oxidation process, hypophosphite [Pi(I)] was oxidized into phosphite [Pi(III)] and phosphate [Pi(V)]. The condensed products included pyrophosphite [PPi(III)], isohypophosphate [PPi(III–V)], and pyrophosphate [PPi(V)]. This showed that overall inorganic reduced P compounds have a tendency to oxidize in air. The mechanism of the reaction can be compared with previously reported work by Kee and colleagues [27]. As suggested, pyrophosphite [PPi(III)] reacts readily with aqueous solutions of orthophosphate [Pi(V)] to give isohypophosphate, PPi(III–V) [27].

Previous studies have shown that hypophosphite [Pi(I)] can be considered a plausible intermediate in oxidation state between phosphide [Pi(0)] and phosphite [Pi(III)] [8]. Moreover, monitoring hypophosphite [Pi(I)] over the course of one day showed that 25% of this compound was oxidized to phosphite under air, independent of other additives in the solution [8]. Phosphite [Pi(III)], however, is more stable than hypophosphite [Pi(I)] over longer periods of time, hence it can be found in ancient rock samples [16] or in meteoritic solutions preserved for years [26]. The stability and longevity of phosphite [Pi(III)] in any solution is mainly dependent on the amount of oxidizing radicals in solution and the metals bonded to phosphite. $CaHPO_3$, for example, is significantly more resistant to oxidation than Na_2HPO_3. It can be stabilized under mildly reducing conditions that can potentially remove oxidants (oxidizing agents) from solutions [1].

Another study suggests a route of conversion of hypophosphite [Pi(I)] and phosphite [Pi(III)] through wet oxidation at 453 K (180 °C) under a partial oxygen pressure of

0.5–5 MPa [45]. The study found only a slight oxidation of hypophosphite [PPi(III)] under a N_2 atmosphere. However, in the presence of an O_2 atmosphere of 1 Mpa, hypophosphite [Pi(I)] was decreased to 5% after 180 min. Furthermore, for the wet-oxidation of phosphite [Pi(III)] under an O_2 pressure of 0.5–5 Mpa, the oxidation of phosphite [Pi(III)] to phosphate [Ppi(V)] was increased with an increase in the partial pressure of O_2. The oxidation of phosphite [Pi(III)] to phosphate [Pi(V)] was also increased when the pH was lowered from 6.05 to 1.04 [45]. We, however, did not observe any impact of pH on the oxidation of phosphite to phosphate during these studies. For example, in all samples containing sodium phosphite [Pi(III)] (pH = 8.5), phosphorous acid [Pi(III)] (pH = 2), and hypophosphite [Pi(I)] (pH = 4–5), around 1–13% phosphate [Pi(V)] was detected, which was not found consistent with any specific pH range.

Phosphite [Pi(III)] is considered to be thermodynamically unstable but kinetically stable on the Earth's surface [1]. The rate-limiting step in the oxidation process of phosphite [Pi(III)] is considered to be the breaking of the P-H (phosphorus-hydrogen) bond, which has a large activation energy of around 370 kJ [1]. This P-H bond, however, can be broken via a radical exchange mechanism (i.e., the reaction of phosphite with an $^{\bullet}OH$ radical to form H_2O and PO_3^{2-}). Further investigation into the free radical formation from phosphide [Pi(0), e.g., schreibersite] has shown that O_2 from air does not participate in forming the radical species [8].

The yields (% abundances) of the organic-P compounds were remarkably improved when urea was used as an additive (Figure 6). Urea seemed to facilitate condensation to form high-energy inorganic P compounds that readily reacted with the organics in contrast to the sample solution containing phosphite and organic without urea. In our reactions, we did not observe any organo-pyrophosphite species (P-O-P). One possibility is the instability and quick hydrolysis of such compounds. These results are also comparable with our recent studies that show one-pot syntheses of organic phosphates and phosphites specifically favored under alkaline conditions and additives such as urea and NH_4^+ ions [33]. The idea that high-energy condensed phosphates (and in this case condensed phosphites) are formed in the presence of urea that readily reacts with organics is also supported by our previous studies [39]; these show that pyrophosphate reacts with uridine to form uridine monophosphates alongside dimer (e.g., uridine-phosphate-uridine) [39]. However, the reaction of 'high-energy phosphorus compounds' formed in the mixture with the organics has not been verified independently e.g., by taking out the high-energy condensed P compounds from the crude reaction mixture and reacting them with organic substrates. Recently, it has been shown that phosphorous acid yields around 32.6% 5′-nucleoside monophosphate, along with di- and tri- organophosphate species, in a single reaction step at room temperature using liquid SO_2 under prebiotic conditions. Simultaneous oxidation results exist for the formation of organophosphates from phosphorus ([Pi(III)]) acid [46].

On the whole, it is quite plausible to envision cycles of de- and rehydration on the early Earth that could have driven nucleic acid polymerization in a volcanic environment enriched in phosphite and sulfur [47]. More research will be needed to find the likely pathway of the plausible oxidation of phosphite [Pi(III)] into phosphate [Pi(V)] under such mild conditions and the plausible formation of condensed P compounds that can readily react with organics to form a one-pot mixture of organic phosphates and phosphites.

Supplementary Materials: The following supporting information can be downloaded at: https://www.mdpi.com/article/10.3390/life13040920/s1, Figure S1: H-coupled ^{31}P-NMR spectrum of Sample P3-NaCl; Figure S2: ^{31}P-NMR (H-coupled) spectrum of sodium phosphite [Pi(III)] solution prior to reaction; Figure S3: H-coupled ^{31}P-NMR of heating sodium hypophosphite and urea solution (mixture); Figure S4: ^{31}P-NMR (H-coupled) spectrum of sodium hypophosphite [Pi(I)] solution prior to reaction showing no impurities or air oxidation phosphate species present in the starting compound; Figure S5: H-coupled ^{31}P-NMR of phosphorylation and phosphonylation reactions of glycerol spiked with standard glycerol phosphate (isomeric) solution; Figure S6: H-coupled ^{31}P-NMR of phosphorylation and phosphonylation reactions of adenosine spiked with standard 5′-AMP; Figure S7: H-coupled ^{31}P-NMR spectrum of Sample P3-thio.

Author Contributions: For Conceptualization, M.A.P., R.K. and M.G.; methodology, M.G.; formal analysis, M.G.; investigation, M.G. and M.A.P.; resources, M.A.P. and R.K.; data curation, M.G., H.A.C. and T.F.; writing—original draft preparation, M.G. and T.F.; writing—review and editing, M.G., T.F., H.A.C., R.K. and M.A.P.; funding acquisition, M.A.P. All authors have read and agreed to the published version of the manuscript.

Funding: This work was supported by NASA Exobiology program No: 80NSSC22K0509.

Institutional Review Board Statement: Not applicable.

Informed Consent Statement: Not applicable.

Data Availability Statement: NMR raw files and all the other relevant research results can be obtained by request from the corresponding author.

Acknowledgments: This work was supported in part by the University of South Florida Interdisciplinary NMR Facility, The Department of Chemistry and the College of Arts and Sciences, Tampa, Florida. Authors thank the USF-NMR facility, especially Benjamin Smith and Kyle, for their help with NMR. The authors also acknowledge Ryan Barkley for their help with the figures and support. Maheen Gull would also like to thank Luna and Nova Barkley for their support.

Conflicts of Interest: The authors declare no conflict of interest.

References

1. Pasek, M.A. Rethinking early Earth phosphorus geochemistry. *Proc. Natl. Acad. Sci. USA* **2008**, *105*, 853–858. [CrossRef] [PubMed]
2. Schwartz, A.W. Phosphorus in prebiotic chemistry. *Philos. Trans. R. Soc. Lond. B Biol. Sci.* **2006**, *361*, 1743–1749. [CrossRef]
3. Kring, D.A.; Cohen, B.A. Cataclysmic bombardment throughout the inner solar system 3.9–4.0 Ga. *J. Geophys. Res. Planets* **2002**, *107*, 4-1. [CrossRef]
4. Strom, R.G.; Malhotra, R.; Ito, T.; Yoshida, F.; Kring, D.A. The origin of planetary impactors in the inner solar system. *Science* **2005**, *309*, 1847–1850. [CrossRef] [PubMed]
5. Maciá, E.; Hernández, M.V.; Oró, J. Primary sources of phosphorus and phosphates in chemical evolution. *Orig. Life Evol. Biosph.* **1997**, *27*, 459–480. [CrossRef]
6. Pasek, M.; Lauretta, D. Extraterrestrial flux of potentially prebiotic C, N, and P to the early Earth. *Orig. Life Evol. Biosph.* **2008**, *38*, 5–21. [CrossRef]
7. Pasek, M.A.; Lauretta, D.S. Aqueous corrosion of phosphide minerals from iron meteorites: A highly reactive source of prebiotic phosphorus on the surface of the early Earth. *Astrobiology* **2005**, *5*, 515–535. [CrossRef]
8. Pasek, M.A.; Dworkin, J.P.; Lauretta, D.S. A radical pathway for organic phosphorylation during schreibersite corrosion with implications for the origin of life. *Geochim. Cosmochim. Acta* **2007**, *71*, 1721–1736. [CrossRef]
9. Ritson, D.J.; Mojzsis, S.J.; Sutherland, J.D. Supply of phosphate to early Earth by photogeochemistry after meteoritic weathering. *Nat. Geosci.* **2020**, *13*, 344–348. [CrossRef]
10. Pasek, M.A. Schreibersite on the early Earth: Scenarios for prebiotic phosphorylation. *Geosci. Front.* **2017**, *8*, 329–335. [CrossRef]
11. Melosh, H.J. *Impact Cratering: A Geologic Process*, 1st ed.; Oxford University Press: New York, NY, USA, 1989.
12. Simonson, B.M.; Glass, B.P. Spherule layers—Records of ancient impacts. *Annu. Rev. Earth Planet Sci.* **2004**, *32*, 329–361. [CrossRef]
13. Hunter, R.H.; Taylor, L.A. Rust and schreibersite in Apollo 16 highland rocks-Manifestations of volatile-element mobility. In Proceedings of the 12th Lunar and Planetary Science Conference, Houston, TX, USA, 16–20 March 1981; Pergamon Press: New York, NY, USA, 1982; Volume 12, pp. 253–259.
14. Yakovlev, O.I.; Dikov, Y.P.; Gerasimov, M.V. Experimental data on the thermal reduction of phosphorus and iron and their significance for the interpretation of the impact reworking of lunar materials. *Geochem. Int.* **2006**, *44*, 847. [CrossRef]
15. Cooper, G.W.; Onwo, W.M.; Cronin, J.R. Alkyl phosphonic acids and sulfonic acids in the Murchison meteorite. *Geochim. Cosmochim. Acta* **1992**, *56*, 4109–4115. [CrossRef] [PubMed]
16. Pasek, M.A.; Harnmeijer, J.P.; Buick, R.; Gull, M.; Atlas, Z. Evidence for reactive reduced phosphorus species in the early Archean Ocean. *Proc. Natl. Acad. Sci. USA* **2013**, *110*, 10089–10094. [CrossRef]
17. Pasek, M.; Block, K. Lightning-induced reduction of phosphorus oxidation state. *Nat. Geosci.* **2009**, *2*, 553–556. [CrossRef]
18. Pech, H.; Henry, A.; Khachikian, C.S.; Salmassi, T.M.; Hanrahan, G.; Foster, K.L. Detection of geothermal phosphite using high-performance liquid chromatography. *Environ. Sci. Technol.* **2009**, *43*, 7671–7675. [CrossRef]
19. Pasek, M.A.; Sampson, J.M.; Atlas, Z. Redox chemistry in the phosphorus biogeochemical cycle. *Proc. Natl. Acad. Sci. USA* **2014**, *111*, 15468–15473. [CrossRef]
20. Pasek, M.A.; Gull, M.; Herschy, B. Phosphorylation on the early earth. *Chem. Geol.* **2017**, *475*, 149–170. [CrossRef]
21. Gulick, A. Phosphorus as a factor in the origin of life. *Am. Sci.* **1955**, *43*, 479–489.
22. Pasek, M.A.; Kee, T.P. On the origin of phosphorylated biomolecules. In *Origins of Life: The Primal Self-Organization*, 1st ed.; Egel, R., Lankenau, D.-H., Mulkidjanian, A.Y., Eds.; Springer: Berlin, Germany, 2011; pp. 57–84.
23. Gull, M. Prebiotic Phosphorylation Reactions on the Early Earth. *Challenges* **2014**, *5*, 193–212. [CrossRef]

24. Gull, M.; Mojica, M.A.; Fernández, F.M.; Gaul, D.A.; Orlando, T.M.; Liotta, C.L.; Pasek, M.A. Nucleoside phosphorylation by the mineral schreibersite. *Sci. Rep.* **2015**, *5*, 17198. [CrossRef]
25. La Cruz, N.L.; Qasim, D.; Abbott-Lyon, H.; Pirim, C.; McKee, A.D.; Orlando, T.; Gull, M.; Lindsay, D.; Pasek, M.A. The evolution of the surface of the mineral schreibersite in prebiotic chemistry. *Phys. Chem. Chem. Phys.* **2016**, *18*, 20160–20167. [CrossRef] [PubMed]
26. Gull, M.; Feng, T.; Pasek, M.A. Results of an Eight-Year Extraction of Phosphorus Minerals within the Seymchan Meteorite. *Life* **2022**, *12*, 1591. [CrossRef]
27. Kee, T.P.; Bryant, D.E.; Herschy, B.; Marriott, K.E.; Cosgrove, N.E.; Pasek, M.A.; Atlas, Z.D.; Cousins, C.R. Phosphate activation via reduced oxidation state phosphorus (P). Mild routes to condensed-P energy currency molecules. *Life* **2013**, *3*, 386–402. [CrossRef]
28. Arrhenius, G.; Sales, B.; Mojzsis, S.; Lee, T. Entropy and charge in molecular evolution—The case of phosphate. *J. Theor. Biol.* **1997**, *187*, 503–522. [CrossRef] [PubMed]
29. Handschuh, G.J.; Lohrmann, R.; Orgel, L.E. The effect of Mg^{2+} and Ca^{2+} on urea-catalyzed phosphorylation reactions. *J. Mol. Evol.* **1973**, *2*, 251–262. [CrossRef] [PubMed]
30. Holm, N.G. The significance of Mg in prebiotic geochemistry. *Geobiology* **2012**, *10*, 269–279. [CrossRef] [PubMed]
31. Holm, N.G.; Baltscheffsky, H. Links between hydrothermal environments, pyrophosphate, Na+, and early evolution. *Orig. Life Evol. Biosph.* **2011**, *41*, 483–493. [CrossRef]
32. Yang, J.; Junium, C.K.; Grassineau, N.V.; Nisbet, E.G.; Izon, G.; Mettam, C.; Martin, A.; Zerkle, A.L. Ammonium availability in the Late Archaean nitrogen cycle. *Nat. Geosci.* **2019**, *12*, 553–557. [CrossRef]
33. Gull, M.; Feng, T.; Bracegirdle, J.; Abbott-Lyon, H.; Pasek, M.A. Organophosphorus Compound Formation through the Oxidation of Reduced Oxidation State Phosphorus Compounds on the Hadean Earth. *J. Mol. Evol.* **2022**, *91*, 60–75. [CrossRef]
34. Toner, J.D.; Catling, D.C. A carbonate-rich lake solution to the phosphate problem of the origin of life. *Proc. Natl. Acad. Sci. USA* **2020**, *117*, 883–888. [CrossRef] [PubMed]
35. Schoonen, M.; Smirnov, A.; Cohn, C. A perspective on the role of minerals in prebiotic synthesis. *AMBIO A J. Hum. Environ.* **2004**, *33*, 539–551. [CrossRef] [PubMed]
36. Lambert, J.B.; Gurusamy-Thangavelu, S.A.; Ma, K. The silicate-mediated formose reaction: Bottom-up synthesis of sugar silicates. *Science* **2010**, *327*, 984–986. [CrossRef] [PubMed]
37. Österberg, R.; Orgel, L.E.; Lohrmann, R. Further studies of urea-catalyzed phosphorylation reactions. *J. Mol. Evol.* **1973**, *2*, 231–234. [CrossRef]
38. Lohrmann, R.; Orgel, L.E. Urea-inorganic phosphate mixtures as prebiotic phosphorylating agents. *Science* **1971**, *171*, 490–494. [CrossRef]
39. Gull, M.; Omran, A.; Feng, T.; Pasek, M.A. Silicate-, magnesium ion-, and urea-induced prebiotic phosphorylation of uridine via pyrophosphate; revisiting the hot drying water pool scenario. *Life* **2020**, *10*, 122. [CrossRef]
40. Gull, M.; Zhou, M.; Fernández, F.M.; Pasek, M.A. Prebiotic phosphate ester syntheses in a deep eutectic solvent. *J. Mol. Evol.* **2014**, *78*, 109–117. [CrossRef]
41. Gull, M.; Pasek, M.A. Catalytic Prebiotic Formation of Glycerol Phosphate Esters and an Estimation of Their Steady State Abundance under Plausible Early Earth Conditions. *Catalysts* **2021**, *11*, 1384. [CrossRef]
42. Yoza, N.; Ueda, N.; Nakashima, S. pH-dependence of 31 P-NMR spectroscopic parameters of monofluorophosphate, phosphate, hypophosphate, phosphonate, phosphinate and their dimers and trimers. *Fresenius J. Anal. Chem.* **1994**, *348*, 633–638. [CrossRef]
43. Pasek, M.A. *Phosphorus NMR of Natural Samples*, 1st ed.; Free Radical Consulting: Seffner, FL, USA, 2018.
44. Gull, M.; Pasek, M.A. The role of glycerol and its derivatives in the biochemistry of living organisms, and their prebiotic origin and significance in the evolution of life. *Catalysts* **2021**, *11*, 86. [CrossRef]
45. Fujita, T.; Kawaguchi, Y.; Fukuta, T.; Matsuda, H.; Kojima, Y.; Yagishita, K. Effect of pH on conversion of hypophosphite and phosphite to phosphate by wet oxidation. *J. Surf. Finish. Soc. Jpn.* **2006**, *57*, 368–372, (In Japanese with English Abstract). [CrossRef]
46. Sydow, C.; Seiband, C.; Siegle, A.F.; Trapp, O. Phosphorylation in liquid sulfur dioxide under prebiotically plausible conditions. *Commun. Chem.* **2022**, *5*, 143. [CrossRef] [PubMed]
47. Lönnberg, T. Sulfurization of H-phosphonate diesters by elemental sulfur under aqueous conditions. *ACS Omega* **2017**, *2*, 5122–5127. [CrossRef] [PubMed]

life

Opinion

Distinguishing Biotic vs. Abiotic Origins of 'Bio'signatures: Clues from Messy Prebiotic Chemistry for Detection of Life in the Universe

Niraja V. Bapat [1,*] and Sudha Rajamani [2,*]

1 Independent Researcher, Hyderabad 500046, Telangana, India
2 Department of Biology, Indian Institute of Science Education and Research (IISER), Pashan, Pune 411008, Maharashtra, India
* Correspondence: nirajavbapat@gmail.com (N.V.B.); srajamani@iiserpune.ac.in (S.R.)

Abstract: It is not a stretch to say that the search for extraterrestrial life is possibly the biggest of the cosmic endeavors that humankind has embarked upon. With the continued discovery of several Earth-like exoplanets, the hope of detecting potential biosignatures is multiplying amongst researchers in the astrobiology community. However, to be able to discern these signatures as being truly of biological origin, we also need to consider their probable abiotic origin. The field of prebiotic chemistry, which is aimed at understanding enzyme-free chemical syntheses of biologically relevant molecules, could particularly aid in this regard. Specifically, certain peculiar characteristics of prebiotically pertinent messy chemical reactions, including diverse and racemic product yields and lower synthesis efficiencies, can be utilized in analyzing whether a perceived 'signature of life' could possibly have chemical origins. The knowledge gathered from understanding the transition from chemistry to biology during the origin of life could be used for creating a library of abiotically synthesized biologically relevant organic molecules. This can then be employed in designing, standardizing, and testing mission-specific instruments/analysis systems, while also enabling the effective targeting of exoplanets with potentially 'ongoing' molecular evolutionary processes for robust detection of life in future explorative endeavors.

Keywords: origin of life; search for extraterrestrial life; prebiotic chemistry; abiosignatures; nonenzymatic reactions; astrobiology

Citation: Bapat, N.V.; Rajamani, S. Distinguishing Biotic vs. Abiotic Origins of 'Bio'signatures: Clues from Messy Prebiotic Chemistry for Detection of Life in the Universe. *Life* **2023**, *13*, 766. https://doi.org/10.3390/life13030766

Academic Editors: Alberto Vázquez-Salazar and Ranajay Saha

Received: 29 December 2022
Revised: 4 March 2023
Accepted: 11 March 2023
Published: 13 March 2023

1. Introduction

The search for life beyond Earth has intrigued humankind for a very long time. Nonetheless, we are still far from conclusively answering the questions related to whether we are alone in the Universe even after decades of enquiry. We only know one example of life as yet, and it is that which exists on Earth. However, with the discovery of a large number of exoplanets, the current count being at 5272 (https://exoplanets.nasa.gov/; accessed on 4 March 2023), the prospects of detecting extraterrestrial life look promising and more exciting than ever before.

An array of potential biosignatures have been tabulated for extraterrestrial life detection strategies [1]; most of these are drawn from our current knowledge of biology on Earth. Biosignatures are usually assigned based on a non-zero possibility of them being of potential biological origin. However, this does not necessarily preclude their chemical origin. Hence, it is important to consider that the detection of one or a few potential biosignatures might not necessarily confirm the possibility/presence of extraterrestrial life. For example, the abiotic production of oxygen might result in a false positive detection of life on candidate exoplanets [2]. In addition, abiotic organic matter may result in the formation of pseudomicrofossils and pseudomicrobialites, which can be mistaken as signatures reminiscent of existing or past biological activities [3]. In such cases, it is worthwhile to consider the abiotic origin/s of biologically important molecules in question. In the recently published report of the Committee on the Planetary Science and Astrobiology Decadal Survey of the National Academies of Sciences, Engineering, and Medicine, the importance of constructing a framework to interpret potential biosignatures, abiosignatures, false positives, and false negatives, as well as that of efforts to better understand those abiosignatures that may mimic the ones originating from biological sources, has been highlighted [4]. This report has marked the research pertaining to the differentiation between abiotic and biotic sources and processes in the context of biosignature detection, as an area of focus for the next decade. In addition, the European Astrobiology roadmap (AstRoMap) has also identified distinction between life and nonlife during biosignature detection, as a key objective that should be addressed in a more detailed and coordinated manner in the future [5].

Biologically important molecules can also arise from different abiotic reactions, thus, leading to an incorrect interpretation of the detected putative biosignatures. A plausible way to overcome this issue is to analyze the nuanced differences stemming from the biotic vs. chemical origin of these potential 'bio'signatures. Simultaneously, it is also necessary to deduce/have some understanding of the geochemical settings on that particular planetary body, as the outcomes of organic reactions would vary based on various geochemical factors, including temperature, pH, reactant concentration, and availability of mineral ions.

To understand the likelihood of producing a given biologically pertinent molecule abiotically, which would result in 'abiosignatures' (i.e., signatures originating from abiotic processes), we need to gain insights from the events that led to the formation of first living cells (protocells) on the early Earth. The origin of life on Earth was an outcome of a complex set of non-trivial processes that allowed for the transition from the 'nonliving' chemistry realm to that of the 'living' biology realm. Since the first experimental demonstration of the synthesis of organic molecules relevant to biology in 1953 [6], researchers have been investigating this transition of complex organic chemistry to biology that eventually led to the formation of protocells. Research in this field of prebiotic chemistry has shed substantial light on the synthesis of many organic molecules relevant to biology, even in the absence of life. Such prebiotic chemical syntheses that are considered to have played an important role during the origin of life on Earth have some peculiar characteristics that are different from those associated with biosynthetic processes. Importantly, these prebiotic syntheses can provide clues regarding the extent to which complex organic molecules can be produced abiotically, thereby helping to rule out any abiotic sources when detecting potential biosignatures [7].

Herein, we discuss in detail the production and peculiar aspects of abiotically synthesized biologically important molecules, and emphasize the need for considering them while designing strategies to achieve convincing evidence of life's presence on extraterrestrial cosmic bodies. This would also help in narrowing down those cosmic bodies where prebiotic chemistry might be currently active and may lead to the formation of living entities in the near or distant future. This would yield potential 'targets' for researchers to effectively direct their efforts of detecting extraterrestrial life. Importantly, the aim of this article is not to question the credibility of any of the actual or potential biosignatures, but to use our understanding of messy prebiotic chemistry to efficiently and productively maneuver our search for life beyond Earth.

2. Chemical Synthesis of Biologically Important Molecules

The main building blocks of life on Earth include monomers of nucleic acids, proteins, sugars, and lipids. Although all of these molecules are important for sustaining life, the two fundamental polymers required for the origin of life as we know it are nucleic acids, which carry information, and proteins, which perform the catalytic functions. Prebiotically relevant abiotic synthesis of amino acids has been known for almost 70 years. The seminal spark discharge experiment, which laid the foundation for research being undertaken in the area of prebiotic chemistry, demonstrated the successful formation of glycine from a mixture of water, methane, ammonia, and hydrogen, in addition to some other amino acids and many hydroxy acids [6]. Re-analysis of the original samples using modern techniques revealed the formation of many other amino acids during the original Urey–Miller synthesis experiment [8]. Furthermore, short chain peptides can be chemically obtained from these amino acids using mineral catalysis, condensing agents, or large amounts of salts [9]. A study has previously demonstrated the formation of up to 20-mer long peptides under alternate wet-and-dry conditions [10]. Moreover, studies have also demonstrated the formation of oligopeptides from diketopiperazines, which are usually obtained as the inhibitory byproducts of abiotic peptide synthesis [10,11]. Once formed, these peptides have also been shown to potentially pass on their information by acting as a template for self-replication [12].

As for nucleic acid monomers, their formation and subsequent polymerization have been achieved using chemical means. Small quantities of adenosine were obtained by direct heating of adenine with β-D-ribose [13]. In an alternative approach, synthesis of pyrimidine ribonucleotides was shown using pentose amino-oxazolines as an intermediate [14]. Once synthesized, the enzyme-free polymerization of such monomers and short oligomers has been shown to occur under varied conditions, including ice-water eutectic phases [15], clay-mineral assisted synthesis [16], alternate cycles of dehydration-rehydration [17], and supramolecular liquid crystalline assembly conditions [18]. Oligomerization of cyclic purine and pyrimidine nucleotides to yield short RNA oligomers has also been demonstrated under various conditions [19,20]. In addition, the abiotic synthesis of relatively long nucleic acids via RNA-catalyzed RNA elongation reactions has been successfully demonstrated [21,22]. The other two important types of biomolecules, viz. amphiphiles (e.g., fatty acids and lipids) and sugars, have been synthesized using chemical reactions such as Fischer–Tropsch Type (FTT) synthesis and formose reaction, respectively [23]. The starting reactants for both of these syntheses are small chain carbon-based molecules, which ultimately yield biologically important complex organic molecules such as fatty acids and pentose sugars.

The aforementioned examples highlight the fact that the mere detection of biologically important molecules would not necessarily confirm or even indicate the presence of active biology on an exoplanet or a planetary body or their satellites (e.g., Enceladus and Europa in our solar system). One has to factor in the possibility of chemical syntheses of such molecules, many of which have been demonstrated to occur under varied conditions, including at very high to even sub-zero level temperatures [24]. Additionally, the type and length of the abiotically produced biologically relevant molecules would also vary depending on the available reactants, chemical evolutionary stage, and geochemical settings on the extraterrestrial body under consideration. Hence, it is necessary to assess the success of abiotic reactions under diverse 'messy' chemical environmental settings. This is crucial for determining the extent to which potential biosignatures can, in fact, be produced in the absence of biology (i.e., prebiotically relevant abiotic signatures), which also could be an indication of an ongoing or past, chemical or molecular evolutionary processes.

3. Homochirality Is Atypical in Prebiotic Reactions

Many theories have been put forth in order to delineate the mechanism/s underlying the prominence of enantioselectivity in extant biology [25]. However, a comprehensive understanding of the same is still lacking. It is not clear whether the presence of homochiral compounds was an absolute necessity for the origin of life, or if it was an outcome of later events that occurred during the course of evolution. Given this, enantiopurity seems especially difficult to achieve using only chemical means. Nonetheless, the selective presence of either D- or L-enantiomers of certain biomolecules in cells is a very important hallmark of extant life on Earth. Efficient enzymatic machinery ensures almost exclusive incorporation of D-sugars and L-amino acids during most biochemical processes. Although this symmetry breaking implies under-utilization of monomers from the available chemical space, it also imparts selectivity to biological processes. Arguably, the simultaneous presence of enantiomers might have compromised the competence of biological enzymes in terms of substrate specificity, thereby resulting in their sub-optimal performance. In addition, the presence of racemic mixtures of monomers has also been shown to affect enzyme-free processes such as nonenzymatic template copying [26].

Prebiotically plausible syntheses of complex organic molecules usually yield both enantiomers. The formose reaction yields a mixture of both D- and L-isomers of different sugars, including ribose, threose, and erythrose [27]. Similar mixtures of sugars can be obtained by irradiating interstellar ice containing water, methanol, and ammonia [28]. Notably, the amino acids from the spark discharge experiments were obtained as racemic mixtures as well [8,29]. Additionally, frozen solutions of NH_4CN kept at sub-zero temperatures for a prolonged period of time yielded both D- and L-isomers of amino acids [30]; such prolonged period of freezing is analogous to the conditions observed on Europa and other potential icy exoplanets. According to a conceptual model, racemization of the amino acids is expected in the hydrothermal production of organics on Enceladus [31]. Even extraterrestrial sources, such as carbonaceous meteorites, are known to contain only an enantiomeric excess of certain L-amino acids [32] and D-sugars [33]; the analyzed samples have never been observed to be entirely enantiopure.

Furthermore, both D- and L-isomers are known to participate in the abiotic reactions pertinent to biomolecules. For example, both D- and L-isomers of amino acids have been shown to get selectively adsorbed onto mineral surfaces [34]. Pertinently, their subsequent polymerization would lead to the formation of homochiral peptides of both handedness in the absence of any external selection pressure. Similarly, nucleotides containing either L-ribose or D-ribose can nonenzymatically polymerize in the absence or presence of a template [26]. This would potentially give rise to nucleic acid polymers with either chirality in the absence of selective pressure(s) exerted by biology. The presence of information-carrying molecules, specifically that of RNA, is considered to be of importance for the origin of life on Earth, especially due to its additional ability to facilitate catalytic reactions [35]. In this context, both right- and left-handed ribozymes have been shown to be capable of carrying out template copying [36], suggesting the presence and propagation of both the chiral forms of RNA on prebiotic Earth.

Thus, pertinent biologically important molecules obtained from abiotic reactions are not usually enantiopure. Observed enantiomeric excess may be an indication of how prebiotically relevant selection pressures might have shaped the evolutionary processes that had implications for life's origins. This has been shown for protocellular systems, wherein relevant environmental pressures have been shown to affect the robustness and survival of certain molecular systems over the others [37,38]. Given these aforementioned aspects, our efforts should be aimed at detecting not just the biologically important complex organic molecules but also characterizing their chirality for discerning their abiotic vs. biotic origin [39]. The presence of messy racemic mixtures of such molecules might hint more at their abiological origin and need not necessarily be a resultant of past or present biological processes. Nonetheless, the presence of racemic mixtures may also represent ongoing chemical evolutionary processes, making the parent cosmic bodies that harbor such molecules interesting in terms of plausible detection of life or life-like entities in the future.

4. Prebiotic Syntheses Yield Heterogeneous Products

Enzyme-catalyzed reactions in living cells are usually optimized to yield a certain product or a set of products, from a predetermined set of substrates. These products usually have a defined chemical structure; for example, all nucleotides in DNA are linked to each other only by a 3′-5′ phosphodiester bond. The extent of product formation is further modulated according to the physiological needs of the cell. Thus, biochemical reactions are generally tailored to yield only the most useful set of products required by the cell at any given time. On the other hand, prebiotically relevant reactions tend to yield a plethora of products.

A case in point is the famous Urey–Miller synthesis that yields a mixture of isomers of different compounds. Using modern analytical techniques, it was found that the original samples from the spark discharge experiments contained a total of 22 different amino acids and 5 different amines [8]. More recent studies have shown that similar electric discharge experiments can also yield all four RNA nucleobases [40]. In addition, meteoritic samples have been shown to contain many nucleobase analogs and non-biological amino acids along with the nucleobases and amino acids that are observed in extant biology [41,42]. The formose synthesis does not yield only the ribose sugar; it instead yields a mixture of tetrose, pentose, hexose, and heptose sugars when starting from a simple carbon compound such as glycolaldehyde or glyceraldehyde [27]. The formose reaction also yields metabolically relevant carboxylic acids, including α-hydroxy acids, when carried out under alkaline conditions [43]. The FTT synthesis reactions, which are thought to be a plausible terrestrial source of fatty acids and lipids on early Earth, yield an array of compounds containing 2 to more than 30 carbon moieties. Usually, a complex mixture of alkanols, alkanoic acids, alkenes, and alkanes is obtained during FTT synthesis [44]. Even meteorites are now known to contain a highly complex mixture of amino acids, sugars and sugar-related compounds, nucleobases, carboxylic acids, and insoluble organic matter, thus hinting at the large chemodiversity present in Space [45].

It is conceivable that, when such diverse monomers get chemically linked to yield polymers, such as peptides and nucleic acids, these polymers would also be diverse in terms of their chemical bonds, sequence, structure, etc. In accordance with this, when co-oligomerization of different amino acids was attempted, varied sequences of peptides were obtained in the resultant product mix [10]. Similarly, a large number of sequence isomers were obtained during the co-oligomerization of amino acids and hydroxyl acids, resulting in the formation of depsipeptides [46]. When activated nucleic acid monomers undergo enzyme-free oligomerization, the resultant products typically contain both 3′-5′ and 2′-5′ phosphodiester bonds, the occurrence of which is not pre-determined [23]. Furthermore, the nonenzymatic replication of nucleic acids has been argued to result in numerous diverse sequences due to the intrinsic low fidelity of this process. On the other hand, prebiotic formation of non-biological nucleic acid monomers is possible using nucleobase analogs [47,48] and alternate sugars [49]. Such non-conventional monomers also seem to get linked to each other by a phosphodiester linkage [48]. Moreover, two chemically different monomers are also known to covalently interact with each other to yield new molecules capable of biologically relevant functions; the examples include peptide nucleic acids (PNAs), which can hybridize with other PNAs, RNAs, and/or DNAs via base pairing [50], and N-acyl amino acids capable of self-assembling to yield vesicles [51]. Given these studies, the existence of nucleic acids with alternate backbones that are capable of duplex formation and information transfer should also be factored in while looking for potential signatures of life. These studies clearly underline the possibility of the presence of a vast variety of peptide- and nucleic acid-like polymers in the absence of substantial biological constraints.

Biology seems to utilize only a limited chemical space when compared to the enormous variety of entities that could have been readily generated from un-intervened chemical syntheses [52]. It is very possible that, in biology, the most 'useful' molecules might have gotten selected very early on during the course of evolution, as is indicated by the presence of certain common set of biomolecules in all domains of known life. Retaining this set through billions of years would have also saved the efforts of re-inventing the fundamental structure of life, while still allowing for evolution. On the other hand, chemical systems do not necessarily show any preference for only a certain set of molecules, unless limited by reactivity or by environmental constraints. Furthermore, it is important to consider that extant biochemistry could be one of the many possible ways of sustaining life. Therefore, while assessing the molecules on or from extraterrestrial cosmic bodies (using, e.g., in situ or remote detection), it is worthwhile to account for the diversity of the chemical space and possibility of the presence of alternate polymers as well.

5. Nonenzymatic Reactions Are Less Efficient than Biotic Syntheses

The two main biopolymers fundamental to life on extant Earth, viz. nucleic acids and proteins, are of sufficiently long length that in turn supports their respective functions. In theory, there are four choices available for each position in a nucleic acid polymer and twenty choices available for each position in a peptide, i.e., 4^n and 20^n sequences possible for nucleic acids and proteins, respectively, for a polymer of length n. Therefore, the number of possible sequences would increase exponentially with the increasing length of these polymers. Increased sequence space would further mean higher chances of finding an optimal sequence for a suitable function. Additionally, a larger number of optimal sequences would help in adding complexity to living cells, thus facilitating their evolution. Concurrent with this is the fact that the functional genomes of multicellular organisms are observed to be much longer than those of prokaryotic cells [53]. Nevertheless, such long functional polymers can be formed and maintained inside living cells only due to the availability of a kinetically efficient enzymatic machinery.

On the contrary, nonenzymatic synthesis of peptides and nucleic acids usually yield short-chain products. This is mainly because of the intrinsic low rate of these condensation reactions as well as the presence of competing reactions such as hydrolysis and/or alternate product formation [54,55]. Higher product yields can be obtained by using catalysts such as clay surfaces [56], by chemically activating monomers to promote polymer formation [23], by concentrating monomers in really small spaces such as in the brine channels in ice-water eutectic phases [15], or by using combinations thereof. However, the length of the resultant polymer products is still not comparable to what is possible using biological enzymes. The length of the RNAs obtained using ribozyme-catalyzed reactions are also not close to that of the functional nucleic acid genomes observed in extant biology. Similar results are observed for FTT synthesis, wherein the yield of the amphiphiles decreases with the increase in the number of carbons [44]. In the formose reaction as well, sugars containing more than six carbons are usually obtained in miniscule quantities [27]. Thus, nonenzymatic chemical syntheses usually mainly yield a plethora of small molecular weight entities, and achieving complex longer-chain-length products using prebiotically relevant organic synthesis approaches has typically proven to be a daunting challenge.

In addition, nonenzymatic reactions follow equilibrium kinetics while extant biochemical reactions are usually observed in a disequilibrium state. Therefore, the distribution of chemical species would differ for products obtained using chemical synthesis vs. biological synthesis. In biology, the product from one reaction is usually used by other reactions as substrates, forming interdependent reaction networks. These tightly interacting networks ensure efficient formation and utilization of biologically important molecules, and might have been essential for the origin of life on Earth [57]. However, during chemical syntheses in a prebiotic pool, such strong interactions and feedback control, which represent an advanced stage of chemical evolution, might not be plausible unless the molecules are concentrated in a smaller volume that would allow for efficient molecular interactions. The well-orchestrated chemistry required to sustain life is, in general, more efficient and dynamic than the simple conversion of reactants to products. Hence, the detection of a simple conversion reaction under equilibrium state is likely to hint at a chemical evolutionary stage, even if the end product is a biologically relevant molecule.

6. Implications for the Search of Life in the Universe

The outcomes of prebiotic reactions studied thus far clearly highlight the possible abiotic origin/s of biologically relevant molecules. This is important to consider when assessing the source of a detected potential biosignature, which can in fact be a prebiotically relevant abiotic signature. If the concerned biologically relevant molecule/s is/are detected in high and enantiopure quantities in the absence of messy background molecules, such signature/s can indeed be considered as likely originating from a past or present biological activity. However, if these molecules are detected on an extraterrestrial body in the presence of a noisy background and in low quantities, such signature may likely have abiotic origin/s.

In such cases, the presence of these complex organic molecules may not hint at the presence of life. Nonetheless, such signatures can indicate the presence of an active/ongoing chemical evolutionary process(es). Diverse prebiotically relevant chemical evolutionary paths are thought to have yielded the first living entities on the prebiotic Earth. Hence, the cosmic bodies where such potential biosignatures originating from abiotic sources (abiosignatures or complex chemical signatures) are detected may not just be interesting but will have direct implications for life's eventual emergence on that particular extraterrestrial body, and hence for its detection in the future. The life detection community can thus focus on such cosmic bodies for planning and execution of targeted life detection endeavors in a concentrated and effective manner.

The accrued knowledge gathered from understanding chemical processes relevant to life's origin on Earth and elsewhere can also be used to design and standardize mission-relevant instruments and related software for robotic/in situ as well as sample return missions. For example, as prebiotic syntheses usually yield a messy isomeric mixture of diverse molecules, it would be worthwhile to design instruments capable of analyzing the stereochemistry of a set of target molecules from the plethora of observable molecules. Understanding the yields of such prebiotic reactions will also help in setting the detection limits for various molecules on these analytical instruments. It is difficult to assess potential biosignatures without considering the planetary history, its geochemical environment, and the stage of chemical/biological evolution [58]. In this case, the products of prebiotic reactions that are carried out in the laboratory under various reaction conditions (mimicking the geochemical environment/s on cosmic bodies), and starting with different reactants (mimicking the reactant space heterogeneity and extent of chemical/biological evolution), may act as control samples for testing the mission instruments. In fact, the researchers engaged in understanding the origin of life need to come together and create a library of such molecules yielded from abiotic syntheses of biologically important molecules. These reactions are to be carried out in the laboratory and in natural analogue sites for delineating the various resultant species, which can then be used by the teams engaged in the detection of life in the Universe. Notably, eliminating the abiotic origin/s of a potential biosignature using such a library would still not necessarily confirm the presence of life because the signature may still have originated from a previously unknown chemistry and/or geochemical setting. Nonetheless, such a library will surely aid in designing more targeted efforts for detecting extraterrestrial life as well as understanding the limits of a planet's habitability.

Certain cross-disciplinary research collaboration networks have recently been established to address important questions related to the emergence of life on Earth and beyond, and detection of life in the Universe. For example, the Prebiotic Chemistry and Early Earth Environments (PCE3) Consortium has been established to bring together early earth geoscientists and prebiotic chemists to better understand the processes involved in the emergence of life on Earth and other cosmic bodies. Researchers in the Network for life detection (NfoLD) are actively involved in undertaking research pertaining to life detection, including biosignature creation and preservation, as well as related technology development. The proposed library of abiotically synthesized biologically important molecules will benefit from the research carried out by scientists involved in collaborations such as the PCE3 consortium, while also being useful for scientists involved in collaborations such as NfoLD for designing effective life detection strategies. Importantly, we recommend that a collaborative data sharing platform be established between researchers from the origins of life and life detection communities (including biosignature assessment and instrument development), for facilitating concrete advancement in our understanding of life's presence and distribution in the Universe. Such collaborations across diverse disciplines would greatly help in initiating a community-level dialogue that would positively support and refine our collective efforts in expanding life detection objectives.

7. Conclusions

With the continuous discovery of new exoplanets and advances in life detection technologies, the search for extraterrestrial life looks unprecedentedly promising. However, researchers should be prudent while detecting and segregating 'complex chemical signatures' or 'abiosignatures' from true 'biosignatures.' Experimental and theoretical studies carried out to decipher the origin of life on Earth could provide the requisite framework in this regard. Prebiotically relevant organic syntheses of biologically relevant molecules usually yield a heterogeneous spread of products containing many low molecular weight isomers. This contrasts with our understanding of how functional biomolecules are produced enzymatically in a living system. It can be expected that a minimal chaotic background would be detected for a 'true' biosignature as opposed to that observed for a prebiotically relevant abiosignature originating due to a divergent pool of chemicals yielded from messy prebiological reactions. To avoid this conundrum, a library of diverse molecules obtained from prebiotic reactions carried out under conditions reflecting various relevant geochemical settings, needs to be created in order to understand the extent to which biologically important complex organic molecules can be obtained in the absence of life. Such a library can be used to efficiently demarcate cosmic bodies with plausible active chemical evolution that is ongoing, while also allowing for designing and standardizing mission-specific instruments for a targeted and effective search for life beyond our pale blue dot.

Author Contributions: N.V.B. conceived the subject and structure of the review with crucial inputs from S.R.; N.V.B. and S.R. wrote and edited the manuscript. All authors have read and agreed to the published version of the manuscript.

Funding: This research received no external funding.

Institutional Review Board Statement: Not applicable.

Informed Consent Statement: Not applicable.

Data Availability Statement: This article has no additional data.

Acknowledgments: N.V.B. would like to acknowledge the support received from the ELSI-EON long-term visitors program, which gave her a chance to attend the EON workshop titled 'Cosmic Perspective of Earth: A Planet Permeated and Shaped by Life—Implications for Astrobiology'; the discussions in this workshop helped in the initial conceptualization of this article. S.R. acknowledges research support provided over the years from IISER Pune, Department of Science and Technology's Science and Engineering Research Board (SERB) (EMR/2015/000434, CRG/2021/001851) and Department of Biotechnology, Govt. of India (BT/PR19201/BRB/10/1532/2016); all this support has allowed her group to contribute productively towards understanding the chemical origins of life.

Conflicts of Interest: The authors declare no conflict of interest.

References

1. Schwieterman, E.W.; Kiang, N.Y.; Parenteau, M.N.; Harman, C.E.; DasSarma, S.; Fisher, T.M.; Arney, G.N.; Hartnett, H.E.; Reinhard, C.T.; Olson, S.L.; et al. Exoplanet biosignatures: A review of remotely detectable signs of life. *Astrobiology* **2018**, *18*, 663–708. [CrossRef]
2. Harman, C.E.; Schwieterman, E.W.; Schottelkotte, J.C.; Kasting, J.F. Abiotic O_2 Levels on Planets around F, G, K, and M Stars: Possible False Positives for Life? *Astrophys. J.* **2015**, *812*, 137. [CrossRef]
3. McMahon, C.; Cosmidis, J. False biosignatures on Mars: Anticipating ambiguity. *J. Geol. Soc.* **2021**, *179*, jgs2021-050. [CrossRef]
4. National Academies of Sciences, Engineering, and Medicine. *Origins, Worlds, and Life: A Decadal Strategy for Planetary Science and Astrobiology 2023–2032*; The National Academies Press: Washington, DC, USA, 2022. [CrossRef]
5. Horneck, G.; Walter, N.; Westall, F.; Grenfell, J.L.; Martin, W.F.; Gomez, F.; Leuko, S.; Lee, N.; Onofri, S.; Tsiganis, K.; et al. AstRoMap European Astrobiology Roadmap. *Astrobiology* **2016**, *16*, 201–243. [CrossRef]
6. Miller, S.L. A production of amino acids under possible primitive Earth conditions. *Science* **1953**, *117*, 528–529. [CrossRef] [PubMed]
7. Hoehler, T.; Brinckerhoff, W.; Davila, A.; Marais, D.D.; Getty, S.; Glavin, D.; Pohorille, A.; Quinn, R.; Bebout, L.; Broddrick, J.; et al. Groundwork for Life Detection. *Bull. AAS* **2021**, *53*. [CrossRef]

8. Johnson, A.P.; Cleaves, H.J.; Dworkin, J.P.; Glavin, D.P.; Lazcano, A.; Bada, J.L. The Miller volcanic spark discharge experiment. *Science* **2008**, *322*, 404. [CrossRef] [PubMed]
9. Rode, B.M. Peptides and the origin of life. *Peptides* **1999**, *20*, 773–786. [CrossRef]
10. Rodriguez-Garcia, M.; Surman, A.J.; Cooper, G.J.T.; Suárez-Marina, I.; Hosni, Z.; Lee, M.P.; Cronin, L. Formation of oligopeptides in high yield under simple programmable conditions. *Nat. Comm.* **2015**, *6*, 8385. [CrossRef]
11. Sasaki, M.; Miyagawa, Y.; Nonaka, K.; Miyanomae, R.; Quitain, A.T.; Kida, T.; Goto, M.; Honma, T.; Furusato, T.; Kawamura, K. Nano-pulsed discharge plasma-induced abiotic oligopeptide formation from diketopiperazine. *Sci. Nat.* **2022**, *109*, 33. [CrossRef]
12. Lee, D.H.; Granja, J.R.; Martinez, J.A.; Severin, K.; Ghadiri, M.R. A self-replicating peptide. *Nature* **1996**, *382*, 525–528. [CrossRef] [PubMed]
13. Fuller, W.D.; Sanchez, R.A.; Orgel, L.E. Studies in prebiotic synthesis: VI. Synthesis of purine nucleosides. *J. Mol. Biol.* **1972**, *67*, 25–33. [CrossRef]
14. Powner, M.W.; Gerland, B.; Sutherland, J.D. Synthesis of activated pyrimidine ribonucleotides in prebiotically plausible conditions. *Nature* **2009**, *459*, 239–242. [CrossRef]
15. Kanavarioti, A.; Monnard, P.A.; Deamer, D. Eutectic phase in ice facilitates nonenzymatic nucleic acid synthesis. *Astrobiology* **2001**, *1*, 271–281. [CrossRef]
16. Ferris, J.P. Montmorillonite-catalysed formation of RNA oligomers: The possible role of catalysis in the origins of life. *Philos. Trans. R. Soc. B* **2006**, *361*, 1777–1786. [CrossRef] [PubMed]
17. Rajamani, S.; Vlassov, A.; Benner, S.; Coombs, A.; Olasagasti, F.; Deamer, D. Lipid-assisted synthesis of RNA-like polymers from mononucleotides. *Orig. Life Evol. Biosph.* **2007**, *38*, 57–74. [CrossRef]
18. Todisco, M.; Fraccia, T.P.; Smith, G.P.; Corno, A.; Bethge, L.; Klussmann, S.; Paraboschi, E.M.; Asselta, R.; Colombo, D.; Zanchetta, G.; et al. Nonenzymatic Polymerization into Long Linear RNA Templated by Liquid Crystal Self-Assembly. *ACS Nano* **2018**, *12*, 9750–9762. [CrossRef]
19. Dagar, S.; Sarkar, S.; Rajamani, S. Geochemical influences on nonenzymatic oligomerization of prebiotically relevant cyclic nucleotides. *RNA* **2020**, *26*, 756–769. [CrossRef] [PubMed]
20. Costanzo, G.; Saladino, R.; Botta, G.; Giorgi, A.; Scipioni, A.; Pino, S.; Di Mauro, E. Generation of RNA molecules by a base-catalysed click-like reaction. *ChemBioChem* **2012**, *13*, 999–1008. [CrossRef]
21. Wochner, A.; Attwater, J.; Coulson, A.; Holliger, P. Ribozyme-catalyzed transcription of an active ribozyme. *Science* **2011**, *332*, 209–212. [CrossRef]
22. Attwater, J.; Wochner, A.; Holliger, P. In-ice evolution of RNA polymerase ribozyme activity. *Nat. Chem.* **2013**, *5*, 1011–1018. [CrossRef] [PubMed]
23. Orgel, L.E. Prebiotic chemistry and the origin of the RNA world. *Crit. Rev. Biochem. Mol. Biol.* **2004**, *39*, 99–123. [CrossRef] [PubMed]
24. Kitadai, N.; Maruyama, S. Origins of building blocks of life: A review. *Geosci. Front.* **2018**, *9*, 1117–1153. [CrossRef]
25. Blackmond, D.G. The origin of biological homochirality. *Cold Spring Harb. Perspect. Biol.* **2010**, *2*, a002147. [PubMed]
26. Joyce, G.F.; Visser, G.M.; Van Boeckel, C.A.; Van Boom, J.H.; Orgel, L.E.; Van Westrenen, J. Chiral selection in poly(C)-directed synthesis of oligo(G). *Nature* **1984**, *310*, 602. [CrossRef]
27. Zweckmair, T.; Böhmdorfer, S.; Bogolitsyna, A.; Rosenau, T.; Potthast, A.; Novalin, S. Accurate analysis of Formose reaction products by LC–UV: An analytical challenge. *J. Chromatogr. Sci.* **2014**, *52*, 169–175. [CrossRef]
28. Meinert, C.; Myrgorodska, I.; De Marcellus, P.; Buhse, T.; Nahon, L.; Hoffmann, S.V.; Le Sergeant d'Hendecourt, L.; Meierhenrich, U.J. Ribose and related sugars from ultraviolet irradiation of interstellar ice analogs. *Science* **2016**, *352*, 208–212. [CrossRef]
29. Ring, D.; Wolman, Y.; Friedmann, N.; Miller, S.L. Prebiotic synthesis of hydrophobic and protein amino acids. *Proc. Natl. Acad. Sci. USA* **1972**, *69*, 765–768. [CrossRef]
30. Levy, M.; Miller, S.L.; Brinton, K.; Bada, J.L. Prebiotic synthesis of adenine and amino acids under Europa-like conditions. *Icarus* **2000**, *145*, 609–613. [CrossRef]
31. Steel, E.L.; Davila, A.; McKay, C.P. Abiotic and biotic formation of amino acids in the Enceladus ocean. *Astrobiology* **2017**, *17*, 862–875. [CrossRef]
32. Pizzarello, S. The Chemistry of life's origin: A carbonaceous meteorite perspective. *Acc. Chem. Res.* **2006**, *39*, 231–237. [CrossRef] [PubMed]
33. Cooper, G.; Rios, A.C. Enantiomer excesses of rare and common sugar derivatives in carbonaceous meteorites. *Proc. Natl. Acad. Sci. USA* **2016**, *113*, E3322–E3331. [CrossRef] [PubMed]
34. Hazen, R.M.; Filley, T.R.; Goodfriend, G.A. Selective adsorption of L- and D-amino acids on calcite: Implications for biochemical homochirality. *Proc. Natl. Acad. Sci. USA* **2001**, *98*, 5487–5490. [CrossRef] [PubMed]
35. Robertson, M.P.; Joyce, G.F. The origins of the RNA world. *Cold Spring Harb. Perspect. Biol.* **2010**, *4*, a003608. [CrossRef]
36. Sczepanski, J.T.; Joyce, G.F. A cross-chiral RNA polymerase ribozyme. *Nature* **2014**, *515*, 440–442. [CrossRef]
37. Sarkar, S.; Dagar, S.; Verma, A.; Rajamani, S. Compositional heterogeneity confers selective advantage to model protocellular membranes during the origins of cellular life. *Sci. Rep.* **2020**, *10*, 4483. [CrossRef]
38. Joshi, M.P.; Steller, L.; Van Kranendonk, M.J.; Rajamani, S. Influence of Metal Ions on Model Protoamphiphilic Vesicular Systems: Insights from Laboratory and Analogue Studies. *Life* **2021**, *11*, 1413. [CrossRef]

39. Glavin, D.P.; Burton, A.S.; Elsila, J.E.; Aponte, J.C.; Dworkin, J.P. The Search for Chiral Asymmetry as a Potential Biosignature in our Solar System. *Chem. Rev.* **2020**, *120*, 4660–4689. [CrossRef]
40. Ferus, M.; Pietrucci, F.; Saitta, A.M.; Knížek, A.; Kubelík, P.; Ivanek, O.; Shestivska, V.; Civiš, S. Formation of nucleobases in a Miller–Urey reducing atmosphere. *Proc. Natl. Acad. Sci. USA* **2017**, *114*, 4306–4311. [CrossRef]
41. Callahan, M.P.; Smith, K.E.; Cleaves, H.J.; Ruzicka, J.; Stern, J.C.; Glavin, D.P.; House, C.H.; Dworkin, J.P. Carbonaceous meteorites contain a wide range of extraterrestrial nucleobases. *Proc. Natl. Acad. Sci. USA* **2011**, *108*, 13995–13998. [CrossRef]
42. Elsila, J.E.; Aponte, J.C.; Blackmond, D.G.; Burton, A.S.; Dworkin, J.P.; Glavin, D.P. Meteoritic amino acids: Diversity in compositions reflects parent body histories. *ACS Cen. Sci.* **2016**, *2*, 370–379. [CrossRef]
43. Omran, A.; Menor-Salvan, C.; Springsteen, G.; Pasek, M. The Messy Alkaline Formose Reaction and Its Link to Metabolism. *Life* **2020**, *10*, 125. [CrossRef] [PubMed]
44. McCollom, T.M.; Ritter, G.; Simoneit, B.R.T. Lipid synthesis under hydrothermal conditions by Fischer- Tropsch-Type reactions. *Orig. Life Evol. Biosph.* **1999**, *29*, 153–166. [CrossRef] [PubMed]
45. Ruf, A.; D'Hendecourt, L.L.S.; Schmitt-Kopplin, P. Data-Driven Astrochemistry: One Step Further within the Origin of Life Puzzle. *Life* **2018**, *8*, 18. [CrossRef] [PubMed]
46. Forsythe, J.G.; Petrov, A.S.; Millar, W.C.; Yu, S.-S.; Krishnamurthy, R.; Grover, M.A.; Hud, N.V.; Fernández, F.M. Surveying the sequence diversity of model prebiotic peptides by mass spectrometry. *Proc. Natl. Acad. Sci. USA* **2017**, *114*, E7652–E7659. [CrossRef]
47. Chen, M.C.; Cafferty, B.J.; Mamajanov, I.; Gállego, I.; Khanam, J.; Krishnamurthy, R.; Hud, N.V. Spontaneous prebiotic formation of a β-Ribofuranoside that self-assembles with a complementary heterocycle. *J. Am. Chem. Soc.* **2014**, *136*, 5640–5646. [CrossRef]
48. Mungi, C.V.; Singh, S.K.; Chugh, J.; Rajamani, S. Synthesis of barbituric acid containing nucleotides and their implications for the origin of primitive informational polymers. *Phys. Chem. Chem. Phys.* **2016**, *18*, 20144–20152. [CrossRef]
49. Fialho, D.M.; Clarke, K.C.; Moore, M.K.; Schuster, G.B.; Krishnamurthy, R.; Hud, N.V. Glycosylation of a model proto-RNA nucleobase with non-ribose sugars: Implications for the prebiotic synthesis of nucleosides. *Org. Biomol. Chem.* **2018**, *16*, 1263–1271. [CrossRef]
50. Frenkel-Pinter, M.; Samanta, M.; Ashkenasy, G.; Leman, L.J. Prebiotic peptides: Molecular hubs in the origin of life. *Chem. Rev.* **2020**, *120*, 4707–4765. [CrossRef]
51. Joshi, M.P.; Sawant, A.A.; Rajamani, S. Spontaneous emergence of membrane-forming protoamphiphiles from a lipid-amino acid mixture under wet-dry cycles. *Chem. Sci.* **2021**, *12*, 2970–2978. [CrossRef]
52. Meringer, M.; Cleaves, H.J. Exploring astrobiology using in silico molecular structure generation. *Phil. Trans. R. Soc. A* **2017**, *375*, 20160344. [CrossRef]
53. Lynch, M.; Conery, J.S. The origins of genome complexity. *Science* **2003**, *302*, 1401–1404. [CrossRef] [PubMed]
54. Bapat, N.V.; Rajamani, S. Templated replication (or lack thereof) under prebiotically pertinent conditions. *Sci. Rep.* **2018**, *8*, 15032. [CrossRef]
55. Mungi, C.V.; Bapat, N.V.; Hongo, Y.; Rajamani, S. Formation of abasic oligomers in nonenzymatic polymerization of canonical nucleotides. *Life* **2019**, *9*, 57. [CrossRef]
56. Huang, W.; Ferris, J.P. One-step, regioselective synthesis of up to 50-mers of RNA oligomers by Montmorillonite catalysis. *J. Am. Chem. Soc.* **2006**, *128*, 8914–8919. [CrossRef] [PubMed]
57. Cronin, L.; Walker, S.I. Beyond prebiotic chemistry. *Science* **2016**, *352*, 1174–1175. [CrossRef]
58. Barge, L.M.; Rodriguez, L.E.; Weber, J.M.; Theiling, B.P. "Determining the Biosignature Threshold" for Life Detection on Biotic, Abiotic, or Prebiotic Worlds. *Astrobiology* **2022**, *22*, 481–493. [CrossRef] [PubMed]

life

MDPI

Article

A New Approach in Prebiotic Chemistry Studies: Proline Sorption Triggered by Mineral Surfaces Analysed Using XPS

Eduardo J. Cueto-Díaz [1], Santos Gálvez-Martínez [1], María Colin-García [2] and Eva Mateo-Martí [1,*]

[1] Centro de Astrobiología (CSIC-INTA), Ctra. Ajalvir, km. 4, Torrejón de Ardoz, 28850 Madrid, Spain
[2] Instituto de Geología, Universidad Nacional Autónoma de México, Ciudad de Mexico 04510, Mexico
* Correspondence: mateome@cab.inta-csic.es

Abstract: The role of minerals in the origin of life and prebiotic evolution remains unknown and controversial. Mineral surfaces have the potential to facilitate prebiotic polymerization due to their ability to adsorb and concentrate biomolecules that subsequently can catalyse reactions; however, the precise nature of the interaction between the mineral host and the guest biomolecule still needs to be understood. In this context, we spectroscopically characterized, using infrared, X-ray photoemission spectroscopy (XPS) and X-ray diffraction (XRD) techniques, the interaction between L-proline and montmorillonite, olivine, iron disulphide, and haematite (minerals of prebiotic interest), by evaluating their interaction from a liquid medium. This work provides insight into the chemical processes occurring between proline, the only cyclic amino acid, and this selection of minerals, each of them bearing a particular chemical and crystal structures. Proline was successfully adsorbed on montmorillonite, haematite, olivine, and iron disulphide in anionic and zwitterionic chemical forms, being the predominant form directly related to the mineral structure and composition. Silicates (montmorillonite) dominate adsorption, whereas iron oxides (haematite) show the lowest molecular affinity. This approach will help to understand structure-affinity relationship between the mineral surfaces and proline, one of the nine amino acids generated in the Miller-Urey experiment.

Keywords: proline; XPS; infrared; prebiotic chemistry; mineral surfaces; montmorillonite; olivine; haematite; iron disulphides; spectroscopies

Citation: Cueto-Díaz, E.J.; Gálvez-Martínez, S.; Colin-García, M.; Mateo-Martí, E. A New Approach in Prebiotic Chemistry Studies: Proline Sorption Triggered by Mineral Surfaces Analysed Using XPS. *Life* **2023**, *13*, 908. https://doi.org/10.3390/life13040908

Academic Editors: Ranajay Saha and Alberto Vázquez-Salazar

Received: 22 February 2023
Revised: 24 March 2023
Accepted: 28 March 2023
Published: 30 March 2023

1. Introduction

Minerals have been considered the key player for prebiotic synthesis [1]. Indeed, nowadays, researchers have confirmed the catalytic properties of many prebiotic mineral catalysts towards the nitrogen and carbon fixation reactions, and based on these features, it is evident that minerals have actively participated in the availability of some key elements on modern and primigenial Earth.

Minerals are conspicuous components of environments where chemical evolution could have occurred, for example, meteorites, comets, the surface of early Earth and other planets. Minerals have been proposed as key elements in prebiotic chemistry studies and could have behaved as catalysts to favour prebiotic chemical reactions [2].

The catalytic abilities of metals have been thoroughly recognized [3]. Biochemistry is extremely linked to metals, metabolism is driven by metal gradients, and they are part of the active sites of enzymes [4]. In fact, one important fraction of known proteins (almost one-third) contains metal cofactors [5]. This link is probably rooted to the environment where life originated [4], and it has been proposed that metal ions could be the first catalyst during the first stages of the Earth [6]. Keeping this in mind, there are studies interested in elucidating the role of metals in prebiotic environments. For example, Pinter et al. (2021) [7] demonstrated that transition metals (Zn^{2+}, Cu^{2+}y Co^{2+}) increase the yield of depsipeptides, which contain histidine. In another study, Muchowska et al. (2017) [8] reported that some non-enzymatic reactions of the rTCA (reverse Krebs cycle) were promoted by metals

(Zn^{2+}, Cr^{3+} and Fe^0). Bray et al. (2018) [9] studied the structure, function, and cation content of ribosomes under conditions of the early Earth; they demonstrated that iron and manganese (Fe^{2+} and Mn^{2+}) are essential cofactors for the structure and function of ribosomes. Minerals could have also behaved as catalysts to expedite the release of chemical energy in water–rock–organic systems, enabling prebiotic chemical reactions [10]. It has been proposed that metal-containing minerals could be the predecessors of enzymes. There are resemblances in both the structure and reactivity of metallic minerals and the active sites of some enzymes [11]. In fact, most reactions involved in the bioenergetics of living beings require the participation of metal clusters in metalloenzymes [12].

Mineral surfaces have the potential to facilitate prebiotic polymerization by adsorbing and concentrating precursors or organic molecules on their surface (for example, Zaia et al., 2012 [13]; Rimola et al., 2019 [14]), which may result in catalytic reactions; however, the precise nature of the interaction between the mineral host and the biomolecule guest must be deeply studied. Over 70 years ago, it was proposed by Bernal [15] that the polymerization of biomolecules actually involved reactions on mineral surfaces, a mechanism later dubbed "polymerization on the rocks" [16,17]. For a long time, few experimental investigations were undertaken to test this hypothesis, so the adsorption and reactivity of amino acids and "small peptides" and their subsequent reactivity on surfaces resembling minerals that were probably present on the early Earth, such as iron disulphides, clays as montmorillonite, silicates as olivine, and iron oxides as haematite, would be a relevant approach to this hypothesis.

Amino acids are the building blocks of proteins, and they are essential units to build up more complex molecules from polymerization reactions [18,19]. Among all the amino acids, proline ($C_5H_9NO_2$) is the only amino acid lacking primary amine groups; in fact, the amine group is located inside the heterocyclic-pyrrolidine group (see Figure 1). In proline, the carbonyl group is more electron-rich than those of other natural amino acids. This fact promotes the formation of hydrogen bridges, which are involved in macromolecular self-organization and currently play an important role in protein structure and function.

Proline is the only cyclic amino acid, and this molecule is of particular interest because it contains a secondary amine (NH) group, fixed rigidly in the pyrrolidine ring, limiting the conformational mobility of the N–H bond with respect to the carboxyl group (COOH), which is itself restricted in mobility by the pyrrolidine cycle [20,21]. This rigidity is one of the factors that determines the important role played by proline in the folding of proteins and peptides, where steric effects arising from the proline side chains help to determine the stabilities and positions of the protein folds [22]. In modern living beings, proline-rich regions (PRRs) are common in both eukaryotic and prokaryotic proteins [23], and these PRRs are involved in protein–protein recognition processes [24]. Furthermore, proline has been synthesized in prebiotic experiments under different conditions [25–30]; it has also been detected on meteorites [31,32].

Therefore, in this complex context, these experiments are focused on understanding how an amino acid, proline, interacts with different mineral surfaces (relevant in prebiotic chemistry) due to their different functionalities or capacities. Surface science tools such as XPS have been used to study pyrite surface-driven molecular chemistry [33,34], and understanding the reactivity of molecule-pyrite surface systems has been improved by using a wide range of studies characterized by powerful and innovative spectroscopy techniques. These techniques have perhaps been the most useful in establishing the structure of the pristine pyrite surface and molecule/pyrite interaction. However, despite their importance, few spectroscopic studies have been performed on molecular interactions on other mineral surfaces. It is well known that the adsorption of organic molecules or biopolymers on mineral surfaces may substantially modify mineral morphology and surface composition and reactivity, thereby tailoring their properties [35]. Therefore, surface science techniques are very promising tools to study these molecular processes on catalytic mineral surfaces to obtain molecular information.

On the early Earth, ultramafic rocks were very abundant. On these rocks, olivine and pyroxene were the most common minerals [36], and this lithology probably hosted hydrothermal systems [37]. Olivine is not a mineral species but a group of them, and the olivine group corresponds to a solid solution with a variable chemical composition ($Fe_{1-x}Mg_xSiO_4$) [38], in which the Fe end member is called fayalite and the Mg end member is forsterite. The Earth crust is composed of more than 60% olivine, and fayalite is the most abundant mineral in the mantle [39].

Clays were the first minerals employed in prebiotic experiments since the proposal of Bernal [15]. Clay minerals are complex layered aluminium hydrous silicates. One sheet is formed by tetrahedral silicates, and the other by octahedral hydroxide [40]. The sheets are linked together by hydrogen bonds, and their assembly determines the type of clay, for example, 2:1 clays (two tetrahedral and one octahedral) or 1:1 clays (one tetrahedral and one octahedral) [40]. Montmorillonite is a 2:1 clay and is the most commonly used clay in prebiotic experiments; between the 2:1 blocks, there is a channel where cations can be stored and released.

Pyrite (FeS_2) is a very common mineral on Earth [41] because it is formed by different geological processes; it can be found in sediments [42] or as a result of hydrothermal activity [43]. On our planet, pyrite allows many geochemical and biogeochemical processes [44], and it has been proposed as crucial in the iron-sulphur world hypothesis [45]. Due to its properties, it has also been used in many prebiotic experiments [33,35,46–48]. It has been demonstrated that pyrite can act as a catalyst for nitrogen fixation [49]. Pyrite and haematite are common minerals in hydrothermal environments [50].

The potential catalytic activity of minerals in the context of prebiotic chemistry remains for the most part unexplored. In this scenario, we have studied the interaction of proline with montmorillonite (Mnt), olivine (Ol), iron disulphide, and haematite (Hem) by evaluating the interaction of the amino acid in a liquid alkaline environment. This strategy was designed to evaluate how mineral compositions modify the molecular interaction between the amino acid and the iron, silica, and sulphur surficial groups of the minerals. The interaction of some functional groups in organics may be governed by electrostatic forces (physisorption), while the interaction of others may be dictated by chemical bonding (chemisorption) to a specific group on the surface. In the present work, we will try to respond to these uncertainties by applying spectroscopic techniques.

In our study, we apply X-ray photoelectron spectroscopy (XPS), which is a surface-sensitive spectroscopic technique that can identify the elements that exist within a material (elemental composition) or on its surface, as well as their chemical state (e.g., oxidation state or chemical group which belong to). XPS is based on the photoelectric effect, in which a free electron is ejected from an atom after it has absorbed the energy of a photon from a source (X-rays). The relationship between the kinetic energy of the free electron (Ek), the binding energy (Eb), the photon energy (hν), and the work function (Ø) can be expressed as:

$$\mathrm{Ek} = h\nu - \mathrm{Eb} - \acute{\text{Ø}} \quad (1)$$

Knowing the energy of the incident photon and measuring Ek, we can obtain Eb. The electron binding energy (Eb) is a parameter that identifies the electron specifically in terms of the element and atomic energy level from which it comes [51–55]. The energy emitted by the photoelectrons is analysed using a spectrometer, and the data are plotted on a graph of intensity versus binding energy. The number of detected electrons in each peak is directly related to the amount of element within the XPS sampling volume. XPS is a powerful measurement technique because it shows not only what elements are present but also what other elements they are bonded to.

The study of the physical properties of minerals is fundamental for understanding the chemical reactions driven by them. Furthermore, developing a picture (at the atomic level) of the molecular structure and reactivity of these surfaces is also critical to understanding the role of these surfaces in the formation of complex biomolecules in a broad environment.

2. Materials and Methods

2.1. Materials

Four different minerals were used for the experiments (see Table 1): Na-rich montmorillonite (SWy-3, County of Crook, WY, USA), chemical composition (%) SiO_2: 62.9, Al_2O_3: 19.6, Fe_2O_3: 3.35, MgO: 3.05, CaO: 1.68; iron disulphide (>99.9%, Thermo Scientific®, Waltham, USA); haematite (NCS DC® 14033), chemical composition (%) SiO_2: 9.82, Al_2O_3: 0.48, CaO: 0.11, TFe: 61.68, FeO: 1.43; and olivine–basalt RM No. 1044-94®. L-proline (≥99.9%) was purchased from Sigma–Aldrich® (Missouri, USA).

Table 1. Minerals employed in this study and their characteristics.

Mineral	Chemical Formula	Mineral Group	Environment
Montmorillonite [1]	$(Na,Ca)_{0.3}(Al,Mg)_2Si_4O_{10}(OH)_2 \cdot nH_2O$	Phyllosilicate	Sedimentary, hydrothermal, diagenetic
Olivine [2]	$(Mg,Fe)_2SiO_4$	Silicate	Basic and ultra-basic igneous rocks, Group species name
Pyrite [1]	FeS_2	Sulphide	Sedimentary, magmatic, metamorphic, and hydrothermal deposits
Haematite [1]	Fe_2O_3	Oxide	Magmatic, hydrothermal, metamorphic, and sedimentary

[1] Mineral species [56], [2] Mineral group [57].

2.2. Interaction Process/Adsorption Experiments

The adsorption experiments were conducted by using the same stock solution of L-proline (1 M; pH 10). None of the natural minerals or FeS_2 were pre-treated or previously purified.

Ten millilitres of the solution of L-proline was mixed with 300 mg of Na-rich montmorillonite in a Teflon-capped reaction vial. Thereafter, the reaction was heated to 80 °C and stirred for 48 h. Then, the mixture was cooled down, and the collected suspension was centrifuged (13,000× *g* rpm, at RT). The supernatant was discarded, and the remaining solid was exhaustively washed with Milli-Q water (3 × 10 mL). The sample (Pro/Mnt) was dried under high vacuum, leading to 280 mg of a pale solid. In the case of the control sample, 10 mL of Milli-Q water was added to 300 mg of Na-rich montmorillonite and treated as previously described. Once the sample was dried under high vacuum, 282 mg of a pale solid was recovered.

The experimental protocol carried out for the Na-Montmorillonite (Mnt) sample case was repeated for the adsorption experiments in haematite (Hem), olivine (Ol), and pyrite (FeS_2), recommended abbreviations for the names of clay minerals and associated phases [58]. The quantities of sorbate/mineral obtained after work-up were roughly similar, leading to approximately 280 mg in each case.

2.3. Characterization by XPS, IR, and XRD Analysis

The XPS spectra of the mineral samples were recorded before and after the interaction among the molecule, from the solution, and the mineral. The XPS analysis of the samples was carried out in an ultra-high vacuum chamber equipped with a hemispherical electron analyser and with the use of an Al Kα X-ray source (1486.6 eV) with an aperture of 7 mm × 20 mm. The base pressure in the chamber was 2×10^{-9} mbar, and the experiments were performed at room temperature. The peak decomposition in different components was shaped, after background subtraction, as a convolution of Lorenztian and Gaussian curves. Binding energies were calibrated against the binding energy of the carbon 285.0 eV samples.

IR analysis and Fourier transform infrared (FTIR) spectroscopy of the mineral and proline/mineral systems were performed on eight samples in a Thermo-Nicolet spectrometer. Spectra (2 cm^{-1} resolution and 128 scans) were collected in the mid-infrared region (400–4000 cm^{-1}) using a DTGS-ATR detector and an XT-KBr beam splitter.

X-ray diffraction analysis and XRD characterization were carried out using a Bruker D8 Advance diffractometer with Cu Kα radiation (λ = 1.542 Å) and a diffractometer using Cu K1α radiation (λ = 1.54056 Å) operating at 45 kV and 40 mA. A Bragg–Brentano configuration geometry was used. The 2θ range was from 0 to 40 or 60° at 0.041 scanning steps.

3. Results and Discussion

In order to study the L-proline-mineral interaction, XPS analysis was performed. InXPS photons with sufficient energy are absorbed by a system, core electrons are ejected from the sample. A typical XPS spectrum is a plot of the number of electrons detected at a specific binding energy. Each element produces a set of characteristic XPS peaks. These peaks correspond to the electron configuration of the electrons within the atoms, e.g., core levels 1s, 2s, 2p, 3s, etc. Therefore, XPS allows confirmation through molecular fingerprints of sorption, and those traits would confirm the presence of proline on the mineral surfaces.

3.1. L-Proline Adsorption on Mineral Surfaces

3.1.1. XPS Studies

First, it is relevant to search for the presence of nitrogen as a molecular fingerprint since it confirms the adsorption and, therefore, the presence of the amino acid on the mineral surfaces. Figure 1 shows the XPS spectra of the N1s core level of proline on all minerals (montmorillonite, iron disulphide, haematite, and olivine). In the four cases, a nitrogen signal was detected, which proves the adsorption of proline on all the minerals tested. Furthermore, the intensity of the signal is related to the amount sorbed, so we can conclude that proline has a higher affinity for montmorillonite, followed by iron disulphide (pyrite), whereas olivine and haematite show low molecular affinity. This behaviour would be in good agreement with the fact that exposure of clay minerals to alkaline fluids, as in our case, induces the dissolution of silica, which can also lead to an increase in the specific surface area [59], favouring an increase in molecular adsorption on the mineral. In addition, montmorillonite has an interlayer space that holds cations [60], which could behave as a specific cation binding site.

One of the main objectives of this research was to deeply understand the sorption process. This objective has to do, on the one hand, with the identification of the functional groups of proline that interact with each mineral and, on the other hand, with determining the state of the amino acid (for example, ion, zwitterion, neutral molecule) when adsorbed on each mineral. The proline molecule can exist in four possible chemical species, namely, neutral (molecular form), cationic (NH group is protonated to give NH_2^+), anionic (COOH group is deprotonated, leading to COO^-), or zwitterionic (with both NH_2^+ and COO^- groups). The main factor that determines the chemical species (of an amino acid) present in a given system is the immediate geochemical environment. The zwitterion form dominates in the solid-state and in aqueous solution (near the isoelectric point, proline has a pKa_1 = 1.99 (carboxyl), pKa_2 = 10.96 (amino)). The anionic and cationic forms are pH-dependent and are abundant in alkaline and acidic media, respectively. The molecular form is present in the gas phase and can be isolated in a low-temperature argon matrix [61], while an anionic form, formed via the oxygen atoms of the carboxylate group and the nitrogen of the imine group, is predominant in the interaction on metallic surfaces such as copper [62].

In our case, L-proline yields a clearly differentiated spectroscopic fingerprint on the core level spectra of N1s, C1s, and O1s, the latter of which are shown in Figure 2. The comparative study of the C1s core level spectra among different minerals confirms that L-proline has a high affinity for montmorillonite, whereas for iron disulphide, olivine, and haematite, it exhibits a lower molecular affinity. Moreover, although the intensity of the signal changes in each experiment, the profile of the spectra is similar for the four samples. In the case of the O1s spectra, the contribution of the mineral structure to this region of the spectrum needs to be considered; in particular, iron oxides, iron silicates, and iron sulphates (all oxygen-bearing minerals) confer relevant changes in the profile of the spectrum related to the chemical composition of each mineral under study.

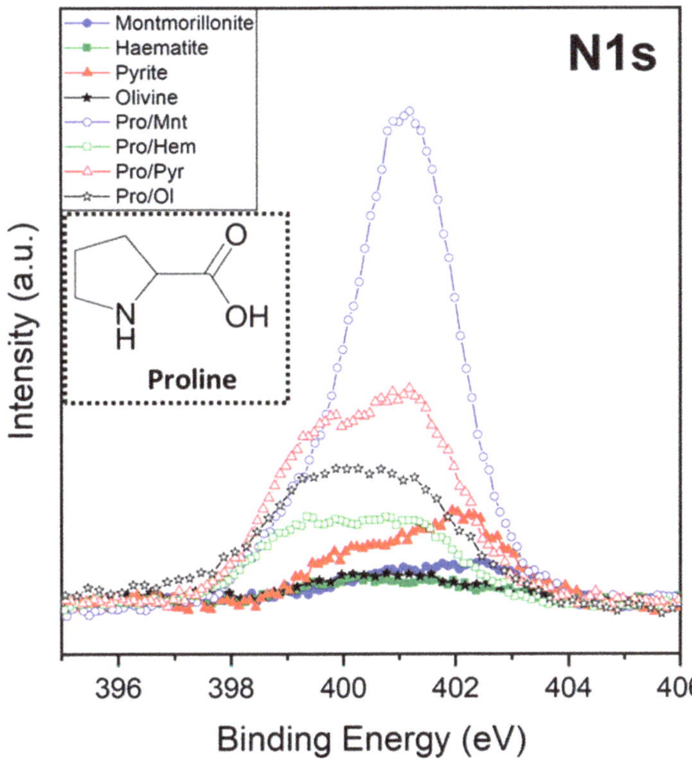

Figure 1. XPS nitrogen signal of the four minerals, before and after the adsorption of proline on montmorillonite, olivine, haematite, and iron disulphide; inset of proline molecular structure.

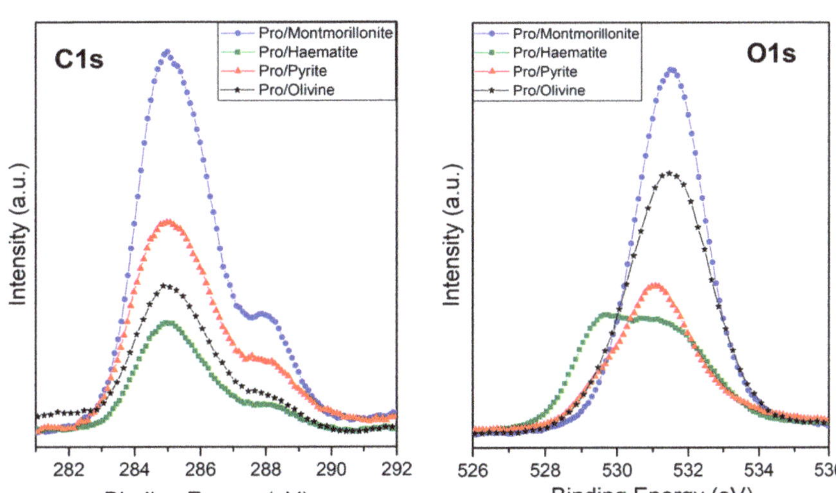

Figure 2. XPS spectra from the carbon (C1s) and oxygen (O1s) regions for the adsorption of proline on montmorillonite, olivine, haematite, and iron disulphide.

A detailed analysis and deconvolution of the N1s, C1s, and O1s components confirm the molecular chemical form of L-proline adsorbed on each mineral surface.

The XPS deconvolution for the proline/montmorillonite system indicates that the best fit for the C1s peak presents three main contributions: the first, at 284.6 eV binding energy (B.E.), assigned to CH plus one adventitious component; the second (at 285.9 eV) related to the C-N groups; and the third one (288.1 eV) assigned to the carboxylate (COO^-) or carboxylic (COOH) group. Regarding the O1s peak, the experimental data adjustment reveals the contribution of one component at 531.0 eV, which has been assigned to the two equivalent oxygen atoms belonging to the so-called resonant state of the deprotonated carboxylic group (carboxylate group), and the adventitious oxygen component due to the atmospheric air sample exposition before measuring XPS [63–65]. The N1s core level presents two components, the first one at 399.5 eV (19%) assigned to the NH species and the second centred at 401.2 eV (81%), which is assigned to the NH_2^+ species [66]. In summary, this spectroscopy analysis indicates that the molecule is adsorbed on the montmorillonite surface mainly in its zwitterionic (81%) and anionic forms (19%).

In the case of the Pro/Pyr system, the XPS analysis shows the same three components for the C1s peak at 284.2 eV, 285.9 eV, and 288.2 eV binding energies (B.E.), assigned to the CH plus adventitious carbon, to the C-N groups, and to the carboxylate (COO^-) or carboxylic (COOH) group, respectively. In the O1s peak, two components were found, one each at 529.6 eV (assigned to the iron oxides) and at 531.0 eV, which has been assigned to the carboxylate group and an adventitious component, respectively [63–65]. The N1s core level also presents two components, at 399.5 eV (65%) assigned to NH species and the second at 401.1 eV (35%) assigned to the NH_2^+ species; accordingly, the proline molecule is present in different chemical forms, the anionic (65%) and the zwitterionic form (35%), and pyrite seems to enhance the anionic adsorption with respect to the zwitterionic form. Cystine also exhibits the same behaviour, and there is a preference of the anionic form to be sorbed on a pyrite surface, as described previously [46]. XPS analysis for the proline/olivine system shows the same three components for the C1s peak, which were observed at 284.5 eV, 285.9 eV, and 287.8 eV binding energy (B.E.) assigned to the CH and adventitious carbon, to the C-N groups, and to carboxylate (COO^-) or carboxylic (COOH) groups, respectively. Regarding the O1s peak, two components were identified at 529.4 eV (assigned to iron oxides) and at 531.0 eV, which have been assigned to the deprotonated carboxylic group (carboxylate group) and adventitious oxygen. The N1s shows two components, at 399.5 eV (57%) assigned to NH species, and the second at 401.3 eV (43%), which is assigned to the NH_2^+ species; therefore, in this case, the molecule also presents two different chemical forms as well as in the case of pyrite. In both surfaces, the sorbed proline is predominantly in the anionic form over the zwitterionic form. However, in the case of iron disulphide, the percentage of anionic adsorption increases slightly with respect to the olivine surface.

XPS analysis for the proline/haematite system shows the same three components for the C1s peak at 284.7 eV, 286.0 eV, and 288.1 eV binding energies (B.E.). Those peaks can be assigned to the CH and adventitious carbon, to the C-N groups, and to carboxylate (COO^-) or carboxylic (COOH) groups, respectively. The O1s deconvoluted spectra show three main components centred at 529.5 eV that can be assigned to iron oxides, at 531.2 eV assigned to the carboxylate group, and at 532.1 eV assigned to the hydroxide species. The N1s shows two components, at 399.3 eV (50%) assigned to NH species, and the second at 401.3 eV (50%), which is assigned to the NH_2^+ species[67]. In this case, L-proline is observed to be present in two different chemical forms, as in the previous cases (anionic and zwitterionic forms), but it is noteworthy that in the case of haematite, there are no preferential adsorption forms, meaning that both molecular forms are equally sorbed. Table 2 shows a summary of the proline species percentages adsorbed on each molecule/mineral studied system.

Table 2. Summary of the proline species percentages adsorbed on each molecule/mineral studied system.

System of Study	Anion	Zwitterion	Mineral Composition
Pro/Mnt	19%	81%	Mg, Ca, Na silicate
Pro/Iron disulphide	65%	35%	Iron disulphide
Pro/Ol	57%	43%	Fe Mg silicate
Pro/Hem	50%	50%	Iron oxide

3.1.2. FT-IR Studies

For a better identification of the vibrational groups and accurate confirmation of the molecular chemical forms adsorbed on the different minerals studied, the IR-ATR spectra of proline adsorption on the four minerals studied were recorded in Table 3, which shows the peak assignments. A careful analysis of the IR spectra was carried out in the 1200–1800 cm^{-1} region, where significant changes were observed between the mineral matrix and the mineral exposed to proline adsorption (Figure 3). Unfortunately, the dark surface of pyrite does not give a good reflection of the infrared beam, and therefore, we could not obtain a good signal in the spectrum.

Table 3. List of peak assignments on the FT-IR spectra.

Wavenumber (cm^{-1})		Vibration Mode	Refs.
Mnt	**Pro/Mnt**		
3625	3614	ν(OH)	[68,69]
-	3069	$\nu_{as}(CH_2)$	[69]
-	2981	$\nu_s(CH_2)$	[69]
1633	-	ν(H-O-H)	[68]
-	1611	$\nu(COO^-)$	[69]
-	1559	$\delta(NH_2^+)$	[69]
-	1407	$\nu(COO^-)$	[69]
-	1372	ν(C-H)	[69]
1114	1115	ν(Si-O) out of plane	[70]
1000	1007	ν(Si-O) in plane	[70]
797	796	Platy form of tridymite	[70]
Hem	**Pro/Hem**		
-	1611	$\nu(COO^-)$	[69]
-	1549	$\delta(NH_2^+)$	[69]
-	1405	$\nu(COO^-)$	[69]
-	1376	ν(C-H)	[69]
1089	1082	$\nu(SiO_4)$ [1]	[71,72]
1047	1030	ν(Si-O-Si) [1]	[73]
908	903	ν(Si-OH) [1]	[73]
799	796	ν(Si-O-Si) [1]	[73]
518	519	δ(Fe-O)	[73]
441	441	δ(Fe-O)	[73]
Ol	**Pro/Ol**		
-	3068	$\nu_{as}(CH_2)$	[69]
-	2975	$\nu_s(CH_2)$	[69]
-	1612	$\nu(COO^-)$	[69]
-	1559	$\delta(NH_2^+)$	[69]
-	1449	$\nu(COO^-)$	[69]
-	1379	ν(C-H)	[69]
1086	1087	ν(Si-O)	[70]
959	959	$\nu(SiO_4)$ tetrahedron structure	[71,72]
929	930	$\nu(SiO_4)$ tetrahedron structure	[71,72]
740	743	$\nu(SiO_4)$ tetrahedron structure	[71,72]

[1] Quartz impurities.

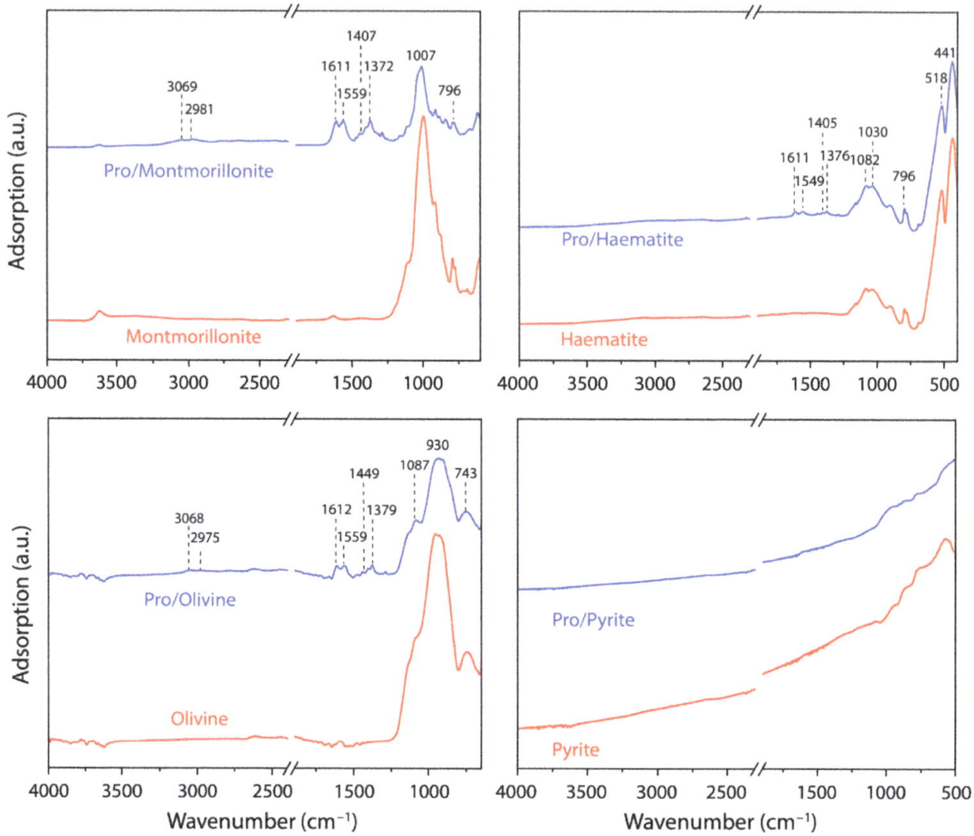

Figure 3. ATR-FTIR spectra for the minerals (red curves) and proline/mineral systems (blue spectra).

The chemical forms of adsorbed L-proline may be distinguished by characteristic vibrational bands associated with the main functional groups present. The zwitterion form is identified on the four minerals studied by both the vibrations associated with the scissor deformation of the NH_2^+ group $\delta(NH_2^+)$, at approximately 1559 cm^{-1}, and the asymmetric and symmetric carboxylate stretches, $\nu(COO^-)$, at approximately 1611 cm^{-1} and 1402 cm^{-1}, respectively. Due to the molecule/mineral interaction, the absence of the strong stretching vibration $\nu(C{=}O)$ at 1790–1760 cm^{-1}, assigned to the COOH group, is noticeable, while the appearance of vibrations attributable to the COO^- functionality is evident [62].

For the proline anionic form, one can expect the presence of characteristic vibrations associated with COO^-, as already described for the zwitterionic form. The NH-functionality vibrations (specific to the anionic form) are identified by the presence of the NH-bending vibration at 1377 cm^{-1}, which is observed in the IR spectrum. Furthermore, vibrations assigned to the other molecular functionalities, such as $\nu(CH_2)_{sciss}$ at 1452 cm^{-1} and ν(CH) bend at 1377 cm^{-1}, were also identified. Additionally, the spectra show less intense bands in the 3100–2800 cm^{-1} region; nevertheless, it is possible to identify the presence of ν(NH) at 3060 cm^{-1}, $\nu(CH_2)_{asym}$ at 2981 cm^{-1}, $\nu(CH_2)_{sym}$ at 2947 cm^{-1}, and ν(CH) at 2870 cm^{-1}, which are also related to molecular vibrations from the proline molecule.

3.1.3. XRD Studies

The XRD patterns of montmorillonite, pro/montmorillonite, haematite, pro/haematite, pyrite, pro/pyrite, olivine, and pro/olivine powder are shown in Figure 4 (below). The diffractograms showed that montmorillonite (Figure 4a, in red) displays a typical diffraction peak at $2\theta = 6.99°$, corresponding to a basal interlayer distance of 1.27 nm according to Bragg's law [74]. A second narrowed peak was observed at $2\theta = 8.83°$, which could be ascribed to the partial collapse of the mineral when exposed to pH = 10 (NaOH). The group of peaks observed between $2\theta = 27–29°$ can be attributed to impurities of quartz and feldspar.

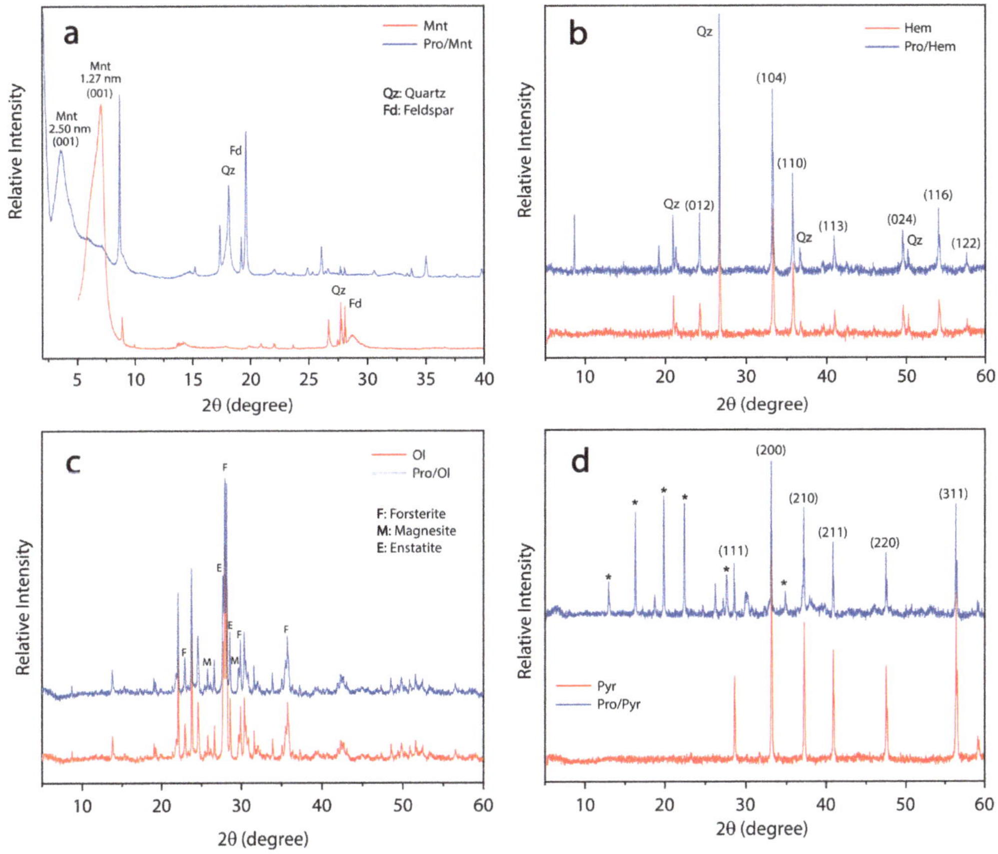

Figure 4. XRD diffractograms of the four cases of study: Pro/Mont (**a**), Pro/hem (**b**), Pro/Ol (**c**), and Pro/Pyr (**d**) in blue and their corresponding control samples in red. The measured diffraction peak positions and intensities are a fingerprint-like of a particular crystalline phase. Sharp and narrowed peaks at the XRD pattern indicate certain degree of crystallinity, whereas broad peaks denote amorphous materials, such as the (001) lattice plane in both montmorillonite cases.

After the adsorption of L-proline (Figure 4a, blue), the interlayer distance (d_{001}) increased to 2.50 nm. This result confirms that the organic molecule was profusely intercalated into the Mnt interlayer spaces. As the interlayer channel usually hosts cations, it can be suggested that the more likely proline species adsorbed here is the zwitterion form, while the sorption of the anionic form could occur on the surface of the clay.

On the other hand, both the haematite (Figure 4b) and olivine (Figure 4c) cases of study showed a similar pattern between the pristine (in red) and the mineral exposed to proline (blue). These results are in agreement with the small amount adsorbed on the surface recorded by XPS and FT-IR spectroscopies on the mineral surface.

Finally, the XRD diffractograms of the pyrite samples (Figure 4d) were recorded. As expected, the sample in the absence of proline displayed the typical pyrite pattern [75] showing six intensive peaks (111), (200), (210), (211), (220), and (311) (Figure 4d, red); however, due to oxidation, six new peaks (denoted with an asterisk) arise and are attributed to Rozenite ($FeSO_4 \cdot 4H_2O$) formation. This is in agreement with the observed XPS spectra of pyrite (S2p). On the other hand, when pyrite is in the presence of proline, the oxidation of FeS_2 is prevented. This is also confirmed using XPS spectroscopy show at the 3.2.2. Iron Disulphide section.

After IR and XPS analysis, we confirmed that the only two chemical species of proline adsorbed onto mineral surfaces are zwitterionic and anionic and that the percentages of each species are different depending on the nature of the mineral. Montmorillonite has a strong affinity for the zwitterionic form, whereas pyrite strongly interacts with the anionic form, whereas olivine and haematite barely discriminate each chemical species. A detailed mineral analysis of relevant elements such as iron and silicates would then be necessary.

3.2. *The Role of the Mineral Surface in the Competitive Adsorption of Zwitterionic and Anionic Species*

3.2.1. Montmorillonite

Clay minerals possess unique chemical and physical properties, such as a small particle size and the presence of charge on their surface, which gives them the ability to retain and release both organic and inorganic molecules. This could be the case for montmorillonite.

The hydration of clays involves the adsorption of water molecules on their surface, which can occur through hydration and interaction of the mineral surface with water molecules and with interlayer cations. The charges developed on the surface of the mineral (usually negative) [76] will promote the adsorption of cations, which behave as the main hydrophilic centres on the mineral. In this case, montmorillonite shows a much higher affinity for the zwitterionic (81%) form than for the anionic (19%) form since the positive charge is missing. Previous work on other amino acids, such as glycine adsorbed onto nontronite, also shows the same behaviour; amino acids are adsorbed on clays mainly in the zwitterionic form, although the anionic form is also sorbed but in a minor amount [77]. These results are consistent with the findings of our current study.

Figure 5 shows the spectrum of silicon (Si2p), the main chemical element in the montmorillonite structure, with a binding energy at 102.9 eV (assigned to the silicate species). After molecular adsorption, the intensity of the spectrum decreases, which can be due to amino acid adsorption on the surface, which attenuates the signal of the mineral. The same behaviour is observed for sodium, as the signal decreases after proline adsorption.

It has been observed that the negative layer charge on the silicate framework is delocalized over Si, O, and Al ions, owing to the systematic shifts of photoelectron binding energies. The chemical shift of the Si2p binding energy to lower binding-energy values may result from an increase in the negative layer charge in the silicate framework because of Si^{4+} replacement by Al^{3+} and/or Fe^{3+} [78]. This was confirmed in our work, where the Si2p peak is shifted towards lower binding energy values after proline adsorption (see Figure 5). Therefore, the interaction process of the positively charged zwitterion form in the negatively charged mineral is favoured over the anionic form.

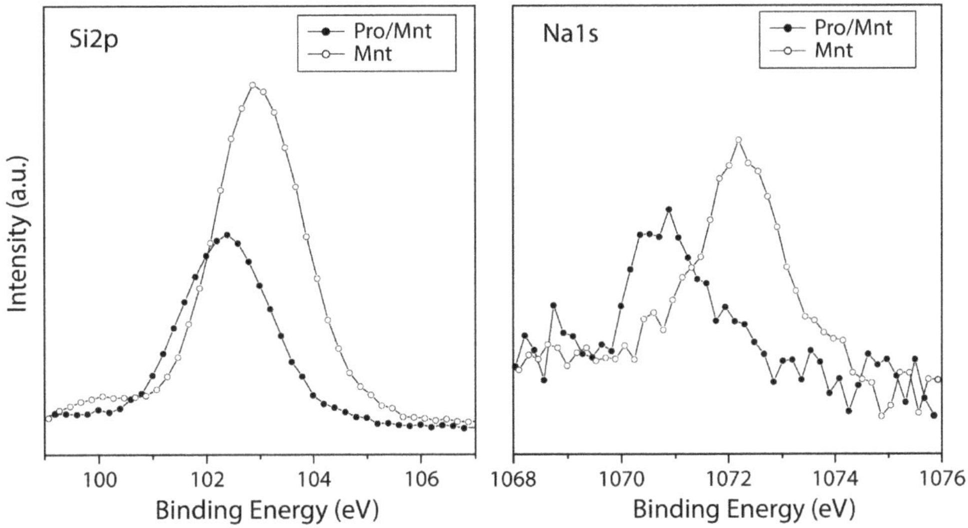

Figure 5. XPS spectra from the Si2p and Na1s regions for the montmorillonite mineral surface (empty circles) and after proline molecular adsorption on the mineral (black circles).

3.2.2. Iron Disulphide

Figure 6 shows the XPS spectra comparison of the pyrite surface before and after the interaction of the molecules. It is clear that pyrite minerals in water solution are highly oxidized, which can be confirmed by the identification of both iron oxide and sulphate species from the Fe2p iron spectra at 709–711 eV FeO and Fe_2O_3. Furthermore, the sulphur S2p spectra show a peak at 168–169 eV characteristic of sulphate species, which confirms the oxidation process. The predominant component is iron disulphide at 706.7–707.7 eV (Fe2p region) and 162.0–163.0 eV (S2p region).

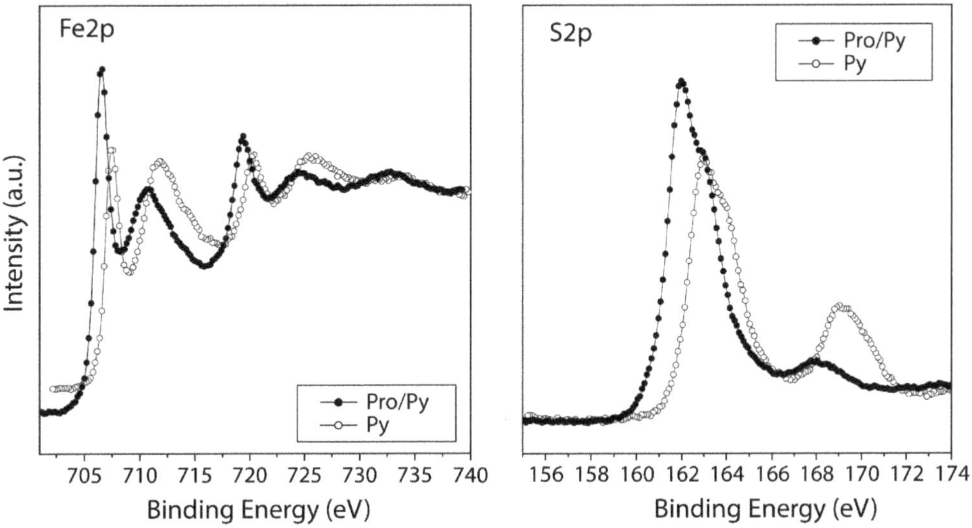

Figure 6. XPS spectra from the Fe2p and S2p regions for the iron disulphide mineral surface (empty circles) and after proline molecular adsorption on the mineral (black circles).

Under solution conditions, the pyrite surface is oxidized, so oxygen promotes an autocatalytic oxidation process on the pyrite surface. Proline anion adsorption has been detected on the oxidized surface, and zwitterion species adsorption is less significant under these conditions. Therefore, anion adsorption might compensate for the positive charge on the oxidized pyrite surface.

A comparison between the iron disulphide surface before and after molecular adsorption shows that iron oxide and iron disulphide species are similar in intensity to the isolated mineral, whereas in Pro/Pyrite systems, the component assigned to iron disulphide is predominant over iron oxides, which means that L-proline partially inhibits surface oxidation during the adsorption process. However, the clear identification of the iron oxide species present in both cases suggests that the interaction of the anionic form is favoured over that of the zwitterion.

3.2.3. Haematite

Figure 7 shows the XPS spectra comparison of the haematite surface before and after the interaction of the molecules. It is evident that the mineral haematite is the least reactive surface among the four tested in this study. First, the spectra of the N1s core level (a signature of the proline molecule), obtained from the comparative mineral study (Figure 1), show the lowest intensity for haematite. Second, the haematite surface after interaction with the proline solution does not show a noticeable decrease in the iron signal compared to the iron spectrum of the mineral alone. When molecular adsorption is successful, one would expect a noticeable decrease in the intensity of the initial surficial elements in the XPS spectrum, and this is not the actual case. However, even with a lower adsorption of proline than in any of the other adsorbed minerals, it can still be concluded that in the case of haematite, the anchoring of both species (zwitterionic and anionic) is equally favoured (Table 2). It seems that the positive (zwitterionic) selection effect as a silicate is compensated by the effect of the oxidation of iron in solution and therefore the need for the presence of the anion, which could compensate for the positive charge on the oxidized iron surface.

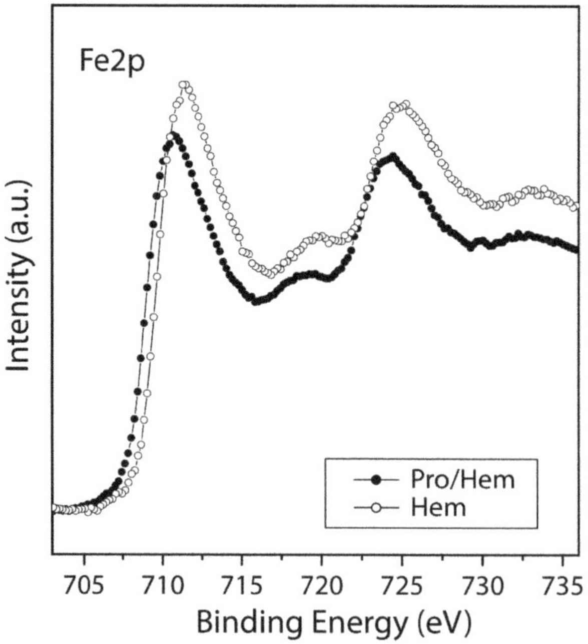

Figure 7. XPS spectra from the Fe2p region for the haematite mineral surface (empty circles) and after proline molecular adsorption on the mineral (black circles).

3.2.4. Basaltic Olivine

Figure 8 shows the spectra of silicon (Si2p) and iron (Fe2p), both chemical elements present in the structure of basaltic olivine, the former with a binding energy at 103.2 eV assigned to the silicate species and the latter at 710.7 eV assigned to iron oxides. After molecular adsorption, the intensity of the silicon spectrum slightly decreases, while that of iron is hardly attenuated at all; this result means that the proline molecule has a preferential affinity for silicon rather than iron atoms. Furthermore, low adsorption occurs on the surface of the olivine mineral, as depicted previously in the comparative mineral study (see Figure 1).

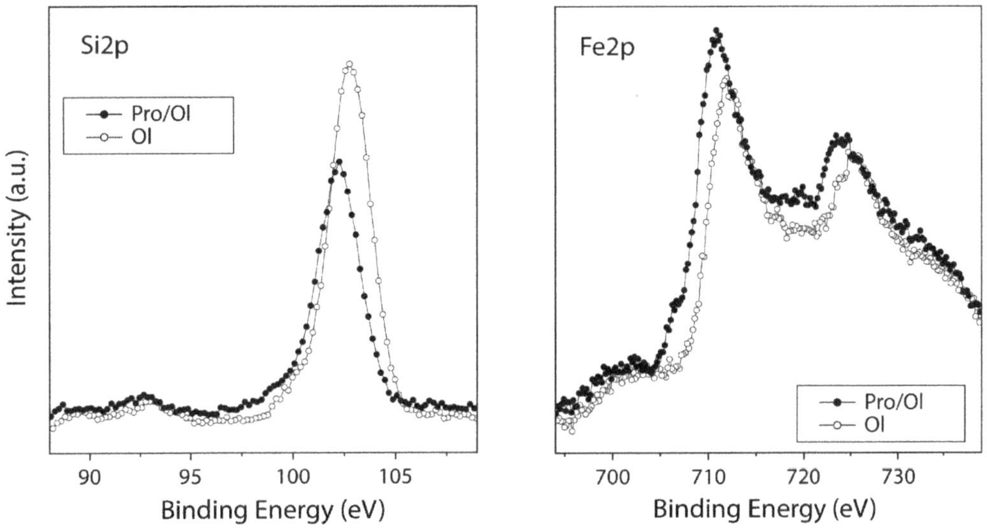

Figure 8. XPS spectra from the Fe2p and Si2p regions for the olivine mineral surface (empty circles) and after proline molecular adsorption on the mineral (black circles).

In this case, olivine shows a similar shift behaviour in the spectrum from 103.2 to 102.7 eV for Si2p towards lower binding energies after proline adsorption; this phenomenon was already described for montmorillonite. Thus, the interaction process of the positively charged zwitterion on the negatively charged mineral is somewhat favoured over the pyrite surface case. The presence of iron oxide species on the surface seems to compensate for the interaction inducing the coexistence of both chemical forms on the surface in almost equal percentages.

A schematic representation of the different chemical forms of proline molecules interacting with different mineral surfaces has been made. Therefore, Figure 9 shows the possible interaction among the species of the amino acid (anion or zwitterion) and the surface. The point of zero charge (pzc) is the pH for which the net surface charge of adsorbent is equal to zero; then, pzc values were taken from different sources: montmorillonite, pyrite, olivine, and haematite [79–82].

Amino acid/mineral systems, relevant in prebiotic chemistry, have been studied from a new approach using complementary spectroscopic techniques (XPS and IR) and a diffraction technique (XRD) to study in detail the mineral phase instead of the solution. It is noteworthy to explain that XPS is not a routine technique in this field, but from our studies, we can probe its potential to study the molecular affinity, the surface sites of preference, the favoured chemical form of interaction, and the spectroscopic fingerprint to follow for molecular preservation. Therefore, the information extracted from our studies allows us to propose montmorillonite as a highly reactive mineral with an enormous capacity for molecular preservation, which would be more likely to be conducive to polymerization/condensation reactions. Furthermore, it preferentially selects the zwitterion form, which provides two crucial and reactive charge functional groups (COO^- and NH_2) and thus two molecular sites to drive molecular chemistry and anchor possible surface reactions, depending on the surrounding conditions. Minerals played a crucial role in the origin of life and in prebiotic chemistry, but it is relevant to open up these studies to tools from other disciplines such as surface science to have a broader view and deeper understanding of surface chemistry, and thus to elucidate interesting implications for discerning catalysed prebiotic chemistry reactions.

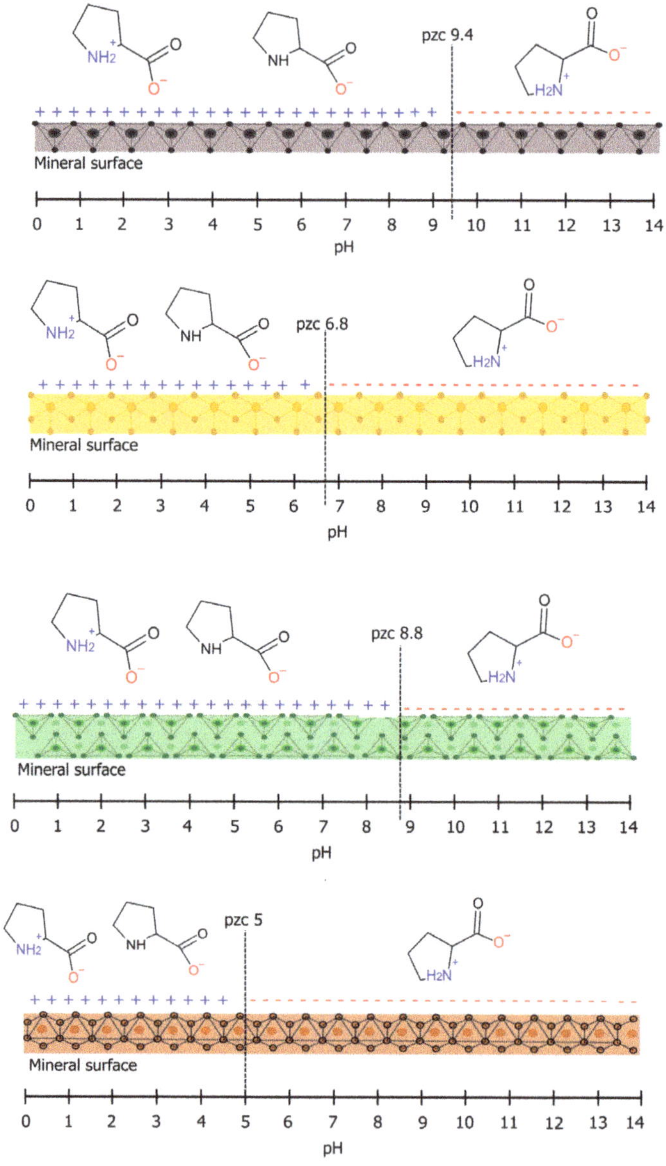

Figure 9. The Scheme shows a summary of the interactions between L-proline (anion or zwitterion) and the different mineral surfaces tested: (from top to bottom) montmorillonite, pyrite, olivine, and haematite.

4. Conclusions

We have performed the first spectroscopic characterization of L-proline adsorption on a battery of mineral surfaces, which is relevant to prebiotic chemistry. XPS analysis is used to accurately explore each molecular adsorption process and to determine the nature of the interaction (chemi/physisorption) of L-proline. In XPS spectroscopy, changes in the spectra of key elements in minerals reveal the interaction of organics with these surfaces. Noticeable differences in these spectra before and after adsorption dictate the mineral affinity towards one of the two preferable molecular forms of L-proline, carboxylate (anionic), or zwitterionic. Montmorillonite has a negative surface charge and shows preferential adsorption of the amino acid zwitterion, that is, a molecule related to the amino group (NH_2^+) that is positively charged. Iron disulphide has a surface that is easily oxidized, which implies the generation of a positive surface charge, favouring the sorption of the anionic form of proline. However, for olivine and haematite, there is a less remarkable preference for the adsorption of one of the chemical species.

Olivine is a magnesium iron silicate, where the positive iron and the negative silicate share surficial environments, and this mineral exhibits a slight preference for proline anion adsorption. In addition, haematite is the only mineral tested with no preferential adsorption of anionic or zwitterionic forms.

The understanding of the specific elemental role of the different mineral surfaces is of interest because it governs the interaction between them and the biomolecule of study. The XPS spectra of the mineral constituent elements show remarkable differences before and after molecular adsorption, proving the success of the molecule's adsorption on the mineral surface. Indeed, XPS spectroscopy demonstrated which chemical form has been anchored, and in this case, the affinity towards one of the two preferred molecular forms of L-proline (anionic or zwitterionic) can be established.

In this research, specific mineral-organic interactions have been described; the adsorbed chemical forms and their charge could force the functional groups more kindred to the mineral surface, thus leaving the groups less reactive with the mineral free and available to activate chemical reactions with other molecules, leading to an increase in chemical complexity. Here, we have examined the physical processes and driving forces of the molecular interaction of four systems, amino acid (proline)/minerals, which will provide insight into the understanding of a wide range of chemical environments that would be prebiotically viable by concentrating precursor molecules and their subsequent reactivity due to the specific structure of each mineral. In addition, it is important to search for minerals with the highest molecular concentration potential, as well as those capable of protecting the molecules in their structure from UV damage. Therefore, the location of these minerals would be the planetary sites most likely to detect organic molecules. This study confirms the potential of clays such as montmorillonite to preserve and concentrate simple biomolecules such as amino acids, as well as pushing the chemistry governing molecular interaction.

Author Contributions: Conceptualization, E.M.-M. and E.J.C.-D.; methodology, E.M.-M. and E.J.C.-D.; software, E.M.-M., E.J.C.-D. and S.G.-M.; validation, E.M.-M., E.J.C.-D., S.G.-M. and M.C.-G.; formal analysis, E.M.-M., E.J.C.-D. and S.G.-M.; investigation, E.M.-M. and E.J.C.-D.; resources, E.M.-M.; data curation, E.M.-M., E.J.C.-D. and S.G.-M.; writing—original draft preparation, E.M.-M. and M.C.-G.; writing—review and editing, E.M.-M., E.J.C.-D., S.G.-M. and M.C.-G.; visualization, E.M.-M., E.J.C.-D., S.G.-M. and M.C.-G.; supervision, E.M.-M., E.J.C.-D., S.G.-M. and M.C.-G.; project administration, E.M.-M.; funding acquisition, E.M.-M. All authors have read and agreed to the published version of the manuscript.

Funding: This research was funded by the projects PID2019-104205GB-C21 from the Spanish Ministry of Science and Innovation. Finally, we acknowledge funding through the Talent Attraction Postdoctoral Fellowship from CAM, ref: 2018-T2/TIC-10616. The support of CONACyT (A1-S-25341) and DGAPA-PAPIIT (IN111720 and IN218323) is also gratefully acknowledged.

Institutional Review Board Statement: Not applicable.

Informed Consent Statement: Not applicable.

Data Availability Statement: Not applicable.

Acknowledgments: The authors extend their appreciation to the Centro de Astrobiología (CAB) CSIC-INTA and the support by the Instituto Nacional de Técnica Aeroespacial "Esteban Terradas" (INTA). Additionally, the authors are grateful to Mª Teresa Fernández for helping during the FTIR and XRD measurements and to K. M. Lerin-Morales for the schematic diagrams showing the interaction molecule minerals (Scheme was drawn with https://inkscape.org/es/, (accessed on 20 March 2023)).

Conflicts of Interest: The authors declare no conflict of interest.

References

1. Walton, C.R.; Shorttle, O.; Jenner, F.E.; Williams, H.M.; Golden, J.; Morrison, S.M.; Downs, R.T.; Zerkle, A.; Hazen, R.M.; Pasek, M. Phosphorus mineral evolution and prebiotic chemistry: From minerals to microbes. *Earth-Sci. Rev.* **2021**, *221*, 103806. [CrossRef]
2. Martin, S.; Alexander, S.; Corey, C. A Perspective on the Role of Minerals in Prebiotic Synthesis. *AMBIO J. Hum. Environ.* **2004**, *33*, 539–551. [CrossRef]
3. Duval, M.; Deboos, V.; Hallonet, A.; Sagorin, G.; Denicourt-Nowicki, A.; Roucoux, A. Selective palladium nanoparticles-catalyzed hydrogenolysis of industrially targeted epoxides in water. *J. Catal.* **2021**, *396*, 261–268. [CrossRef]
4. Belmonte, L.; Mansy, S.S. Metal Catalysts and the Origin of Life. *Elements* **2016**, *12*, 413–418. [CrossRef]
5. Bartnikas, T.B.; Gitlin, J.D. How to make a metalloprotein. *Nat. Struct. Biol.* **2001**, *8*, 733–734. [CrossRef]
6. Aithal, A.; Dagar, S.; Rajamani, S. Metals in Prebiotic Catalysis: A Possible Evolutionary Pathway for the Emergence of Metalloproteins. *ACS Omega* **2023**, *8*, 5197–5208. [CrossRef]
7. Frenkel-Pinter, M.; Sargon, A.B.; Glass, J.B.; Hud, N.V.; Williams, L.D. Transition metals enhance prebiotic depsipeptide oligomerization reactions involving histidine. *RSC Adv.* **2021**, *11*, 3534–3538. [CrossRef]
8. Muchowska, K.B.; Varma, S.J.; Chevallot-Beroux, E.; Lethuillier-Karl, L.; Li, G.; Moran, J. Metals promote sequences of the reverse Krebs cycle. *Nat. Ecol. Evol.* **2017**, *1*, 1716–1721. [CrossRef]
9. Bray, M.S.; Lenz, T.K.; Haynes, J.W.; Bowman, J.C.; Petrov, A.S.; Reddi, A.R.; Hud, N.V.; Williams, L.D.; Glass, J.B. Multiple prebiotic metals mediate translation. *Proc. Natl. Acad. Sci. USA* **2018**, *115*, 12164–12169. [CrossRef]
10. Shock, E.L.; Boyd, E.S. Principles of Geobiochemistry. *Elements* **2015**, *11*, 395–401. [CrossRef]
11. Russell, M.J.; Martin, W. The rocky roots of the acetyl-CoA pathway. *Trends Biochem. Sci.* **2004**, *29*, 358–363. [CrossRef] [PubMed]
12. Duval, S.; Zuchan, K.; Baymann, F.; Schoepp-Cothenet, B.; Branscomb, E.; Russell, M.; Nitschke, W. Minerals and the Emergence of Life. In *Metals in Life Sciences*; Kroneck, P., Sosa Torres, M.E., Eds.; De Gruyter: Berlin, Germany, 2021; pp. 135–157.
13. Zaia, D.A.M. Adsorption of amino acids and nucleic acid bases onto minerals: A few suggestions for prebiotic chemistry experiments. *Int. J. Astrobiol.* **2012**, *11*, 229–234. [CrossRef]
14. Rimola, A.; Sodupe, M.; Ugliengo, P. Role of Mineral Surfaces in Prebiotic Chemical Evolution. In Silico Quantum Mechanical Studies. *Life* **2019**, *9*, 10. [CrossRef]
15. Bernal, J.D. The Physical Basis of Life. J. D. Bernal. London: Routledge and Kegan Paul, 1951. 80 pp. 6s.; The Structure and Mechanical Properties of Metals. Bruce Chalmers. New York: Wiley, 1951. 132 pp. $3.50.; Selective Toxicity with Special Reference to Chemotherapy. Adrien Albert. New York: Wiley; London: Methuen, 1951. 228 pp. $1.75. *Science* **1952**, *115*, 50. [CrossRef]
16. Liu, R.; Orgel, L.E. Polymerization on the Rocks: β-amino Acids and Arginine. *Orig. Life Evol. Biosph.* **1998**, *28*, 245–257. [CrossRef] [PubMed]
17. Orgel, L.E. Polymerization on the Rocks: Theoretical Introduction. *Orig. Life Evol. Biosph.* **1998**, *28*, 227–234. [CrossRef] [PubMed]
18. Marti, E.M.; Quash, A.; Methivier, C.; Dubot, P.; Pradier, C.M. Interaction of S-histidine, an amino acid, with copper and gold surfaces, a comparison based on RAIRS analyses. *Colloids Surf. A Physicochem. Eng. Asp.* **2004**, *249*, 85–89. [CrossRef]
19. Mateo-Martí, E.; Rogero, C.; Gonzalez, C.; Sobrado, J.M.; de Andrés, P.L.; Martin-Gago, J.A. Interplay between Fast Diffusion and Molecular Interaction in the Formation of Self-Assembled Nanostructures of S-Cysteine on Au(111). *Langmuir* **2010**, *26*, 4113–4118. [CrossRef] [PubMed]
20. Kayushina, R.; Vainshtein, B. Kristallografiya 10 (1965) 333. *Sov. Phys. Crystallogr.* **1966**, *10*, 698.
21. Reva, I.D.; Stepanian, S.G.; Plokhotnichenko, A.M.; Radchenko, E.D.; Sheina, G.G.; Blagoi, Y.P. Infrared matrix isolation studies of amino acids. Molecular structure of proline. *J. Mol. Struct.* **1994**, *318*, 1–13. [CrossRef]
22. Sapse, A.M.; Mallah-Levy, L.; Daniels, S.B.; Erickson, B.W. The.gamma. turn: Ab initio calculations on proline and N-acetylproline amide. *J. Am. Chem. Soc.* **1987**, *109*, 3526–3529. [CrossRef]
23. Williamson, M.P. The structure and function of proline-rich regions in proteins. *Biochem. J.* **1994**, *297*, 249–260. [CrossRef] [PubMed]
24. Kay, B.K.; Williamson, M.P.; Sudol, M. The importance of being proline: The interaction of proline-rich motifs in signaling proteins with their cognate domains. *FASEB J.* **2000**, *14*, 231–241. [CrossRef] [PubMed]
25. Fox, S.W.; Windsor, C.R. Synthesis of Amino Acids by the Heating of Formaldehyde and Ammonia. *Science* **1970**, *170*, 984–986. [CrossRef] [PubMed]

26. Janda, M.; Morvova, M.; Machala, Z.; Morva, I. Study of plasma induced chemistry by DC discharges in CO2/N2/H2O mixtures above a water surface. *Orig. Life Evol. Biosph.* **2008**, *38*, 23–35. [CrossRef]
27. Marshall, W.L. Hydrothermal synthesis of amino acids. *Geochim. Et Cosmochim. Acta* **1994**, *58*, 2099–2106. [CrossRef]
28. Plankensteiner, K.; Reiner, H.; Rode, B.M. Amino acids on the rampant primordial Earth: Electric discharges and the hot salty ocean. *Mol. Divers.* **2006**, *10*, 3–7. [CrossRef]
29. Ring, D.; Wolman, Y.; Friedmann, N.; Miller, S.L. Prebiotic Synthesis of Hydrophobic and Protein Amino Acids. *Proc. Natl. Acad. Sci. USA* **1972**, *69*, 765–768. [CrossRef]
30. Ruiz-Bermejo, M.; Menor-Salván, C.; Osuna-Esteban, S.; Veintemillas-Verdaguer, S. The effects of ferrous and other ions on the abiotic formation of biomolecules using aqueous aerosols and spark discharges. *Orig. Life Evol. Biosph.* **2007**, *37*, 507–521. [CrossRef]
31. Pizzarello, S.; Schrader, D.L.; Monroe, A.A.; Lauretta, D.S. Large enantiomeric excesses in primitive meteorites and the diverse effects of water in cosmochemical evolution. *Proc. Natl. Acad. Sci. USA* **2012**, *109*, 11949–11954. [CrossRef]
32. Shimoyama, A.; Ogasawara, R. Dipeptides and diketopiperazines in the Yamato-791198 and Murchison carbonaceous chondrites. *Orig. Life Evol. Biosph.* **2002**, *32*, 165–179. [CrossRef] [PubMed]
33. Galvez-Martinez, S.; Escamilla-Roa, E.; Zorzano, M.-P.; Mateo-Marti, E. Defects on a pyrite(100) surface produce chemical evolution of glycine under inert conditions: Experimental and theoretical approaches. *Phys. Chem. Chem. Phys.* **2019**, *21*, 24535–24542. [CrossRef]
34. Galvez-Martinez, S.; Escamilla-Roa, E.; Zorzano, M.-P.; Mateo-Marti, E. Ar+ ion bombardment dictates glycine adsorption on pyrite (100) surface: X-ray photoemission spectroscopy and DFT approach. *Appl. Surf. Sci.* **2020**, *530*, 147182. [CrossRef]
35. Pérez-Fernández, C.; Ruiz-Bermejo, M.; Gálvez-Martínez, S.; Mateo-Martí, E. An XPS study of HCN-derived films on pyrite surfaces: A prebiotic chemistry standpoint towards the development of protective coatings. *RSC Adv.* **2021**, *11*, 20109–20117. [CrossRef] [PubMed]
36. Berndt, M.E.; Allen, D.E.; Seyfried, W.E., Jr. Reduction of CO_2 during serpentinization of olivine at 300 °C and 500 bar. *Geology* **1996**, *24*, 351–354. [CrossRef]
37. Russell, M.J.; Hall, A.J.; Martin, W. Serpentinization as a source of energy at the origin of life. *Geobiology* **2010**, *8*, 355–371. [CrossRef] [PubMed]
38. de Souza, C.M.D.; Carneiro, C.E.A.; Baú, J.P.T.; da Costa, A.C.S.; Ivashita, F.F.; Paesano, A.; di Mauro, E.; de Santana, H.; Holm, N.G.; Neubeck, A.; et al. Interaction of forsterite-91 with distilled water and artificial seawater: A prebiotic chemistry experiment. *Int. J. Astrobiol.* **2013**, *12*, 135–143. [CrossRef]
39. Deer, W.L.H.; Howie, R.A.; Zussman, J. *Disilicates and Ring Silicates (Rock-Forming Minerals)*; Geological Society of London: London, UK, 1986.
40. Murray, H.H. Chapter 2 Structure and Composition of the Clay Minerals and their Physical and Chemical Properties. In *Developments in Clay Science*; Murray, H.H., Ed.; Elsevier: Amsterdam, The Netherlands; Oxford, UK, 2006; Volume 2, pp. 7–31.
41. Mateo-Martí, E.; Briones, C.; Rogero, C.; Gomez-Navarro, C.; Methivier, C.; Pradier, C.M.; Martín-Gago, J.A. Nucleic acid interactions with pyrite surfaces. *Chem. Phys.* **2008**, *352*, 11–18. [CrossRef]
42. Berner, R.A. Sedimentary pyrite formation: An update. *Geochim. Cosmochim. Acta* **1984**, *48*, 605–615. [CrossRef]
43. Keith, M.; Häckel, F.; Haase, K.M.; Schwarz-Schampera, U.; Klemd, R. Trace element systematics of pyrite from submarine hydrothermal vents. *Ore Geol. Rev.* **2016**, *72*, 728–745. [CrossRef]
44. Luther, G.W.; Kostka, J.E.; Church, T.M.; Sulzberger, B.; Stumm, W. Seasonal iron cycling in the salt-marsh sedimentary environment: The importance of ligand complexes with Fe(II) and Fe(III) in the dissolution of Fe(III) minerals and pyrite, respectively. *Mar. Chem.* **1992**, *40*, 81–103. [CrossRef]
45. Wächtershäuser, G. On the chemistry and evolution of the pioneer organism. *Chem. Biodivers.* **2007**, *4*, 584–602. [CrossRef]
46. Sanchez-Arenillas, M.; Mateo-Marti, E. Spectroscopic study of cystine adsorption on pyrite surface: From vacuum to solution conditions. *Chem. Phys.* **2015**, *458*, 92–98. [CrossRef]
47. Sanchez-Arenillas, M.; Mateo-Marti, E. Pyrite surface environment drives molecular adsorption: Cystine on pyrite(100) investigated by X-ray photoemission spectroscopy and low energy electron diffraction. *Phys. Chem. Chem. Phys.* **2016**, *18*, 27219–27225. [CrossRef] [PubMed]
48. Xian, H.; Zhu, J.; Tan, W.; Tang, H.; Liu, P.; Zhu, R.; Liang, X.; Wei, J.; He, H.; Teng, H.H. The mechanism of defect induced hydroxylation on pyrite surfaces and implications for hydroxyl radical generation in prebiotic chemistry. *Geochim. Cosmochim. Acta* **2019**, *244*, 163–172. [CrossRef]
49. Mateo-Marti, E.; Galvez-Martinez, S.; Gil-Lozano, C.; Zorzano, M.-P. Pyrite-induced uv-photocatalytic abiotic nitrogen fixation: Implications for early atmospheres and Life. *Sci. Rep.* **2019**, *9*, 15311. [CrossRef] [PubMed]
50. Colín-García, M.; Heredia, A.; Cordero, G.; Camprubí, A.; Negrón-Mendoza, A.; Ortega-Gutiérrez, F.; Beraldi, H.; Ramos-Bernal, S. Hydrothermal vents and prebiotic chemistry a review. *Boletín De La Soc. Geológica Mex.* **2016**, *68*, 599–620. [CrossRef]
51. Attard, G.; Barnes, C. *Surfaces*; Oxford University Press: Oxford, UK, 1998.
52. Hüfner, S. *Photoelectron Spectroscopy*; Springer: Berlin/Heidelberg, Germany, 1995; Volume 82, p. 27.
53. Moulder, J.F.; Stickle, W.F.; Sobol, W.M.; Bomben, K.D. Handbook of X-Ray Photoelectron Spectroscopy. *Perkin-Elmer Corp.* **1992**, *40*, 221.

54. Siegbahn, K. ESCA atomic, molecular and solid state structure studies by means of electron spectroscopy. In *Nova acta Regiae Societatis Scientiarum Upsaliensis*; Almqvist & Wiksell: Uppsala, Sweden, 1967.
55. Siegbahn, K.M. Electron Spectroscopy for Atoms, Molecules and Condensed Matter. *Rev. Mod. Phys.* **1982**, *54*, 709. [CrossRef]
56. Commission on New Minerals, Nomenclature and Classification. Available online: http://cnmnc.units.it/ (accessed on 20 March 2023).
57. Mineralogy Database. Available online: http://webmineral.com/ (accessed on 20 March 2023).
58. Warr, L.N. Recommended abbreviations for the names of clay minerals and associated phases. *Clay Miner.* **2020**, *55*, 261–264. [CrossRef]
59. Jozefaciuk, G.; Bowanko, G. Effect of Acid and Alkali Treatments on Surface Areas and Adsorption Energies of Selected Minerals. *Clays Clay Miner.* **2002**, *50*, 771–783. [CrossRef]
60. Yotsuji, K.; Tachi, Y.; Sakuma, H.; Kawamura, K. Effect of interlayer cations on montmorillonite swelling: Comparison between molecular dynamic simulations and experiments. *Appl. Clay Sci.* **2021**, *204*, 106034. [CrossRef]
61. Wu, R.; McMahon, T.B. Stabilization of the Zwitterionic Structure of Proline by an Alkylammonium Ion in the Gas Phase. *Angew. Chem. Int. Ed.* **2007**, *46*, 3668–3671. [CrossRef]
62. Mateo Marti, E.; Barlow, S.M.; Haq, S.; Raval, R. Bonding and assembly of the chiral amino acid S-proline on Cu(110): The influence of structural rigidity. *Surf. Sci.* **2002**, *501*, 191–202. [CrossRef]
63. Clark, D.; Peeling, J.; Colling, L. An experimental and theoretical investigation of the core level spectra of a series of amino acids, dipeptides and polypeptides. *Biochim. Et Biophys. Acta (BBA)-Protein Struct.* **1976**, *453*, 533–545. [CrossRef]
64. Uvdal, K.; Bodö, P.; Ihs, A.; Liedberg, B.; Salaneck, W.R. X-ray photoelectron and infrared spectroscopy of glycine adsorbed upon copper. *J. Colloid Interface Sci.* **1990**, *140*, 207–216. [CrossRef]
65. Uvdal, K.; Bodö, P.; Liedberg, B. l-cysteine adsorbed on gold and copper: An X-ray photoelectron spectroscopy study. *J. Colloid Interface Sci.* **1992**, *149*, 162–173. [CrossRef]
66. Gao, F.; Wang, Y.; Burkholder, L.; Tysoe, W.T. Chemistry of l-proline on Pd(111): Temperature-programmed desorption and X-ray photoelectron spectroscopic study. *Surf. Sci.* **2007**, *601*, 3579–3588. [CrossRef]
67. Scheglov, A.; Helmke, A.; Loewenthal, L.; Ohms, G.; Vioel, W. XPS and ATR-FTIR study on chemical modifications of cold atmospheric plasma (CAP) operated in air on the amino acids L-proline and trans-4-Hydroxy-l-proline. *Plasma Process. Polym.* **2018**, *15*, 1800078. [CrossRef]
68. Madejová, J. FTIR techniques in clay mineral studies. *Vib. Spectrosc.* **2003**, *31*, 1–10. [CrossRef]
69. Mary, Y.S.; Ushakumari, L.; Harikumar, B.; Varghese, H.T.; Panicker, C.Y. FT-IR, FT-raman and SERS spectra of L-proline. *J. Iran. Chem. Soc.* **2009**, *6*, 138–144. [CrossRef]
70. Patel, H.A.; Somani, R.S.; Bajaj, H.C.; Jasra, R.V. Nanoclays for polymer nanocomposites, paints, inks, greases and cosmetics formulations, drug delivery vehicle and waste water treatment. *Bull. Mater. Sci.* **2006**, *29*, 133–145. [CrossRef]
71. Paques-Ledent, M.T.; Tarte, P. Vibrational studies of olivine-type compounds—I. The i.r. and Raman spectra of the isotopic species of Mg2SiO4. *Spectrochim. Acta Part A Mol. Spectrosc.* **1973**, *29*, 1007–1016. [CrossRef]
72. Manfred, K.; Alexandr, A.; Thomas, O.; Cathcart, J.M.; Boris, M. The influence of wetting and drying cycles on mid-infrared attenuated total-reflection spectra of quartz: Understanding spectroscopy of disturbed soil. In Proceedings of the Detection and Remediation Technologies for Mines and Minelike Targets IX, Orlando, FL, USA, 12–16 April 2004; pp. 629–637.
73. Tadic, M.; Panjan, M.; Tadic, B.V.; Lazovic, J.; Damnjanovic, V.; Kopani, M.; Kopanja, L. Magnetic properties of hematite (−FeO) nanoparticles synthesized by sol-gel synthesis method: The influence of particle size and particle size distribution. *J. Electr. Eng.* **2019**, *70*, 71–76. [CrossRef]
74. Yu, C.; Ke, Y.; Deng, Q.; Lu, S.; Ji, J.; Hu, X.; Zhao, Y. Synthesis and Characterization of Polystyrene-Montmorillonite Nanocomposite Particles Using an Anionic-Surfactant-Modified Clay and Their Friction Performance. *Appl. Sci.* **2018**, *8*, 964. [CrossRef]
75. Yuniati, M.D.; Hirajima, T.; Miki, H.; Sasaki, K. Silicate Covering Layer on Pyrite Surface in the Presence of Silicon–Catechol Complex for Acid Mine Drainage Prevention. *Mater. Trans.* **2015**, *56*, 1733–1741. [CrossRef]
76. Au, P.-I.; Leong, Y.-K. Surface Chemistry and Rheology of Slurries of Kaolinite and Montmorillonite from Different Sources. *KONA Powder Part. J.* **2016**, *33*, 17–32. [CrossRef]
77. Gil-Lozano, C.; Fairén, A.G.; Muñoz-Iglesias, V.; Fernández-Sampedro, M.; Prieto-Ballesteros, O.; Gago-Duport, L.; Losa-Adams, E.; Carrizo, D.; Bishop, J.L.; Fornaro, T.; et al. Constraining the preservation of organic compounds in Mars analog nontronites after exposure to acid and alkaline fluids. *Sci. Rep.* **2020**, *10*, 15097. [CrossRef] [PubMed]
78. Elmi, C.; Guggenheim, S.; Gieré, R. Surface Crystal Chemistry of Phyllosilicates Using X-Ray Photoelectron Spectroscopy: A Review. *Clays Clay Miner.* **2016**, *64*, 537–551. [CrossRef]
79. Rath, R.K.; Subramanian, S.; Pradeep, T. Surface Chemical Studies on Pyrite in the Presence of Polysaccharide-Based Flotation Depressants. *J. Colloid Interface Sci.* **2000**, *229*, 82–91. [CrossRef]
80. Balomenou, G.; Stathi, P.; Enotiadis, A.; Gournis, D.; Deligiannakis, Y. Physicochemical study of amino-functionalized organosilicon cubes intercalated in montmorillonite clay: H-binding and metal uptake. *J. Colloid Interface Sci.* **2008**, *325*, 74–83. [CrossRef] [PubMed]

81. Mustafa, S.; Tasleem, S.; Naeem, A.; Safdar, M. Solvent effect on the electrophoretic mobility and adsorption of Cu on iron oxide. *Colloids Surf. A Physicochem. Eng. Asp.* **2008**, *330*, 8–13. [CrossRef]
82. Oelkers, E.H.; Golubev, S.V.; Chairat, C.; Pokrovsky, O.S.; Schott, J. The surface chemistry of multi-oxide silicates. *Geochim. Et Cosmochim. Acta* **2009**, *73*, 4617–4634. [CrossRef]

life

MDPI

Article

"Sea Water" Supplemented with Calcium Phosphate and Magnesium Sulfate in a Long-Term Miller-Type Experiment Yields Sugars, Nucleic Acids Bases, Nucleosides, Lipids, Amino Acids, and Oligopeptides

Robert Root-Bernstein [1,*], Andrew G. Baker [2], Tyler Rhinesmith [3], Miah Turke [4], Jack Huber [1] and Adam W. Brown [5]

1 Department of Physiology, Michigan State University, East Lansing, MI 48824, USA
2 Department of Chemical Engineering and Biotechnology, University of Cambridge, Cambridge CB3 0AS, UK
3 MRC Laboratory of Molecular Biology, Francis Crick Avenue, Cambridge CB2 0QH, UK
4 Department of Chemistry, University of Chicago, Chicago, IL 60637, USA
5 Department of Art, Art History and Design, Michigan State University, East Lansing, MI 48824, USA
* Correspondence: rootbern@msu.edu

Abstract: The standard approach to exploring prebiotic chemistry is to use a small number of highly purified reactants and to attempt to optimize the conditions required to produce a particular end product. However, purified reactants do not exist in nature. We have previously proposed that what drives prebiotic evolution are complex chemical ecologies. Therefore, we have begun to explore what happens if one substitutes "sea water", with its complex mix of minerals and salts, for distilled water in the classic Miller experiment. We have also adapted the apparatus to permit it to be regassed at regular intervals so as to maintain a relatively constant supply of methane, hydrogen, and ammonia. The "sea water" used in the experiments was created from Mediterranean Sea salt with the addition of calcium phosphate and magnesium sulfate. Tests included several types of mass spectrometry, an ATP-monitoring device capable of measuring femtomoles of ATP, and a high-sensitivity cAMP enzyme-linked immunoadsorption assay. As expected, amino acids appeared within a few days of the start of the experiment and accumulated thereafter. Sugars, including glucose and ribose, followed as did long-chain fatty acids (up to C_{20}). At three-to-five weeks after starting the experiment, ATP was repeatedly detected. Thus, we have shown that it is possible to produce a "one-pot synthesis" of most of the key chemical prerequisites for living systems within weeks by mimicking more closely the complexity of real-world chemical ecologies.

Keywords: prebiotic chemistry; "dirty experiments"; ATP; cAMP; amino acids; peptides; sugars; fatty acids; steroids; chemical ecology; sea water; minerals

Citation: Root-Bernstein, R.; Baker, A.G.; Rhinesmith, T.; Turke, M.; Huber, J.; Brown, A.W. "Sea Water" Supplemented with Calcium Phosphate and Magnesium Sulfate in a Long-Term Miller-Type Experiment Yields Sugars, Nucleic Acids Bases, Nucleosides, Lipids, Amino Acids, and Oligopeptides. *Life* **2023**, *13*, 265. https://doi.org/10.3390/life13020265

Academic Editors: Ranajay Saha and Alberto Vázquez-Salazar

Received: 20 December 2022
Revised: 13 January 2023
Accepted: 15 January 2023
Published: 18 January 2023

1. Introduction

One of the general principles of science appears to be that order emerges from complexity within bounds set by system–level constraints [1,2]. Miller [3] demonstrated that given the input of energy in the form of heat and electrical discharges, even relatively simple systems composed of ammonia, methane, and hydrogen gases supplemented by water vapor could give rise to compounds of interest to the understanding of prebiotic chemistry and the origins of living systems. Following in Miller's footsteps, most chemists have attempted to simplify or modify the original chemical conditions to better yield essential molecules of living systems, and with great success (e.g., [4–6]). However, in light of the order-from-complexity concept, one must wonder what would happen if Miller-type experiments were complexified rather than simplified. This possibility has led us to approach the question of prebiotic evolution as a problem in the evolution of complex chemical ecologies rather than the optimization of specific chemical pathways [2,7,8].

In consequence, we have chosen not only what Vincent et al. [9] have recently described as a "synthesized mixtures" approach over an "assembled mixtures of reagent grade chemicals" approach (see also [10]) to prebiotic experimentation but have begun to explore the effects of using what might be called "complex" or "dirty synthesized chemical mixtures" that more closely approach the messiness of real-world conditions [11,12]. Even "synthesized mixtures" generally originate from the use of reagent-grade gases and distilled, deionized water in most Miller-like experiments [3,6], but real atmospheres are much more complex, and water sources are never pure.

The purpose of these experiments was to explore whether novel products resulted from Miller-like experiments modified to last weeks to months by regassing the atmosphere and employing "dirty" chemical conditions resulting from the use of "sea water" salts and common geological minerals. The experiments were preliminary ones that employed some arbitrarily chosen conditions designed to explore possible avenues for future optimization and variation. The rationale for these experiments was that increasing the complexity of the environment in which the reactions take place might also increase the probability that novel reactions would take place, yielding compounds of importance to the origins of cellular life, such as nucleic acids, sugars, and fatty acids or lipids, as well as amino acids. Of course, it would not be possible to produce phosphate-containing compounds, such as cAMP or ATP, without a source of phosphates, such as hydroxyapatite, and the presence of magnesium was likely required in order to stabilize such phosphates as occurs in biotic systems [13]. Thus, we chose to supplement the "sea water" with soluble amounts of calcium phosphate and magnesium sulfate. The sulfate was also thought to be important for providing a possible source of sulfur for the synthesis of amino acids, such as cystine and methionine. We also speculated that the hydroxyapatite, magnesium sulfate, and/or various trace elements, such as iron and iron sulfate present in the "sea water" (or resulting from the presence of the sulfur in the magnesium sulfate), might function as such catalysts for sugar synthesis, as sugar synthesis in prebiotic conditions has previously been demonstrated in aqueous solutions only in the presence of mineral catalysts. Notably, Reid and Orgel were able to produce sugars in prebiotic conditions in the presence of hydroxyapatite [14], and other investigators have succeeded by using other catalysts, such as iron and titanium oxides (e.g., [15–18]). Amino acids are common products of Miller-type experiments, and we reasoned further that if alpha fatty acids, such as butyric acid, fumaric acid, and succinic acid, could be produced, these might participate in ester-mediated amide bond formation [19] while production of longer-chain lipids could catalyze peptide formation [20].

An additional feature of the experiments was to run the syntheses for five to eight weeks by regassing our apparatus and adding aliquots of "sea water" at weekly intervals so as to maintain a reasonably constant or increasing supply of reactants over extended periods of time. Such long-term Miller-type experiments have been relatively rare (reviewed in [4]), so it is not known what effect increasing concentrations of products, such as amino acids, might have on the probability of supporting polymerization into peptides or the emergence of other classes of compounds.

This paper reports the results of these preliminary experiments. As summarized in Figure 1, in addition to the amino acids produced in classic Miller-type experiments, we have evidence for the production of peptides, alpha acids, fatty acids, steroids, sugars, nucleic acid bases, and nucleosides.

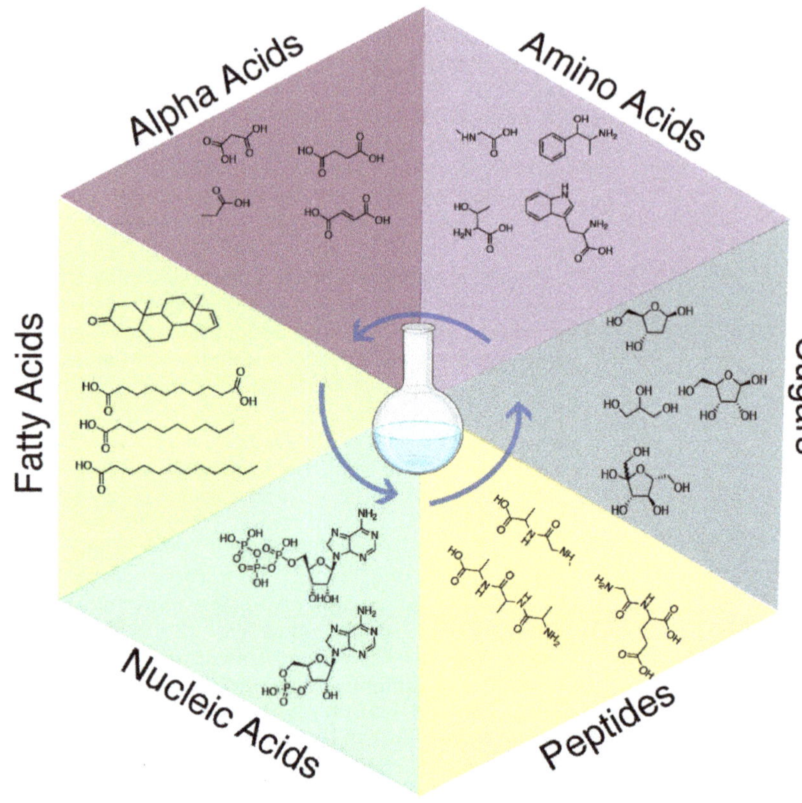

Figure 1. Overview of the molecules produced in the experiments. Multiple regassing cycles, as well as the addition of "sea water" supplemented with calcium phosphate and magnesium sulfate, led to the production of a diverse set of prebiotic molecules in a "one pot" synthesis.

2. Materials and Methods

2.1. Apparatus

A modified version of the original Urey–Miller apparatus (Figure 2) was constructed in a way that permitted regassing of the apparatus, integrated sample ports for ease of repeated sampling, and had electrodes to produce high-energy sparks. The overall design of the apparatus was the same as the Urey–Miller one, with a flask to hold water brought to a boil by means of a heating mantel, a 5-L flask with ports through which electrodes were placed, a refrigerated condenser unit to act as a heat sink, and a u-tube to collect the condensed material and feed it back into the flask. Three ports between the flask and the spherical gas element were provided with fittings controlled by valves that permitted the apparatus to be evacuated to near vacuum or filled with desired mixtures of ga to permit monitoring of the gas pressure/vacuum and to permit the "atmosphere" within the apparatus to be released under pressure. Two additional ports for sampling the liquid in the flask and u-tube were also provided and sealed with rubber sampling septa. The entire apparatus was fabricated from borosilicate glass (a point that is important in terms of the long-term use of the apparatus).

The electrodes were activated by a Marx generator powered by a variable voltage (up to 12 V) DC power supply with ten capacitors that yielded an output charge of approximately 250 kilovolts. The power supply was variable so that the frequency of discharge could be controlled, and it was set to produce sparks approximately every five to ten seconds.

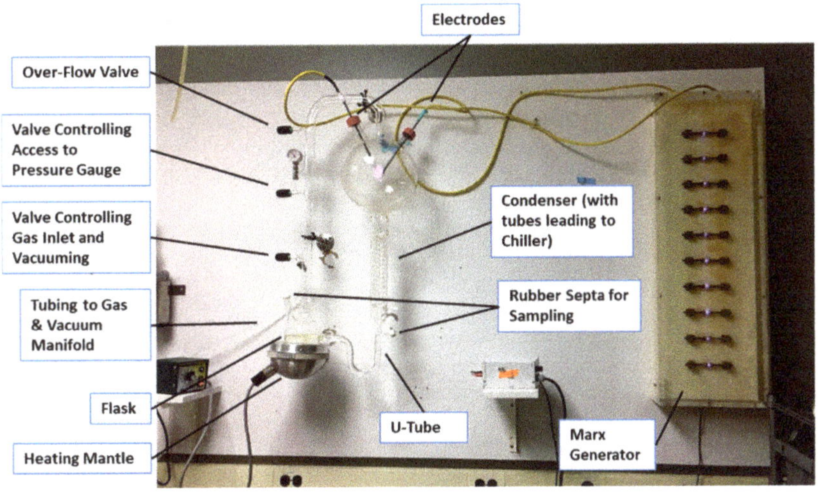

Figure 2. The modified Miller apparatus used in the experiments.

All components of the apparatus that could contact the gases or water sources were rinsed with 70% ethanol followed by reverse-osmosis deionized, autoclaved water, and then sterilized in an autoclave at 140 °C under 1.5 psi pressure for 40 min before each experiment. The apparatus was tested for possible leaks by being evacuated using the vacuum pump until the pressure gauge read "zero," and this reading was maintained for a minimum of 24 h. The water used in the experiments was also sterilized by autoclaving and reverse osmosis deionized. The absence of ATP was confirmed using the ATP assay system described below. Prior to the start of each experiment, a total of 350 mL of this water was introduced through the septa into the flask and u-tube of the apparatus using a sterile 100 mL syringe and non-coring sampling needle. The apparatus was then subjected three times to as complete a vacuum as possible (with the pressure gauge again reading "zero"), which was held for at least one hour each time in order to purge the water of as much gas (particularly oxygen) as possible.

2.2. "Sea Water"

Initially, as noted above, deionized, sterilized water was introduced into the flask and u-tube at the beginning of each experiment. Each time a 10 mL sample was taken from the flask or u-tube, it was replaced with an equal amount of "sea water" so that the concentration of salts and minerals slowly increased over time. The rationale for this procedure was based on our ignorance of the concentrations of salts that might have been present in diverse types of water across the globe and the fear that too high a concentration of some elements or minerals might poison prebiotic reactions. Slowly ramping up the salt/mineral concentrations permitted us to be able to observe a range of conditions with any single experiment. While not optimal in terms of controlling the conditions of the experiments, it was, nonetheless, easily possible to monitor the concentration of phosphates as a general measure of the increases in other mineral concentrations, and the procedure could be reproduced from one experiment to the next or varied, as results might suggest.

"Sea water" was created by dissolving 35.5 g Alessi Sea Salt (evaporated from the Mediterranean Sea) in 2.0 L of deionized, sterilized water, creating a concentration of salts and minerals one-half that of normal sea water. The rationale for using this decreased concentration of salts was the assumption that the concentrations of salts in ocean waters were less than at present because the leaching of minerals from geological deposits and run-offs from rivers had been going on for far less time. This solution was augmented with 1.0 g of hydroxyapatite [$Ca_{10}(PO_4)_6(OH)_2$] (Sigma Aldrich, St. Louis, MO, USA) and

1.0 g magnesium sulfate heptahydrate (Epsomite or Epsom salt [$MgSO_4{\cdot}7H_2O$]) (Sigma Aldrich). These two minerals were chosen as a starting point for our experiments because they are both abundant worldwide and would have been present in many locations in which prebiotic chemistries were taking place. The concentrations of hydroxyapatite and magnesium sulfate were chosen conservatively to ensure their solubility and were, otherwise, arbitrary. The resulting solution was tested during every experiment using the mass spectrometry and ATP testing detailed below to ensure that it contained no detectable biological molecules.

As a consequence of the supplemented "sea water" being added in small aliquots over time to the reaction mixture, it is likely that the concentrations of minerals represent a lower bound of what may have been present during many prebiotic chemical environments and may have reached far higher concentrations at least in some locations on Earth. As noted in the Introduction, these were designed as preliminary experiments to explore possible avenues for future optimization and variation and not as attempts to model any particular environment or location in the prebiotic Earth.

2.3. *Regassing*

After the water had been degassed, the gases comprising the "atmosphere" within the apparatus were added by means of a gas manifold that permitted each gas to be introduced in a controlled manner. The gas proportions were 40% ammonia, 40% methane, and 20% hydrogen, the total producing 1 atmosphere of pressure and mimicking the original conditions employed by Miller [3]. Although there is much controversy over the make-up of the prebiotic Earth atmosphere, we chose to stick with Miller's "recipe" in order to limit the number of novel variables introduced into the experiments.

The apparatus was regassed approximately every seven days using the following procedure: the valve to the vacuum pump was opened, the vacuum pump turned on, and the gas inside the apparatus was evacuated until the liquid inside began to bubble. This process retained most of the gases dissolved in the liquid and did not yield a complete vacuum but did remove most of the existing "atmosphere," resulting in about a 90% vacuum (0.1-atmosphere pressure). The vacuum was then shut off, the gas line switched on, and ammonia, methane, and hydrogen were added proportionally, as above, to return the 0.1-atmosphere "vacuum" to 1.0-atmosphere pressure.

2.4. *Sampling*

The heater was turned off for at least an hour prior to sampling so as to permit the liquid in the flask and the u-tube to cool. Then, 5 mL samples were obtained from the flask and the u-tube via their sampling ports using a sterile technique (sterile gloves were donned, rinsed with 70% ethanol, and dried before the apparatus was touched) employing sterile, individually wrapped syringes, and autoclaved, non-coring sampling needles. A sterile technique was then used to replace the 10 mL of liquid withdrawn with 10 mL of augmented "sea salt", keeping the volume of water in the apparatus constant throughout the duration of the experiment and slowly raising the concentration of salts and minerals in solution.

ATP, pH, and phosphate tests were carried out immediately following sampling (see below). Samples for mass spectrometry were placed in RNA-free Nunc vials, sealed, and, if not immediately prepared, then refrigerated (for no more than two days) prior to preparation. Some samples were also frozen for future use (though none of the results reported here involve such frozen samples).

2.5. *Ultraviolet (UV) Spectroscopy*

An initial set of experiments were run to determine whether regassing increased the concentrations of compounds produced. During this initial set of experiments, and only in these, a constant concentration of "sea water" at 1/10 dilution was used, and no additional "sea water" was added following each sampling. Then, 100 μL samples were extracted from the flask portion of the apparatus every few days, and the UV spectrum from 190 to 340 nm was obtained on a SpectraMaxPlus scanning UV-Visible light spectrometer. Samples were subjected to spectroscopy immediately following their removal from the apparatus, and the resulting absorption curves were then plotted to provide a rough estimate of changes in the total concentration of compounds produced as a function of time.

2.6. Mass Spectrometry

Samples were prepared for gas-chromatogram/mass spectrometry (GC/MS) by trimethylsilylation using N-Methyl-N-trimethylsilyl-trifluoroacetamide (MSTFA) (Sigma Aldrich P/N 394866-10X1ML). Then, 100 μL samples were placed in 2 mL amber autosampler vials with micro-inserts and evaporated to dryness using vacuum centrifugation (General Electric high capacity, thermally-protected vacuum pump) in a Savant Speed Vac Concentrator. The samples were then redissolved in 140 μL of a mixture of MSTFA and pyridine (Sigma Aldrich ACS reagent, ≥99.0%, Product #360570), three to four by volume, sealed with vial caps, and placed in a sealed, light-opaque container with desiccant. The reaction was permitted to proceed overnight (at least 12 h) at room temperature. After the reaction was complete, the liquid portion of each sample was transferred to glass inserts, leaving any solid material behind, and the inserts were placed back into the mass spectrometry vials from which they came. The vials were recapped, placed back into the desiccant container, and the samples were analyzed immediately thereafter using an Agilent A Gas Chromatograph-Mass Spectrometer with an Agilent CP9013 J&W VF-5ms GC Column, 30 m, 0.25 mm, 0.25 μm, 10 m EZ-Guard, 7-inch cage. The GC–MS was tuned using facility-supplied standards as well as experiment-appropriate standards, such as urea (PHR1406-1G, Pharmaceutical Secondary Standard), glutamic acid (G0355000, European Pharmacopoeia (EP) Reference Standard), lactic acid (PHR1215 Pharmaceutical Secondary Standard; Certified Reference Material), alanine (A0325000 European Pharmacopoeia (EP) Reference Standard), glycine (G7126 ReagentPlus®, ≥99% (HPLC)), serine (S0450000 European Pharmacopoeia (EP) Reference Standard), asparagine (Y0000305 European Pharmacopoeia (EP) Reference Standard), and/or aspartic acid (A1330000 European Pharmacopoeia (EP) Reference Standard), all from Sigma-Aldrich (St. Louis, MO, USA).

Identification of products was performed using Agilent Chemstation software version LTS 01.11 to visualize mass chromatograms and produce mass spectra for matching to the NIST database (versions 11.1 through 20.0). The quality of match (QOM) of each spectrum was aggregated (see Supplementary Material, which reports on all compound identifications above QOM of 50). In general, a QOM of 90 (out of a possible perfect match of 100) was utilized as a criterion for inclusion in the data reported below; however, for compounds of particular interest to origins of life chemistry, such as some amino acids, oligopeptides, sugars, and nucleic acid-related compounds, QOM as low as 50 percent were reported, with the clear understanding that these matches were highly questionable and in need of further investigation in the future. In most of the cases in which low QOM are reported, either the compound is related to another compound identified with a better QOM or other experimental methods suggest the presence of the compound.

2.7. ATP Assay (Luciferin/Luciferase)

The presence of ATP was determined via an AccuPoint® Advanced ATP Reader (Neogen Corporation, Lansing, MI, USA, Item No. 9903), which uses a photomultiplier system to sense photon production in a luciferin/luciferase system in the presence of adenosine triphosphate (ATP). Samples were tested by applying c. 10 μL to AccuPoint® Advanced Samplers, Water (Neogen Corporation, Lansing, MI, USA, Item No. 9906), which have a sensitivity of 10 fmole/sample (10 relative light units equals 1 fmole). All positive values were confirmed by multiple measurements (between 2 and 6 trials) and accepted only if control materials (deionized, sterilized water, and the original "sea water" sources) tested negative for ATP on an equivalent number of tests. Results were interpreted in relation to a standard curve of known ATP concentrations.

2.8. cAMP Assay (Enzyme-Linked Immunoadsorption Assay—ELISA)

A high-sensitivity competitive ELISA assay was used to measure cyclic adenine mononucleotide (cAMP) (ENZO Life Sciences, ADI-900-067A, sensitive to 0.027 pmol/mL). Assays were carried out according to the manufacturer's instructions and plotted against a standard curve of concentrations of the appropriate mononucleotide.

2.9. Phosphate Assay

Approximate phosphate concentrations were determined using Bartovation Phosphorus and Phosphate Detection Test Strips (Part # PWQ09V50), sensitive to 0–100 ppm (Bartovation, Queens, New York, NY, USA).

2.10. pH Measurements

Approximate pH was determined using UltraCruz® pH Indicator Strips (Catalogue # sc-3667), Santa Cruz Biotechnology, Inc., Dallas, TX, USA.

3. Results

3.1. Increased Yields of Compounds through Regassing

The first experiment performed was simply to determine whether the use of the regassing procedure yielded greater concentrations of products over time. The concentration of "sea water" was kept constant in this particular set of experiments so as to limit the number of variables. The increase in product concentration was confirmed via UV spectroscopy (Figure 3). If the absorbance is approximately determined by the concentration of compounds in the solution, then running the apparatus after regassing for three weeks resulted in about four times the concentration of products were found after about one week. Later experiments were extended to five or six weeks with regassing each week, resulting in proportionally greater concentrations of products. During this initial experiment, no attempt was made at this time to identify the molecules produced, although the peak around 310 nm is suggestive of aromatic compounds such as tryptophan, nucleic acid bases, and/or steroids, which were later identified by mass spectrometry (see below).

3.2. Initial Compound Identification

The second set of experiments was designed to explore the effects of adding "sea water" supplemented with calcium phosphate (hydroxyapatite) and magnesium sulfate (Epsom salt, Epsomite). As per the Methods, samples were tested weekly for pH, the concentration of phosphates, ATP, and two sets of cAMP and cGMP; ELISA tests were carried out. All samples were subjected to GC–MS in order to identify the synthesis of other compounds. Notably, the pH of the flask and u-tube portions of the apparatus each started out with the same pH, which was about 9.5–10.0, and while the u-tube pH remained at this value throughout the experiments, the flask pH generally fell into the range of 6.5–7.5, presumably due to the accumulation of amphiphilic compounds. Due to the procedure used for adding the supplemented "sea water" in 5 mL aliquots to replace samples, the phosphate level rose from undetectable to 100 ppm over a period of five weeks and reached 200 ppm at eight weeks. No measure of the magnesium concentration was carried out, but given that the "sea water" was augmented with the same concentration of magnesium sulfate as calcium phosphate, it can be assumed that the magnesium concentration was in the same range as the phosphate concentration.

As expected, amino acids appeared within a few days of beginning each experiment. Confirmation that sample preparation was properly carried out was provided by using control samples of key compounds, such as glycine, alanine, asparagine, serine, and urea, and comparing the spectra of the pure controls with those of the compounds identified in the experimental preparations (e.g., Figure 4).

Among the first compounds to appear in these experiments were what might be called "reactant" molecules, from which many other compounds could be synthesized, which were found during the first week of all the experiments, and these included urea, formamide, acetic acid or acetamide, oxalic acid, and glycerol (Figure 5 and Table 1). Figures 6 and 7 provide GC–MS spectra, identifying these products from the reaction mixtures. Additional spectra are shown in Appendix A.

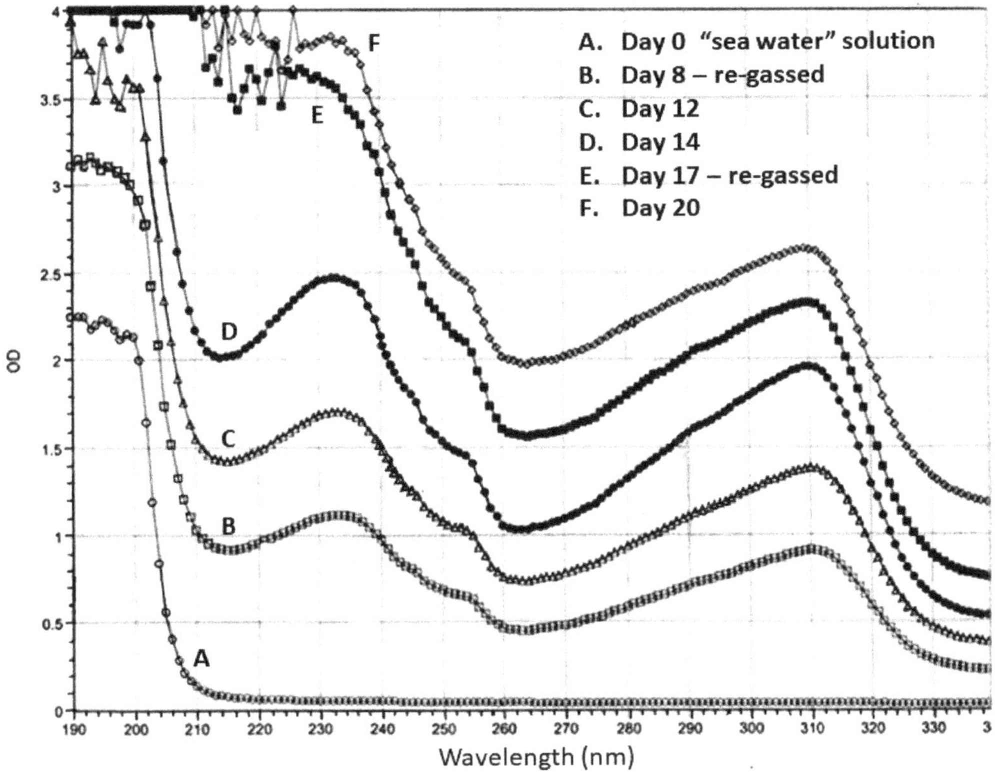

Figure 3. Ultraviolet spectroscopy of samples from the flask of the apparatus taken prior to turning the apparatus on (A) and at days 8 (B), 12 (C), 14 (D), 17 (E), and 20 (F). The apparatus was regassed on days 8 (C) and 17 (E) after the samples were taken. The original Miller experiment ended about day 8 (B) [3], and very few other long-term Miller-type experiments have been run since (reviewed in [4].

Table 1. "Reactant compounds", from which many more complex prebiotic molecules can be synthesized, were found in all of the experiments. "RT" is retention time in minutes (min). "Mol. Wt." is the molecular weight of the identified compound in grams per mole. "QOM" is the quality of match out of 100, according to the NIST database matching software.

Reactant Compounds	RT (min)	Hit Name	Mol. Wt.	QOM
Acetic Acid	3.851	Acetic acid, [(trimethylsilyl)oxy]-, trimethylsilyl ester	220.41	80
Formamide	3.982	Formamide, (trimethylsilyl)-	189.4	87
Glycerol	4.751	Glycerol, tris(trimethylsilyl) ether	308.64	90
Oxalic Acid	4.783	Oxalic acid, di(trimethylsilyl)	234.40	87
Urea	4.327	Urea, (trimethylsilyl)-	426.68	96

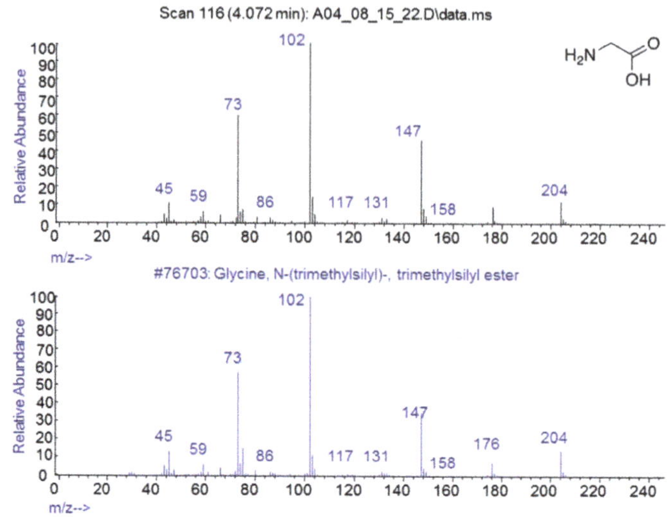

Figure 4. GC–MS spectra of MSTFA-pyridine prepared pure glycine (bottom) and glycine identified in sample from the apparatus after nine days of incubation.

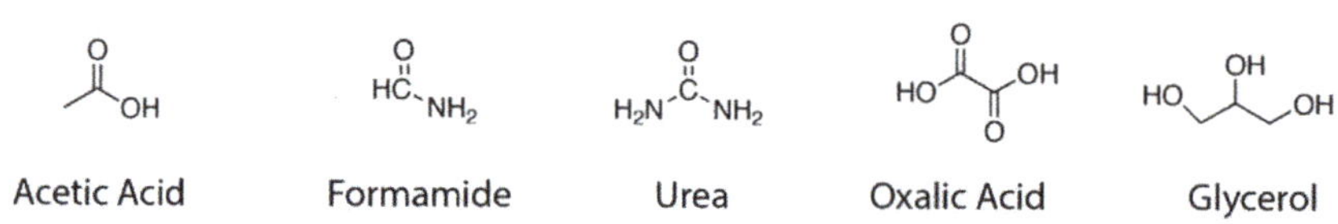

Figure 5. "Reactant compounds", from which many more complex prebiotic molecules can be synthesized, were found in all of the experiments.

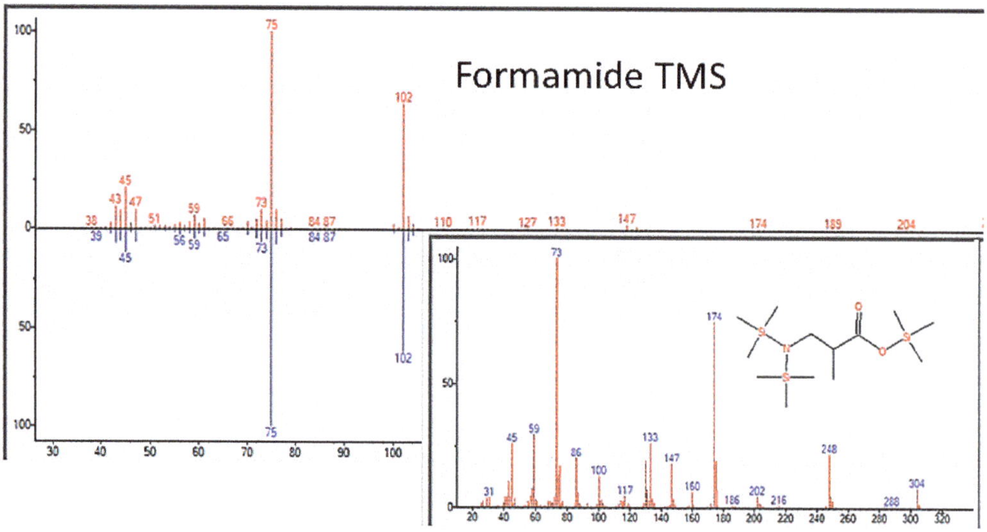

Figure 6. GC–MS identification of formamide from reaction mixture using the NIST database recognition software.

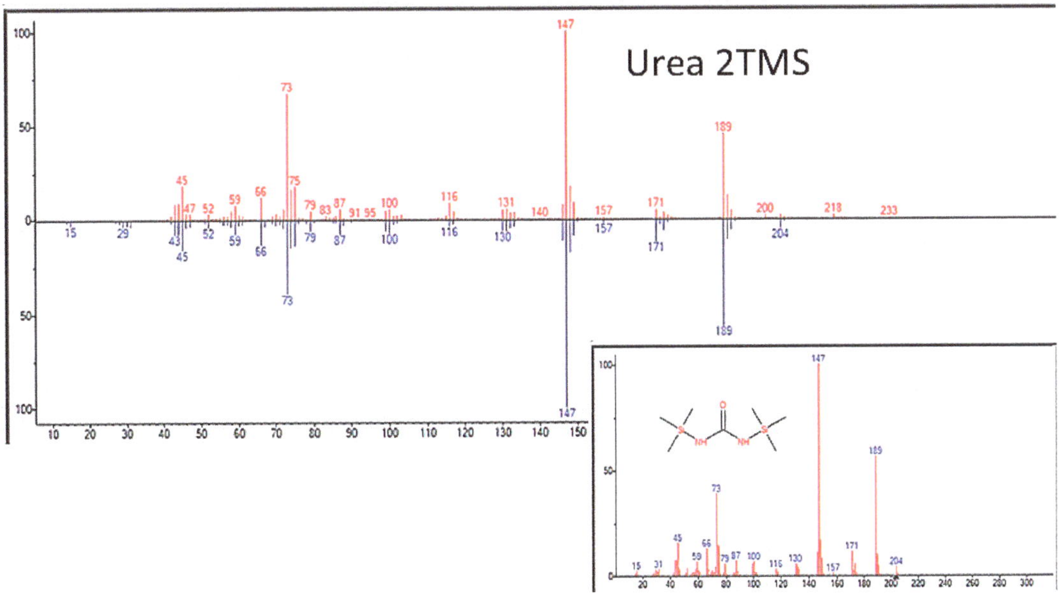

Figure 7. GC–MS identification of urea from reaction mixture using the NIST database recognition software.

3.3. Amino Acids

"Reactant" molecules were always accompanied during the first week of the experiments by amino acids, most commonly glycine, alanine, serine, aspartic acid, leucine, and tryptophan, although most of the biotic amino acids were found repeatedly by the end of four-to-six week experiments (Figures 8–11 and Table 2). Probably due to the presence of magnesium sulfate in the "sea water", both cysteine, methionine, and their metabolites were also repeatedly observed by GC–MS (Table 2 and Appendix B). We assume that these amino acids were racemic mixtures since there is no chiral selection mechanism present in our experiments, and the mass spectrometry methods do not distinguish L- from D- amino acids. Additional spectra are in Appendix B.

Table 2. Summary of amino acids identified by GC–MS repeatedly in at least two independent experiments. "QOM" is quality of match out of 100 as determined by the NIST database. "RT" is retention time in minutes (min). "Mol. Wt." is the molecular weight of the identified compound in grams per mole. Note that the QOMs for Cystathione and proline are much lower than for the other compounds identified in the table, and the reliability of the identification is correspondingly lower. However, the presence of both cysteine and proline with higher QOM suggests that these identifications are plausible.

Amino Acids	RT (min)	Hit Name	Mol. Wt.	QOM
Alanine	4.080	Alanine, N-(trimethylsilyl)-, trimethylsilyl ester	233.45	93
Asparagine	6.525	Asparagine, N,N2-bis(trimethylsilyl)-, trimethylsilyl ester	348.66	99
Aspartic Acid	5.821	Aspartic acid, N-(trimethylsilyl)-, bis(trimethylsilyl) ester	349.64	98
Beta-alanine	5.867	Beta-alanine, N-(trimethylsilyl)-, bis(trimethylsilyl) ester	305.64	75
Cystathionine	9.777	Cystathionine, N-(trimethylsilyl)-, bis(trimethylsilyl) ester	366.63	51
Cysteine	8.622	Cysteine, N,N′-bis(trimethylsilyl)-, bis(trimethylsilyl) ester	337.70	76

Table 2. *Cont.*

Amino Acids	RT (min)	Hit Name	Mol. Wt.	QOM
Glutamic Acid	6.313	Glutamic acid, N-(trimethylsilyl)-, bis(trimethylsilyl) ester	363.67	98
Glutamine	6.924	Glutamine, tris(trimethylsilyl)	362.69	64
Glycine	4.929	Glycine, N,N-bis(trimethylsilyl)-, trimethylsilyl ester	291.61	86
Hydroxytryptophan	4.581	5-Hydroxytryptophan, tetramethylsilylester	508.95	87
Isoleucine	8.588	Isoleucine, N-(trimethylsilyl)-, trimethylsilyl ester	275.53	81
Leucine	4.42	Leucine, trimethylsilyl ester	203.35	95
Methionine	10.311	Methionine-(trimethylsilyl)	221.39	72
Oxyproline	5.965	Proline, 5-oxo-1-(trimethylsilyl)-, trimethylsilyl ester	275.49	74
Phenylalanine	6.415	Phenylalanine, N,O-Bis-(trimethylsilyl)	309.55	91
Phenylpropanolamine	4.080	Phenylpropanolamine, bis(trimethylsilyl)	295.57	80
Proline	10.651	Proline, trimethylsilyl)-, trimethylsilyl ester	259.49	57
Sarcosine	4.165	Sarcosine, Bis(trimethylsilyl)	233.45	81
Serine	5.201	Serine, N,O-bis(trimethylsilyl)-, trimethylsilyl ester	321.63	87
Threonine	5.320	Threonine, N,O,O-Tris(trimethylsilyl)-	335.66	91
Tryptophan	9.420	Tryptophan, bis(trimethylsilyl)-	348.6	89
Tyramine	5.707	Tyramine, tri(trimethylsilyl)-	353.72	90
Valine	4.581	Valine, N-(trimethylsilyl)-, trimethylsilyl ester	261.51	90
Norvaline	4.581	Norvaline, N-(trimethylsilyl)-, trimethylsilyl ester	261.51	83

Additionally, beginning in the fifth week of the experiments, dipeptides began to appear, and alanyl–glycine (Figure 12), alanyl–alanine, leucyl–alanine, and glycyl–glutamic acid were observed in several independent experiments (Table 3). The tripeptide alanyl–alanyl–alanine was also observed in one experiment after the fifth week, although the quality of the identification by mass spectrometry was low (Table 3).

Table 3. Summary of peptides identified by GC–MS in independent experiments. "RT" is retention time in minutes (min). "Mol. Wt." is the molecular weight of the identified compound in grams per mole. "QOM" is quality of match out of 100 as determined by the NIST database.

Peptides	RT	Hit Name	Mol. Wt.	QOM
Alanyl-beta-alanine	5.201	Alanyl-beta-alanine, N,O-bis(trimethylsilyl)-, trimethylsilyl ester	232.35	87
Alanyl–alanyl–alanine	9.768	Alanyl–alanyl–alanine methyl ester	245.28	59
Alanyl–glycine	6.0242	Alanyl–glycine, bis(trimethylsilyl) ester	218.33	87
Glycyl-glutamic acid	17.286	Glycyl-glutamic acid, bis(trimethylsilyl) ester	348.54	86
Leucyl–alanine	8.588	Leucyl–alanine, bis(trimethylsilyl) ester	376.69	81

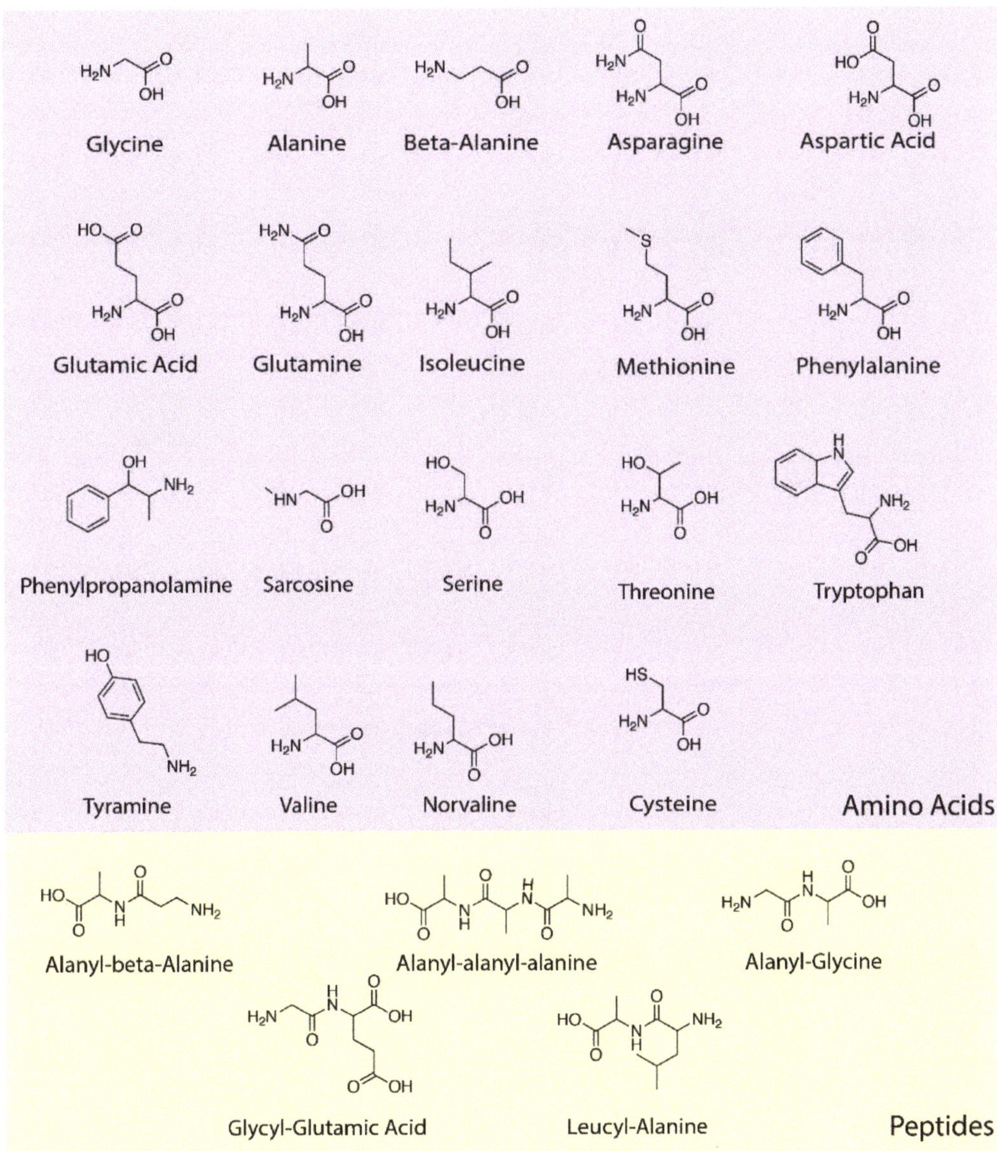

Figure 8. Overview of amino acids and peptides produced in this study. These compounds were undoubtedly racemic, and thus, the use of a non-chiral structural formalism here.

3.4. Sugars

In addition to amino acids, a wide range of monosaccharides were produced in the experiments (Figure 13 and Table 4), including fructose, glucose, mannose, ribose, and many of their metabolites, such as levo-glucosan, gluconic acid, ribonic acid, ribo-hexos-3-ulose, and ribono-1,4-lactone (Figures 14 and 15 and Appendix C). Two disaccharides, maltose and trehalose, were repeatedly observed, though the identification of trehalose was consistently of poor quality and, therefore, questionable (Table 5).

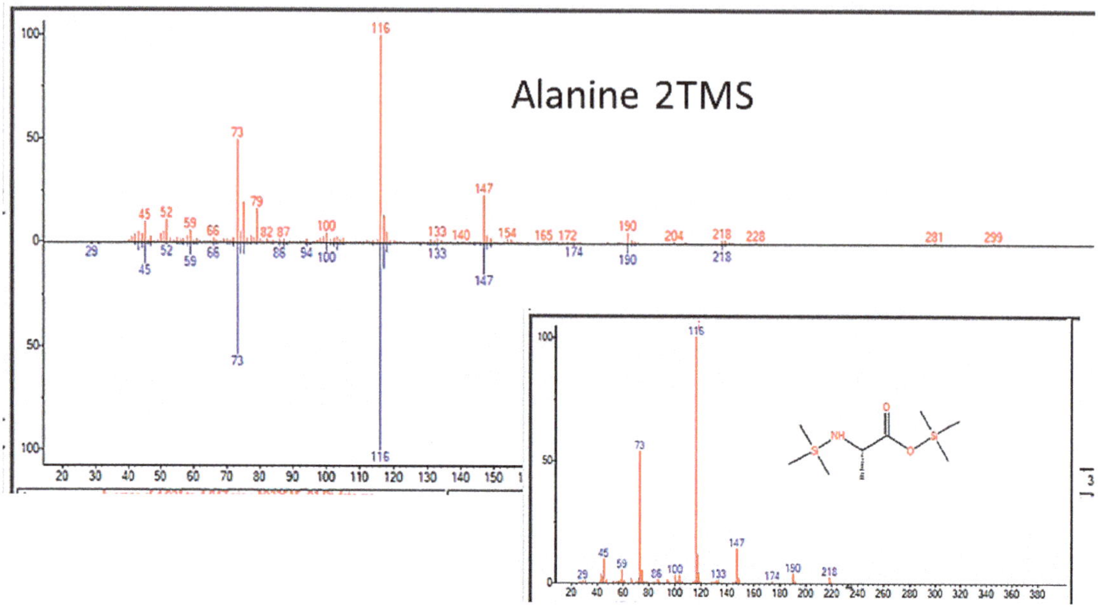

Figure 9. GC–MS identification of alanine using the NIST database recognition software.

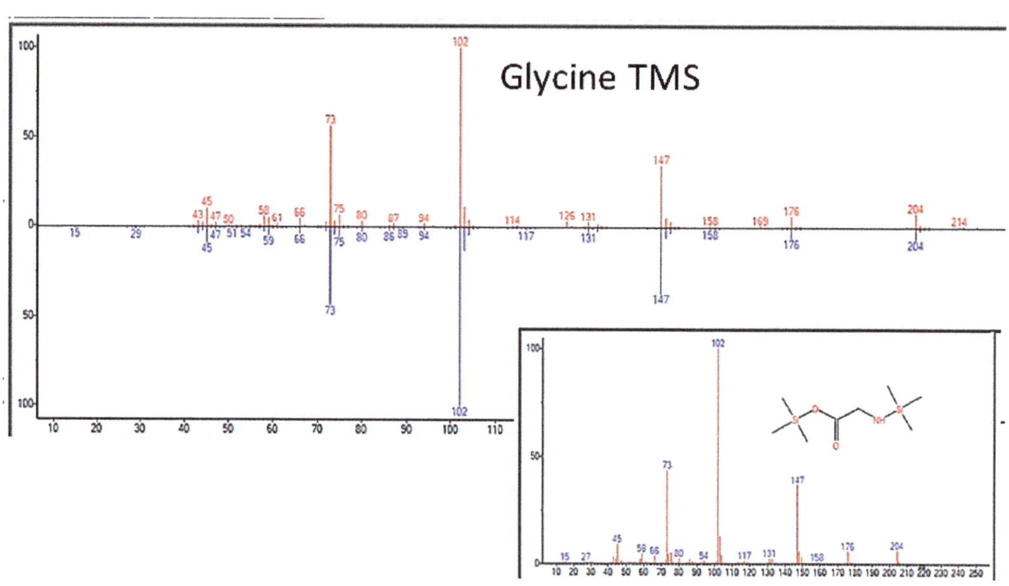

Figure 10. GC–MS identification of glycine from reaction mixture using the NIST database recognition software.

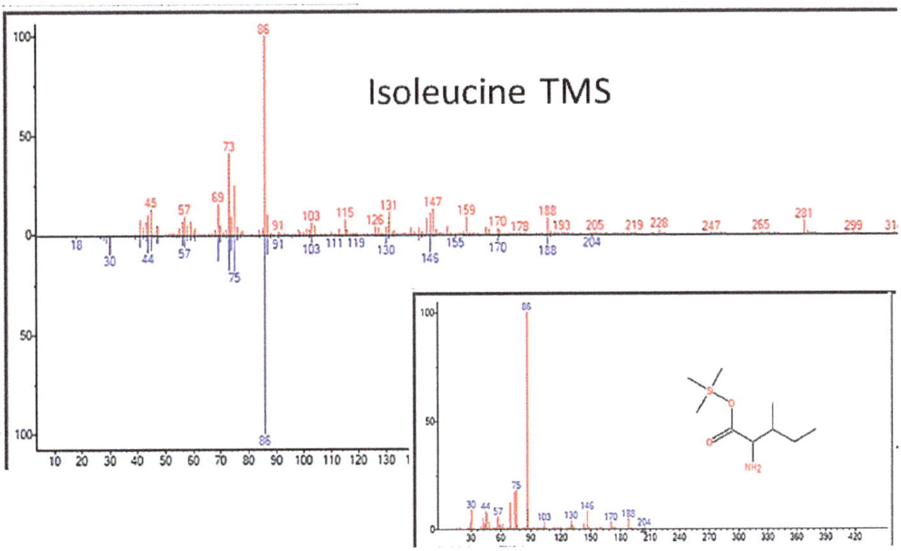

Figure 11. GC–MS identification of isoleucine from reaction mixture using the NIST database recognition software.

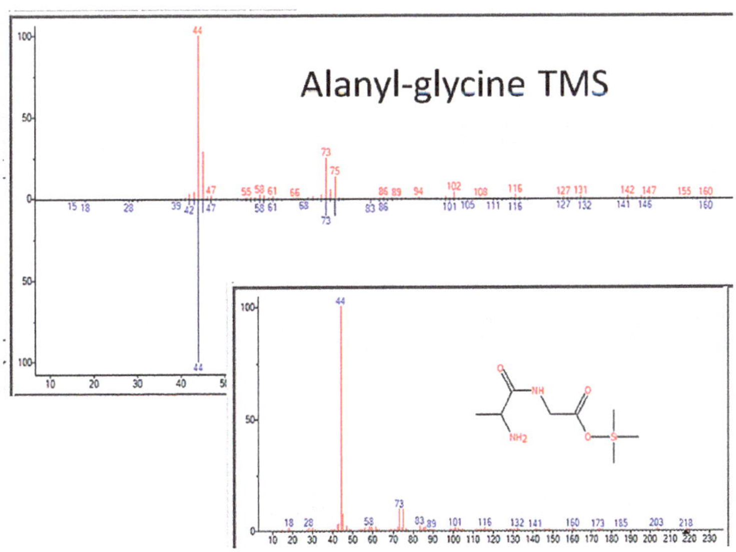

Figure 12. GC–MS identification of Alanyl–glycine from reaction mixture using the NIST database recognition software.

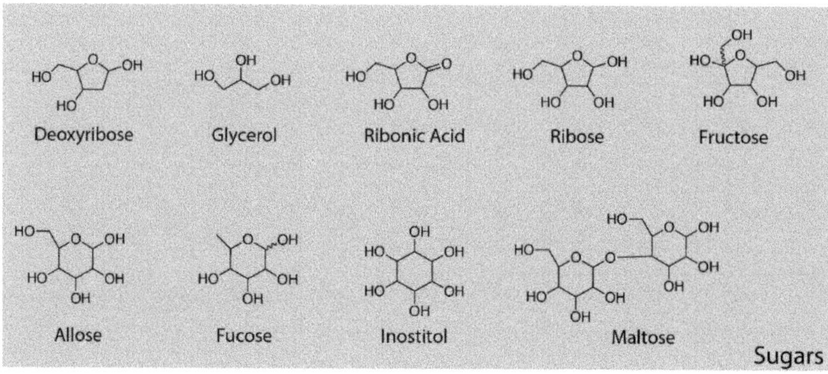

Figure 13. Sugars repeatedly identified by GC–MS in the experiments. Both D- and L- sugars were undoubtedly produced, and so was the use of a non-chiral structural formalism in these structures.

Table 4. Summary of sugars repeatedly identified by GC–MS in the experiments. "RT" is retention time in minutes (min). "Mol. Wt." is the molecular weight of the identified compound in grams per mole. "QOM" is the quality of match out of 100 according to the NIST database matching software.

Monosaccharides	RT (min)	Hit Name	Mol. Wt.	QOM
Allose	7.4	Allose, pentakis(trimethylsilyl) ether, methyloxime (anti)	570.10	90
Deoxyribose	4.072	2-Deoxy-ribose, tris(trimethylsilyl) ether	437.87	53
Fructose	7.221	Fructose, pentakis(trimethylsilyl) ether, methyloxime (anti)	570.10	97
Fucose	7.552	Fucose, tetrakis(trimethylsilyl) ether	481.90	91
Galactopyranose	7.552	Galactopyranose, pentakis(trimethylsilyl) ether (isomer 2)	541.06	90
Galactose	7.4	Galactose, 2,3,4,5,6-pentakis-O-(trimethylsilyl)-, o-methyloxyme, (1E)-	570.10	91
Glucose	7.4	Glucose, 2,3,4,5,6-pentakis-O-(trimethylsilyl)-, o-methyloxyme, (1Z)-	628.30	87
Gluconic Acid	8.792	Gluconic acid, 2,3,4,6-tetrakis-O-(trimethylsilyl)-, .delta.-lactone	466.86	77
Inositol	7.994	Inositol, 1,2,3,4,5,6-hexakis-O-(trimethylsilyl)-, cis-	613.24	90
Levoglucosan	6.67	Levoglucosan, tris(trimethylsilyl)-	378.68	87
Lyxose	7.221	Lyxose, tetrakis(trimethylsilyl) ether, trimethylsilyloxime (isomer 1)	526.05	90
Mannose	7.4	Mannose, 2,3,4,5,6-pentakis-O-(trimethylsilyl)-, o-methyloxyme, (1Z)-	570.10	91
Myo-Inositol	8.002	Myo-Inositol, 1,2,3,4,5,6-hexakis-O-(trimethylsilyl)-	613.24	99
Ribonic Acid	5.091	Ribonic acid, 5-deoxy-2,3-bis-O-(trimethylsilyl)-, .gamma.-lactone	276.48	89
Ribose	6.355	Ribose, 2,3,4,5-tetrakis-O-(trimethylsilyl)-	438.9	87
Ribo-hexos-3-ulose	4.700	Ribo-hexos-3-ulose, 2,4,5,6-tetrakis-O-(trimethylsilyl)-, bis(O-methyloxime)	525.0	73
Ribono-1,4-lactone	9.352	Ribono-1,4-lactone, tris(trimethylsilyl) ether	364.65	63
Sorbitol	7.434	Sorbitol, hexakis(trimethylsilyl) ether	615.26	89

Table 4. *Cont.*

Monosaccharides	RT (min)	Hit Name	Mol. Wt.	QOM
Sorbose	7.221	Sorbose, pentakis(trimethylsilyl) ether, trimethylsilyloxime (isomer 1)	628.26	90
Tagatose	7.221	Tagatose, pentakis(trimethylsilyl) ether, trimethylsilyloxime	628.26	91
Talose	7.4	Talose, pentakis(trimethylsilyl) ether, methyloxime (syn)	570.10	91

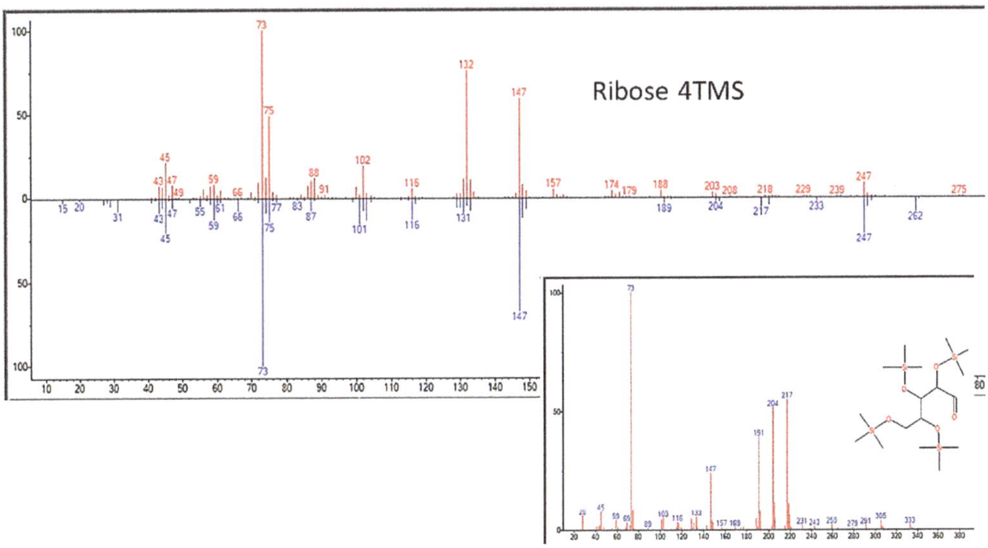

Figure 14. GC–MS identification of ribose from reaction mixture using the NIST database recognition software.

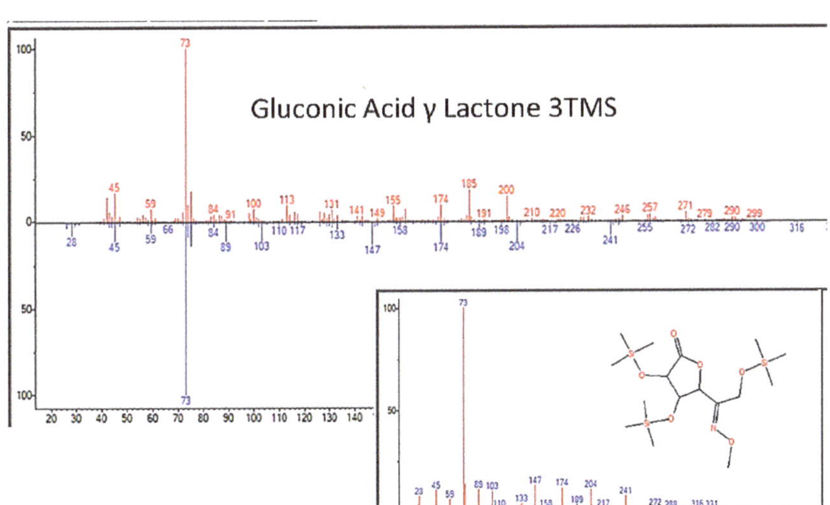

Figure 15. GC–MS identification of one of several glucose metabolites from reaction mixture using the NIST database recognition software.

Table 5. Summary of disaccharides identified in reaction mixtures using the NIST database recognition software. "RT" is retention time in minutes (min). "Mol. Wt." is the molecular weight of the identified compound in grams per mole. "QOM" is the quality of match out of 100 according to the NIST database matching software.

Disaccharides	RT (min)	Hit Name	Mol. Wt.	QOM
Maltose	9.836	Maltose, octakis(trimethylsilyl) ether, methyloxime (isomer 1)	948.78	93
Trehalose	9.836	Trehalose, octakis(trimethylsilyl) ether	919.75	64

3.5. Nucleic Acid Bases, Nucleosides, and Nucleotides

GC–MS also identified nucleic acid bases repeatedly in the experiments, including purine metabolites, adenosine, guanosine, and dihydrouracil (though the quality of the match for guanosine and uracil was marginal) (Figure 16 and Table 6). We assume that we made not only D-adenosine and D-guanosine but L-forms as well, but mass spectrometry does not differentiate these. Figures 17 and 18 and Appendix D illustrate some of the identified compounds listed in the table.

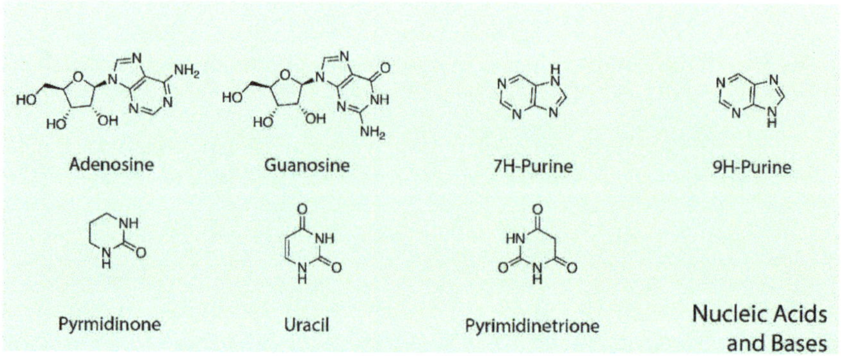

Figure 16. Nucleic acid bases identified by GC–MS in the experiments. The figures shown here illustrate D-adenosine and D-guanosine, but both D- and L- isomers of these molecules were presumably produced since there is no chiral selection mechanism involved in our experiments.

Table 6. Summary of nucleic acid bases and precursors repeatedly identified by GC–MS in the experiments. "RT" is retention time in minutes (min). "Mol. Wt." is the molecular weight of the identified compound in grams per mole. "QOM" is the quality of match out of 100 according to the NIST database matching software.

Nucleic Acid Precursors	RT (min)	Hit Name	Mol. Wt.	QOM
Adenine	6.551	Adenine, N,7-bis(trimethylsilyl)-	279.49	60
Adenosine	8.461	Adenosine, N-(4-hydroxy-3-methyl-2-butenyl)-, (E)-	351.15	80
Adenosine	9.598	Adenosine-tetrakis(trimethylsilyl)-	555.97	93
Guanine	7.824	Thioguanine	167.19	59
Guanosine	9.972	Guanosine,N-Methyl penta(trimethylsilyl)-	644.10	56
7H-Purine	7.799	7-(Trimethylsilyl)-2,6-bis[(trimethylsilyl)oxy]-7H-purine	368.65	99
7H-Purine	7.714	7-(Trimethylsilyl)-2,6-bis[(trimethylsilyl)oxy]-7H-purine	368.65	91

Table 6. *Cont.*

Nucleic Acid Precursors	RT (min)	Hit Name	Mol. Wt.	QOM
9H-Purine	6.271	9H-Purine, 9-(trimethylsilyl)-2,6-bis[(trimethylsilyl)oxy]-	368.65	78
Pyrimidinetrione	7.484	2,4,6(1H,3H,5H)-Pyrimidinetrione, 5-[2-(methoxyimino)-3-[(trimethylsilyl)]-	399.6	93
Pyrimidine	7.077	Pyrimidine, 2,4,6-tris[(trimethylsilyl)oxy]-	344.63	76
Pyrimidine	7.391	1,2,4-Triazolo[1,5-a]pyrimidine, 5,7-dimethyl-2-phenyl-	224.26	64
Pyrmidinone	5.345	5-Methyldihydro-2,4(1H,3H)-pyrimidinedione diTMS	272.49	70
Pyrmidinone	8.962	2(1H)-Pyrimidinone, 5-(4-methylphenoxy)-4-(4-nitrophenyl)-6-phenyl-	399.40	64
Uracil	5.779	Dihydro-uracil-di(trimethylsilyl)-	258.46	69
Uracil	5.209	6-Azauracil, bis(tert-butyldimethylsilyl) deriv.	341.6	72

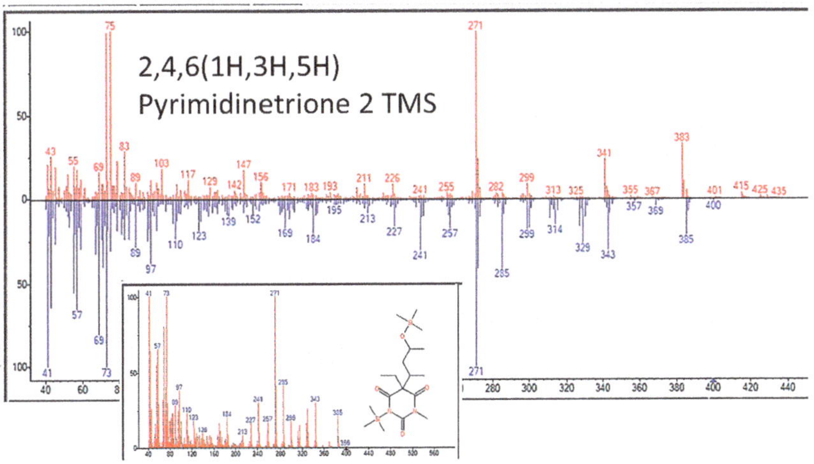

Figure 17. GC–MS identification of one of several pyrimidine metabolites from reaction mixture using the NIST database recognition software.

Because we found that mass spectrometry (both GC and liquid chromatography–MS) was unable to detect concentrations of nucleic acids below 10 nM, additional non-spectroscopic methods were employed. The presence of nucleic acids was further verified by ELISA, which demonstrated increasing concentrations of cyclic adenosine monophosphate (cAMP) reaching more than 1 pmol/mL (Figure 19) during each of the two separate experiments. The cAMP results replicated for each experiment were repeatedly negative for the original "sea water" solution. The reliability of the cAMP data was further confirmed using a luciferin–luciferase reaction to demonstrate the presence of adenosine triphosphate (ATP) in four separate experiments (Figure 20) beginning as early as week four of the experiments but most often beginning to appear in week five when the phosphate (and presumably magnesium) concentration had reached 100 ppm. Once again, however, it is important to note that ATP appeared in only four of eight experiments that ran for six weeks or more, and the reasons for the failure to produce ATP in four of these experiments are unknown but may involve undetected oxygen leaks or simply effects of variations in initial conditions by altering the synthetic pathways to ribose and the nucleic acid bases, and/or by oxidizing the ATP itself as quickly as it is formed. The possibility of contamination is extremely unlikely for reasons that will be addressed below.

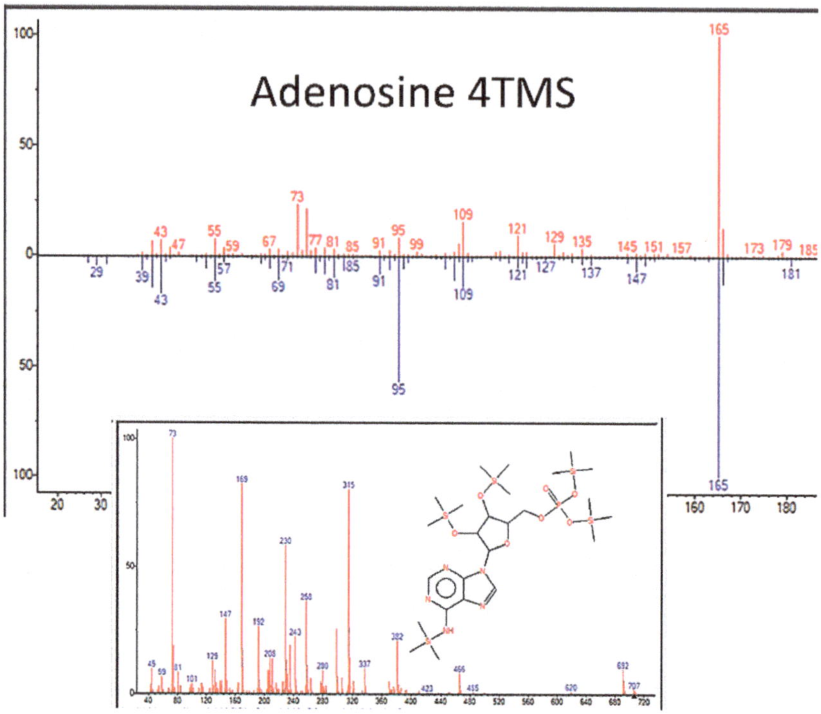

Figure 18. GC–MS identification of one of adenosine from reaction mixture using the NIST database recognition software.

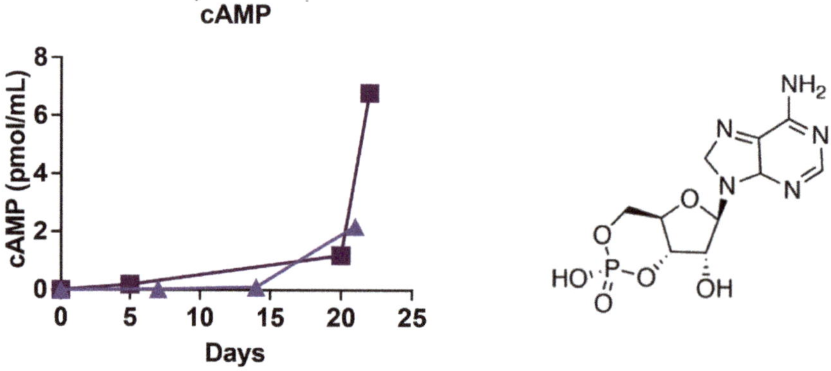

Figure 19. High sensitivity cyclic adenosine monophosphate ELISA results for two independent experiments, indicated by solid squares or triangles. Since this was an antibody assay, it presumably reacted only to the D-enantiomer of the compound, although there is every reason to believe the reaction mixture contained the L-enantiomer as well.

3.6. Fatty Acids and Steroids

In addition to amino acids, sugars, and nucleic acids, the GC/MS experiments also repeatedly yielded a range of fatty acids ranging from C4 through C20, as well as a few steroids (but never cholesterol or phospholipids, a point that will be amplified in the Discussion section) (Figure 21 and Table 7). The GC–MS spectra and NIST identifications of some of these compounds are illustrated in Figures 22–24 and in Appendix E.

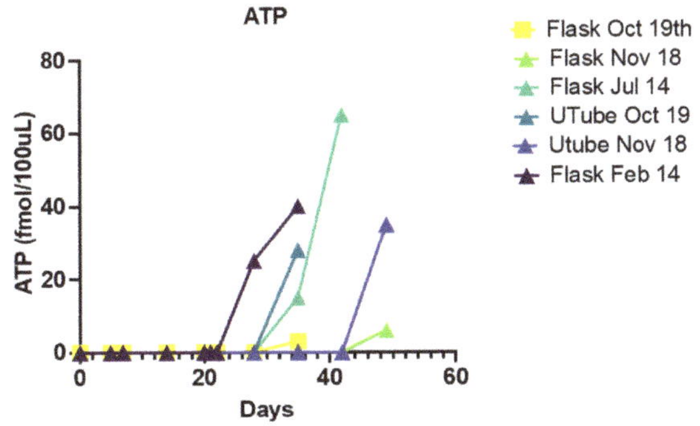

Figure 20. Adenosine triphosphate (ATP) readings from four independent experiments (two comparing values from the flask to the u-tube sources). No ATP was observed from any experiment until 28 days or thereafter, and all values were confirmed by multiple measurements (between 2 and 6) and accepted only if control materials (deionized, sterilized water, and the original "sea water") sources tested repeatedly negative for ATP. Since this was an enzyme assay, it presumably reacted only to the D-enantiomer of the compound, although there is every reason to believe the reaction mixture contained the L-enantiomer as well.

Table 7. Summary of steroidal and fatty acid compounds identified by GC–MS in the experiments. "RT" is retention time in minutes (min). "Mol. Wt." is the molecular weight of the identified compound in grams per mole. "QOM" is quality of match out of 100 as determined by the NIST database.

Steroids	**RT**	**Hit Name**	**Mol. Wt.**	**QOM**
Androstan-17-one	16.462	Androstan-17-one, 3-[(trimethylsilyl)oxy]-, (3.alpha.,5.alpha.)-	362.6	97
Androstan-17-one	6.848	5.alpha.-Androstan-17-one, 11.beta.-hydroxy-3.alpha.-(trimethylsiloxy)-	378.6	98
Androstan-17-one	8.461	5.beta.-Androstan-17-one, 11.beta.-hydroxy-3.alpha.-(trimethylsiloxy)-	378.6	95
Cholestane	5.88	Cholestane, 2,3-epoxy-, (2.alpha.,3.alpha.,5.alpha.)-	386.7	92
Deoxycorticosterone	8.919	4-Pregnen-21-ol-3,20-dione glucoside	492.6	95
ALPHA AND FATTY ACIDS	RT	Hit Name	Mol. Wt.	QOM
Propanedioic Acid, Malonic Acid (C3)	9.53	Propanedioic acid, (1H-indole-3-ylmethylene)-, diethyl ester	287.31	93
Butenoic Acid, Succinic Acid (C4)	6.186	2-Butenoic acid, 3-methyl-2-[(trimethylsilyl)oxy]-, trimethylsilyl ester	260.48	76
Butanedioic Acid, Fumeric Acid (C4)	5.048	Butanedioic acid, bis(trimethylsilyl) ester	262.45	97
Butyric Acid (C4)	6.024	2,3,4-Trihydroxybutyric acid tetrakis(trimethylsilyl) deriv., (, (R*,R*)-)	424.8	87
Aminohexanoic Acid (C6)	4.581	N,O,O′-Tris-(trimethylsilyl)-6-hydroxy-2-aminohexanoic acid	363.71	90
Sebacic Acid (C10)	7.349	Sebacic acid, bis(trimethylsilyl) ester	346.61	87
Decanoic Acid (C10)	9.318	Decanoic acid, 2-[(trimethylsilyl)oxy]-1-[[(trimethylsilyl)oxy]methyl]ethyl ester	390.7	62
Lauric Acid (C12)	12.985	12-Methylaminolauric acid	229.36	59
Myristic Acid (C14)	7.061	Myristic acid, trimethylsilyl ester	300.6	89

Table 7. *Cont.*

Steroids	RT	Hit Name	Mol. Wt.	QOM
Pentadecanoic Acid (C15)	7.394	Pentadecanoic acid, trimethylsilyl ester	314.6	80
Palmitelaidic Acid (C16)	7.773	Palmitelaidic acid, trimethylsilyl ester	326.6	62
Octadecenoic Acid, Stearic Acid (C18)	7.484	9-Trimethylsilyloxy-12-octadecenoic acid, methyl ester	384.7	80
Arachidonic Acid (C20)	7.604	Arachidonic acid, trimethylsilyl ester	376.6	91
5,8,11-Eicosatrienoic acid (C20)	7.255	cis-5,8,11-Eicosatrienoic acid, trimethylsilyl ester	378.7	91

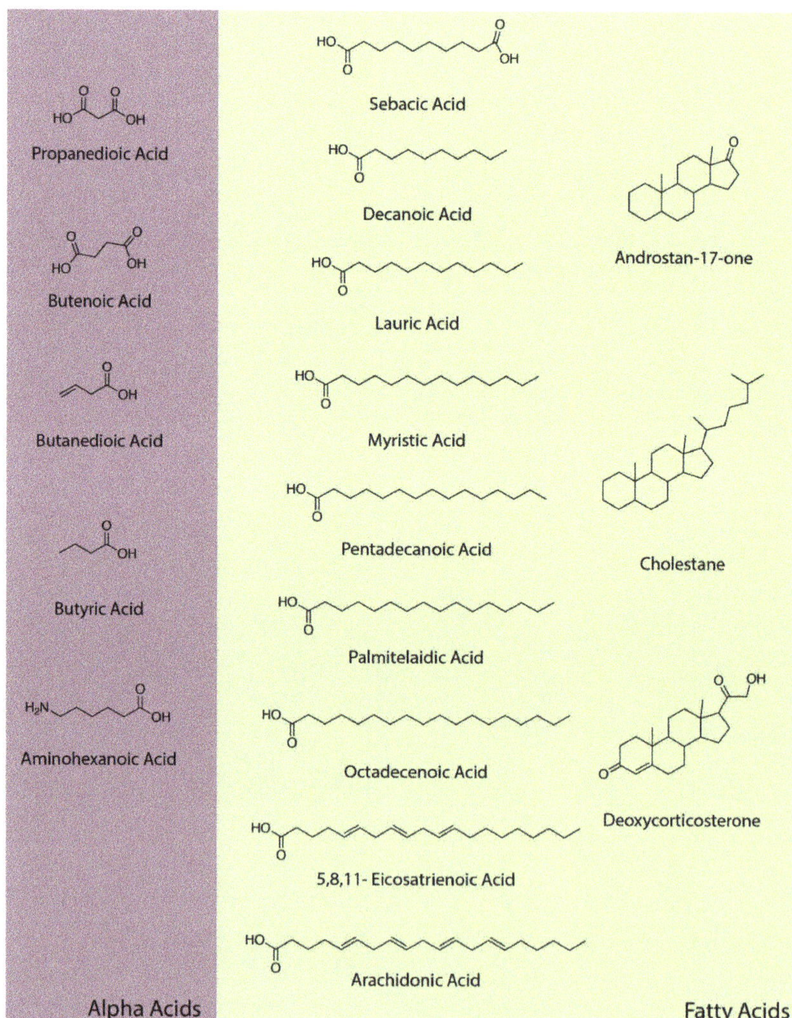

Figure 21. Alpha acids, fatty acids, and steroidal compounds identified by GC–MS in the experiments.

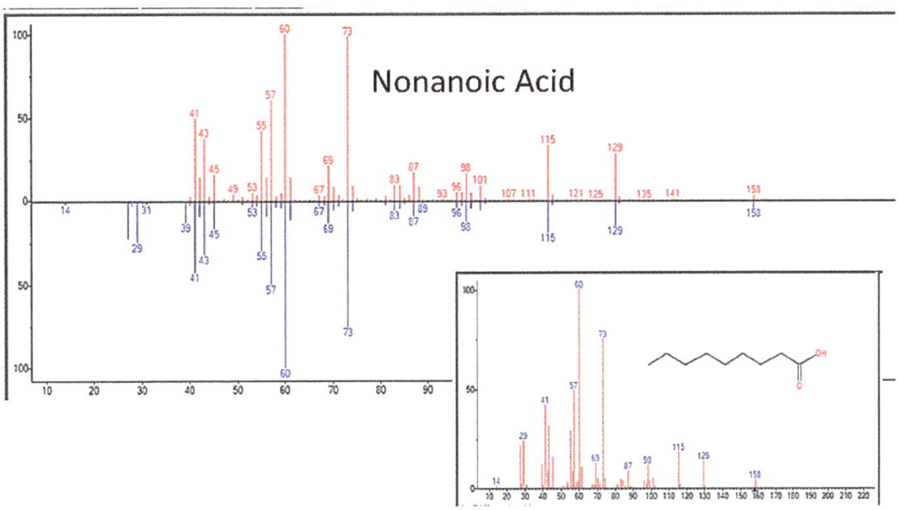

Figure 22. GC–MS identification of nonanoic acid from reaction mixture using the NIST database recognition software.

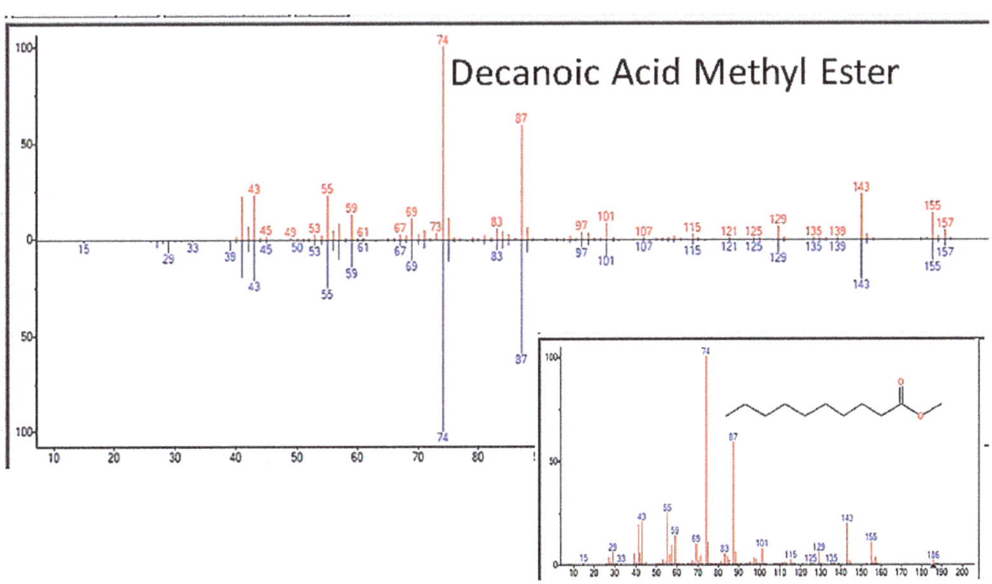

Figure 23. GC–MS identification of decanoic acid from reaction mixture using the NIST database recognition software.

Figure 25 illustrates a GC–MS chromatogram for the end of week three of the synthetic process, after two regassings of the apparatus and with phosphate and magnesium concentrations of approximately 200 ppm. The chromatogram has been annotated with the NIST identifications of the compounds present in each peak and clearly shows the presence of amino acids, sugars, fatty acids, as well as the nucleic acid base adenine and adenosine.

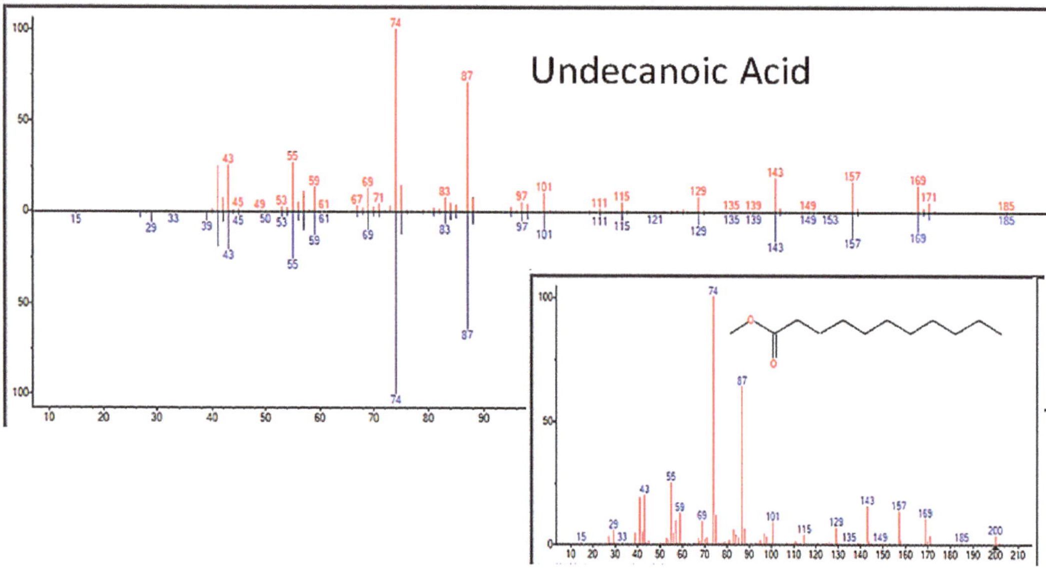

Figure 24. GC–MS identification of undecanoic acid from reaction mixture using the NIST database recognition software.

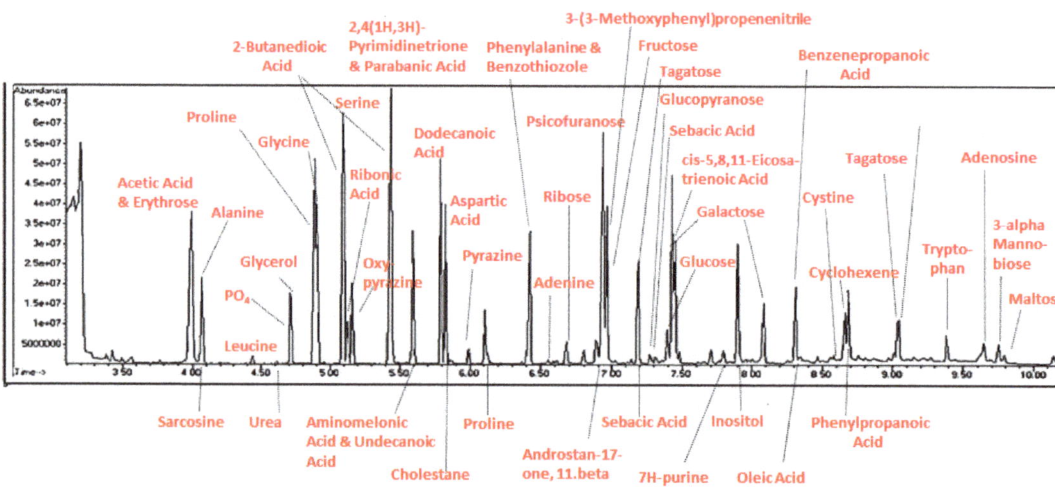

Figure 25. Chromatogram of end of week three sample (u-tube) labeled with the NIST-identified products.

3.7. Controls for Possible Sources of Contamination

Long-term experiments with repeated sampling, such as the ones reported here, are potentially subject to contamination. We, therefore, emphasize that certain types of molecules that would have been expected to be present if extremophile bacteria or human contact contaminated our experiments were never identified in any experiment.

We begin with possible bacterial contamination. Because of the conditions within the apparatus (anaerobic with a continuous flask temperature of 100 °C), only select extremophile bacteria could possibly have become contaminants. However, many types of compounds that such extremophiles produce were never observed. These absent compounds included the amino acids arginine, glutamine, and tyrosine. Common bacterial peptides, such as the cell–wall component muramyl dipeptide and the tripeptide glutathione, also were not observed. No polysaccharides or starches were identified. The most common extremophile bacterial lipids are tetra-ether lipids, glycerol-ester lipids, and phospholipids that characterize lipid membranes [21,22], none of which were identified in any experiment. Nor were the nucleic acids, cytosine, or thymine ever found. Thus, although we did not test directly for the presence of such extremophile organisms, using, for example, a polymerase chain reaction assay for mitochondrial genes, the absence of so many key molecules from our results provides a strong case against their presence and, therefore, their participation in producing the molecules that we did identify.

Human contamination is also unlikely due to the strict use of sterile techniques at all times while handling, loading, and sampling the apparatus. The effectiveness of our sterile technique was evident in the absence of peptides, such as glutathione, and disaccharides, such as sucrose, lactose, and maltose. Phospholipids, glycerophspholipids, phosphotidylcholine, and cholesterol, which are the most abundant lipids present in finger grease [23], were never observed in any experiment. It is further unlikely that the repeated pattern of finding ATP about four or five weeks into the experiments was due to human error contaminating the ATP test solely around that time in each of the four experiments. Furthermore, if our observation of adenosine was the result of contamination, then cytosine, uridine, and inosine should also have been observed, but they were not. Finally, contamination during sample processing for GC–MS was extremely unlikely as no contaminant compound was ever observed in the spectra of the control compounds (urea, amino acids, etc., e.g., Figure 4) that were run at the beginning of every spectrometry series.

One final type of evidence also argues against human or microbial contamination as a cause of our results and particularly addressed whether the abiotic synthesis of sugars and nucleic acids occurred during the experiments. We noted in our Methods that our apparatus was constructed from borosilicate glass, as it must be in order to resist the high heat conditions necessary for its operation and autoclaving. However, after several years of running the apparatus, we noticed that our mass spectrometry samples were increasingly contaminated with boric acid (Figure 26), an observation that, after eliminating all other possible sources of boron from components of the apparatus, such as the rubber septa, the "sea water" and added minerals, and the reagents used for preparing mass spectrometry samples, we attributed it to a very slow deterioration of the glass caused by long periods of time exposed to ammonia at high heat. Two striking modifications in the products of the experiments accompanied this boric acid contamination. One was the production of boronate sugars (Figures 27 and 28), which could not possibly have originated from human contamination. The other modification was the loss of adenine and adenosine (Table 6 and Figure 18) from the samples and the appearance, instead of uracil, dihydrouracil, and azauracil (Appendix F). The observation that samples contained either adenine-related compounds or uracil-related compounds, but never both (and never cytosine- or thymidine-related compounds), argues, firstly, against contamination as a source of these nucleic acids and, secondly, for the possibility that boron-containing minerals may play a role in the catalysis of uracil-related compounds.

To summarize, the results of the experiments using "dirty" water and repeated regassing of the Miller-style apparatus yielded greater concentrations of compounds and a wider range of products than the original Miller experiment. In particular, sugars, amino acids, some peptides, several nucleic acids, alpha acids, and fatty acids were produced. The typical time course of the appearance of these compounds is summarized in Table 8. Notably, except for some of the sugars, the rest of the compounds only began to appear after regassing at least twice, which may explain why some of these compounds have not been observed in previous, shorter-term studies.

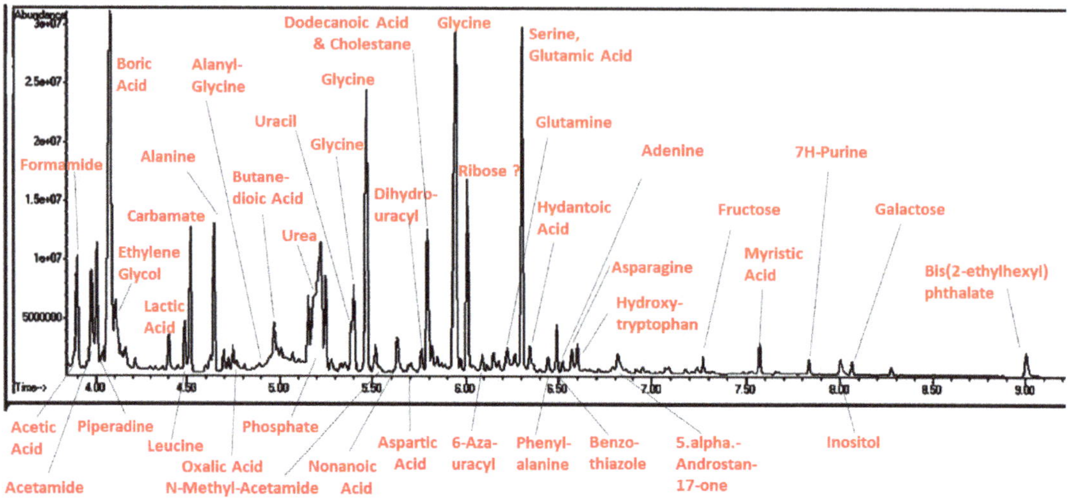

Figure 26. Chromatogram of end of week three sample (u-tube) labeled with the NIST-identified products.

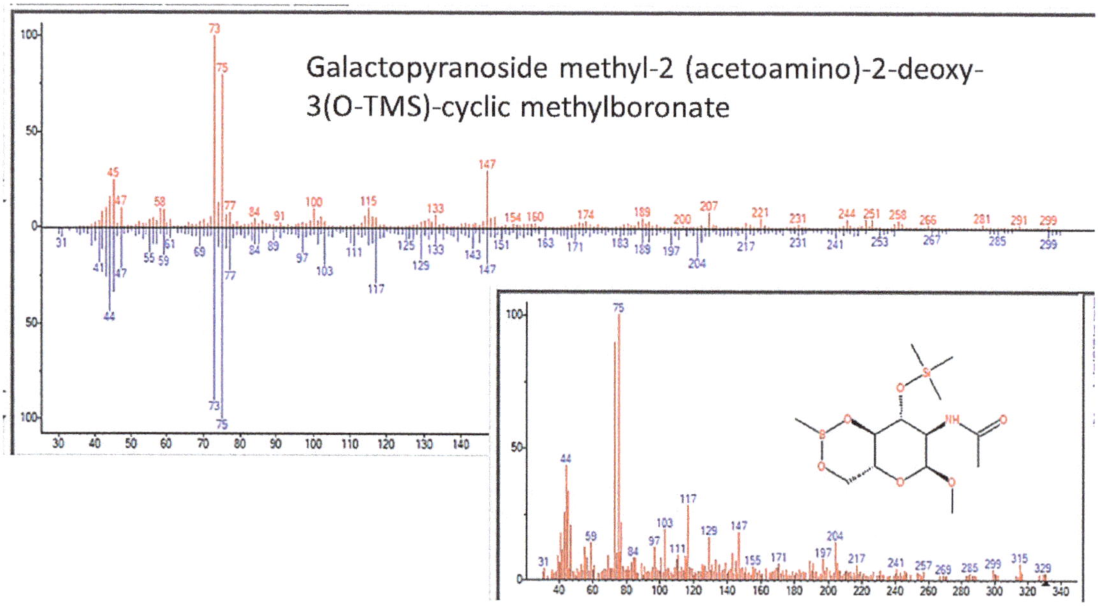

Figure 27. GC–MS identification of 6-azauracil from reaction mixture using the NIST database recognition software.

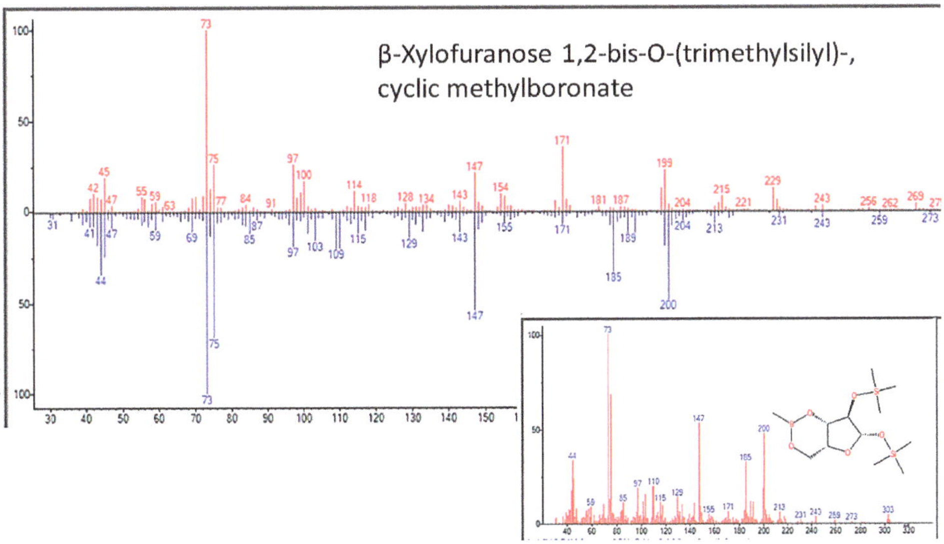

Figure 28. GC–MS identification of 6-azauracil from reaction mixture using the NIST database recognition software.

Table 8. Time course of appearance of classes of compounds. All compounds listed here were observed at least twice (and usually more often) in independent experiments and had high–quality identification scores from GC–MS or other assays. Exceptions are noted with a question mark following the compound name, indicating that the identification of the compound was not repeatable or had a low identification quality. "Times" refers to how many regassings had occurred (usually done every seven or eight days).

	Day 0	1 Time	2 Times	3 Times	4 Times	5 Times
Phosphates	0	25 ppm	50 ppm	75 ppm	100 ppm	150 ppm
Sugars		Fructose, Galactose, Glucose, Mannose, Sorbose, Tagatose	Xylose, Fucose, Maltose Trehalose?	Ribose, Deoxyribose?		
Amino Acids		Ala, Asn, Asp, Gln, Glu, Gly, Ile, Leu, Ser, Trp	Phenylpropanolamine (amphetamine), Cys, Met, Val, Nor-Val		Ala-Gly Gly-Glu	Ala-Ala Ala-Ala-Ala Leu-Ala
Nucleic Acids			Adenosine, Guanosine? 7H-Purine, 9H-Purine, Pyrimidinones	cAMP, cGMP?	Uracil	ATP
Fatty Acids			Aminohexanoic acid (C6), Butanoic Acid (C4), (Succinic Acid), Butyric Acid (C4)	Sebacic acid (C10), Decanoic Acid (C10), Arachidonic Acid (C20), Eicosatrianoic Acid (C20)	Lauric Acid (C14), Myristic Acid (C14), Octadecenoic Acid (C18), Steroids	Hexadecanoic Acid (C16) Palmitalaidic Acid (C16)

4. Discussion

In practice, the experiments reliably produced well-studied precursors to many prebiotic molecules, such as urea, formamide and glycerol, most (though not all) of the biotic and some abiotic amino acids, the key biotic sugars, a range of fatty acids and steroids, and, in some experiments, evidence of nucleic acids, including adenosine, cAMP, and ATP. Among the amino acids were two that were not found in Miller's original experiments [3], which were cysteine and methionine. These amino acids were later produced by Miller through the addition of H_2S in the gases [6,23], but in this case, they presumably resulted from the presence of a source of sulfur from the magnesium sulfate. Notably, by regassing the apparatus and running the experiments for many weeks, the yield of total products increased (Figure 2), resulting in the presence of sufficient amino acids and their derivatives to permit polymerization into di- and tri-peptides. Similarly, the production of key sugars, including ribose, in an environment including phosphates and nucleic acid bases, appears to have made possible the production of cAMP, ATP, and, possibly, guanosine. Whether the presence of magnesium permitted their stabilization will require additional research, and, if it does, more experiments will be needed to determine the concentrations of phosphates and magnesium required to optimize nucleoside production and stability. The observation that boron appears to alter the types of nucleic acid bases produced may also be a clue of importance for understanding what minerals are needed to catalyze the range of products characterizing living systems and what minerals might have "poisoned" essential reactions. Notably, borate compounds have been found to stabilize nucleic acids [24], making the results reported here concerning the shift from adenine-like compounds to uracil-like ones possibly significant.

As would be expected, precursors molecules (amino acids, monosaccharides, nucleic acid bases, short-chain fatty acids) appear a week or two prior to the more complex molecules (peptides, disaccharides, nucleotides, longer-chain fatty acids) (Figure 29 and Table 8). We emphasize that each one of these classes of compounds has independently been synthesized previously in Miller-type experiments, for example, amino acids [3,6,24], peptides [25], nucleic acid bases [26–28], sugars [14–18], and fatty acids [29,30]. In general, these previous Miller-type experiments have previously been explored as means to generate one particular class of compounds, such as amino acids or nucleic acids, and conditions have generally been optimized for the production of that particular compound class. Our experiment differs in having no particular class of compounds as a goal and functioning as an exploratory rather than an optimizing exercise. We must point out, however, that Saladino, et al. [31] have reported the synthesis of amino acids, lipids (including arachidonic and eicosatrienoic acids), and nucleic acid bases (but not nucleosides or sugars), starting with pure formamide solution heated in the presence of meteorite particles so that it is known that several classes of prebiotic compounds can be synthesized under Miller-like conditions with an appropriate mix of mineral catalysts present. It does not appear that anyone has previously demonstrated the production of amino acids, peptides, fatty acids, sugars, nucleic acids, and nucleosides under one set of conditions.

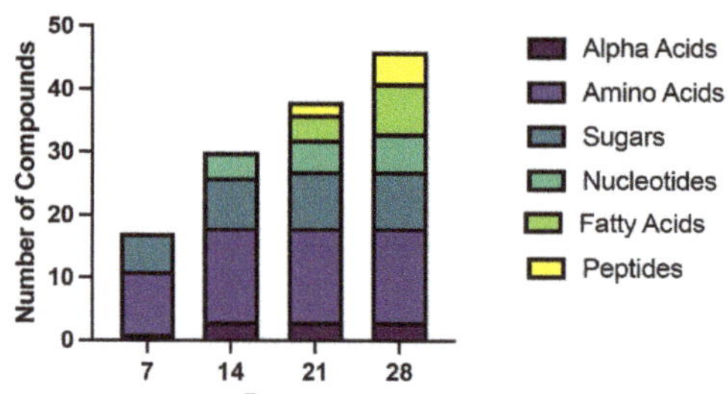

Figure 29. Time course of appearance of different classes of compounds. The *y* axis is the number of compounds where each color is type.

In short, by increasing the time that the experiments ran with a reasonably consistent atmosphere and by adding sea salts augmented with calcium phosphate and magnesium sulfate, we believe that we have produced the conditions necessary for a one-pot synthesis of all major classes of compounds required for the origins of living systems. Better optimized conditions may help to define the environments, in which living systems are most likely, or least likely, to have originated. In this context, it is likely that the addition of clays or minerals, other than calcium phosphate or magnesium sulfate, will alter the distribution and types of products. Iron and titanium oxides and ferrous sulfate, are obvious examples of minerals that should be explored since iron oxides have been employed as catalysts for sugar production in prebiotic experiments [14–18], and iron–sulfur clusters can coordinate with and be stabilized by cysteine-containing peptides and mediate the assembly of iron–sulfur cluster peptide complexes that can drive enzymatic reactions [32]. Iron complexes have also been implicated in the production and breakdown of universal metabolic precursor compounds [31]. A listing of the many other mineral catalysts of prebiotic reactions would be too long to include here, but some representative publications include [33–36], and a brief review can be found in [37]. More complex atmospheres (including CO_2, CO, H_2S, NO, etc.) may also increase or significantly alter the molecular complexity, types, or ratios, of products. Such experiments could also be designed to more closely mimic non-Earth environments, in which prebiotic chemistries can take place [38–42]. Exploring combinations of these various atmospheres with sets of minerals might more accurately mimic real-world environments in which prebiotic chemistries occurred. Finally, the production of nucleic acid precursors along with amino acids opens up new possible pathways within complex Miller-like experiments for performing the type of amino-acid-adenylate-mediated peptide syntheses carried out in purer experimental conditions [43,44].

5. Conclusions

We report a possible advance in mimicking real-world environments through our "dirty" prebiotic experiments and possible "one-pot syntheses" of all the essential classes of molecules required for the emergence of living systems. We are, however, fully aware of the severe limitations of the present study, many of which can be addressed by further experimentation. We do not have data on amounts of each compound formed in each sample, though some sense of relative concentrations can be gained from the peak heights in the original chromatograms (Figures 25 and 26). One difficulty is that compound spectra overlap, creating serious difficulties in determining how much of each observed peak is due to the contribution of any particular compound. Additionally, for each compound identified, of which there are many, titration spectra based on the pure compound will be needed with which to compare the peak heights derived from the chromatograms; this will involve a great deal of additional work. The use of the atmosphere (ammonia, methane, hydrogen) originally employed by Miller [3] is certainly questionable given subsequent research on the atmosphere of the primordial earth (e.g., [38–40], nor is it representative of many other planetary atmospheres (e.g., [41,42], so that other mixtures of gases should be explored using the "dirty" approach employed here. Similarly, the use of Mediterranean sea salt at low concentrations relative to those present in modern sea waters is similarly open to modification in order to mimic freshwater, sea, lake, and pond waters, as well as hydrothermal vents, hot springs, etc. Many other minerals, other than or in addition to calcium phosphate and magnesium sulfate, should be explored to augment "dirty" conditions (e.g, [32–36]). We must rethink the assumption that the borosilicate glass used as part of the experimental apparatus is entirely non-reactive and consider instead that it may play an essential role in the synthetic environment as a reactive surface. Much longer experiments should also be performed to determine whether it is possible to produce more and longer peptides, di- and polysaccharides, di- and polynucleic acids, etc. We also believe that additional energy sources, such as ultraviolet light cycles, freeze–thaw cycles, wet–dry cycles, etc., should be introduced into these "dirty" experiments not only to drive novel reactions but also to act as selective agents to limit the chemical combinatorial explosion that can, otherwise, be expected to result [37].

In sum, it appears that a possible way to evolve living systems is first to evolve chemical ecosystems complex enough to support the range of necessary chemical reactions.

Increasingly "dirty" experiments are, however, only a first step in that direction. Prebiotic evolution will also require increasingly complex environmental selection pressures, such as light–dark, wet–dry, and freeze–thaw cycles, in order to control the "explosion" of chemical species that complex environments will engender [36]. Real progress will occur when we can model both the chemical complexity of prebiotic environments and also their range of physicochemical selection pressures.

Supplementary Materials: The following supporting information can be downloaded at: https://www.mdpi.com/article/10.3390/life13020265/s1, Summary of Mass Spectrometry Data (Excel File).

Author Contributions: Conceptualization, R.R.-B. and A.W.B.; methodology, R.R.-B., T.R. and A.G.B.; validation, R.R.-B., T.R., A.G.B., M.T. and J.H.; formal analysis, R.R.-B., T.R., A.G.B., M.T. and J.H.; investigation, R.R.-B., T.R., A.G.B., M.T. and J.H.; resources, R.R.-B. and A.W.B.; data curation, R.R.-B.; writing—original draft preparation, R.R.-B.; writing—review and editing, R.R.-B., T.R., A.G.B., M.T. and J.H.; visualization, R.R.-B., A.W.B., A.G.B.; supervision, R.R.-B.; project administration, R.R.-B.; funding acquisition, R.R.-B. and A.W.B.; apparatus design and construction: A.W.B. All authors have read and agreed to the published version of the manuscript.

Funding: This research was partially funded by the National Science Foundation EAGER Grant Program, (Biology, Chemistry and Informal Science Directorates) NSF DRL-1212365 "Exploring Public Engagement with Real-Time Experimentation in Different Public Venues" Robert Root-Bernstein, Primary Investigator; Adam Brown, co-investigator, September 2012–August 2014. Additional funding was provided as unrestricted gifts from Maurine Bernstein.

Data Availability Statement: Original data are available by application to the corresponding author.

Acknowledgments: We thank Scott Bankroff (The Glass Lab, MSU Chemistry Department) for glasswork; Barry Tignor (Electronics Facility, MSU Physics and Astronomy) for electronics; the staff of the Machine Shop (Physics and Astronomy); Scott Smith, Lijun Chen and Cassandra Johnny of Research Technology Support Facility Mass Spectrometry and Metabolomics Core, Michigan State University, for assistance and advice with mass spectrometry techniques.

Conflicts of Interest: The authors declare no conflict of interest.

Appendix A. Additional "Metabolite" Spectra

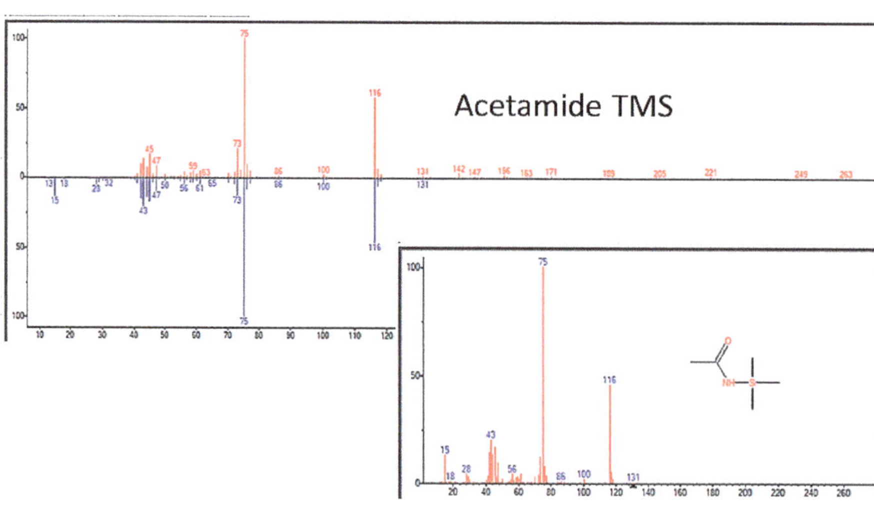

Figure A1. GC–MS identification of acetamide from reaction mixture using the NIST database recognition software.

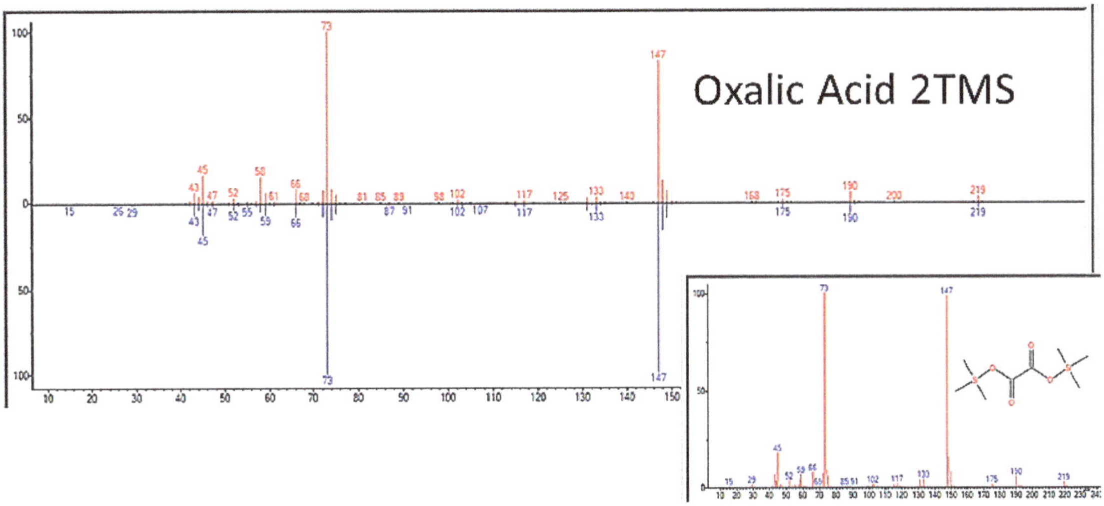

Figure A2. GC–MS identification of oxalic acid from reaction mixture using the NIST database recognition software.

Appendix B. Additional Amino Acid Spectra

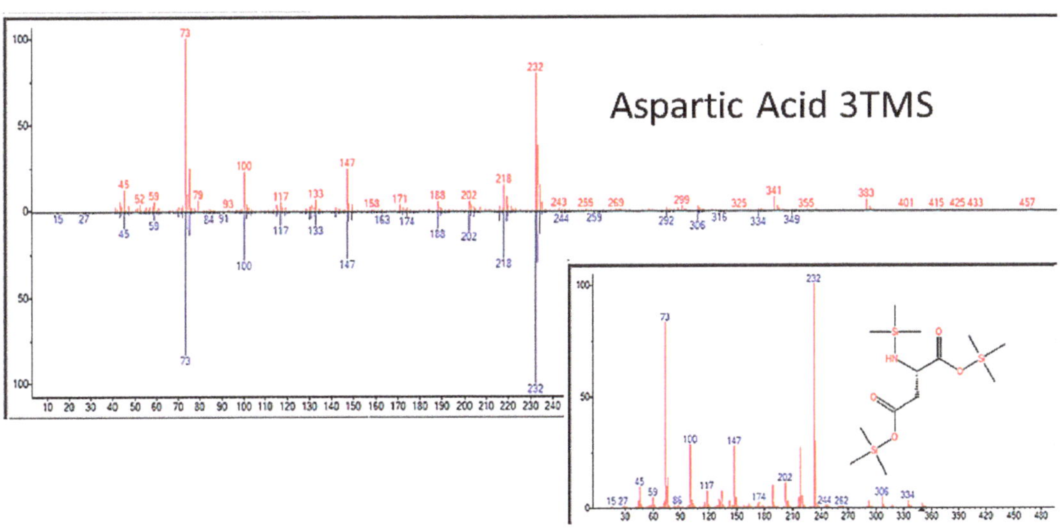

Figure A3. GC–MS identification of aspartic acid from reaction mixture using the NIST database recognition software.

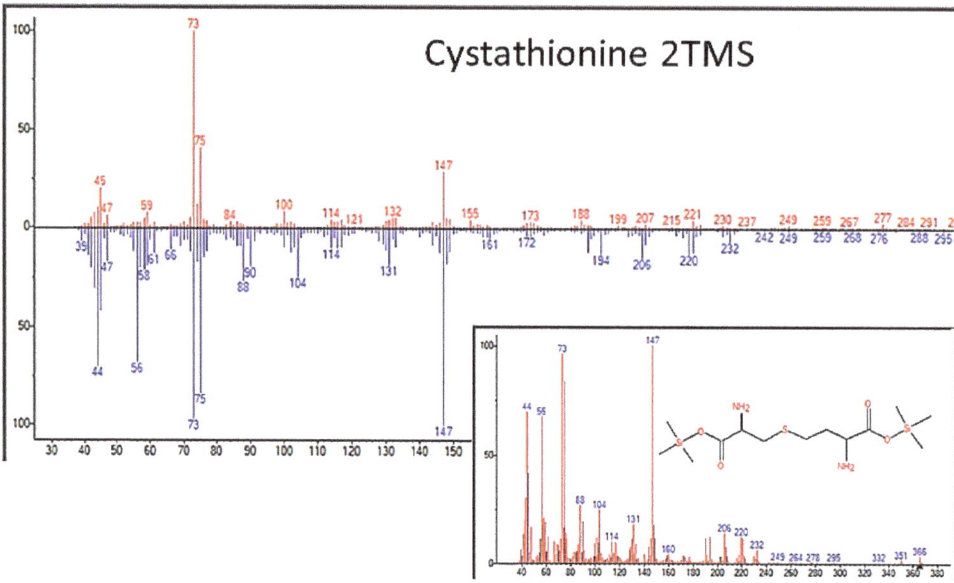

Figure A4. GC–MS identification of cystathionine, a metabolite of cysteine, from reaction mixture using the NIST database recognition software.

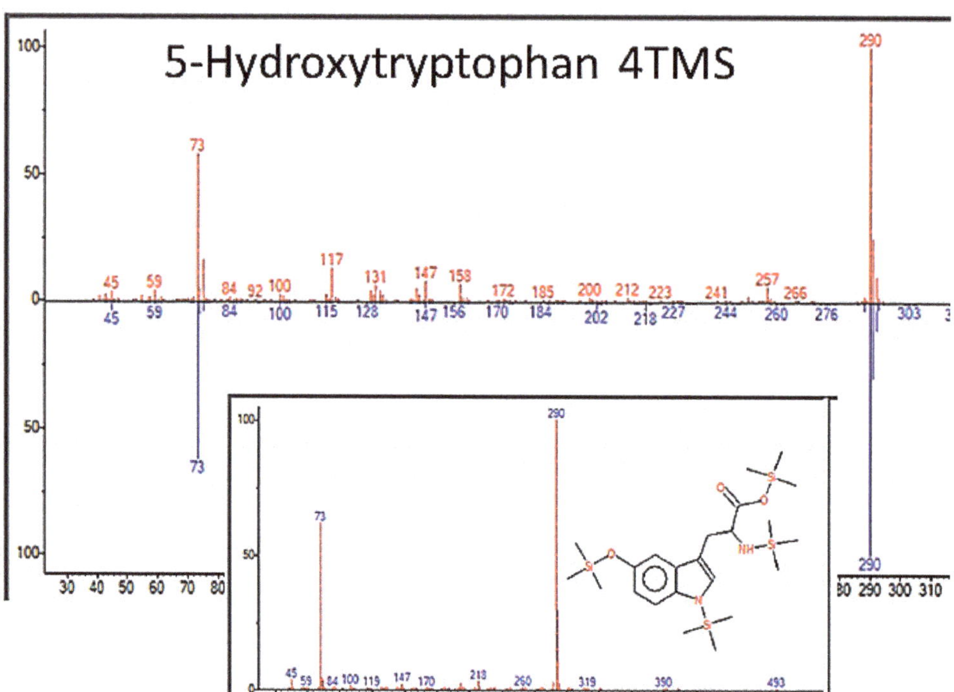

Figure A5. GC–MS identification of 5-hydroxytryptophan from reaction mixture using the NIST database recognition software.

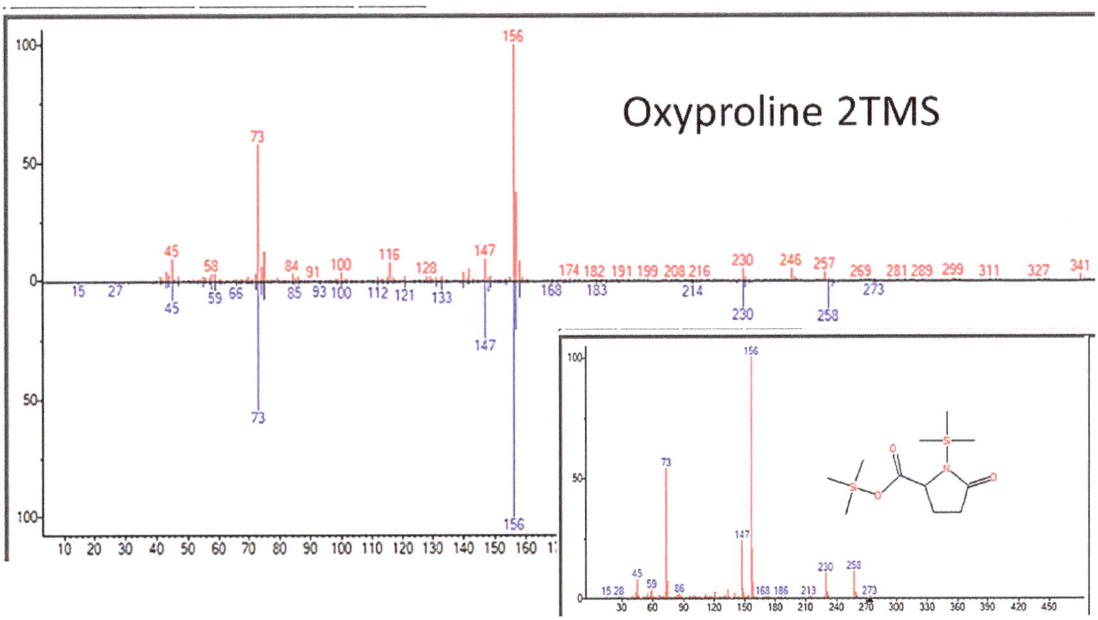

Figure A6. GC–MS identification of oxyproline from reaction mixture using the NIST database recognition software.

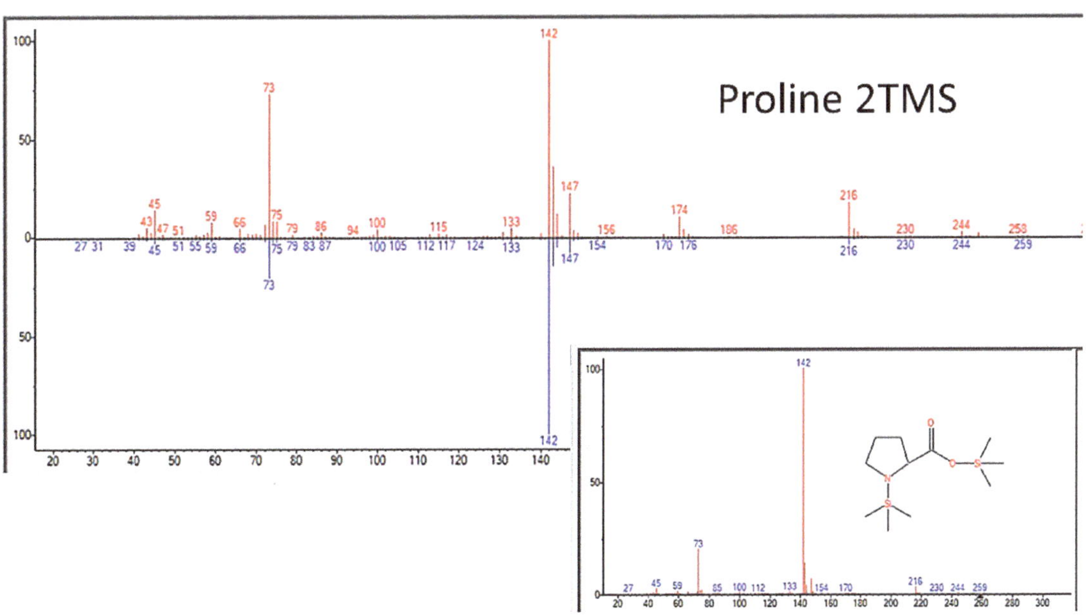

Figure A7. GC–MS identification of proline from reaction mixture using the NIST database recognition software.

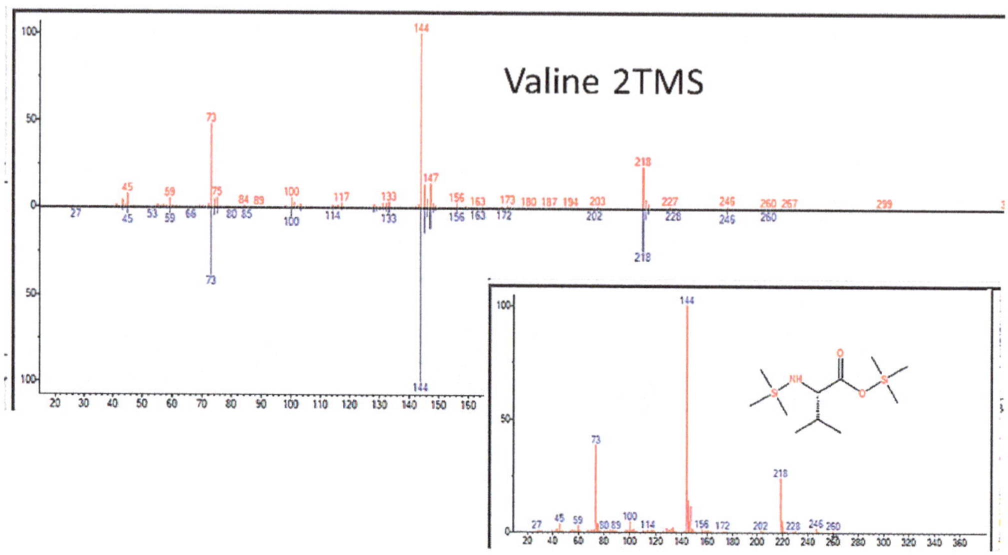

Figure A8. GC–MS identification of valine from reaction mixture using the NIST database recognition software.

Appendix C. Additional Sugar Spectra

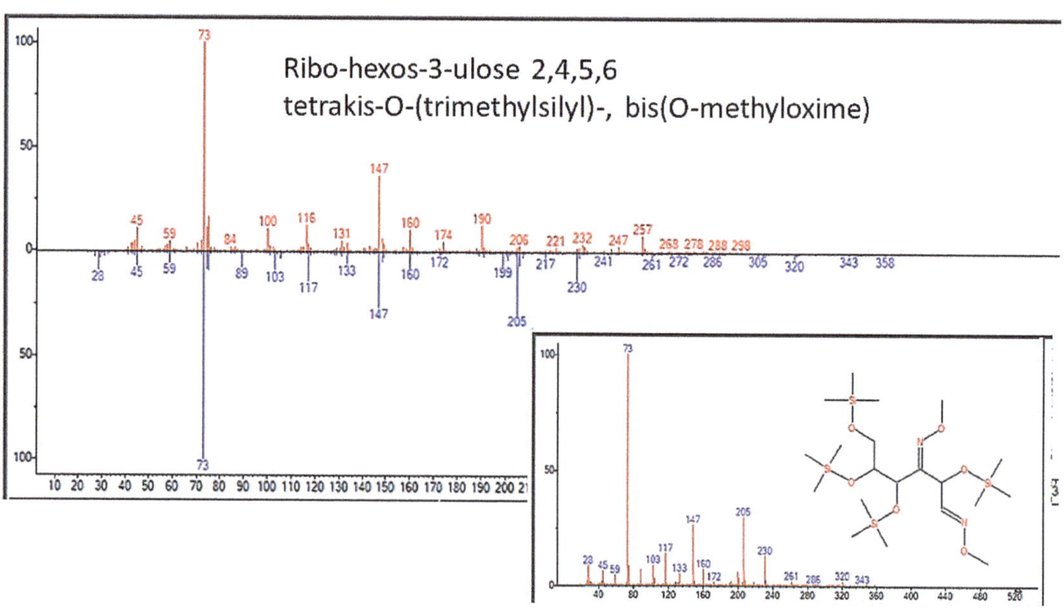

Figure A9. GC–MS identification of one of several ribose metabolites from reaction mixture using the NIST database recognition software.

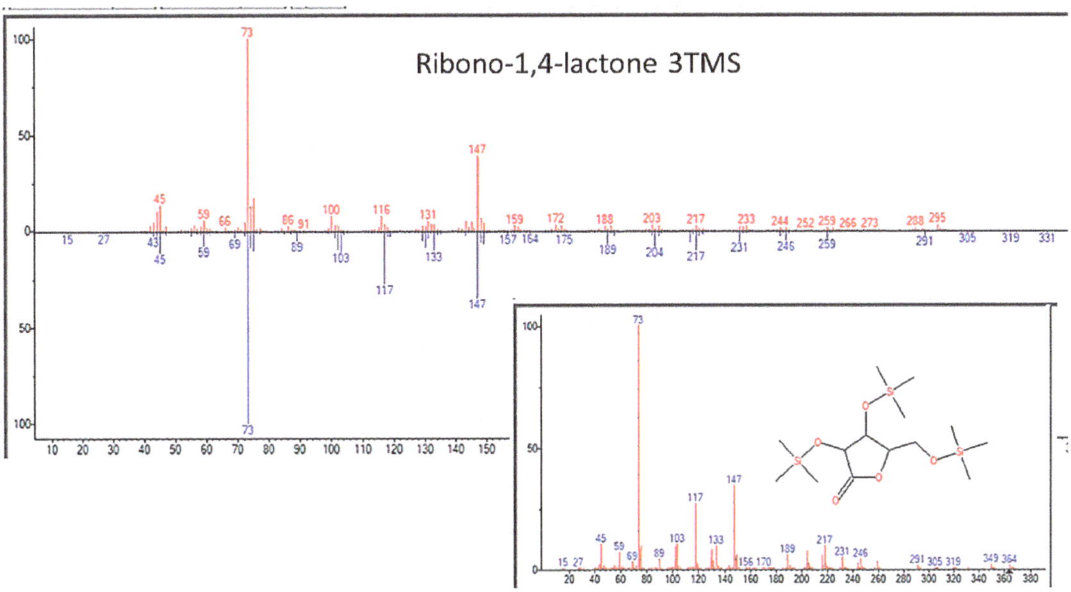

Figure A10. GC–MS identification of one of several ribose metabolites from reaction mixture using the NIST database recognition software.

Appendix D. Additional Nucleic Acid Spectrum

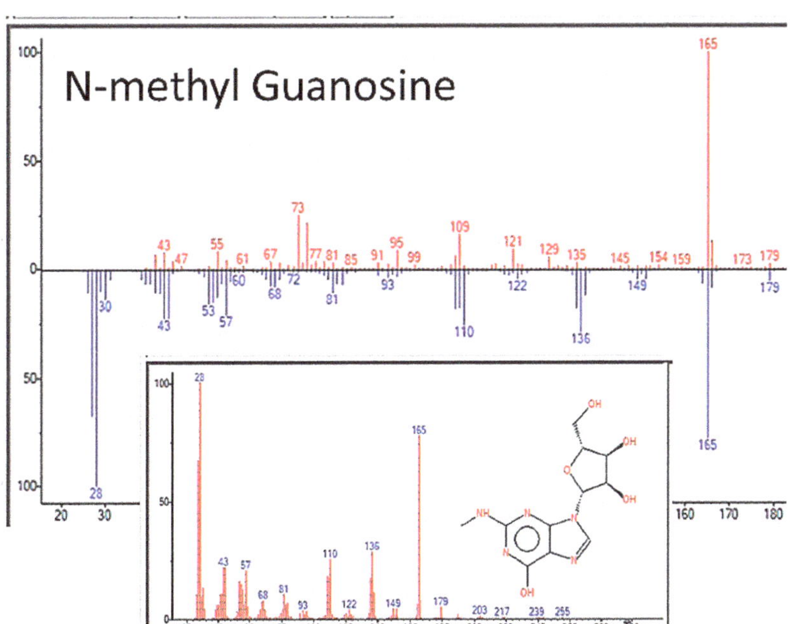

Figure A11. GC–MS identification of N-methyl guanosine from reaction mixture using the NIST database recognition software. As noted in the text and in Table 6, while this is the best match the NIST program found for this compound, it has a relatively low quality of match score compared with the other compounds illustrated.

Appendix E. Additional Fatty Acid Spectra

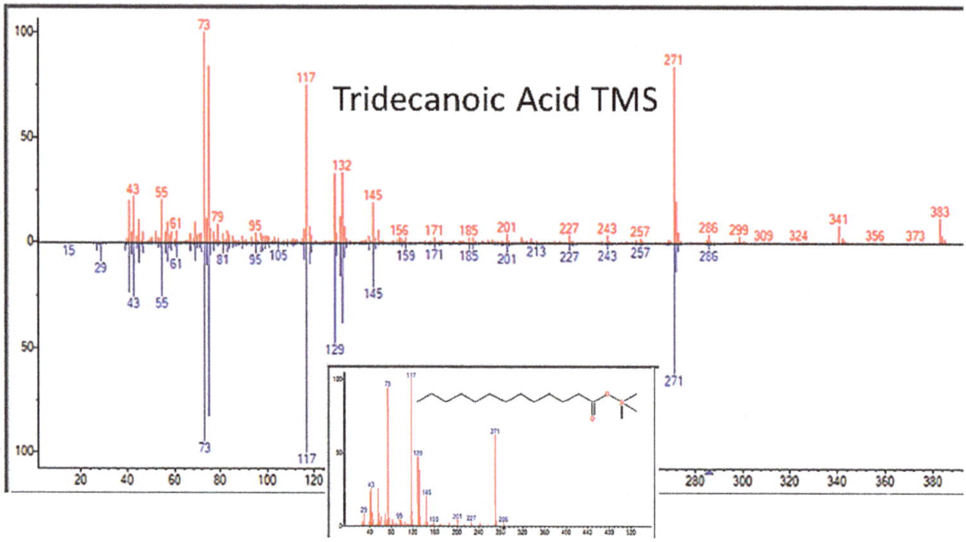

Figure A12. GC–MS identification of tridecanoic acid from reaction mixture using the NIST database recognition software.

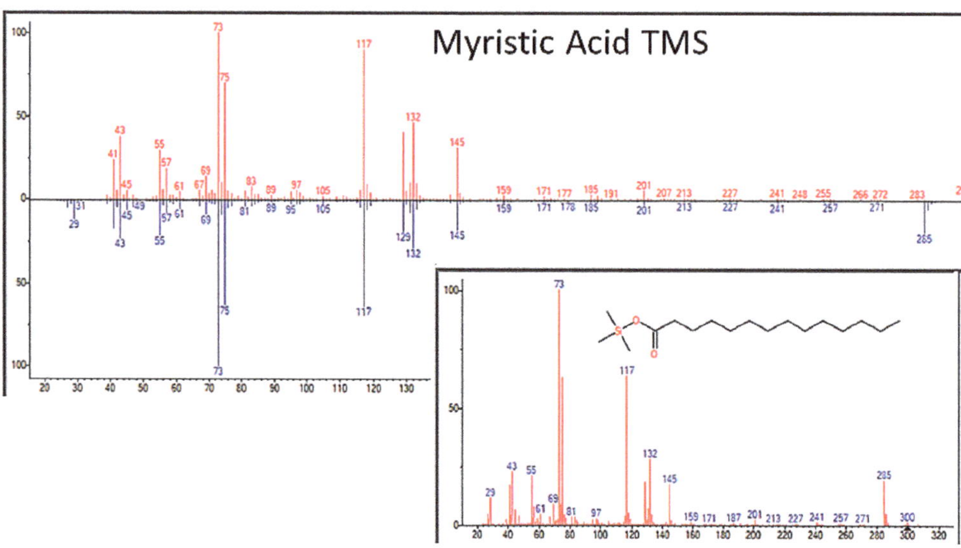

Figure A13. GC–MS identification of myristic acid from reaction mixture using the NIST database recognition software.

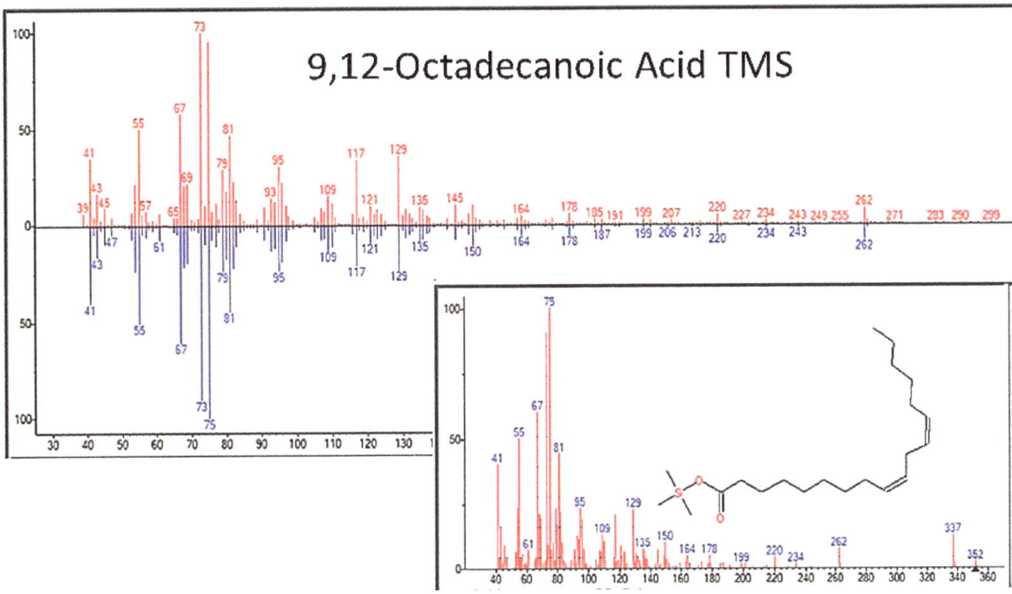

Figure A14. GC–MS identification of octadecanoic acid from reaction mixture using the NIST database recognition software.

Appendix F. Uracil Compound Spectra

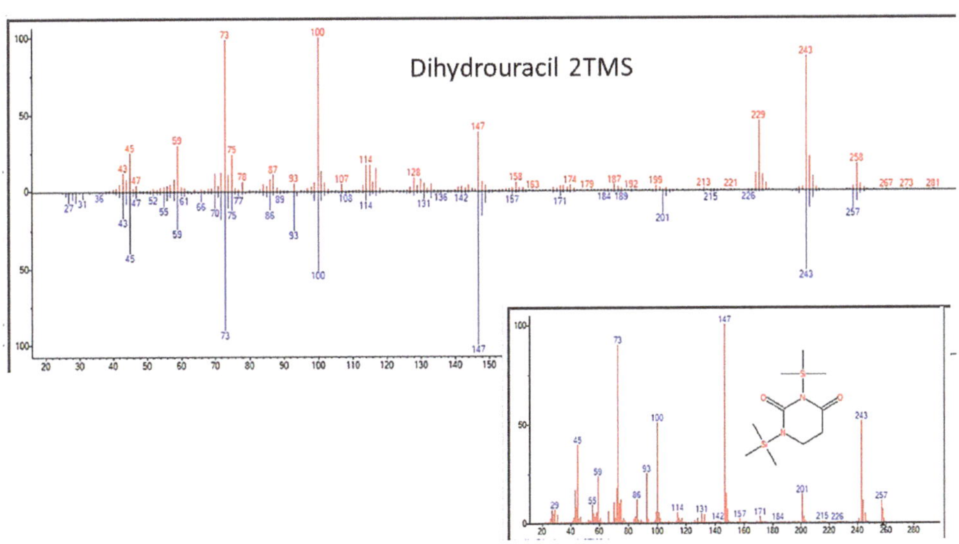

Figure A15. GC–MS identification of dihydrouracil from reaction mixture using the NIST database recognition software.

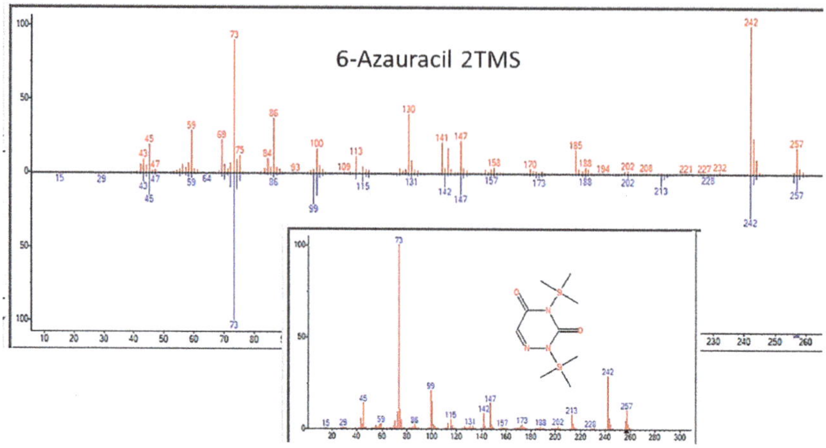

Figure A16. GC–MS identification of 6-azauracil from reaction mixture using the NIST database recognition software.

References

1. Kauffmann, S.A. The Origins of Order. In *Self-Organization and Selection in Evolution*; Oxford University Press: Oxford, UK, 1993.
2. Root-Bernstein, R.S.; Dillon, P.F. Molecular complementarity I, the complementarity theory of the origin and evolution of life. *J. Theor. Biol.* **1997**, *188*, 447–479. [CrossRef]
3. Miller, S.L. Production of amino acids under possible primitive earth conditions. *Science* **1953**, *117*, 528–529. [CrossRef]
4. Keseru, G.M.; Soos, T.; Kappe, C.O. Anthropogenic reaction parameters—the missing link between chemical intuition and the available chemical space. *Chem. Soc. Rev.* **2014**, *43*, 5387–5399. [CrossRef]
5. Cleaves, H.J. Prebiotic chemistry: What we know, what we don't. *Evo. Edu. Outreach.* **2012**, *5*, 342–360. [CrossRef]
6. Bada, J.L. New insights into prebiotic chemistry from Stanley Miller's spark discharge experiments. *Chem. Soc. Rev.* **2013**, *42*, 2186–2196. [CrossRef]
7. Hunding, A.; Kepes, F.; Lancet, D.; Minsky, A.; Norris, V.; Raine, D.; Sriram, K.; Root-Bernstein, R. Compositional complementarity and prebiotic ecology in the origin of life. *Bioessays* **2006**, *28*, 399–412. [CrossRef]
8. Root-Bernstein, R. A modular hierarchy-based theory of the chemical origins of life based on molecular complementarity. *Acc. Chem. Res.* **2012**, *45*, 2169–2177. [CrossRef]
9. Vincent, L.; Colón-Santos, S.; Cleaves, H.J., 2nd; Baum, D.A.; Maurer, S.E. The prebiotic kitchen: A guide to composing prebiotic soup recipes to test origins of life hypotheses. *Life* **2021**, *11*, 1221. [CrossRef]
10. Zaia, D.A.; Zaia, C.T.; De Santana, H. Which amino acids should be used in prebiotic chemistry studies? *Orig. Life Evol. Biosph.* **2008**, *38*, 469–488. [CrossRef]
11. Guttenberg, N.; Virgo, N.; Chandru, K.; Scharf, C.; Mamajanov, I. Bulk measurements of messy chemistries are needed for a theory of the origins of life. *Philos. Trans. A Math. Phys. Eng. Sci.* **2017**, *375*, 20160347. [CrossRef]
12. Asche, S.; Cooper, G.J.T.; Keenan, G.; Mathis, C.; Cronin, L. A robotic prebiotic chemist probes long term reactions of complexifying mixtures. *Nat. Commun.* **2021**, *12*, 3547. [CrossRef]
13. Wikipedia. Magnesium in Biology. 2022. Available online: https//en.wikipedia.org/wiki/Magnesium_in_biology (accessed on 10 August 2022).
14. Reid, C.; Orgel, L.E. Synthesis in sugars in potentially prebiotic conditions. *Nature* **1967**, *216*, 455. [CrossRef]
15. Weber, A.L. Prebiotic sugar synthesis: Hexose and hydroxy acid synthesis from glyceraldehyde catalyzed by iron(III) hy-droxide oxide. *J. Mol. Evol.* **1992**, *35*, 1–6. [CrossRef]
16. Gabel, N.W.; Ponnamperuma, C. Model for origin of monosaccharides. *Nature* **1967**, *216*, 453–455. [CrossRef]
17. Saladino, R.; Neri, V.; Crestini, C. Role of clays in the prebiotic synthesis of sugar derivatives from formamide. *Philos Mag.* **2010**, *90*, 2329–2337. [CrossRef]
18. Civiš, S.; Szabla, R.; Szyja, B.M.; Smykowski, D.; Ivanek, O.; Knížek, A.; Kubelík, P.; Šponer, J.; Ferus, M.; Šponer, J.E. TiO2-catalyzed synthesis of sugars from formaldehyde in extraterrestrial impacts on the early Earth. *Sci. Rep.* **2016**, *6*, 23199. [CrossRef]
19. Forsythe, J.G.; Yu, S.S.; Mamajanov, I.; Grover, M.A.; Krishnamurthy, R.; Fernández, F.M.; Hud, N.V. Ester-mediated amide bond formation driven by wet-dry cycles: A possible path to polypeptides on the prebiotic earth. *Angew. Chem. Int. Ed. Engl.* **2015**, *54*, 9871–9875. [CrossRef]
20. Raine, D.J.; Norris, V. Lipid domain boundaries as prebiotic catalysts of peptide bond formation. *J. Theor. Biol.* **2007**, *246*, 176–185. [CrossRef]

21. Yamauchi, K.; Kinoshita, M. Highly stable lipid membranes from archaebacterial extremophiles. *Prog. Polym. Sci.* **1993**, *18*, 763–804. [CrossRef]
22. Driessen, A.J.M.; van de Vossenberg, J.A.C.M.; Konings, W.N. Membrane composition and ion-permeability in extremophiles. *FEMS Microbio. Rev.* **1996**, *18*, 139–148. [CrossRef]
23. van Meer, G.; Voelker, D.R.; Feigenson, G.W. Membrane lipids: Where they are and how they behave. *Nat Rev Mol Cell Biol.* **2008**, *9*, 112–124. [CrossRef]
24. Parker, E.T.; Zhou, M.; Burton, A.S.; Glavin, D.P.; Dworkin, J.P.; Krishnamurthy, R.; Fernández, F.M.; Bada, J.L. A plausible simultaneous synthesis of amino acids and simple peptides on the primordial Earth. *Angew. Chem. Int. Ed. Engl.* **2014**, *53*, 8132–8136. [CrossRef]
25. Cossetti, C.; Crestini, C.; Saladino, R.; Mauro, E.D. Borate Minerals and RNA Stability. *Polymers* **2010**, *2*, 211–228. [CrossRef]
26. Yuasa, S.; Flory, D.; Basile, B.; Oró, J. Abiotic synthesis of purines and other heterocyclic compounds by the action of electrical discharges. *J. Mol. Evol.* **1984**, *21*, 76–80. [CrossRef]
27. Ruiz-Bermejo, M.; Menor-Salván, C.; Osuna-Esteban, S.; Veintemillas-Verdaguer, S. Prebiotic microreactors, a synthesis of purines and dihydroxy compounds in aqueous aerosol. *Orig. Life Evol. Biosph.* **2007**, *37*, 123–142. [CrossRef]
28. Ferus, M.; Pietrucci, F.; Saitta, A.M.; Knížek, A.; Kubelík, P.; Ivanek, O.; Shestivska, V.; Civiš, S. Formation of nucleobases in a Miller-Urey reducing atmosphere. *Proc. Natl. Acad. Sci. USA* **2017**, *114*, 4306–4311. [CrossRef]
29. Allen, W.V.; Ponnamperuma, C. A possible prebiotic synthesis of monocarboxylic acids. *Curr. Mod. Biol.* **1967**, *1*, 24–28. [CrossRef]
30. Bossard, A.R.; Raulin, F.; Mourey, D.; Toupance, G. Organic synthesis from reducing models of the atmosphere of the primitive earth with UV light and electric discharges. *J. Mol. Evol.* **1982**, *18*, 173–178. [CrossRef]
31. Saladino, R.; Carota, E.; Botta, G.; Kapralov, M.; Timoshenko, G.N.; Rozanov, A.Y.; Krasavin, E.; Di Mauro, E. Meteorite-catalyzed syntheses of nucleosides and of other prebiotic compounds from formamide under proton irradiation. *Proc. Natl. Acad. Sci. USA* **2015**, *112*, E2746–E2755. [CrossRef]
32. Muchowska, K.B.; Varma, S.J.; Moran, J. Synthesis and breakdown of universal metabolic precursors promoted by iron. *Nature* **2019**, *569*, 104–107. [CrossRef]
33. Lawless, J.G.; Levi, N. The role of metal ions in chemical evolution, polymerization of alanine and glycine in a cation-exchanged clay environment. *J. Mol. Evol.* **1979**, *13*, 281–286. [CrossRef] [PubMed]
34. Hansma, H.G. Potassium at the origins of life: Did biology emerge from biotite in micaceous clay? *Life* **2022**, *12*, 301. [CrossRef] [PubMed]
35. Baú, J.; Carneiro, C.; da Costa, A.; Valezi, D.F.; di Mauro, E.; Pilau, E.; Zaia, D. The effect of goethites on the polymerization of glycine and alanine under prebiotic chemistry conditions. *Orig. Life Evol. Biosph.* **2021**, *51*, 299–320. [CrossRef] [PubMed]
36. Shanker, U.; Bhushan, B.; Bhattacharjee, G. Oligomerization of glycine and alanine catalyzed by iron oxides, implications for prebiotic chemistry. *Orig. Life Evol. Biosph.* **2012**, *42*, 31–45. [CrossRef]
37. Root-Bernstein, R.; Brown, A.W. Novel Apparatuses for Incorporating Natural Selection Processes into Origins-of-Life Experiments to Produce Adaptively Evolving Chemical Ecosystems. *Life* **2022**, *12*, 1508. [CrossRef]
38. Miyakawa, S.; Yamanashi, H.; Kobayashi, K.; Cleaves, H.J.; Miller, S.L. Prebiotic synthesis from CO atmospheres, implications for the origins of life. *Proc. Natl. Acad. Sci. USA* **2002**, *99*, 14628–14631. [CrossRef]
39. Kasting, J.F. The evolution of the prebiotic atmosphere. *Orig. Life* **1984**, *14*, 75–82. [CrossRef]
40. Mazankova, V.; Torokova, L.; Krcma, F.; Mason, N.J.; Matejcik, S. The influence of CO_2 admixtures on the product composition in a nitrogen-methane atmospheric glow discharge used as a prebiotic atmosphere mimic. *Orig. Life Evol. Biosph.* **2016**, *46*, 499–506. [CrossRef]
41. Raulin, F.; Brassé, C.; Poch, O.; Coll, P. Prebiotic-like chemistry on Titan. *Chem. Soc. Rev.* **2012**, *41*, 5380–5393. [CrossRef]
42. Cruikshank, D.P.; Materese, C.K.; Pendleton, Y.J.; Boston, P.J.; Grundy, W.M.; Schmitt, B.; Lisse, C.M.; Runyon, K.D.; Keane, J.T.; Beyer, R.A.; et al. Prebiotic Chemistry of Pluto. *Astrobiology* **2019**, *19*, 831–848. [CrossRef]
43. Paecht-Horowitz, M.; Berger, J.; Katchalsky, A. Prebiotic synthesis of polypeptides by heterogeneous polycondensation of amino-acid adenylates. *Nature* **1970**, *228*, 636–639. [CrossRef] [PubMed]
44. Liu, Z.; Beaufils, D.; Rossi, J.C.; Pascal, R. Evolutionary importance of the intramolecular pathways of hydrolysis of phosphate ester mixed anhydrides with amino acids and peptides. *Sci. Rep.* **2014**, *4*, 7440. [CrossRef] [PubMed]

life

MDPI

Review

Prebiotic Chemistry Experiments Using Microfluidic Devices

Karen Melissa Lerin-Morales [1,*], Luis F. Olguín [2], Eva Mateo-Martí [3] and María Colín-García [4,*]

1 Posgrado en Ciencias de la Tierra, Universidad Nacional Autónoma de México, Ciudad de Mexico 04510, Mexico
2 Laboratorio de Biofisicoquímica, Facultad de Química, Universidad Nacional Autónoma de México, Ciudad de Mexico 04510, Mexico
3 Centro de Astrobiología (CAB), CSIC-INTA, Carretera de Ajalvir Km 4, Torrejón de Ardoz, 28850 Madrid, Spain
4 Instituto de Geología, Universidad Nacional Autónoma de México, Ciudad de Mexico 04510, Mexico
* Correspondence: mel.lerin@comunidad.unam.mx (K.M.L.-M.); mcolin@geologia.unam.mx (M.C.-G.); Tel.: +52-(55)-5622-4300 (ext. 164) (M.C.-G.)

Abstract: Microfluidic devices are small tools mostly consisting of one or more channels, with dimensions between one and hundreds of microns, where small volumes of fluids are manipulated. They have extensive use in the biomedical and chemical fields; however, in prebiotic chemistry, they only have been employed recently. In prebiotic chemistry, just three types of microfluidic devices have been used: the first ones are Y-form devices with laminar co-flow, used to study the precipitation of minerals in hydrothermal vents systems; the second ones are microdroplet devices that can form small droplets capable of mimic cellular compartmentalization; and the last ones are devices with microchambers that recreate the microenvironment inside rock pores under hydrothermal conditions. In this review, we summarized the experiments in the field of prebiotic chemistry that employed microfluidic devices. The main idea is to incentivize their use and discuss their potential to perform novel experiments that could contribute to unraveling some prebiotic chemistry questions.

Keywords: microfluidic devices; prebiotic chemistry; hydrothermal vents; mineral membranes; cellular compartmentalization

Citation: Lerin-Morales, K.M.; Olguín, L.F.; Mateo-Martí, E.; Colín-García, M. Prebiotic Chemistry Experiments Using Microfluidic Devices. *Life* **2022**, *12*, 1665. https://doi.org/10.3390/life12101665

Academic Editors: Ranajay Saha and Alberto Vázquez-Salazar

Received: 30 September 2022
Accepted: 18 October 2022
Published: 21 October 2022

1. Introduction

Microfluidic devices are, according to George Whitesides, "the science and technology of systems that process or manipulate small (10^{-9} to 10^{-18} L) amounts of fluids, using channels with dimensions of tens to hundreds of micrometers" [1]. It can also be said that microfluidics is the design and construction of small devices, including channels and chambers in microscale size, where the flow and mixing of fluids can be accurately controlled [1–3].

The multidisciplinary field of microfluidics emerged from a conjunction of technologies and principles of physics, biology, microtechnology, material science, chemistry, and microelectronics, among others [4,5]. Precedents to microfluidics were microanalytical methods, such as gas phase chromatography (GPC), high-pressure liquid chromatography (HPLC), and capillary electrophoresis (CE), all in the capillary format; these techniques entirely changed chemical analysis [1].

Microfluidic devices are used in biological [6–10] and chemical essays [11,12], clinical and forensics [13–15], molecular and medical diagnosis [16–19], pollution monitoring [20–22], drug design and therapeutic activity [23–26], and biohazard detection [7,27–29], among others disciplines.

There are multiple advantages to using microfluid devices. The most evident is that miniaturization reduces reagent consumption, thus cost and waste are diminished too. Other advantages are high precision of the mixing regimes and control of fluids [4], as well as the possibility to manipulate multiphase flows (i.e., liquid, gases, particles) [1].

Diffusive mixing is fast and increases the speed of reactions. In microfluidics devices, sample processing is fast [17], and it is possible to work under supercritical conditions at the microscale [11]. The dynamics of interfaces in complex systems can also be explored [30]. Compared to macroscopic systems, there is an improvement in thermal and mass transfer due to the high surface-to-volume ratios on these devices [31]. Microfluidic devices enable quick temperature adjustments and accurate temperature control. The concentration of the molecules both in space and time is feasible [1]. Additionally, the huge design flexibility of these devices is remarkable [32]. In a single-designed integrated device, such as lab-on-chip (LOC), a full experiment could be performed, including sampling, processing, and measurements, thanks to the combination of microfluidic channels and active or passive components [33].

1.1. Components

A microfluidic system consists of different components: (i) a device for introducing samples and reactants; (ii) a method for pushing, mixing, and combining the fluids; and (iii) other components (detectors, purification tools, etc.) [1].

- Accessories for introducing samples. Samples can be loaded in different ways: (i) manually, (ii) by integrated capillaries that transfer the fluid directly by vacuum, (iii) by capillaries connected via clamping or screwing; or (iv) by dispensing systems activated by short pressure pulses [34].
- Methods for pushing, mixing, and combining fluids. The dispositive needs components, either active or passive structures, arranged in such a manner that guide liquids through channels, channel networks, or chambers. Depending on the nature of the experiments, some adaptations can be made, including (but not limited to) microvalves installation (for blocking/unblocking channels), pumps (for promoting/increasing fluid flow), and micromixers [34]; micromixers can be active (requiring external activation) and passive mixers [35].
- Other components. All needed components can be added for processing or analyzing the samples. For example, samples can be sorted according to their size for filtration; in classic filtration, components transported on the flow are retained [34]. Filtration can also be accomplished in membranes [36] or by centrifugal forces in centrifugal platforms [37]. The inclusion of solid-phase chromatography extraction elements is also applied [38]. In microfluidic devices, sometimes it is necessary to measure and control the temperature [39].

1.2. Materials for Microfluidic Devices Construction

Microfluidic devices can be fabricated from different materials and following different techniques; there is continuous design and improvement in this regard [5]. The material is critical because it affects the flow, biocompatibility, absorptivity, and function of the components [40]. Some materials that have been used comprise thermoplastic polymers (polycarbonate), elastomers (polydimethylsiloxane), hydrogels, inorganic substrates (glass, silicon), paper, resins, ceramics, or a combination of these materials [41]. For more detailed information, consult the work of Niculescu [5].

1.3. Basic Principles of Microfluidics

The manipulated volume of fluid in microfluidic devices is not so relevant. The most significant feature of these devices is the length scale in the channels that allows laminar flow of the fluid [42]. In laminar flow, the fluid moves smoothly as if layers slip over each other [43]. The Reynolds number (*Re*) is a dimensionless number that shows the flow regime as a function of the density (ρ) and viscosity (μ) of the flow, the flow velocity (ν), and the duct diameter (D) where the fluid flows. The Reynolds number can be expressed as follows:

$$Re = \frac{\nu D \rho}{\mu} \tag{1}$$

Flows with high Reynolds numbers tend to be turbulent, whereas flows with low Reynolds numbers are laminar. For practical applications of flows in ducts, if the Reynolds number is below 2000, the flow regime is laminar; if the Reynolds number is above 4000, it is turbulent. Regimes at intermediate values of Reynolds numbers between 2000 and 4000 are named transition regimes [44]. In microfluidic devices, the Reynolds number is usually smaller, with values less than 1 [45].

1.4. Types of Microfluidic Devices Configurations

In microfluidic devices, the channel design will depend on the device's function, and devices can come up with various types of channels. Some types of channels are straight, Y-form, T-junction, spiral, cross-junction, flow-focusing, division, serpentine, and microchambers (Figure 1). Devices with spiral and Y channels are commonly used for separations (although Y channels are used for combining fluids as well). To carry out the mixing of fluids, a serpentine design is employed, and the division channels are used for splitting fluids. T-junction, cross-junctions, and flow-focusing are commonly used in microdroplet devices, and in microchambers, physical, chemical, and biological reactions are performed.

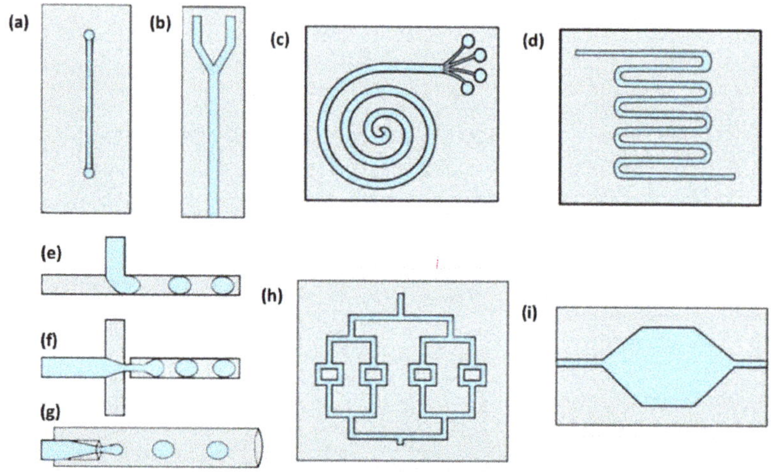

Figure 1. Types of microfluidic devices channel configurations: (**a**) straight, (**b**)"Y" shaped, (**c**) spiral, (**d**) serpentine, (**e**) "T"-junction, (**f**) cross-junction, (**g**) flow-focusing, (**h**) division, and (**i**) microchamber.

The necessity to reproduce conditions occurring in some environments has led researchers to conceive experiments that permit the integration of various factors and/or simplify the way such experiments are carried out. Microfluidic devices facilitate the reproduction of certain prebiotic environments, and for that reason, they have been used to perform prebiotic chemistry experiments. Three types of microfluidic devices have been used for this purpose: Y-form devices with laminar co-flow, microdroplet devices, and devices with microchambers. The first is commonly employed to study mineral precipitation, under conditions simulating those of hydrothermal systems; the second type is used to imitate cell compartmentalization; and the last is used to recreate rock pores in hydrothermal environments. In some cases, the type of device does not match the mentioned purpose. In this review, the prebiotic chemistry experiments performed with microfluidic devices are summarized and their use to accomplish novel experiments that contribute to the development of this field is analyzed.

2. Microfluidic Devices and Prebiotic Chemistry

2.1. Mineral Precipitation and Mineral Membranes on Hydrothermal Systems

Since its discovery in the late seventies [46], submarine hydrothermal vent systems have been proposed as environments where life could have emerged [47–52]. This idea arose after considering the physical and chemical gradients of hydrothermal vents as energy sources for the abiotic synthesis of organic molecules [47,53–55]. There are two types of submarine hydrothermal systems designated as prebiotic environments: black and white smokers. Each one has certain attributes, but they can coexist in the same hydrothermal field [56,57].

Black smokers are systems found near oceanic ridges, and they form when oceanic waters seep into the oceanic crust and are heated by magma [57]. As the oceanic water heats up, they begin to return to the ocean, and due to elevated temperatures (400 °C) and acidic pH, metals are dissolved from the crust rocks and carried with water [57]. When the water comes up from the crust and contacts the low-temperature ocean, sulfides precipitate, and they are the main responsible for the black smoke appearance [56,58]. On the other hand, white smokers are more recently discovered hydrothermal systems found farther away from the heat source than black smokers. White smokers' temperature is rather warm (40–90 °C), their pH is alkaline (9–11), and within them are mainly carbonate precipitates [59,60]. Due to the fewer extreme characteristics of white smokers, they are pointed out as favorable places for the beginning of life [55,61,62].

In the study of hydrothermal vents systems (HVS) as prebiotic environments, laboratory simulations, to recreate the conditions in these systems, have been scarcely employed. These experiments must include factors such as pH, pressure, temperature, and the use of minerals identified in these types of environments [63]. One way to simulate the reactions on HVS is by bringing two fluids in contact; one mimics hydrothermal fluids, and the other simulates oceanic water, just as it happens in hydrothermal systems. The solution simulating the hydrothermal fluids is alkaline, and the solution simulating oceanic water is acidic. Both solutions may contain soluble metal salts and sodium silicate compounds [64–67].

Different methods can be employed to put the two solutions in contact, e.g., by adding drops of one solution to the other [54,68] or by injecting the hydrothermal fluid into a reservoir containing the oceanic solution [64–67]. The contact of the two solutions creates hollow structures called chemical gardens [69,70]. Chemical gardens are thus inorganic structures formed by the reaction of a soluble metallic salt and an aqueous solution of anions [71].

As an approach to studying chemical gardens more easily, Batista and Steinbock [72] created microfluidic devices for mineral membrane precipitation. Such devices are fabricated with low-cost materials such as acrylic, Parafilm®, polystyrene, glass, or Teflon®. Fabrication of microfluidic devices for mineral precipitation is simple; for example, a cut Y-pattern parafilm is sandwiched between two acrylic sheets with drilled inlets. Barb fittings are glued to the perforations to connect the tubing [72]. The Y-shaped channel configuration (Figure 2) permits the flux of two parallel streams without mixing. Although the streams do not mix, they are in contact with each other, allowing the creation of conditions far from equilibrium (pH, temperature, and concentration gradients) [45]. For instance, Möller et al. [73] prove the existence of pH gradients up to six units at micrometric scales in the microfluidic device channel.

In many of the experiments of mineral membrane precipitation, as in chemical gardens, two reactants are used to simulate the ocean and the hydrothermal fluids. One is an acidic solution that contains metals, and the other is a NaOH alkaline solution (Table 1). Different compositions of solutions with different metals in the form of chlorides have been used. Some examples are chlorides of Mg^{2+}, Mn^{2+}, $Fe^{2+/3+}$, Co^{2+}, and Cu^{2+} [72]. When the two solutions come into contact, in the interface, a mineral membrane is formed by precipitation. Its composition depends on the metal salt employed. In experiments with cobalt salt, $Co(OH)_2$ and cobalt oxide are produced [74], while membranes formed with manganese chloride are composed of $Mn(OH)_2$ [72]. Further, when membranes were produced in the presence of iron chloride and phosphate, the composition of the membrane consisted of the goethite and vivianite minerals [75].

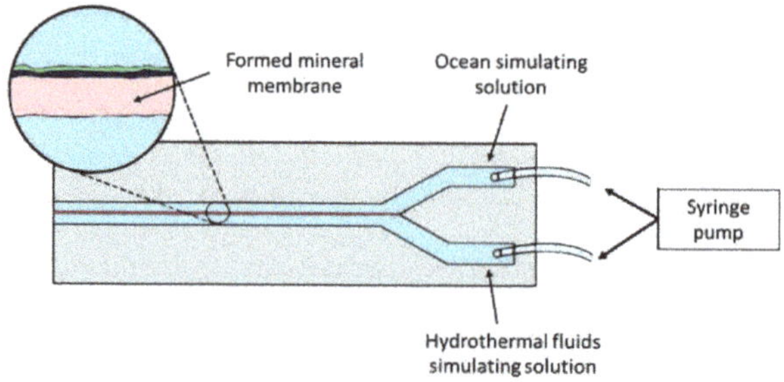

Figure 2. "Y"-shaped channel microfluidic devices design for simulating mineral precipitation in hydrothermal systems.

Some mineral membrane characteristics have been studied. For example, Wang and Steinbock [76] explore the effect of temperature in the formation of $Ni(OH)_2$ membranes, finding that the membrane growth rate is independent of the temperature in the interval of 10 to 40 °C. In another experiment, Ding et al. [77] study mineral membranes formed by $Mn(OH)_2$ and observe the formation of waves on the membrane surface. Such waves facilitate ion transport in the membrane; thus, they may be relevant in studying membrane transport in protocells.

The catalytic properties of these membranes have been also explored. Wang et al. [75], using membranes formed at different pH gradients and with different metal salts, explored if the membranes could synthesize pyrophosphate from acyl-phosphate. They found that membranes formed under high pH gradients and containing Fe^{2+} favored the yield of pyrophosphate. In two independent experiments, Sojo et al. [78] and Vasiliadou et al. [79] probed the capacity of (Fe, Ni)S membranes to reduce CO_2 in the presence of H_2, obtaining negative results, concluding that step pH gradients are not enough to reduce CO_2. They suggested that the reaction could be achieved at pressures greater than atmospheric. In a further experiment by Hudson et al., using the same factors as Sojo et al. [78], except for the H_2 pressure (1.5 bar in this study), the CO_2 reduction to formate, promoted by a pH gradient, was observed [80].

Mineral membrane precipitation experiments in microfluidic devices with Y-shaped channels are recent and have the potential to be used in the fields of prebiotic chemistry and material synthesis [45]. The exploration of various experimental conditions, such as different reactants and their concentration, and different physical and chemical conditions, such as temperature or pH, and the presence of prebiotic molecules is still missing. An important matter is to evaluate the potential of mineral membranes as catalysts and concentrators of relevant organic molecules in the prebiotic chemistry field. Additionally, this kind of microfluidic device has traits that allow the generation of physical and chemical gradients. Those gradients can be useful in other types of experiments necessary to examine non-equilibrium processes, which are important in prebiotic environments.

Table 1. Prebiotic chemistry experiments in microfluidic devices with Y-shaped channels.

Experiment Type	Device Description	Experimental	Findings	Reference
Mineral precipitation in hydrothermal systems	Parallel laminar flow, Y-shaped channel, external temperature control.	$Ni(OH)_2$ mineral membranes form from NaOH and $NiCl_2$ solutions, at different temperatures.	From T-10 to 40 °C, the effective diffusion coefficient is temperature independent.	[76]
Mineral precipitation in hydrothermal systems	Parallel laminar flow, Y-shaped channel.	Membrane precipitation by alkaline inorganic phosphorous, acyl-phosphate solution, and acidic solution including cations (Fe^{2+}, Fe^{3+}, Ca^{2+}, Mn^{2+}, Co^{2+}, Cu^{2+}, Zn^{2+}, or Ni^{2+}). Precipitated membrane was incubated in a water bath for 1 h at 38 °C and analyzed.	Fe^{2+}, other divalent cations and Fe^{3+} promote the formation of pyrophosphate from inorganic phosphorus and acyl-phosphate.	[75]
Mineral precipitation in hydrothermal systems	Parallel laminar flow, Y-shaped channel, heated by a heating plate.	Membranes formed from alkaline (Na_2S, $Na_6Si_2O_7$, pH 11) and acidic ($FeCl_2$, $NiCl_2$, and $NaHCO_3$, pH 6) solutionsH_2 were introduced into the alkaline solution.	Fe(Ni)S mineral membrane and a pH gradient of 5 units formed. The reaction between H_2 and CO_2 at the mineral was not possible at atmospheric pressure.	[79]
Mineral precipitation in hydrothermal systems	Parallel laminar flow, Y-shaped channel, heated by a plate.	Membranes formed by alkaline (pH 11, Na_2S, K_2HPO_4, Na_2MoO_4, H_2 at atmospheric pressure) and acidic (pH 6, $FeCl_2$, $NiCl_2$, and CO_2 at atmospheric pressure) solutions. The reaction times were 0, 0.5,1, 2, 5, 12 and 24 h.	CO_2 reduction was not achieved in the presence of the Fe(Ni)S membrane. High pressures of H_2 are required to achieve CO_2 reduction.	[78]
Mineral precipitation in hydrothermal systems	Parallel laminar flow, Y-shaped channel, microfluidic pumps driven by pressure.	n alkaline (pH 12.3, Na_2S, $Na_2Si_3O_7$, degassed water, 1.5 bar of H_2) and acidic (pH 3.9, $FeCl_2$, $NiCl_2$, 1.5 bar of CO_2) solutions.	CO_2 reduced to formate in a Fe(Ni)S mineral membrane promoted by a pH gradient. H_2 oxidation, and movement of electrons across the mineral membrane to reduceCO_2.	[80]
Mineral precipitation in hydrothermal systems	Parallel laminar flow, Y-shaped channel.	Membranes from NaOH and $MnCl_2$ solutions at different concentrations. A theoretical model of electron transport was made.	The waviness enhances the diffusion across the $Mn(OH)_2$ membrane.	[77]
Pores in rocks in hydrothermal systems	Parallel laminar flow, Y-shaped channel, platinum electrodes for electric potential measurement.	Na_2S at pH 11.8 and a $FeCl_2$ at pH 5.8 solutions.	A chemical gradient, 6 pH units of difference, at a micrometric scale. Precipitation reaction stabilizes the pH gradient and makes it larger.	[73]

2.2. Prebiotic Chemistry Experiments with Droplet-Based Microfluidic Devices

There are three minimum characteristics for a system to be considered alive: (i) metabolism, which consists of a reactions network capable of synthesizing useful molecules; (ii) self-replication, which refers to a system capable of producing macromolecules with a template, by polycondensation of molecules generated by metabolism; and (iii) membranes, structures that delimitate and protect the living system from the exterior; the membrane must be permeable and have the capacity to grow using the molecules produced by metabolism [81,82]. It is believed that membranes were the first protobiological structures on early Earth [83]. Cellular life probably emerged when self-assembling membranes captured in their interior catalytic and informational polymers [83]. Moreover, compartmentalization is essential for the emergence of Darwinian evolution since, in this way, different catalytic/informational systems are well delimited one for another [84].

In prebiotic chemistry, various methods of compartmentalization have been considered. The principal is encapsulation by vesicles made of amphiphilic molecules. Amphiphilic molecules possess polar (hydrophilic) and non-polar (hydrophobic) moieties. The polar component consists of hydroxyl, carboxyl, amine, phosphate, or sulfate groups, while the non-polar part comprises a hydrophobic hydrocarbon chain [83]. At specific conditions of temperature, pH, salt concentration, or biopolymer presence, the amphiphilic molecules self-assemble into vesicles with an aqueous interior separated from the medium by one or two bilayers [85]. The self-assembly process of amphiphilic molecules to vesicles has been studied under various conditions, including the interaction with molecules that can be encapsulated such as amino acids, sugars, nitrogen bases, and even with macromolecules as peptides [86]. Droplet-based microfluidic devices are an alternative to studying the compartmentalization phenomenon since they have been used to perform confined chemical and biochemical reactions [87–89]. These devices have geometries that allow them to intersect two non-miscible fluids, and thanks to this intersection, monodisperse microdroplets are created (Figure 3) [87,90]. Microdroplets are discrete units with specific microenvironments that favor reactions that probably occurred in early Earth compartments [91]. There are few studies describing the use of droplet-based microfluidic devices to address prebiotic chemistry problems and their principal objective is to investigate processes that can lead to Darwinian evolution (Table 2). The studies are described below.

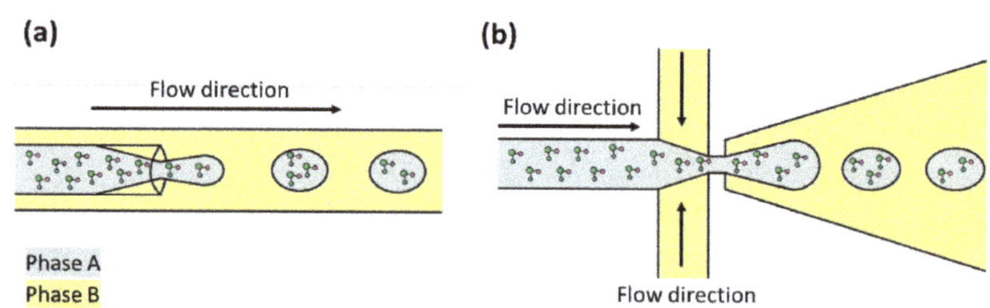

Figure 3. Encapsulation of molecules with droplet-based microfluidic devices. Phases A and B are immiscible liquids. (**a**) Flow-focusing geometry. (**b**) Cross-junction geometry.

Doran et al. [92] designed a microfluidic device consisting of a droplet generator, an incubation microchamber, and a droplet size sorter. In the study, the authors assume that larger droplets with high osmolarity are the most suitable to suffer evolutive processes. So, in the device, droplets with catalytic networks would be created, then incubated, and those with the highest polymer concentration (i.e., the biggest) would be selected for another incubation cycle. To test the device two kinds of droplets were placed in the incubation chamber: one with water and the other with glycyl-glycine. The results show that glycyl-glycine droplets grew at the expense of water droplets.

In a different experiment, Ameta et al. [93] used a device to create microdroplets loaded with RNA networks that catalyze their own formation. They analyzed the accumulation of products (growth) and the fraction of the networks' catalytic species (composition) as a function of their topology. Furthermore, the authors studied the reproduction (interpreted as species accumulation) and the variation (changes in the fraction of species). As a result, they found that strong variations surge from the new catalytic species that perturb networks with weak connections and that growth increases with the global connectivity of networks.

There is another way to create droplets without amphiphilic molecules or non-miscible fluids. The use of solutions of two polyelectrolytes, with opposed charges, leads to spontaneous liquid–liquid phase separation and the formation of droplets called complex coacervates [94]. This kind of coacervates can be created by microfluidic devices, as shown by van Swaay et al. [95]. Complex coacervates were made from poly diallyl dimethylammonium chloride (PDDA) and ATP or PDDA and carboxymethyl-dextran (CM-dextran). In the experiment, DNA oligonucleotides were added to the coacervates flow. The aim was to detect if the genetic information (DNA oligonucleotides) could be immobilized in the coacervates or be transferred. The authors showed that two populations of coacervates, with different DNA oligonucleotides, can coexist for up to 48 h without information exchange.

The synthesis of prebiotic molecules can be achieved in microdroplets. A study by Ju et al. [96] shows that microdroplets favor the phosphorylation of adenosine, guanosine, uridine, and cytidine in the presence of KH_2PO_4 (as a phosphate source), at ambient conditions and without a catalyst. Normally, this synthesis cannot be achieved in bulk aqueous solution, but the formation of microdroplets gave a negative ΔG, allowing the synthesis to take place. Further, under the same conditions in microdroplets, the polymerization of nucleotides in dimers is possible.

In the studies presented here, the microdroplets do not have properly a membrane, although it is possible to create it in microfluidic devices. Many studies are focused on the formation of liposomes, including encapsulated molecules with pharmaceutical applications [97]. Majumder et al. [89] have described droplet microfluidic methods to create membranes for the creation of artificial cells for the biomedical field. The same method can be applied to fabricate prebiotic membranes. For example, it would be interesting to include amphiphilic molecules with prebiotic relevance. Another possibility is the simultaneous use of amphiphilic molecules and another variety of prebiotic molecules that can be encapsulated (e.g., amino acids, nitrogenous bases; carboxylic acids, sugar precursors, or their polymers). In this way, it would be possible to analyze the occurrence of synergies between organic molecules that make up protocells. These synergies are important for protocell formation, protometabolism, replication, and evolution of these structures. By employing microfluidic devices, various systems with different combinations of molecules can be tested [86]. At the same time, microdroplets offer the potential to analyze reaction processes in confined areas, including microenvironments and different conditions, both from each other and from the outside medium as demonstrated by Ameta et al. [93]. These processes are important in prebiotic chemistry since they can give rise to Darwinian evolution.

Compartmentalization is a feature of living systems, and some authors suggest that it is important for the concentration and/or protection of organic molecules synthesized in prebiotic environments. Experimentally, compartmentalization can be spawned by fabricating microdroplets with encapsulated molecules. There are several methods to prepare them without the use of microdroplet devices. These methods already have extensive use not only in prebiotic chemistry but also in other fields and are older than microdroplet devices. This is a probable reason why compartmentalization in a prebiotic chemistry context is generally studied by using "traditional methods". Droplet microfluidics usage is being explored recently by only a few groups. However, there are some advantages of microfluidic devices over traditional methods, such as: (i) a massive production of monodisperse microdroplets, (ii) the precise generation and repeatability of droplets operation, and (iii) the possibility of encapsulating molecules into the droplets and use them as microreactors [87,88,98].

Table 2. Prebiotic chemistry experiments in droplet-based microfluidic devices.

Experiment Type	Device Description	Experimental	Findings	Reference
Cellular compartmentalization models	Double droplet generator, flow-focusing channel arrangement, microchamber, droplet size sorting (pressurized air), droplet spitter, and fuser.	Water and glycyl-glycine droplets created and incubated in the microchamber.	Osmotic exchange between water and glycyl-glycine droplets. Glycyl-glycine grows at expense of water droplets.	[92]
Cellular compartmentalization models	Flow-focusing geometry.	DNA oligonucleotides labeled (fluorescein or cyanine) added to a flow of coacervates (poly(dialyl dimethylammon ium chloride)), ATP or carboxymethyl-dextran].	Two populations of DNA oligonucleotides coexist near each other without genetic information exchange for up to 48 h.	[95]
Reaction networks evolution	Four devices: (i) "T" junction, (ii) droplet fusion by electrocoalescence and incubation, (iii) droplet split, and (iv) droplet fusion by electrocoalescence and incubation.	Microdroplets with catalytic RNAs fragments and hairpin RNA reporters, incubated (48 °C/1 h) and split. Labeled with barcoded DNA and incubated (60°/1 h). RNA fraction of each droplet measured by next-generation DNA barcoded sequencing.	The final fraction of RNA species depends on the composition of the network. Networks with greater yield show fewer perturbations.	[93]
Prebiotic synthesis	Two devices: (i) sparyer similar to ESSI [a], connected MS [b], (ii) cylindrical chamber, ceramic atomizer, heating tape.	Nebulization of a solution (adenosine, guanosine, uridine, cytidine, KH_2PO_4) with the two devices.	Produced microdroplets have negative ΔG, allowing ribonucleosides phosphorylation and polymerization under ambient conditions.	[96]

[a] Electrosonic spray ionization, [b] mass spectrometry.

2.3. Prebiotic Chemistry Experiments in Microfluidic Devices with Microchambers

Proposed terrestrial environments for life appearance are diverse, from the primitive ocean, hydrothermal vents (submarine and subaerial), subaerial exposures, water bodies, and oceanic ice [99]. Each environment favors particular and relevant mechanisms for prebiotic chemistry but disfavors others. Nevertheless, the presence of microenvironments is key for prebiotic reactions, since they exhibit unique traits that make them different from the whole environment.

Stüeken et al. [99] highlight the possible microenvironments in the early Earth and how these environments could have contributed to the synthesis and accumulation of organic compounds. They propose that each microenvironment has attributes that allow processes different from the others. In addition, the authors say that life could have required the contribution of all those microenvironments to arise. This idea implies that global transport was needed to communicate the microenvironments and interchange the feedstock for the emergence of life.

Microfluidic devices with microchambers (millimeter size) represent an approach to model processes that could occur in microenvironments. Dieter Braun's research group is using these devices to study the microscale effects that may have occurred in pore rocks at hydrothermal environments. It is worth mentioning the series of papers in which the effects of thermophoresis (particle movement driven by thermal gradients) on DNA are analyzed. To study this phenomenon the authors constructed devices called "thermal traps" (TT), which are made of various materials and allow the creation of thermal gradients (Table 3). The thermal gradient is created by: (i) heating a capillary with an IR laser [100]; (ii) heating one side and cooling the other side of the capillary [101], or (iii) heating the two sides of a microchamber at different temperatures [102]. In TTs, convection currents are generated in which the flow is upward on the hot side and downward on the cold side of the chamber. If dissolved molecules are present in a thermal trap, they are pushed toward the cold side and accumulate at the bottom of the chamber by the effects of both thermophoresis and gravity (Figure 4). This effect is called thermogravitational accumulation [103]. If the chamber contains gas besides the solution of molecules, the molecules accumulate in the gas-water interface (Figure 4) [104]. By placing DNA molecules in solution in TTs, Braun's group was able to observe their accumulation [100], replication [105], elongation [106] and the replication and subsequent selection of long DNA chains over short [101].

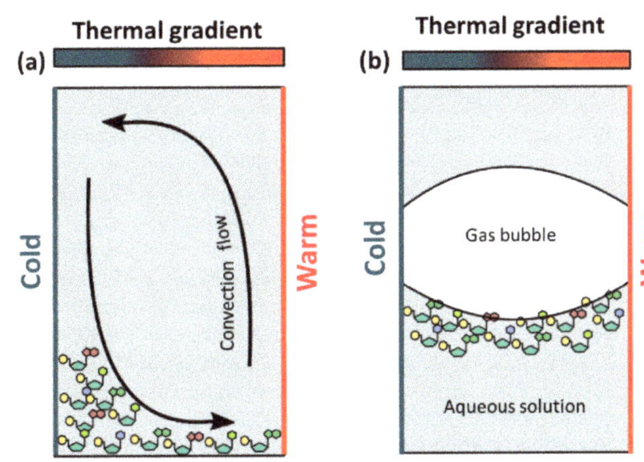

Figure 4. Accumulation of molecules in thermal traps: (**a**) thermal trap filled with an aqueous solution: molecules accumulated in the bottom of the capillary (by gravitational forces) and on the cold side (by convection induced by the thermal gradient); (**b**) thermal trap filled with an aqueous solution and gas: molecules tend to accumulate at the gas–water interface.

In more recent works, Braun's group introduced the existence of interfaces and other molecules, besides DNA, in their experiments. For example, Ianeselli et al. [107] employed a Teflon microchamber as a TT, loaded it with gas and water, and then applied heat. This induced the formation of a microscale analog of the water cycle and the consequent formation of gas–liquid interfaces in the microchamber. The authors performed further experiments with a DNA–saline buffer solution instead of pure water and observed fluctuations in the salt concentration in the gas–water interface that promote DNA string division below its melting point temperature. Further experiments showed that DNA replication was possible in the microenvironments created in the devices. Ianeselli et al. [108] observed the creation of two ideal settings for DNA replication in a microfluidic chamber: one favorable for the denaturation of DNA or RNA (droplets with lower pH and salt concentration) and the other with characteristics for replication (higher salt concentration and neutral pH). On the other hand, Morasch et al. [104] used a corrugated microchamber as a TT to create air–water interfaces. In these experiments, besides DNA, they included other molecules such as RNA, ribose aminooxazoline, and cytidine nucleotides. Various relevant mechanisms in the prebiotic context were observed: an increase in RNA catalytic activity in the gas–water interface, DNA and RNA accumulation up to the formation of hydrogels, ribose aminooxazoline crystallization, and phosphorylation of cytidine nucleotides.

In addition to DNA and RNA molecules, Braun's group has also studied larger structures such as vesicles and coacervates in thermal gradients. Morasch et al. [104] introduced oleic acid vesicles in a microchamber filled with a DNA solution and gas. They noted the vesicle aggrupation in clusters and the DNA encapsulation inside the vesicles when applying a thermal gradient. In another experiment, Ianeselli et al. [102] explored the behavior of coacervates under a thermal gradient. Coacervates tend to accumulate and grow by fusion on the interface gas–water and the forces in the chamber (heat and gas bubbles movement) promote the coacervate segregation in two different populations: oligonucleotide-polypeptide coacervates, in the aqueous bulk; and sugar-oligonucleotide-polypeptide coacervates, in the gas–water interface.

Besides studying thermal gradients, microfluidic devices have been used to approach the chirality bias in biomolecules in a novel way. Sun et al. [109] used a microfluidic device, consisting of ten pairs of inclined chambers, to imitate rock micropores in hydrothermal systems, within which microvortices can form. The authors concluded that laminar microvortices could induce enantioselectivity in supramolecular systems composed of non-chiral molecules.

Compartmentalization was also explored using microfluidic devices with microchambers. In a device composed of serial microchambers and microchannels, Sugiyama et al. [110] caught and selected liposomes by size. Afterwards, the liposomes were exposed to a uranine and fructose solution, resulting in the encapsulation of the molecules in the liposomes, even against the concentration gradient. The authors argue that this mechanism can provide clues in understanding the continuous development of protocells in early Earth.

The experiments performed by Braun's group are relevant because they highlight the importance of prebiotic microenvironments, where processes not observed at large scales can occur. All the works of this research group are within the RNA world theory framework, nevertheless, it would be interesting to perform experiments with other types of molecules such as amino acids, carboxylic acids, sugars, or their precursors. This will allow us to observe their behavior in microenvironments and to test if it is possible to observe processes, such as concentration, polymerization, or catalysis. On the other hand, the experiments performed by Sun et al. [109] and Sugiyama et al. [110] show us how microfluidic with microchamber devices also can be used in novel ways to approach various issues related to prebiotic chemistry such as chirality and compartmentalization.

Table 3. Prebiotic chemistry experiments in microfluidic devices with microchambers.

Experiment Type	Device Description	Experimental	Findings	Reference
Simulation of Pores in rocks in hydrothermal systems	Thermal trap. A microchamber heated by an infrared laser.	Fluorescent-stained DNA heated in the capillary, under a temperature gradient.	DNA thermal diffusion coefficient was measured. DNA accumulates in the lower part of the chamber, near the heating spot (from nmol/L to µmol/L).	[100]
Pores in rocks in hydrothermal systems	Thermal trap. Borosilicate capillary embedded in immersion oil and inserted between a silicon plate and a sapphire cover. An infrared laser as heat source.	PCR [a] solution and DNA oligonucleotide templates (random sequences) stained with fluorescent dye heated by a temperature difference of 27 K.	Temperature gradients trigger replication and accumulation of short DNA by thermophoresis and convection.	[105]
Pores in rocks in hydrothermal systems	Thermal trap. Borosilicate capillary embedded in immersion oil, inserted between silicon plate and sapphire cover. Infrared laser as heat source.	Double-chain DNA segments capable of reversible union by hybridization heated.	Thermal gradient promotes DNA accumulation and polymerization	[106]
Pores in rocks in hydrothermal systems	Thermal trap. Borosilicate capillary inserted between two metallic plates, temperature controlled heated on one side and cooling the other side.	DNA, Taq polymerase, fluorescent dye heated with temperature gradients (38 °C to 71 °C). DNA in PCR buffer (6 µm/s flow) heated with gradients (36–73 °C; 61–94 °C).	Temperature gradients promote replication of DNA oligonucleotides with a sequence length. Long-over short sequences are preferred.	[101]
Pores in rocks in hydrothermal systems	Three inlets, two outlets, ten pairs of asymmetric inclined microchambers, allowing microvortice formation.	BTAC [b] and DMF [c] introduced (1 mL/h) in the central inlet. DMF/H_2O is introduced (30 mL/h; 40 °C) at the side inlets. $TPPS_4$ [d], H_2SO_4, and H_2O injected (30 mL/h) in the side inlets, and in central inlet, a C_2mim^{+} [e] and HCl dissolution (1 mL/h).	Chiral microvortices created in the microchambers induce hydrodynamic selection of enantiomers in supramolecular systems composed of non-chiral molecules.	[109]
Pores in rocks in hydrothermal systems	Thermal trap. Microchamber made of Teflon and placed between a sapphire plate (heated) and a silicon plate (cooled).	Microchamber filled with air and dissolution of DNA chains (labeled with a chromophore) on a salt buffer (EDTA and NaCl), a temperature gradient from 9 °C to 15 °C was applied.	Formation of a mini water cycle analog that induces fluctuations in salt concentrations in the air–water interface promoting periodic separation of DNA strands below their melting temperature.	[107]
Pores in rocks in hydrothermal systems	Thermal traps. Corrugated microchambers (PETG plastic, UV-curable resin, or Teflon) sandwiched between sapphire (heated) and silicon (cooled) plates.	Temperature gradients applied. Devices filled with gas and solutions of DNA, RNA, ribozymes, ribose aminooxazoline, cytidine nucleosides, and monoammonium phosphate or vesicles (oleic acid or 1,2-Dioleoyl-sn-glycero-3-phosphocholine and oligonucleotides).	DNA and RNA form hydrogels and ribozymes, increase catalytic activity at the gas–water interface. Nucleotide encapsulation in vesicles, ribose aminooxazoline crystallization, and cytidine nucleosides phosphorylation.	[104]

Table 3. *Cont.*

Experiment Type	Device Description	Experimental	Findings	Reference
Cellular compartmentalization	Arrangement of serial channels and microchambers.	Liposomes of phospholipids, cholesterol, and fluorescent dye doped with fructose trapped in the microchambers, exposed to a uranine/fructose and fluorescein-12-adenosine triphosphate solution (ATP analog) different flows and pH.	Fructose, uranine, and ATP analog accumulation in the liposomes, even against concentration gradient (between liposome and exterior).	[110]
Pores in rocks in hydrothermal systems	Thermal traps. Triangular PTFE plastic sheets placed between sapphire (heated) and silicon (cooled) plates.	Different coacervates compositions (CM-Dex [f] or ATP, with pLys [g] or PDDA [h] in Na^{++}bicine or tris buffer), temperature gradient, gas volume, and microchamber thickness. Some experiments included RNA.	The fusion, accumulation, and division of coacervates occurred at the gas–water interface. Two coacervate populations can be separated: in the gas–water coacervates of RNA, CM-Dex, and pLys; in the bulk, coacervates of RNA and pLys.	[102]
Pores in rocks in hydrothermal systems	Thermal trap. Cut Teflon sheet placed between a silicon plate (covered with Teflon) and a sapphire plate. Thermal gradient by differential heating of plates.	Microchamber was filled with CO_2 at different pressures and solutions Lysosensor Yellow/blue dye, RNA, $MgCl_2$, and buffer Tris; and DNA nucleotides, Taq polymerase, complementary primers, $MgCl_2$, Tris Buffer, KCl, and $(NH_4)_2SO_4$. Temperature gradients (5 °C to 17 °C).	Dew cycle generation. In dewRNA or DNA, melting is favored, in the bulk solution. Emergence of larger DNA strands was observed.	[108]

[a] Polymerase chain reaction, [b] tris(ethyl cinnamate) benzene-1,3,5-tricarboxamide, [c] N,N-dimethylformamide, [d] tetra-(4-sulfonatophenyl) porphyrin, [e] 1-ethyl-3-methylimidazolium cations, [f] carboxymethyl-dextran, [g] polylysine, [h] polydiallyldimethylammonium chloride.

3. Conclusions

Although prebiotic chemistry experiments using microfluidic devices are few, they have been increasing in recent years. Thanks to the characteristics of microfluidic devices, it is possible that they can be used to simulate different prebiotic environments, such as submarine hydrothermal vents systems in early Earth or rock pores. Additionally, other questions can be approached, as demonstrated by using microfluidic devices to study the chirality problem, compartmentalization phenomena, and catalytic networks that can derive Darwinian evolution. The great potential of microfluidic devices to be used in prebiotic chemistry experiments has yet to be explored. It should be interesting to study more conditions or factors for the experiments that have already been performed. For example, using Y-shaped microfluidic devices, it is possible to study the mineral membrane formation with different physical and chemical conditions such as pH gradient, temperature, chemical composition, or reactant concentrations. In addition, factors such as the presence of relevant organic molecules in mineral membrane formation to evaluate their role as concentrators or catalyzers should be explored. Microdroplet devices that have several applications in the biomedical field also can be used in prebiotic chemistry studies, as shown in the experiments involving reaction networks and compartmentalization. There is great potential to use droplet devices to perform more experiments involving different molecules or conditions. For example, using amphiphilic molecules with prebiotic relevance, to study the probable formation or behavior of what may have been the first membranes. As for studies using microchambers, there is no doubt about their usefulness to study the processes that nucleic acids and some other molecules can undergo in hydrothermal microenvironments. However, it would be also interesting to investigate the processes that other prebiotic molecules, or their precursors, could undergo in those and other kinds of environments.

Perhaps a disadvantage of using microfluidic devices is the individual technical training and laboratory equipment required to manufacturing, operate, and monitor the devices (e.g., manufacture equipment, pumps, and microscopes for observation). However, there are companies dedicated to fabric custom microfluidic devices for diverse purposes. In addition, new methods with low-cost fabrication and materials have already been proposed (e.g., [72,111–115]). The potential use of microfluidic devices constitutes a relatively new approach to prebiotic chemistry, although the present study proves it is rather useful. It would be interesting to use different types of microfluidic devices to respond to key prebiotic chemistry questions. This review pretends to encourage researchers to use microfluidic devices and think about how to use them to perform prebiotic chemistry experiments in novel ways.

Author Contributions: Conceptualization, K.M.L.-M. and M.C.-G.; writing—original draft preparation, K.M.L.-M.; writing—review and editing, M.C.-G., E.M.-M. and L.F.O.; supervision, M.C.-G.; funding acquisition, M.C.-G., E.M.-M. and L.F.O. All authors have read and agreed to the published version of the manuscript.

Funding: This research was funded by CONACyT (Projects A1-S-25341 and 284240), DGAPA-PAPIIT (IN111720), and by the Spanish Ministry of Science and Innovation/State Agency of Research MCIN/AEI (Project PID2019-104205GB-C21).

Institutional Review Board Statement: Not applicable.

Informed Consent Statement: Not applicable.

Data Availability Statement: Not applicable.

Acknowledgments: This research is part of K.M.L.-M.'s dissertation. She acknowledges CONACyT for the Ph.D. scholarship granted (355850) and the Posgrado en Ciencias de la Tierra for the academic support.

Conflicts of Interest: The authors declare no conflict of interest.

References

1. Whitesides, G.M. The Origins and the Future of Microfluidics. *Nature* **2006**, *442*, 368–373. [CrossRef] [PubMed]
2. Nguyen, N.-T.; Shaegh, S.A.M.; Kashaninejad, N.; Phan, D.-T. Design, Fabrication and Characterization of Drug Delivery Systems Based on Lab-on-a-Chip Technology. *Adv. Drug Deliv. Rev.* **2013**, *65*, 1403–1419. [CrossRef] [PubMed]
3. Tsui, J.H.; Lee, W.; Pun, S.H.; Kim, J.; Kim, D.-H. Microfluidics-Assisted in Vitro Drug Screening and Carrier Production. *Adv. Drug Deliv. Rev.* **2013**, *65*, 1575–1588. [CrossRef] [PubMed]
4. Hamdallah, S.I.; Zoqlam, R.; Erfle, P.; Blyth, M.; Alkilany, A.M.; Dietzel, A.; Qi, S. Microfluidics for Pharmaceutical Nanoparticle Fabrication: The Truth and the Myth. *Int. J. Pharm.* **2020**, *584*, 119408. [CrossRef]
5. Niculescu, A.-G.; Chircov, C.; Bîrcă, A.C.; Grumezescu, A.M. Fabrication and Applications of Microfluidic Devices: A Review. *Int. J. Mol. Sci.* **2021**, *22*, 2011. [CrossRef]
6. Beebe, D.; Mensing, G.; Walker, G. Physics and Applications of Microfluidics in Biology. *Annu. Rev. Biomed. Eng.* **2002**, *4*, 261–286. [CrossRef]
7. Holmes, D.; Gawad, S. The Application of Microfluidics in Biology. In *Microengineering in Biotechnology*; Springer: Berlin/Heidelberg, Germany, 2010; pp. 55–80.
8. Jakeway, S.C.; de Mello, A.J.; Russell, E.L. Miniaturized Total Analysis Systems for Biological Analysis. *Fresenius J. Anal. Chem.* **2000**, *366*, 525–539. [CrossRef]
9. Mu, X.; Zheng, W.; Sun, J.; Zhang, W.; Jiang, X. Microfluidics for Manipulating Cells. *Small* **2013**, *9*, 9–21. [CrossRef]
10. Weibel, D.B.; Whitesides, G.M. Applications of Microfluidics in Chemical Biology. *Curr. Opin. Chem. Biol.* **2006**, *10*, 584–591. [CrossRef]
11. Marre, S.; Roig, Y.; Aymonier, C. Supercritical Microfluidics: Opportunities in Flow-through Chemistry and Materials Science. *J. Supercrit. Fluids* **2012**, *66*, 251–264. [CrossRef]
12. Kurniawan, Y.S.; Imawan, A.C.; Rao, S.R.; Ohto, K.; Iwasaki, W.; Miyazaki, M. Jumina Microfluidics Era in Chemistry Field: A Review. *J. Indones. Chem. Soc.* **2019**, *2*, 7. [CrossRef]
13. Verpoorte, E. Microfluidic Chips for Clinical and Forensic Analysis. *Electrophoresis* **2002**, *23*, 677–712. [CrossRef]
14. Horsman, K.M.; Bienvenue, J.M.; Blasier, K.R.; Landers, J.P. Forensic DNA Analysis on Microfluidic Devices: A Review. *J. Forensic Sci.* **2007**, *52*, 784–799. [CrossRef]
15. Bruijns, B.; Van Asten, A.; Tiggelaar, R.; Gardeniers, H. Microfluidic Devices for Forensic DNA Analysis: A Review. *Biosensors* **2016**, *6*, 41. [CrossRef] [PubMed]
16. Rivet, C.; Lee, H.; Hirsch, A.; Hamilton, S.; Lu, H. Microfluidics for Medical Diagnostics and Biosensors. *Chem. Eng. Sci.* **2011**, *66*, 1490–1507. [CrossRef]
17. Sackmann, E.K.; Fulton, A.L.; Beebe, D.J. The Present and Future Role of Microfluidics in Biomedical Research. *Nature* **2014**, *507*, 181–189. [CrossRef]
18. Van den Berg, A.; Segerink, L. *Microfluidics for Medical Applications*; Nanoscience & Nanotechnology Series; The Royal Society of Chemistry: London, UK, 2014; p. 303. ISBN 978-1-84973-637-4.
19. Yamada, K.; Shibata, H.; Suzuki, K.; Citterio, D. Toward Practical Application of Paper-Based Microfluidics for Medical Diagnostics: State-of-the-Art and Challenges. *Lab A Chip* **2017**, *17*, 1206–1249. [CrossRef]
20. Alhalaili, B.; Popescu, I.N.; Rusanescu, C.O.; Vidu, R. Microfluidic Devices and Microfluidics-Integrated Electrochemical and Optical (Bio) Sensors for Pollution Analysis: A Review. *Sustainability* **2022**, *14*, 12844. [CrossRef]
21. Kudr, J.; Zitka, O.; Klimanek, M.; Vrba, R.; Adam, V. Microfluidic Electrochemical Devices for Pollution Analysis–A Review. *Sens. Actuators B: Chem.* **2017**, *246*, 578–590. [CrossRef]
22. Poenar, D.P. Microfluidic and Micromachined/MEMS Devices for Separation, Discrimination and Detection of Airborne Particles for Pollution Monitoring. *Micromachines* **2019**, *10*, 483. [CrossRef]
23. Dittrich, P.S.; Manz, A. Lab-on-a-Chip: Microfluidics in Drug Discovery. *Nat. Rev. Drug Discov.* **2006**, *5*, 210–218. [CrossRef] [PubMed]
24. Riahi, R.; Tamayol, A.; Shaegh, S.A.M.; Ghaemmaghami, A.M.; Dokmeci, M.R.; Khademhosseini, A. Microfluidics for Advanced Drug Delivery Systems. *Curr. Opin. Chem. Eng.* **2015**, *7*, 101–112. [CrossRef] [PubMed]
25. Sun, J.; Warden, A.R.; Ding, X. Recent Advances in Microfluidics for Drug Screening. *Biomicrofluidics* **2019**, *13*, 061503. [CrossRef] [PubMed]
26. Jaradat, E.; Weaver, E.; Meziane, A.; Lamprou, D.A. Microfluidics Technology for the Design and Formulation of Nanomedicines. *Nanomaterials* **2021**, *11*, 3440. [CrossRef] [PubMed]
27. Fang, X.; Chen, H.; Yu, S.; Jiang, X.; Kong, J. Predicting Viruses Accurately by a Multiplex Microfluidic Loop-Mediated Isothermal Amplification Chip. *Anal. Chem.* **2011**, *83*, 690–695. [CrossRef] [PubMed]
28. Yamaguchi, N.; Torii, M.; Uebayashi, Y.; Nasu, M. Rapid, Semiautomated Quantification of Bacterial Cells in Freshwater by Using a Microfluidic Device for on-Chip Staining and Counting. *Appl. Environ. Microbiol.* **2011**, *77*, 1536–1539. [CrossRef]
29. Nilghaz, A.; Mousavi, S.M.; Li, M.; Tian, J.; Cao, R.; Wang, X. Paper-Based Microfluidics for Food Safety and Quality Analysis. *Trends Food Sci. Technol.* **2021**, *118*, 273–284. [CrossRef]
30. Baret, J.-C. Surfactants in Droplet-Based Microfluidics. *Lab A Chip* **2012**, *12*, 422–433. [CrossRef]
31. Chován, T.; Guttman, A. Microfabricated Devices in Biotechnology and Biochemical Processing. *Trends Biotechnol.* **2002**, *20*, 116–122. [CrossRef]

32. Vaccari, L.; Birada, G.; Grenci, G.; Pacor, S.; Businaro, L. Synchrotron Radiation Infrared Microspectroscopy of Single Living Cells in Microfluidic Devices: Advantages, Disadvantages and Future Perspectives. *J. Phys. Conf. Ser.* **2012**, *359*, 012007. [CrossRef]
33. Bragheri, F.; Martínez Vázquez, R.; Osellame, R. Chapter 12.3—Microfluidics. In *Three-Dimensional Microfabrication Using Two-Photon Polymerization*, 2nd ed.; Baldacchini, T., Ed.; Micro and Nano Technologies; William Andrew Publishing: Oxford, UK, 2020; pp. 493–526. ISBN 978-0-12-817827-0.
34. Neumann, C.; Rapp, B.E. Fluidic Platforms and Components of Lab-on-a-Chip Devices. In *Lab-on-a-Chip Devices and Micro-Total Analysis Systems*; Springer: Berlin/Heidelberg, Germany, 2015; pp. 83–139.
35. Lee, C.-Y.; Chang, C.-L.; Wang, Y.-N.; Fu, L.-M. Microfluidic Mixing: A Review. *Int. J. Mol. Sci.* **2011**, *12*, 3263–3287. [CrossRef] [PubMed]
36. De Jong, J.; Lammertink, R.G.; Wessling, M. Membranes and Microfluidics: A Review. *Lab A Chip* **2006**, *6*, 1125–1139. [CrossRef] [PubMed]
37. Strohmeier, O.; Emperle, A.; Roth, G.; Mark, D.; Zengerle, R.; von Stetten, F. Centrifugal Gas-Phase Transition Magnetophoresis (GTM)–a Generic Method for Automation of Magnetic Bead Based Assays on the Centrifugal Microfluidic Platform and Application to DNA Purification. *Lab A Chip* **2013**, *13*, 146–155. [CrossRef] [PubMed]
38. Huh, Y.S.; Choi, J.H.; Park, T.J.; Hong, Y.K.; Hong, W.H.; Lee, S.Y. Microfluidic Cell Disruption System Employing a Magnetically Actuated Diaphragm. *Electrophoresis* **2007**, *28*, 4748–4757. [CrossRef]
39. Kutter, J.P.; Mogensen, K.B.; Klank, H.; Geschke, O. Chapter 4— Microfluidics–Components. In *Microsyst. Eng. Lab-A-Chip Devices*; Geschke, O., Klank, H., Telleman, P., Eds.; Wiley-VCH Verlag GmbH & Co. KGaA: Weinheim, Germany, 2003; pp. 39–77. ISBN 9783527601653.
40. Nielsen, J.B.; Hanson, R.L.; Almughamsi, H.M.; Pang, C.; Fish, T.R.; Woolley, A.T. Microfluidics: Innovations in Materials and Their Fabrication and Functionalization. *Anal. Chem.* **2019**, *92*, 150–168. [CrossRef]
41. Campbell, S.B.; Wu, Q.; Yazbeck, J.; Liu, C.; Okhovatian, S.; Radisic, M. Beyond Polydimethylsiloxane: Alternative Materials for Fabrication of Organ-on-a-Chip Devices and Microphysiological Systems. *ACS Biomater. Sci. Eng.* **2021**, *7*, 2880–2899. [CrossRef]
42. Nguyen, N.-T.; Wereley, S.T.; Shaegh, S.A.M. *Fundamentals and Applications of Microfluidics*, 3rd ed.; Artech House: Boston, MA, USA, 2019; ISBN 978-1-63081-365-9.
43. Newman, J. *Physics of the Life Sciences*; Springer Science & Business Media: Berlin/Heidelberg, Germany, 2008; ISBN 978-0-387-77258-5.
44. Mott, R.L. *Mecánica de Fluidos Aplicada*, 4th ed.; Prentice Hall Hispanoamericana: Edo. Mex, Mexico, 1996; ISBN 978-968-880-542-8.
45. Wang, Q.; Steinbock, O. Materials Synthesis and Catalysis in Microfluidic Devices: Prebiotic Chemistry in Mineral Membranes. *ChemCatChem* **2020**, *12*, 63–74. [CrossRef]
46. Corliss, J.B.; Dymond, J.; Gordon, L.I.; Edmond, J.M.; von Herzen, R.P.; Ballard, R.D.; Green, K.; Williams, D.; Bainbridge, A.; Crane, K.; et al. Submarine Thermal Springs on the Galápagos Rift. *Science* **1979**, *203*, 1073–1083. [CrossRef]
47. Baross, J.A.; Hoffman, S.E. Submarine Hydrothermal Vents and Associated Gradient Environments as Sites for the Origin and Evolution of Life. *Orig. Life Evol Biosph.* **1985**, *15*, 327–345. [CrossRef]
48. RUSSELL, M.J.; HALL, A.J. The Emergence of Life from Iron Monosulphide Bubbles at a Submarine Hydrothermal Redox and PH Front. *J. Geol. Soc.* **1997**, *154*, 377–402. [CrossRef]
49. Corliss, J.B.; Baross, J.; Hoffman, S. An Hypothesis Concerning the Relationships between Submarine Hot Springs and the Origin of Life on Earth. *Oceanol. Acta Spec. Issue* **1981**, 59–69.
50. Russell, M.J.; Daniel, R.M.; Hall, A.J. On the Emergence of Life via Catalytic Iron-Sulphide Membranes. *Terra Nova* **1993**, *5*, 343–347. [CrossRef]
51. Wächtershäuser, G. Before Enzymes and Templates: Theory of Surface Metabolism. *Microbiol. Rev.* **1988**, *52*, 452–484. [CrossRef]
52. Wächtershäuser, G. Pyrite Formation, the First Energy Source for Life: A Hypothesis. *Syst. Appl. Microbiol.* **1988**, *10*, 207–210. [CrossRef]
53. Omran, A.; Pasek, M. A Constructive Way to Think about Different Hydrothermal Environments for the Origins of Life. *Life* **2020**, *10*, 36. [CrossRef]
54. Barge, L.M.; Flores, E.; Baum, M.M.; VanderVelde, D.G.; Russell, M.J. Redox and PH Gradients Drive Amino Acid Synthesis in Iron Oxyhydroxide Mineral Systems. *Proc. Natl. Acad. Sci. USA* **2019**, *116*, 4828–4833. [CrossRef]
55. Cartwright, J.H.E.; Russell, M.J. The Origin of Life: The Submarine Alkaline Vent Theory at 30. *Interface Focus* **2019**, *9*, 20190104. [CrossRef]
56. Colín-García, M.; Villafañe-Barajas, S.; Camprubí, A.; Ortega-Gutiérrez, F.; Colás, V.; Negrón-Mendoza, A. *Prebiotic Chemistry in Hydrothermal Vent Systems*; Routledge Handbooks Online: London, UK, 2018; ISBN 978-1-138-06512-3.
57. Little, S.A.; Stolzenbach, K.D.; Von Herzen, R.P. Measurements of Plume Flow from a Hydrothermal Vent Field. *J. Geophys. Res. Solid Earth* **1987**, *92*, 2587–2596. [CrossRef]
58. Feely, R.A.; Lewison, M.; Massoth, G.J.; Robert-Baldo, G.; Lavelle, J.W.; Byrne, R.H.; Von Damm, K.L.; Curl Jr., H. C. Composition and Dissolution of Black Smoker Particulates from Active Vents on the Juan de Fuca Ridge. *J. Geophys. Res. Solid Earth* **1987**, *92*, 11347–11363. [CrossRef]
59. Kelley, D.S.; Karson, J.A.; Früh-Green, G.L.; Yoerger, D.R.; Shank, T.M.; Butterfield, D.A.; Hayes, J.M.; Schrenk, M.O.; Olson, E.J.; Proskurowski, G.; et al. A Serpentinite-Hosted Ecosystem: The Lost City Hydrothermal Field. *Science* **2005**, *307*, 1428–1434. [CrossRef]

60. Kelley, D.S.; Karson, J.A.; Blackman, D.K.; Früh-Green, G.L.; Butterfield, D.A.; Lilley, M.D.; Olson, E.J.; Schrenk, M.O.; Roe, K.K.; Lebon, G.T.; et al. An Off-Axis Hydrothermal Vent Field near the Mid-Atlantic Ridge at 30° N. *Nature* **2001**, *412*, 145–149. [CrossRef] [PubMed]
61. Altair, T.; Borges, L.G.F.; Galante, D.; Varela, H. Experimental Approaches for Testing the Hypothesis of the Emergence of Life at Submarine Alkaline Vents. *Life* **2021**, *11*, 777. [CrossRef]
62. Martin, W.; Russell, M.J. On the Origin of Biochemistry at an Alkaline Hydrothermal Vent. *Philos. Trans. R. Soc. B Biol. Sci.* **2007**, *362*, 1887–1926. [CrossRef] [PubMed]
63. Villafañe-Barajas, S.A.; Colín-García, M. Submarine Hydrothermal Vent Systems: The Relevance of Dynamic Systems in Chemical Evolution and Prebiotic Chemistry Experiments. *Int. J. Astrobiol.* **2021**, *20*, 427–434. [CrossRef]
64. Barge, L.M.; Doloboff, I.J.; Russell, M.J.; VanderVelde, D.; White, L.M.; Stucky, G.D.; Baum, M.M.; Zeytounian, J.; Kidd, R.; Kanik, I. Pyrophosphate Synthesis in Iron Mineral Films and Membranes Simulating Prebiotic Submarine Hydrothermal Precipitates. *Geochim. Cosmochim. Acta* **2014**, *128*, 1–12. [CrossRef]
65. Burcar, B.T.; Barge, L.M.; Trail, D.; Watson, E.B.; Russell, M.J.; McGown, L.B. RNA Oligomerization in Laboratory Analogues of Alkaline Hydrothermal Vent Systems. *Astrobiology* **2015**, *15*, 509–522. [CrossRef]
66. Herschy, B.; Whicher, A.; Camprubi, E.; Watson, C.; Dartnell, L.; Ward, J.; Evans, J.R.G.; Lane, N. An Origin-of-Life Reactor to Simulate Alkaline Hydrothermal Vents. *J. Mol. Evol.* **2014**, *79*, 213–227. [CrossRef]
67. Mielke, R.E.; Russell, M.J.; Wilson, P.R.; McGlynn, S.E.; Coleman, M.; Kidd, R.; Kanik, I. Design, Fabrication, and Test of a Hydrothermal Reactor for Origin-of-Life Experiments. *Astrobiology* **2010**, *10*, 799–810. [CrossRef]
68. Mattia Bizzarri, B.; Botta, L.; Pérez-Valverde, M.I.; Saladino, R.; Di Mauro, E.; García-Ruiz, J.M. Silica Metal Oxide Vesicles Catalyze Comprehensive Prebiotic Chemistry. *Chem.-A Eur. J.* **2018**, *24*, 8126–8132. [CrossRef]
69. Barge, L.M.; Abedian, Y.; Doloboff, I.J.; Nuñez, J.E.; Russell, M.J.; Kidd, R.D.; Kanik, I. Chemical Gardens as Flow-through Reactors Simulating Natural Hydrothermal Systems. *JoVE J. Vis. Exp.* **2015**, *105*, e53015. [CrossRef]
70. Borrego-Sánchez, A.; Gutiérrez-Ariza, C.; Sainz-Díaz, C.I.; Cartwright, J.H.E. The Effect of the Presence of Amino Acids on the Precipitation of Inorganic Chemical-Garden Membranes: Biomineralization at the Origin of Life. *Langmuir* **2022**, *38*, 10538–10547. [CrossRef] [PubMed]
71. Cartwright, J.H.E.; García-Ruiz, J.M.; Novella, M.L.; Otálora, F. Formation of Chemical Gardens. *J. Colloid Interface Sci.* **2002**, *256*, 351–359. [CrossRef]
72. Batista, B.C.; Steinbock, O. Growing Inorganic Membranes in Microfluidic Devices: Chemical Gardens Reduced to Linear Walls. *J. Phys. Chem. C* **2015**, *119*, 27045–27052. [CrossRef]
73. Möller, F.M.; Kriegel, F.; Kieß, M.; Sojo, V.; Braun, D. Steep PH Gradients and Directed Colloid Transport in a Microfluidic Alkaline Hydrothermal Pore. *Angew. Chem. Int. Ed.* **2017**, *56*, 2340–2344. [CrossRef] [PubMed]
74. Wang, Q.; Bentley, M.R.; Steinbock, O. Self-Organization of Layered Inorganic Membranes in Microfluidic Devices. *J. Phys. Chem. C* **2017**, *121*, 14120–14127. [CrossRef]
75. Wang, Q.; Barge, L.M.; Steinbock, O. Microfluidic Production of Pyrophosphate Catalyzed by Mineral Membranes with Steep PH Gradients. *Chem.–A Eur. J.* **2019**, *25*, 4732–4739. [CrossRef]
76. Wang, Q.; Steinbock, O. Chemical Garden Membranes in Temperature-Controlled Microfluidic Devices. *Langmuir* **2021**, *37*, 2485–2493. [CrossRef]
77. Ding, Y.; Batista, B.; Steinbock, O.; Cartwright, J.H.E.; Cardoso, S.S.S. Wavy Membranes and the Growth Rate of a Planar Chemical Garden: Enhanced Diffusion and Bioenergetics. *Proc. Natl. Acad. Sci. USA* **2016**, *113*, 9182–9186. [CrossRef]
78. Sojo, V.; Ohno, A.; McGlynn, S.E.; Yamada, Y.M.A.; Nakamura, R. Microfluidic Reactors for Carbon Fixation under Ambient-Pressure Alkaline-Hydrothermal-Vent Conditions. *Life* **2019**, *9*, 16. [CrossRef]
79. Vasiliadou, R.; Dimov, N.; Szita, N.; Jordan, S.F.; Lane, N. Possible Mechanisms of CO2 Reduction by H2 via Prebiotic Vectorial Electrochemistry. *Interface Focus* **2019**, *9*, 20190073. [CrossRef]
80. Hudson, R.; de Graaf, R.; Rodin, M.S.; Ohno, A.; Lane, N.; McGlynn, S.E.; Yamada, Y.M.A.; Nakamura, R.; Barge, L.M.; Braun, D.; et al. CO2 Reduction Driven by a PH Gradient. *Proc. Natl. Acad. Sci. USA* **2020**, *117*, 22873–22879. [CrossRef] [PubMed]
81. Ganti, T. *The Principles of Life*; OUP Oxford: Oxford, UK, 2003; ISBN 978-0-19-850726-0.
82. Patel, B.H.; Percivalle, C.; Ritson, D.J.; Duffy, C.D.; Sutherland, J.D. Common Origins of RNA, Protein and Lipid Precursors in a Cyanosulfidic Protometabolism. *Nat. Chem* **2015**, *7*, 301–307. [CrossRef] [PubMed]
83. Pohorille, A.; Deamer, D. Self-Assembly and Function of Primitive Cell Membranes. *Res. Microbiol.* **2009**, *160*, 449–456. [CrossRef] [PubMed]
84. Szostak, J.W. An Optimal Degree of Physical and Chemical Heterogeneity for the Origin of Life? *Philos. Trans. R. Soc. B Biol. Sci.* **2011**, *366*, 2894–2901. [CrossRef] [PubMed]
85. Chen, I.A.; Walde, P. From Self-Assembled Vesicles to Protocells. *Cold Spring Harb Perspect Biol.* **2010**, *2*, a002170. [CrossRef]
86. Lopez, A.; Fiore, M. Investigating Prebiotic Protocells for a Comprehensive Understanding of the Origins of Life: A Prebiotic Systems Chemistry Perspective. *Life* **2019**, *9*, 49. [CrossRef] [PubMed]
87. Zhu, P.; Wang, L. Passive and Active Droplet Generation with Microfluidics: A Review. *Lab A Chip* **2017**, *17*, 34–75. [CrossRef] [PubMed]
88. Teh, S.-Y.; Lin, R.; Hung, L.-H.; Lee, A.P. Droplet Microfluidics. *Lab Chip* **2008**, *8*, 198–220. [CrossRef] [PubMed]

89. Majumder, S.; Wubshet, N.; Liu, A.P. Encapsulation of Complex Solutions Using Droplet Microfluidics towards the Synthesis of Artificial Cells. *J. Micromech. Microeng.* **2019**, *29*, 083001. [CrossRef]
90. Zeng, S.; Liu, X.; Xie, H.; Lin, B. Basic Technologies for Droplet Microfluidics. In *Microfluidics: Technologies and Applications*; Lin, B., Ed.; Topics in Current Chemistry; Springer: Berlin/Heidelberg, Germany, 2011; pp. 69–90. ISBN 978-3-642-23050-9.
91. Jia, T.Z.; Caudan, M.; Mamajanov, I. Origin of Species before Origin of Life: The Role of Speciation in Chemical Evolution. *Life* **2021**, *11*, 154. [CrossRef]
92. Doran, D.; Rodriguez-Garcia, M.; Turk-MacLeod, R.; Cooper, G.J.T.; Cronin, L. A Recursive Microfluidic Platform to Explore the Emergence of Chemical Evolution. *Beilstein. J. Org. Chem.* **2017**, *13*, 1702–1709. [CrossRef] [PubMed]
93. Ameta, S.; Arsène, S.; Foulon, S.; Saudemont, B.; Clifton, B.E.; Griffiths, A.D.; Nghe, P. Darwinian Properties and Their Trade-Offs in Autocatalytic RNA Reaction Networks. *Nat. Commun.* **2021**, *12*, 842. [CrossRef] [PubMed]
94. Deng, N.-N. Complex Coacervates as Artificial Membraneless Organelles and Protocells. *Biomicrofluidics* **2020**, *14*, 051301. [CrossRef] [PubMed]
95. van Swaay, D.; Tang, T.-Y.D.; Mann, S.; de Mello, A. Microfluidic Formation of Membrane-Free Aqueous Coacervate Droplets in Water. *Angew Chem. Int. Ed. Engl.* **2015**, *54*, 8398–8401. [CrossRef] [PubMed]
96. Ju, Y.; Zhang, H.; Wang, W.; Liu, Q.; Yu, K.; Kan, G.; Liu, L.; Jiang, J. Aqueous-Microdroplet-Driven Abiotic Synthesis of Ribonucleotides. *J. Phys. Chem. Lett.* **2022**, *13*, 567–573. [CrossRef] [PubMed]
97. Carugo, D.; Bottaro, E.; Owen, J.; Stride, E.; Nastruzzi, C. Liposome Production by Microfluidics: Potential and Limiting Factors. *Sci. Rep.* **2016**, *6*, 25876. [CrossRef]
98. Han, W.; Chen, X. A Review on Microdroplet Generation in Microfluidics. *J. Braz. Soc. Mech. Sci. Eng.* **2021**, *43*, 247. [CrossRef]
99. Stüeken, E.E.; Anderson, R.E.; Bowman, J.S.; Brazelton, W.J.; Colangelo-Lillis, J.; Goldman, A.D.; Som, S.M.; Baross, J.A. Did Life Originate from a Global Chemical Reactor? *Geobiology* **2013**, *11*, 101–126. [CrossRef]
100. Braun, D.; Libchaber, A. Trapping of DNA by Thermophoretic Depletion and Convection. *Phys. Rev. Lett.* **2002**, *89*, 188103. [CrossRef]
101. Kreysing, M.; Keil, L.; Lanzmich, S.; Braun, D. Heat Flux across an Open Pore Enables the Continuous Replication and Selection of Oligonucleotides towards Increasing Length. *Nat. Chem* **2015**, *7*, 203–208. [CrossRef]
102. Ianeselli, A.; Tetiker, D.; Stein, J.; Kühnlein, A.; Mast, C.B.; Braun, D.; Dora Tang, T.-Y. Non-Equilibrium Conditions inside Rock Pores Drive Fission, Maintenance and Selection of Coacervate Protocells. *Nat. Chem.* **2022**, *14*, 32–39. [CrossRef] [PubMed]
103. Agerschou, E.D.; Mast, C.B.; Braun, D. Emergence of Life from Trapped Nucleotides? Non-Equilibrium Behavior of Oligonucleotides in Thermal Gradients. *Synlett* **2017**, *28*, 56–63. [CrossRef]
104. Morasch, M.; Liu, J.; Dirscherl, C.F.; Ianeselli, A.; Kühnlein, A.; Le Vay, K.; Schwintek, P.; Islam, S.; Corpinot, M.K.; Scheu, B.; et al. Heated Gas Bubbles Enrich, Crystallize, Dry, Phosphorylate and Encapsulate Prebiotic Molecules. *Nat. Chem.* **2019**, *11*, 779–788. [CrossRef]
105. Mast, C.B.; Braun, D. Thermal Trap for DNA Replication. *Phys. Rev. Lett.* **2010**, *104*, 188102. [CrossRef] [PubMed]
106. Mast, C.B.; Schink, S.; Gerland, U.; Braun, D. Escalation of Polymerization in a Thermal Gradient. *Proc. Natl. Acad. Sci. USA* **2013**, *110*, 8030–8035. [CrossRef]
107. Ianeselli, A.; Mast, C.B.; Braun, D. Periodic Melting of Oligonucleotides by Oscillating Salt Concentrations Triggered by Microscale Water Cycles Inside Heated Rock Pores. *Angew. Chem. Int. Ed.* **2019**, *58*, 13155–13160. [CrossRef]
108. Ianeselli, A.; Atienza, M.; Kudella, P.W.; Gerland, U.; Mast, C.B.; Braun, D. Water Cycles in a Hadean CO_2 Atmosphere Drive the Evolution of Long DNA. *Nat. Phys.* **2022**, *18*, 579–585. [CrossRef]
109. Sun, J.; Li, Y.; Yan, F.; Liu, C.; Sang, Y.; Tian, F.; Feng, Q.; Duan, P.; Zhang, L.; Shi, X.; et al. Control over the Emerging Chirality in Supramolecular Gels and Solutions by Chiral Microvortices in Milliseconds. *Nat. Commun.* **2018**, *9*, 2599. [CrossRef]
110. Sugiyama, H.; Osaki, T.; Takeuchi, S.; Toyota, T. Hydrodynamic Accumulation of Small Molecules and Ions into Cell-Sized Liposomes against a Concentration Gradient. *Commun. Chem.* **2020**, *3*, 1–10. [CrossRef]
111. Abu-Dawas, S.; Alawami, H.; Zourob, M.; Ramadan, Q. Design and Fabrication of Low-Cost Microfluidic Chips and Microfluidic Routing System for Reconfigurable Multi-(Organ-on-a-Chip) Assembly. *Micromachines* **2021**, *12*, 1542. [CrossRef]
112. Agustini, D.; Bergamini, M.F.; Marcolino-Junior, L.H. Low Cost Microfluidic Device Based on Cotton Threads for Electroanalytical Application. *Lab. Chip.* **2016**, *16*, 345–352. [CrossRef] [PubMed]
113. Annabestani, M.; Esmaeili-Dokht, P.; Fardmanesh, M. A Novel, Low Cost, and Accessible Method for Rapid Fabrication of the Modifiable Microfluidic Devices. *Sci. Rep.* **2020**, *10*, 16513. [CrossRef]
114. Nguyen, H.-T.; Thach, H.; Roy, E.; Huynh, K.; Perrault, C.M.-T. Low-Cost, Accessible Fabrication Methods for Microfluidics Research in Low-Resource Settings. *Micromachines* **2018**, *9*, 461. [CrossRef] [PubMed]
115. Tiwari, S.K.; Bhat, S.; Mahato, K.K. Design and Fabrication of Low-Cost Microfluidic Channel for Biomedical Application. *Sci. Rep.* **2020**, *10*, 9215. [CrossRef] [PubMed]

life

Article

Role of Stress in the Origin of Life

Vladimir Kompanichenko [1,*] and Oleg Kotsyurbenko [2,3]

1 Institute for Complex Analysis of Regional Problems RAS, 679016 Birobidzhan, Russia
2 Institute of Oil and Gas, School of Ecology, Yugra State University, 628012 Khanty-Mansiysk, Russia
3 Network of Researchers on the Chemical Evolution of Life, Leeds LS7 3RB, UK
* Correspondence: kompanv@yandex.ru

Abstract: The article shows the compatibility of the concept of thermodynamic inversion (TI) of the origin of life with the theory of stress in (micro)biology. According to the proposed TI concept, the first microorganisms on Earth were formed through an effective (intensified and purposeful) response of organic microsystems to incessant oscillations of physicochemical parameters (i.e., to periodic stress) in a hydrothermal environment. This approach allows us to explain the ability of contemporary microorganisms to respond to stress at the individual and population levels. The ability of microorganisms to effectively react to environmental stress factors is corroborated by a number of molecular and other mechanisms that are described in the article.

Keywords: origin of life; population of microorganisms; hydrothermal system; oscillations in parameters; prebiotic microsystem; stress

Citation: Kompanichenko, V.; Kotsyurbenko, O. Role of Stress in the Origin of Life. *Life* **2022**, *12*, 1930. https://doi.org/10.3390/life12111930

Academic Editor: Paul Higgs

Received: 25 October 2022
Accepted: 17 November 2022
Published: 18 November 2022

1. Introduction

Over the past few decades, prebiotic chemistry has achieved significant success. The initial models, such as coacervates [1] or proteinoid microspheres [2], were quite simple. Now, many of the modern models created and studied within the framework of contemporary prebiotic chemistry are much more complicated. In particular, there are some prebiotic models based on a combination of RNA and lipid vesicles, in which the latter mimic the cellular cytoplasmic membrane [3–5]. Another version of such models is based on the introduction of pre-DNA and pre-protein into lipid vesicles with their subsequent self-replication [6,7].

The culmination of laboratory experiments in the field of prebiotic chemistry is the production of "protocells" and the more complex "artificial cells". Many of them bear some resemblance to living microorganisms. Such models can exhibit catalytic activity and the ability to divide, such that the chains of their nucleic acids can reproduce themselves. However, it should be borne in mind that the DNA chains that are spontaneously synthesized in the models are abiotic and do not contain biological information for billions of years accumulated in the genes of living microorganisms. Sometimes, experimenters designate the processes in such laboratory models as "almost life", although the above-mentioned initial signs of a living state are not self-sustaining, and we therefore cannot consider them biological.

So far, no laboratory experimenters have obtained a population of actually living subcells from prebiotic microsystems, which are characterized by the following key biological features: (1) an extremely high rate of biochemical reactions; (2) behavior adequate to certain environmental conditions or changes, that is, purposeful behavior; (3) closes interaction between cells, both within cell populations and between cell populations in the community. Even advanced scenarios of prebiotic evolution do not explain how populations of microbial cells became active with respect to the environment and how they acquired a tendency towards expansion. This apparent gap between prebiotic models and living cells is a "mysterious leap" in evolution or a missing link [7]. The three modern approaches to the origin of life—complex (or primary metabolism), the RNA world (or primary gene), and the cellular (universal common ancestor)—do not explain the nature of this missing link.

An attempt to consider the nature of the gap between nonliving prebiotic microsystems (or protocells) and living cells on the basis of a different approach was undertaken within the framework of the concept of thermodynamic inversion (TI), or simply the inversion concept of the origin of life [8–13]. It differs from other concepts in that it postulates the need for an effective response of a prebiotic system (a cluster of organic microsystems) to oscillations of physicochemical parameters in the hydrothermal environment. Oscillations should occur with pauses in the high-frequency range, reaching their extreme values from time to time. Such periodic changes are stress factors for prebiotic systems. If the reaction of clusters of organic microsystems to such external influences as periodic stress becomes intensified and purposeful (which is possible under the conditions described below), they transform into the simplest living systems—primary populations of subcells.

The purpose of this article is to correlate the theoretical concepts of TI with respect to the role of periodic stress in the emergence of initial sub-microorganisms from organic microsystems, on the one hand, and modern factual data on the role of stress for the existence and development of a population of microorganisms, on the other.

2. Main Provisions of the Inversion Concept of the Origin of Life

The proposed concept of thermodynamic inversion (TI) focuses on the key moment of nonliving prebiotic microsystems' transition to the first life forms (subcells). Its main theses are summarized in this section. A detailed presentation of the latest version of the inversion concept is given in other works [11,12].

The meaning of thermodynamic inversion as a key thermodynamic transformation during the transition of a prebiotic system to a living state is as follows. It is well known that inanimate natural systems eventually evolve in the direction of increasing entropy, which follows from the second law. It is also known that biological evolution paradoxically proceeds in a thermodynamically opposite (negentropic) direction—with an increase in free energy and information in the system due to a relative decrease in entropy. Within the framework of the concept of thermodynamic inversion, attention has been drawn to the fact that the presence of these contradictory trends means the transition of a nonliving chemical system from existence under conditions of the prevailing contribution of entropy to existence under conditions of the prevalence of the contribution of free energy and information at the moment (or a short period) of the appearance of living state in it, regardless of the constituent molecules. This transition is called thermodynamic inversion, or thermodynamic reversal. It is as if we observe a car moving first in one direction and then in the opposite direction; in that case, we understand that it has turned around somewhere (even if we did not see it). After thermodynamic inversion, the transformed system can no longer exist as an ordinary chemical system. It must build its internal chemical and information processes, as well as interaction with the environment, in such a way as to generate more free energy than can be neutralized by entropy. This way of organizing the system can be designated as "negentropic", and it occurs at the moment of the origin of life. From the point of view of the TI concept, chemical evolution in itself, as a process of the complication of organic matter, is insufficient for the origin of life, since it does not include the transition of the system to the qualitatively different (negentropic) thermodynamic state. This requires a prebiotic chemical revolution—a combination of the presence of evolving organic systems of optimal composition (which has yet to be clarified) and thermodynamic inversion. During the latter, the system is periodically exposed to stress factors (changes in the environment), and through an effective stress response, it builds a negentropic network of (bio)chemical reactions and information processes aimed at maintaining and developing such a state.

The transformation of a cluster of nonliving prebiotic microsystems into a primary population of living subcells takes place at the moment of thermodynamic inversion, when the system obtains the total contribution of free energy prevailing over the contribution of entropy (in the same way as the contribution of information becomes prevailing over the contribution of information entropy). An inanimate prebiotic system passes into the primary form of life (a living subcell) through a thermodynamically intermediate state, which is schematically shown in Figure 1. The optimal composition of prebiotic microsystems most suitable for TI should be clarified by the joint efforts of experimenters and theorists involved in the study of the origin of life. As an option, a model of a three-dimensional organic microsystem was previously proposed, consisting of the main biologically important molecules—lipids, polyamino acids (proteins), and nucleotides [11]. However, consideration of this model is not the purpose of this article.

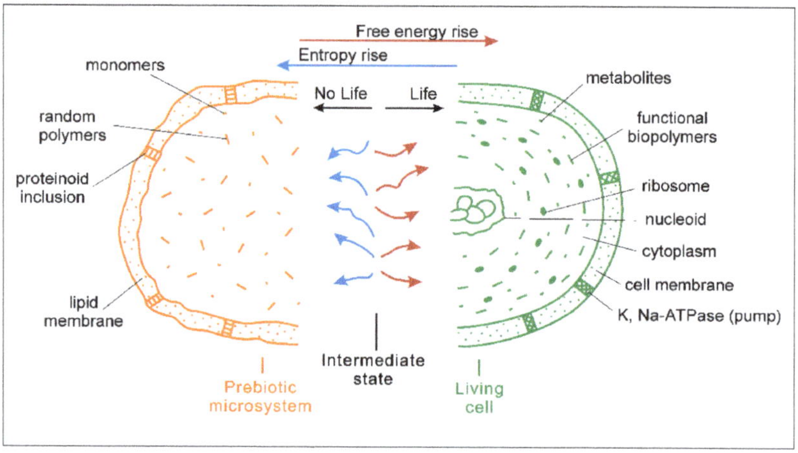

Figure 1. Scheme of change in the thermodynamic state of an inanimate prebiotic microsystem during its transition to the primary living state through an intermediate state.

According to the TI concept, during the transition to life, prebiotic systems are in conditions far from equilibrium, which are initiated by incessant oscillations in physicochemical parameters in the environment, such as pressure, temperature, concentrations of components, and electric potential. An important factor is the periodic change in the scale of oscillations, which leads to a periodic replacement of extreme conditions by pauses. Such regimes are typical of hydrothermal systems, but not typical of the ocean or ice layers. Therefore, it is the hydrothermal fluid that migrates in the Earth's crust up to the surface or ocean floor, which is considered the most suitable area for the emergence of life.

On the left in the Figure 1, in yellow, is a non-living prebiotic microsystem (its half) composed of random polymers, monomers, and simple molecules. In the center is a thermodynamically intermediate state of the microsystem between non-life and life, with relative equality of contributions from chemical reactions producing free energy (red) and entropy (blue). On the right, in green, is a primary living microorganism (its half). The picture shows its evolutionarily advanced form: a modern prokaryotic cell with basic cellular structures, including a nucleoid (circular DNA), ribosomes, and a cell membrane.

Under the described nonequilibrium conditions, thermodynamic inversion occurs due to the enhanced and targeted response of microsystems to high-frequency oscillations ("pumping") of the physicochemical parameters of the environment. Clusters of organic microsystems consisting of lipids, polyamino acids, and other components are formed in the ascending hydrothermal flow due to self-assembly. Continuous interaction between microsystems is supported by different-mode oscillations in the environment (Figure 2). It is well known that abrupt changes occurring in chemical systems under highly nonequilibrium conditions induce extremely fast chemical reactions [14]. Their role in responding to external influences in connection with the process of the beginning of life remains to be studied. The thermodynamic inversion of a cluster of organic microsystems existing in an oscillatory mode occurs in the upper part of hydrothermal channels. Here, the environmental conditions become less extreme (the absolute values of pressure and temperature decrease, as well as the scale of physicochemical fluctuations). Subsequently, the primary living systems that are formed—i.e., populations of the simplest subcells—move out into the ocean, where their biological evolution continues. The proposed mechanism for the emergence of life through thermodynamic inversion with certain variations can be common for different inhabited planets, since the basic terms used—entropy, free energy, and information—are the most fundamental attributes of the Universe.

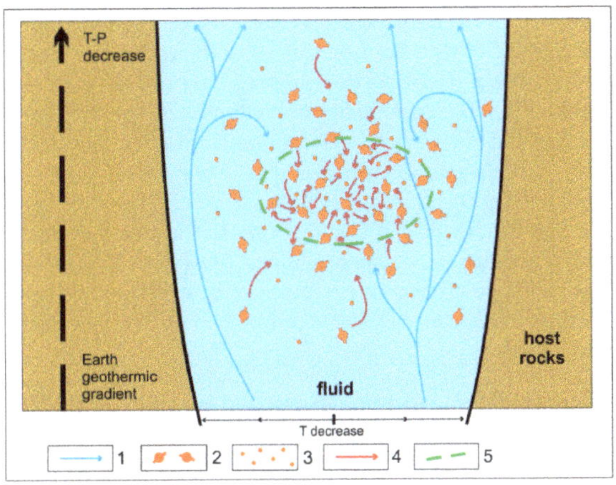

Figure 2. Scheme of the conversion of a cluster of prebiotic microsystems into initial population of living subcells in the upper zone of hydrothermal channel: 1—laminar and convective fluid currents to the surface; 2—interacting organic assemblies (microsystems), composed mainly of lipids, proteins, and nucleotides; 3—disseminated organic molecules; 4—directions of free energy transfer; 5—primary population of living subcells.

A laboratory experiment scheme is proposed to test this approach. It consists of studying the transformations that take place in prebiotic microsystems of different types when they are "pumped" by oscillations of physicochemical parameters of different ranks, including pressure, temperature, pH, Eh, electric potential, etc. [12]. In the course of such experiments, it is necessary to reveal the presence of such a reaction of organic microsystems to physicochemical "pumping", which brings them closer to a living state (strengthening the role of active transport, increasing the degree of homochirality, etc.).

Let us emphasize the main difference between the proposed TI concept and existing alternative concepts. In the latter concepts, it is assumed that living systems arose at some stage of chemical (prebiotic) evolution directed at the consistent chemical complication of organic microsystems (microstructures), with different variations. The need for TI is not provided for in them. According to the approach outlined in the article, it is at the moment of thermodynamic inversion that active metabolic processes are launched in prebiotic microsystems that were inactive earlier (with respect to the environment). According to this understanding, biochemical reactions do not arise by themselves in the course of chemical prebiotic evolution. They are instead launched and integrated through the effective response of a cluster of organic microsystems (or a separate microsystem) to continuous oscillations of physicochemical parameters in the environment, occurring in an optimal mode for the emergence of life.

This difference can be illustrated by the following example. There is a large group of microorganisms belonging to the domains of Archaea and Bacteria living in harsh environments such as extreme temperature and/or pH or high concentrations of metals and/or salt. Such microorganisms are also capable of responding to sudden changes in environmental conditions. A group of Italian scientists has studied and generalized the molecular mechanisms responsible for the survival and adaptation of this type of microorganism, which is related to thermophilic species. The main focus was on their adaptation to toxic metals, with particular emphasis on As (V), As (III), and Cd (II) [15]. The four main mechanisms of heavy metal resistance were found in these microorganisms: (1) an extracellular barrier, forming a selectively permeable system; (2) efflux of metal ions; and (3) enzymatic reduction of metal ions; and (4) intracellular sequestration by small molecule complexing agents or metal-chelating proteins. In more complex mesophilic microorganisms, some of the mechanisms described are already absent. The thermophilic Archaea and Bacteria are known to be very primitive and are at the base of the phylogenetic tree of life on Earth [16,17]. Within the framework of the inversion concept, these facts can naturally be interpreted as follows. At the earliest stages of biological evolution, microorganisms developed their defense mechanisms by effectively resisting extreme environmental conditions. In mesophilic microorganisms, the need for some of the previously formed defense mechanisms had already disappeared, since they existed and exist in less extreme conditions. If we consider these facts in the framework of the concepts of the emergence of life as a result of only chemical evolution without the manifestation of thermodynamic inversion, it becomes unclear at the expense of what and how these protective mechanisms developed in thermophilic microorganisms. At least in scenarios of the origin of life through sequential chemical evolution, such facts are not taken into account, and, accordingly, they do not explain them.

3. Correlation between the Inversion Concept and the Stress Theory in Biology

The TI concept correlates with the fundamental theses of biology in the field of the stress theory, the founder of which is Hans Selye [18]. The key points of the stress theory can be summarized as follows. External influences on a living organism (positive or negative effect of environmental stress factors) cause tension in it (stress itself), followed by a response to stress. Moderate stress factors are necessary to maintain the vitality of the organism, as they allow it to develop an effective response to stress ("stress response") that overcomes environmental pressure. The action of very strong stress factors (distress) leads to the degradation of the organism. Lack of exposure to stress factors (stress = 0) also initiates the degradation of the organism in the long run.

The inversion concept explains these regularities as follows. The fluctuations of the physicochemical parameters of the environment discussed above affect a prebiotic microsystem in the same way as the stress factors of the environment on a living microorganism. When exposed to moderate (optimal) force, the prebiotic microsystem has the ability to develop an enhanced and targeted counteraction (effective stress response), which can transform it into a primary form of life. In the case of a very strong external influence, the prebiotic microsystem is under the overwhelming pressure of entropy (weakened stress response) and cannot be transformed into the primary form of life. In the absence of an external influence (absence of environmental stress factors), a stress response does not develop in the microsystem at all. It also does not have the ability to transition to life.

As a result of fluctuations in physicochemical parameters in the medium, the degree of external pressure (thermodynamically related to the flow of entropy) on the primary form of life will vary all the time. If the environmental pressure begins to exceed the ability of the primary (sub)microorganism to actively exist, then it is forced to return to its original prebiotic state. However, such a reverse transition should initiate the counteraction of an already incipient living microsystem that possesses the sparks of a purposeful, survival-oriented response. It can be assumed that in such conditions of arising "distress", the primary form of life will purposefully reorganize its internal structure and functions in such a way as to survive an unfavorable period and preserve the possibilities for transition to active existence when external pressure subsides. Such "passive" resistance to external pressure can be compared with the state of suspended animation, which is studied by modern microbiology [19–23]. Consequently, it follows from the considered approach to the origin of life on Earth that the microorganism could have initially existed in two states: (a) passive, if the "pressure" from the environment (the action of stress factors) exceeded its ability to effectively counteract; (b) active, if its response to environmental stressors was enhanced and targeted.

Below are three main provisions arising from the inversion concept, which correlate with the fundamental knowledge of modern microbiology.

1. A microorganism is able to pass from a passive state (suspended animation) to an active state (free living form) and vice versa. The current state depends on the conditions in the environment (its "pressure", which within the framework of thermodynamics can be expressed through flow of entropy) and the internal resources of the microorganism.
2. When unfavorable conditions in the environment are approaching, a free-living microorganism is able to purposefully rebuild its structure and functions, consistently preparing for the transition to a passive state (suspended animation) to maintain potential viability.
3. To get out of suspended animation, the (resting) microorganism must receive an impulse (exposure to a stress factor) from the environment, indicating a decrease in the external pressure. In response to this impulse, conserved vital functions are activated, and many reverse sequential transformations are triggered, ensuring its transition to an active state.

4. Microbial Stress Responses

The evolutionary history of the Earth, especially its early period, was associated with various cataclysms, accompanied by a continuous change in geological, water, and atmospheric conditions. During these global transformations, life originated on the planet. In the process of evolution, ancient microorganisms developed mechanisms to resist the adverse environmental factors inherited by subsequent generations of life forms.

Sudden unfavorable changes in conditions either lead to cell death or activate powerful adaptation mechanisms, including damage repair mechanisms, in particular repair of DNA damage, and global reorganization of metabolism. All these transformations are associated with large time and energy resources. This indicates that cells exposed to adverse factors are under stress [24,25].

Moreover, in reality, bacteria, irrespective of natural habitat, are exposed to constant fluctuations in their growth conditions which are usually not optimal. Variations in any environmental parameters can affect the maximum growth rate and, thus, can represent an environmental stress for the microbe. As a result, most bacteria live in a constant state of stress. Furthermore, some microorganisms (extremophiles) view stresses such as extremes in pH or temperature as a lifestyle choice. The ability of microbes to sense and respond (correctly) to such alterations in the environment is crucial to their survival.

The global activation and rearrangement of cellular metabolism in response to stress factors are controlled by the corresponding genes and are accompanied by a global change in the type of gene expression involving cellular hormones called alarmones, which trigger a cascade of sequential reactions using signal transduction systems initiating stress-responsive pathways, some of them being very conserved.

The response to the imposed stress is accomplished by changes in the patterns of gene expression for those genes whose products are required to combat the deleterious nature of the stress. The up-regulation of the transcription of stress responsive genes is achieved by the activation of transcription factors that interact with RNA polymerase to co-ordinate gene expression. One family of transcription factors that play a role in stress resistance is a subunit of RNA polymerase, the sigma factor, which is essential for initiation transcription, playing a key role in promoter recognition. The environmental stress response is controlled by a supramolecular complex known as the stressosome.

The cell is forced to synthesize those enzymes that are necessary under stressful conditions, and hence, genes encoding such enzymes must be activated. Mobile genetic elements (MGE, in prokaryotes these are conjugative plasmids, transposons, and integrons) are involved in the activation of the genetic regulatory system of the organism [26]. MGE is a molecular tool by which the genome can be considered as a special cellular organ designed to detect deviations in the cell from its normal functioning. Moreover, as a result of such detections, the genome can reorganize itself, depending on the needs of the cell under specific conditions. In general, the genome can respond to stress by inducing interspecific transfer of MGE. In this case, the organism experiences the so-called "genomic shock" and the subsequent restructuring of the genome. Plasmid DNA is capable of independent reproduction in the cell and has a special machinery for its transfer to neighboring cells (lateral or horizontal gene transfer), which may cause the formation of new biological species. At the same time, under normal environmental conditions and cellular homeostasis, organisms developed a mechanism for controlling the transfer of MGEs.

Thus, bacteria have developed stress responses, which aim to temporarily increase tolerance limits. These stress responses are often very specific, each specialized for a particular kind of stress. Depending on the nature of the stressor and the type of the damage caused, the cell response may be different. Some stress responses facilitate bacterial transition from a free-living organism to a host-invading pathogen. Bacterial adaptive responses include the development of spores and competence, the activation of motility to more favorable locations, the synthesis of antibiotics and proteases, and changes in energy production systems. The fine-tuning of respiratory electron transfer routes and energy coupling mechanisms play important roles in the ability of bacteria to cope with variations in oxygen and nutrient supply.

The main goal of adaptation mechanisms is to maintain homeostasis in a cell compensating a negative influence of stress factors. For these purposes, the cell synthesizes heat and cold shock proteins and performs a temperature-dependent change in the degree of lipid unsaturation (temperature stress). It uses osmoprotectants and mechanisms of osmoregulations such as K^+ ion influx/glutamate-biosynthesis-coupled systems (osmotic stress) and the modulation of the primary proton pumps as well as the K^+/H^+ and Na^+/H^+ antiporters (acid tolerance), and activates mechanisms of prevention of transporting harmful compounds such as xeno- or antibiotics in the cell. In an oxidative stress response, (2Fe-2S) enzyme centers are usually involved as sensors changing their oxidation state when challenged with oxidative stress. Fe and sulfur compounds are considered to be important components in the metabolic system of the last universal common ancestor (LUCA). This illustrates an ancient nature of some of stress response systems in microorganisms. Another indication of the evolutionary relation of general stress responses is the stress-dependent sigma factor regulating transcription and probably operated in the original ancestor.

In facultative aerobic microorganisms, the transition between aerobic and anaerobic metabolism is accompanied by alterations in the rate, route, and efficiency of pathways of electron flow. The catabolism of fuel molecules is associated with the reduction of NAD^+ to NADH. During the transition to oxygen limited growth and an increased level of NADH built up, as it is less efficiently reoxidized to NAD^+ as a result of reduced aerobic respiration. These roles of NADH and NAD^+ provide a link between energy homeostasis and gene regulation.

Furthermore, microorganisms can control the production of the various respiratory pathway enzymes in response to the availability of alternate electron acceptors. When several e-acceptors are present simultaneously, the more energetically favored acceptor will be used first. Such diversity of metabolic pathways is also an adaptation of a cell to unfavorable conditions. Thus, stress activates the defense mechanisms of the cell, which in turn lead to a rearrangement of the cell metabolism and, often, to a decrease in its level, which contributes to the survival of the cell in conditions that do not allow balanced metabolic activity.

One of the mechanisms of adaptation to stress is the dissociation of bacteria—the splitting of a homogeneous population into variants that differ in morphological, physiological, biochemical and biological properties. The properties of different species and even strains may be different. This creates a phenotypic variety of forms on a single genetic basis. Species that are most adapted to specific environmental conditions survive and develop. The described dissociation is especially typical for pathogenic bacteria. They rearrange their genetic apparatus using various sensory and regulatory mechanisms as a response to their transition to the external environment from a human or animal organism, or after a sharp change in the environmental conditions. This allows them to maintain their viability and to change their virulence and antigenic properties.

Thus, biosystems of various levels of organization can be subject to stress. Accordingly, the methodological approaches to studying the effect of stress and the response of biosystems to it are also different [27]. At the level of the microbial community, the methodology includes incubation experiments with varying physicochemical parameters, the use of various inhibitors and selective substrates to identify the potential of a particular microbial group, and the use of radioactive substrates to study the pathways of decomposition of organic matter [28]. At the microorganism level, experiments include cultivation, the study of enzymatic activity, and genomic and transcriptomic studies and regulatory mechanisms of gene expression under various conditions, followed by bioinformatic processing of the obtained data array and interpretation of the results [29,30]. Shifts in community composition occur due to different biogeochemical capabilities of organisms to resist stress. Thus, the influences of all types of stress operate at both physiological and community composition levels with a linkage between environmental conditions and biogeochemical processes. While the physiological effects likely regulate short-term responses of soil communities and processes, shifts in community composition are likely to regulate them over longer periods [31].

Microbial community is a complex biological system consisting of different microbial groups trophically connected to each other [27]. To cope with stressors in the environment, the microbial community should maintain its basic functional structure according to Le Chatelier's law [27]. It rearranges the trophic interactions and the key microbial groups in order to compensate negative impacts of the environmental changes. It in turn results in redirection of organic matter flows in the community. For example, an anaerobic microbial community producing methane can change its main pathways of organic matter degradation at lower temperature and pH. As a result of such changes, either hydrogen-dependent and acetoclastic methanogenic archaea or homoacetogenic bacteria become the key terminal microbial group, resulting in different product composition.

The diversity of metabolic pathways and taxonomic groups in the community is an important mechanism for withstanding stressors and maintaining their function. The more extreme the external conditions, the less microbial diversity and the more difficult it is for the microbial community to maintain its functionality when the external conditions change, and, hence, the lower is its adaptive potential. Nevertheless, under quite stable conditions, a microbial system consisting of extremophilic microorganisms is formed. Its sustainability is primarily determined by specific conditions of the ecological niche, in which unique non-competitive adaptive survival mechanisms of extremophiles and their proper functionalities are in demand. Thus, microbial community has basic mechanisms of coping with environmental stress factors aiming at keeping its functionality in the environment.

An additional mechanism of resistance to stress can operate in associative mutually beneficial relations between bacteria and plants, for example, in the rhizosphere. In relation to xenobiotics, this mechanism results not only in increasing tolerance to the harmful compound, but also in its joint active removal from the environment [32].

Thus, microorganisms with their huge number in populations, have the mechanisms of physiological variation and transfer of genetic determinants developed by evolution and are in a state of constant adaptive progress in accordance with changing environmental conditions. Activation of the resistance mechanism system in microorganisms, some elements of which are related to very ancient mechanisms of stress response, occurs by signal transduction systems with the participation of transcription factors. These factors bind to specific sites in the DNA molecule and induce the expression of the genes responsible for the cell defense system. The result of this response is a global restructuring, both at the level of an individual cell and at the level of the community, and the activation of speciation processes as a mechanism for increasing the number of more adapted organisms to changing environmental conditions.

5. Conclusions

In a number of publications devoted to the elaboration of the TI concept [9–13], it was substantiated that chemical evolution in itself is insufficient for the emergence of primary life forms. This also requires short-term oscillations in physicochemical parameters in the environment, which exert periodic stress on prebiotic systems. When the stress response of these systems is effective (which is possible under certain conditions), they evolve towards life. From such an understanding of the process of the emergence of life, the conclusion follows that all subsequent (more complex) microorganisms must necessarily respond to stress, and in the event of an effective response, they achieve the opportunity for further development. Sections 3 and 4 of this article show that the ability of microorganisms to respond to stress at the individual and population levels is indeed universal. This confirms the objectivity of the TI concept and its potential to explain the prehistory of the stress phenomenon observed in the modern world of microorganisms. Additionally, within the framework of this approach, the conclusion was substantiated that the main stages of the origin of primary microorganisms are fixed in the anabiotic cycle of modern bacteria and each time they are repeated during the exit of the bacterial population from the anabiotic state [13].

From this approach, it follows that there is a need to move to the next stage of laboratory research on the problem of the origin of life. So far, numerous experiments in prebiotic chemistry have been carried out mainly under stable conditions. Some experiments, which are summarized in [12], were carried out under conditions of reversible oscillations of physicochemical parameters (temperature, humidity) in the medium. These works have demonstrated the progressive complication of organic macromolecules under oscillatory conditions, in comparison with stable ones. As part of the experiments of the next stage, it is proposed to dynamically study the development of the response of clusters of prebiotic microsystems to multi-mode oscillations in parameters in a non-equilibrium environment, including the appearance of induced extremely fast chemical reactions. Through a combination of spontaneous and induced chemical reactions, real proto-biochemical pathways to the emergence of life can be identified. A general approach to experiments of this kind is outlined in [12].

Author Contributions: V.K. wrote Sections 1–3 and 5 of the manuscript. O.K. wrote Section 4 of the manuscript. All authors have read and agreed to the published version of the manuscript.

Funding: This research received no external funding.

Acknowledgments: The authors are grateful to Alla Voronina and Elena Erofeeva for their kind assistance during the manuscript preparation.

Conflicts of Interest: The authors declare no conflict of interest.

References

1. Oparin, A.I. *The Origin of Life*, 2nd ed.; Dover Publishing: New York, NY, USA, 1953.
2. Fox, S.; Balin, P.; Pappelis, A.; Yu, B. Experimental retracement of terrestrial origin of an excitable cell: Was it predictable? In *Chemical Evolution: Physics of the Origin and Evolution of Life*; Chela-Flores, J., Raulin, F., Eds.; Kluwer: Alphen aan den Rijn, The Netherlands, 1996; pp. 21–32.
3. Budin, I.; Szostak, J.W. Expanding roles for diverse physical phenomena during the origin of life. *Annu. Rev. Biophys.* **2010**, *39*, 245–263. [CrossRef] [PubMed]
4. Deamer, D.W. *Assembling Life*; Oxford University Press: Oxford, UK, 2019.
5. Damer, B.; Deamer, D. The Hot Spring Hypothesis for an Origin of Life. *Astrobiology* **2020**, *20*, 429–452. [CrossRef] [PubMed]

6. Kurihara, K.; Tamura, M.; Shohda, K.; Toyota, T.; Suzuki, K.; Sugawara, T. Self-reproduction of supramolecular giant vesicles combined with the amplification of encapsulated DNA. *Nat. Chem.* **2011**, *3*, 775–781. [CrossRef]
7. Sugawara, T.; Kurihara, K.; Suzuki, K. Constructive approach toward protocells. In *Engineering of Chemical Complexity*; Mikhailov, A., Ed.; World Scientific Review: Singapore, 2012; pp. 1–17.
8. Kompanichenko, V.N. Three stages of the origin-of-life process: Bifurcation, stabilization and inversion. *Int. J. Astrobiol.* **2008**, *7*, 27–46. [CrossRef]
9. Kompanichenko, V.N. Inversion concept of the origin of life. *Orig. Life Evol. Biosph.* **2012**, *42*, 153–178. [CrossRef] [PubMed]
10. Kompanichenko, V.N. Thermodynamic Jump from Prebiotic Microsystems to Primary Living Cells (Version 3, Approved). *Science* **2020**, *2*, 14. [CrossRef]
11. Kompanichenko, V.N. *Thermodynamic Inversion: Origin of Living Systems*; Springer International Publishing: Cham, Switzerland, 2017.
12. Kompanichenko, V. The Rise of A Habitable Planet: Four Required Conditions for the Origin of Life in the Universe. *Geosciences* **2019**, *9*, 92. [CrossRef]
13. Kompanichenko, V.; El-Registan, G. Advancement of the TI concept: Defining the origin-of-life stages based on the succession of a bacterial cell exit from anabiosis. *AIMS Geosci.* **2022**, *8*, 398–437. [CrossRef]
14. Eigen, M. Selforganization of matter and the evolution of biological macromolecules. *Naturwissenschaften* **1971**, *58*, 465–523. [CrossRef]
15. Gallo, G.; Puopolo, R.; Limauro, D.; Bartolucci, S.; Fiorentino, G. Metal-Tolerant Thermophiles: From the Analysis of Resistance Mechanisms to their Biotechnological Exploitation. *Open Biochem. J.* **2018**, *12*, 149–160. [CrossRef]
16. Stetter, K.O. Hyperthermophiles in the history of life. *Philos. Trans. R. Soc. B* **2006**, *361*, 1837–1841. [CrossRef]
17. Xue, H.; Tong, K.-L.; Marck, C.; Grosjean, H.; Wong, J.T.-F. Transfer RNA paralogs: Evidence for genetic code-amino acid biosynthesis coevolution and an archaean root of life. *Gene* **2003**, *310*, 59–66. [CrossRef]
18. Selye, H. *Stress without Distress*; JB Lippincott Company: Philadelfia, PA, USA; New York, NY, USA, 1974.
19. Mazur, P. Principles of cryobiology. In *Life in the Frozen State*; Fuller, B.J., Lane, N., Benson, E.E., Eds.; CRC Press: Boca Raton, FL, USA, 2004; pp. 3–65.
20. El-Registan, G.I.; Mylyukin, A.L.; Nikolaev, Y.u.A.; Suzina, N.E.; Gal'cheko, V.F.; Duda, V.I. Adaptogenic functions of extracellular autoregulators of microorganisms. *Microbiology* **2006**, *75*, 380–389. [CrossRef]
21. Parry, B.R.; Surovtsev, I.V.; Cabeen, M.T.; O'Hem, C.S.; Dufresne, E.R.; Jacobs-Wagner, C. Bacterial lytoplasm Has Glass-like Properties and Fluidized by Metabolic Activity. *Cell* **2014**, *156*, 183–194. [CrossRef] [PubMed]
22. Fonesca, F.; Meneghel, J.; Cenard, S.; Passot, S.; Morris, G.J. Determination of Intracellular Vitrification Temperatures for Unicellular Microorganisms under Condition Relevant for Cryopreservation. *PLoS ONE* **2016**, *11*, e0152939. [CrossRef]
23. Loiko, N.G.; Suzina, N.E.; Soina, V.S.; Smirnova, T.A.; Zubasheva, M.V.; Azizbekyan, R.R.; Sinitsyn, D.O.; Tereshkina, K.B.; Nikolaev, Y.A.; Krupyanskii, Y.F.; et al. Biocrysteline Structure in the Nucleoids of the Stationary and Dormant Prokaryotic Cells. *Microbiology* **2017**, *86*, 703–719. [CrossRef]
24. Moat, A.G. Microbial stress responses. In *Microbial Physiology*, 4th ed.; Albert, G.M., John, W.F., Michael, P.S., Eds.; Wiley-Liss Inc.: New York, NY, USA, 2002; pp. 582–611.
25. Gottesman, S. Trouble is coming: Signaling pathways that regulate general stress responses in bacteria. *J. Biol. Chem.* **2019**, *294*, 11685–11700. [CrossRef]
26. Cheresiz, S.V.; Yurchenko, N.N.; Ivannikov, A.V.; Zakharov, I.K. Transposable elements and stress. *Vestn. VOGiS* **2008**, *12*, 216–241. (In Russian)
27. Kotsyurbenko, O.R.; Glagolev, M.V.; Sabrekov, A.F.; Terentieva, I.E. Systems approach to the study of microbial methanogenesis in West-Siberian wetlands. *Environ. Dyn. Glob. Clim. Chang.* **2020**, *11*, 54–68. [CrossRef]
28. Kotsyurbenko, O.R.; Glagolev, M.V. Protocols for measuring methanogenesis. In *Hydrocarbon and Lipid Microbiology Protocols (Springer Protocols Handbooks)*; Terry, J.M., Kenneth, N.T., Balbina, N., Eds.; Springer: Berlin/Heidelberg, Germany, 2015; pp. 227–243.
29. Kumar, R.; Joshi, S.R. Microbial Ecology of Soil: Studying the diversity of microorgansims in the most complex of the environments—A review. *Adv. Appl. Microbiol.* **2015**, *19*, 267–279.
30. Paliy, O.; Shankar, V. Application of multivariate statistical techniques in microbial ecology. *Mol. Ecol.* **2016**, *25*, 1032–1057. [CrossRef] [PubMed]
31. Schime, J.; Balser, T.C.; Wallenstein, M. Microbial stress-response physiology and its implications for ecosystem function. *Ecology* **2007**, *88*, 1386–1394. [CrossRef] [PubMed]
32. Pishchik, V.; Vorob'ev, N.; Provorov, N.; Khomyakov, Y. Mechanisms of plant and microbial adaptation to heavy metals in plant-microbial systems. *Microbiology* **2016**, *85*, 257–271. [CrossRef]

life

MDPI

Article

Stability of DL-Glyceraldehyde under Simulated Hydrothermal Conditions: Synthesis of Sugar-like Compounds in an Iron(III)-Oxide-Hydroxide-Rich Environment under Acidic Conditions

Claudio Alejandro Fuentes-Carreón [1,2,*], Jorge Armando Cruz-Castañeda [2], Eva Mateo-Martí [3] and Alicia Negrón-Mendoza [2]

1 Posgrado en Ciencias de la Tierra, Instituto de Ciencias Nucleares, Universidad Nacional Autónoma de México, Mexico City 04510, Mexico
2 Instituto de Ciencias Nucleares, Universidad Nacional Autónoma de México, Mexico City 04510, Mexico
3 Centro de Astrobiología (CAB) CSIC-INTA, Ctra. de Ajalvir km 4, 28850 Torrejón de Ardoz, Spain
* Correspondence: alejandro.fuent@correo.nucleares.unam.mx

Abstract: Researchers have suggested that the condensation of low-molecular-weight aldehydes under basic conditions (e.g., pH > 11) is the prebiotic reaction responsible for the abiotic formation of carbohydrates. It has also been suggested that surface hydrothermal systems were ubiquitous during the early Archean period. Therefore, the catalysis of prebiotic carbohydrate synthesis by metallic oxide minerals under acidic conditions in these environments seems considerably more probable than the more widely hypothesized reaction routes. This study investigates the stability of DL-glyceraldehyde and its reaction products under the simulated conditions of an Archean surface hydrothermal system. The Hveradalur geothermal area in Iceland was selected as an analog of such a system. HPLC-ESIMS, UV–Vis spectroscopy, Raman spectroscopy and XPS spectroscopy were used to analyze the reaction products. In hot (323 K) and acidic (pH 2) solutions under the presence of suspended iron(III) oxide hydroxide powder, DL-glyceraldehyde readily decomposes into low-molecular-weight compounds and transforms into sugar-like molecules via condensation reactions.

Keywords: chemical evolution; sugar-like compound synthesis; hydrothermal systems

Citation: Fuentes-Carreón, C.A.; Cruz-Castañeda, J.A.; Mateo-Martí, E.; Negrón-Mendoza, A. Stability of DL-Glyceraldehyde under Simulated Hydrothermal Conditions: Synthesis of Sugar-like Compounds in an Iron(III)-Oxide-Hydroxide-Rich Environment under Acidic Conditions. *Life* **2022**, *12*, 1818. https://doi.org/10.3390/life12111818

Academic Editors: Ranajay Saha and Alberto Vázquez-Salazar

Received: 19 September 2022
Accepted: 6 November 2022
Published: 8 November 2022

1. Introduction

A requisite for the emergence of life on the early Earth was the presence of water and organic compounds—primarily amino acids, nucleobases, sugars, and their respective precursors [1]. Sugars are considered one of the most essential molecules for all living organisms as they are vital in contemporary metabolism and the synthesis of other compounds, such as amino acids and nucleotides [2]. The previous functions of these biomolecules suggest that they were synthesized during the initial stages of the origin of life; therefore, to gain insight into the chemical processes that might have taken place on early Earth, investigating the mechanisms of abiotic sugar production is essential. The formose reaction, which involves the condensation of formaldehyde under basic conditions, is the most well-known of the reaction pathways in sugar synthesis and is dependent on the presence of inorganic catalysts, most commonly $CaCO_3$ and $Ca(OH)_2$ [3]. Alternative proposals for sugar synthesis include (A) anaerobic formose-like reactions with low-molecular-weight aldehydes in the presence of thiols [4] and ammonia catalysts [2,5], and (B) aldehyde condensation reactions catalyzed by mineral surfaces [3,6,7] Nevertheless, sugars are only by-products of these reactions, with the main organic compounds involving a wide variety of straight-chain and branched sugars and aldols. Since most of these reactions occur in environments that promote the breakdown of these organic compounds, studies must be carried out to propose alternate sugar synthesis reactions under plausible prebiotic conditions [1,8].

Researchers have proposed glyceraldehyde ($C_3H_6O_3$) as an alternative to formaldehyde in the formose reaction for the synthesis of sugars, acting as a source of energy and monomers in aldehyde condensation reactions. In addition, glyceraldehyde can lead to sugar and sugar-like molecules through condensation reactions catalyzed by mineral surfaces [6] and possible nucleobases via reactions with ammonia under anaerobic conditions [9]. Alternatively, when glyceraldehyde is exposed to ionizing radiation, it acts as a source of molecules relevant to abiotic synthesis [10,11]. The decomposition products include glycolaldehyde, formaldehyde, malondialdehyde, and pyruvaldehyde. Further, glyceraldehyde can originate from both internal and external sources. Dihydroxyacetone, the ketone form of this aldehyde, may have reached early Earth through exogenous delivery from cometary and meteoritic debris [12,13]. Conversely, a by-product of formose-like processes, glyceraldehyde is produced internally and may have originated from endogenous sources in the Earth's early environment [4].

All the previously discussed organic precursors could have been synthesized on Earth when the planet became habitable, as early as 10–20 million years after the Moon-forming impact [14]. Abiotic synthesis might have taken place in aqueous solutions in hydrothermal systems, in the atmosphere, or at the water–mineral interface of the Earth's surface environments [14] during the Hadean (~4.54–4.0 Ga) and early Archean (~3.9 Ga) eons, when the planet's atmosphere may have predominantly comprised CO_2 [15,16]. Among these possibilities, hydrothermal systems have physicochemical traits that are believed to be crucial for the development of life [1]. Hydrothermal fluids are rich in ions, which can act as catalysts for chemical reactions and form metal complexes with organic molecules in solution [17]. Additionally, studies have revealed that hydrothermal systems provide concentration, redox, and temperature gradients that can be utilized for the abiotic synthetic processes of biomolecule precursors [1]. The thermal energy of these systems could also be used to start chemical reactions. Therefore, these geological environments are considered prime candidates for simulation experiments of abiotic synthesis under plausible early Earth geochemical conditions.

The Kverkfjöll Volcano and the Hveradalur Geothermal Area, Central Iceland: Analog of an Ancient Hydrothermal System

The Kverkfjöll volcano, located in Central Iceland, consists of a mountain massif of rocks of evolved basaltic composition, with an area of surface geothermal activity covering approximately 25 km^2. This area displays a wide range of temperatures, pH values, and chemical compositions and has been extensively investigated [18]. Its western part, the Hveradalur geothermal area [19], is home to hydrothermal fluids of low pH (1–5) and low temperatures (<50 °C). This zone is of particular interest to the present work, acting as an analog of a low-temperature primitive early Archean (~3.9 Ga) hydrothermal area (Figure 1). Although representing a novel concept, using analog environments in prebiotic chemistry experiments shows great potential [13], as it allows researchers to reproduce the physicochemical conditions of geological environments that may have existed on ancient Earth. Analog environments also provide insight into the possible mineralogy of early Earth environments by allowing investigations into the distribution of mineral phases under the physicochemical conditions of interest.

Goethite (FeOOH), an iron(III) oxyhydroxide, is one of the mineral phases that potentially existed on primitive Earth and is of special relevance in studies involving the abiotic synthesis of organic molecules. It can act as a catalyst for aldehyde condensation reactions [6] and amino acid surface polymerization [20] and, therefore, possibly plays an important role in the abiotic synthesis of biomolecule precursors. Research has indicated that the sedimentary deposits of the early hydrothermal systems on Earth may have contained this mineral [21,22] Further, reference [19] reported the formation of goethite by hydrothermal modification of an iron meteorite under simulated hydrothermal conditions, thus demonstrating the mineral's stability under conditions of low pH and high temperatures.

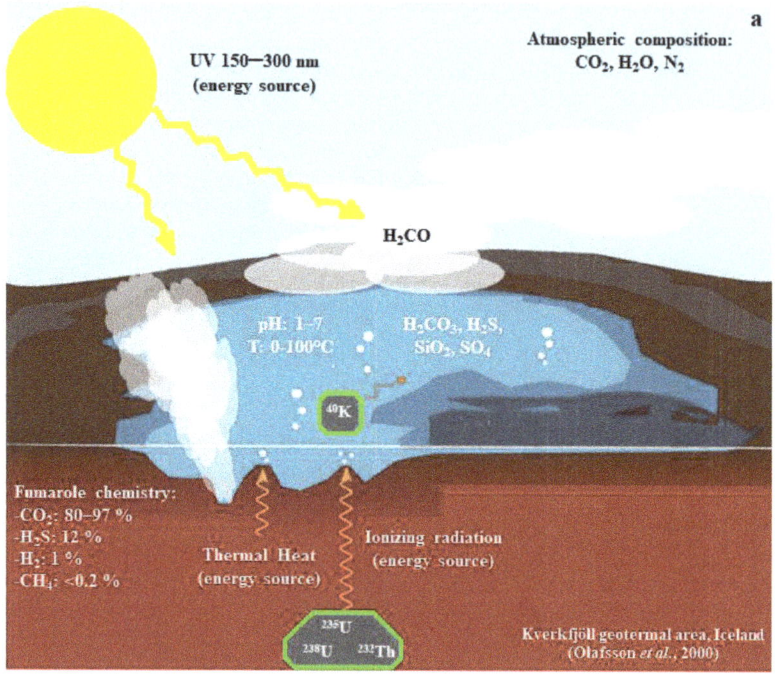

Figure 1. Schematic representation of an early Archean hydrothermal area. These surface hydrothermal environments could have been abundant during the Archean period (~3.9 Ga) in crustal zones with a high incidence of volcanic activity [18].

In the present work, we took quantitative measurements of remnant DL-glyceraldehyde in sample solutions after exposure to the simulated physicochemical conditions of an analog of a primitive Earth hydrothermal area, the Kverkfjöll geothermal field, using goethite as a catalyst for the mineral–water interface reactions. Through these measurements, we aimed to investigate the stability of glyceraldehyde and identify its decomposition products.

2. Materials and Methods

2.1. Reagents

All the reagents were of the highest commercial purity available, obtained from Sigma-Aldrich®, St. Louis, MO, USA, and included the following: DL-glyceraldehyde (>90%), acetaldehyde (>99.5%), pyruvaldehyde (40 wt. % in H_2O), glyoxal (40 wt. % in H_2O), glycolaldehyde (>90%), orcinol (>99%), D-glucose (99.9%), D-ribose (99.9%), and $FeCl_3$ (95%). Methanol-free formaldehyde was prepared from paraformaldehyde according to the method described by [23]. Acetonitrile (HPLC), ethanol (HPLC), H_2SO_4 (95–97%), HCl (37 wt. % in H_2O), HNO_3 (70%), KOH (>85%), and 2,4-dinitrophenylhydrazine (DNPH) (97%) were obtained from Merck Co.®, Kenilworth, NJ, USA. Natural iron(III) hydroxide oxide samples were collected in Durango, Mexico, and were characterized by X-ray diffraction, Raman, and X-ray Photoemission spectroscopy (XPS). To eliminate organic contamination, the goethite samples were treated with a 3% HNO_3 solution (acid wash), followed by two additional washes with Milli-Q® (Merck Millipore, Burlington, MA, USA) water (deionized) and a 3% KOH solution. After the acidic and basic washes, the mineral samples were rinsed with Milli-Q® water and dried in a desiccator at room temperature for 24 h.

2.2. Preparation of Samples

To simulate the Hveradalur geothermal site's physicochemical conditions, a 1×10^{-2} mol/L, aqueous DL-glyceraldehyde solution was prepared using deionized water; the solution was acidified with 36 mM H_2SO_4 (pH 2). The samples were degasified with Ar to remove dissolved O_2 in the solution. For the mineral–water interface reactions in the simulated environment, the system was mixed in 15-mL polyallomer centrifuge tubes with mineral (Goethite) powder (Beckman Coulter®, Brea, CA, USA). Goethite powder (180 mg), previously prepared from natural mineral samples by mechanical grinding in an agate mortar to an approximate grain size of 125 nm, was combined with 5 mL aliquots of the degasified acidic DL-glyceraldehyde standard solutions. The system was saturated with argon to create an anoxic atmosphere, and the polyallomer tubes were sealed to preserve the system's atmospheric integrity. For the XPS measurements, small Goethite sheets (5 × 5 mm) were prepared. The sheets' surfaces were polished with a SiC paste and cleaned with an additional acidic and basic wash.

2.3. Sorption Experiments

The DL-glyceraldehyde/goethite samples were heated to 50 °C by placing the polyallomer tubes inside a water recirculation system. The experimental system was exposed to this thermal treatment at defined time intervals (120, 240, 360, and 480 min). After each time interval had elapsed, the mineral was separated from the aqueous phase using an Allegra XL-90 centrifuge (Beckman Coulter®, Brea, CA, USA) at 25,000 rpm at room temperature for 15 min. The goethite powder was desiccated and preserved for further analysis. The collected supernatants were filtered using 12-µm Acrodisc syringe filters (Whatman, Chicago, IL, USA) and stored at 0 °C for the subsequent HPLC measurements. The formation of new organic compounds in the sample supernatants was monitored by UV–Vis spectrophotometry, HPLC–UV, and HPLC-ESIMS; this sample analysis was repeated for each time interval used in the sorption experiments. For the sorption on goethite sheets, the same experimental procedure was repeated, combining the goethite sheet with 5 mL aliquots of DL-glyceraldehyde standard solutions, modifying the time intervals for the thermal treatment (300 and 600 min).

2.4. Carbonyl Compound Measurements

Using HPLC–UV, the aldehydes, as their 2,4-dinitrophenylhydrazine (DNPH) derivatives, were identified. The derivatization of the carbonyl compounds required reacting 5 mL aliquots of the sample supernatants with 5 mL aliquots of the DNPH reagent (0.4 mg DNPH dissolved in 2 mL H_2SO_4, 3 mL H_2O, and 25 mL of ethanol) for 12 h. The carbonyl derivatives precipitated as yellow-orange crystals, which were filtered, dried, recrystallized, and redissolved in acetonitrile for subsequent analysis. The elution times of the carbonyl DNPH derivatives of formaldehyde, acetaldehyde, glyoxal, pyruvaldehyde, and DL-glyceraldehyde (as standard compounds) were measured.

2.5. Analysis of Samples

2.5.1. HPLC–UV Analysis

For the HPLC–UV analysis of the carbonyl compounds, we used a Knauer®, Berlin, Germany) Azura P 4.1S HPLC pump equipped with a Knauer® Azura UVD 2.1S UV detector. The derivatives of the carbonyl compounds were monitored by measuring their absorbance at 360 nm. The derivatives were separated in a Beckman® Ultrasphere C-18 ODS column (250 × 4.6 mm) using isocratic elution at 1.0 mL/min with a mobile phase comprising 70% acetonitrile and 30% water. The elution times of the standard carbonyl–DNPH derivatives are presented in Table 1.

Table 1. Elution times of standard carbonyl-DNPH derivatives.

Beckman® Ultrasphere C-18 ODS Column.	
Compound	Elution time (min)
Glyceraldehyde-DNPH	2.44, 2.82, 4.59, 8.61 *
Glycolaldehyde-DNPH	3.0
Formaldehyde-DNPH	3.62
Acetaldehyde-DNPH	4.13
Glyoxal-diDNPH	7.25 ± 0.7
Pyruvaldehyde-diDNPH	10.64

* Multiple retention times of the DL-glyceraldehyde standard are possibly associated with multiple compounds formed by the keto-enol tautomerism of the molecule in water [24].

The elution times in the sample supernatants, which were unknown, were identified by comparing them with those of the standard aldehydes and carbohydrates. The quantitation of the remnant DL-glyceraldehyde in the experimental samples was determined by measuring the absorbance of DL-glyceraldehyde–DNPH at 360 nm as a function of time, employing the Beer–Lambert law to calculate the sample concentration. The calibration curve was constructed using the glyceraldehyde–DNPH standards, with a concentration interval of 1×10^{-3} to 1×10^{-5} mol/L.

2.5.2. UV–Vis Spectroscopy

For the UV–Vis spectroscopic analysis, we used a Cary 100 spectrophotometer (Varian, Palo Alto, CA, USA). The supernatants of the DL-glyceraldehyde solutions after sorption were analyzed, monitoring their absorbance at 285 nm, which corresponds to DL-glyceraldehyde in solution. A concentrated (0.2 mol/L) solution was used to increase the sensitivity of the analyte to UV detection.

2.5.3. X-ray Diffraction Analysis

The purity of the natural goethite samples was measured by X-ray diffraction (XRD) using an Empyrean diffractometer (Malvern Panalytical®, Malvern, Worcs, UK) equipped with a PIXcel3D detector, with filtered Fe radiation of 60 kV at 2θ angles of 4° to 70°. The samples were prepared by grinding the goethite with an agate mortar to a particle size of less than <75 μm. The mineral powder was then pressed onto the diffractometer's aluminum sample holder. The software HighScore (Malvern Panalytical®, Malvern, Worcs, UK), the Inorganic Crystal Structure Database (ICSD), and the International Center for Diffraction Data (ICDD) database were used for the mineral phase identification of the measured samples.

2.5.4. Raman Spectroscopy Analysis

To study goethite and DL-glyceraldehyde, Raman spectroscopy was used to record their spectra. This analytical technique was also used to study the adsorption of organic compounds on the mineral surface. The experimental samples, in powder form, were pressed between two NaCl discs, and the Raman probe was used to collect the spectra of the samples. The Raman spectra were recorded using an Optosky ATR 3000 portable Raman spectrometer. The Raman probe containing a Class IIIB laser was positioned at an operating distance from the sample of 6 mm; the laser power was kept at a constant power value of 400 mW to avoid damage to the sample. The Raman frequencies were accurate to ± 5 cm^{-1}.

2.5.5. XPS Analysis

The XPS spectra of the goethite mineral sample were recorded before glyceraldehyde adsorption and after adsorption for 300 and 600 min from the solution. The XPS analysis of the samples was carried out in an ultra-high vacuum chamber equipped with a hemispherical electron analyzer using an Al Kα X-ray source (1486.6 eV) with an aperture of 7 mm × 20 mm. The base pressure in the chamber was 5×10^{-8} mbar, and the experiments were performed at room temperature. The peak decomposition in different components was shaped, after background subtraction, as a convolution of Lorentzian and Gaussian curves. For deconvolution, we applied the criterion of using the lowest number of components for the fit. Binding energies were calibrated against the adventitious carbon component set to 285.0 eV for the goethite samples. The following core-level peaks were recorded under the same experimental conditions: C (1s), Fe (2 $p_{3/2}$), and O (1s). We did not observe any beam radiation damage to the samples' surfaces during the data acquisition.

2.5.6. HPLC-ESIMS Analysis (Sugar-like Compound Measurements)

We define sugar-like compounds as molecules with five or six carbon atoms and multiple OH/CO substituents and isomers of carbohydrates. For the analysis of these molecules, HPLC–ESIMS was used. The analysis was carried out with a Waters® 515 HPLC pump coupled with a Waters® SQ-2 Single Quadrupole Mass Detector system, with electrospray ionization in negative (ESI-) mode. The cone voltage was 23 V, the capillary voltage was 1.46 kV, and the desolvation temperature was 350 °C. The sugar-like compounds were separated in a GL Sciences®, Torrance, CA, USA Inerstil NH_2 5 μm column (4.6 × 250 mm), specifically designed for the separation of carbohydrates, using isocratic elution at 0.8 mL/min with a mobile phase comprising 80% acetonitrile and 20% water.

3. Results

3.1. Goethite and DL-Glyceraldehyde Analysis

The XRD pattern (Figure 2) indicates, according to the ICSD and ICDD databases used for identification, the presence of two mineral phases in the collected samples: Goethite (91%) and hematite (9%), where goethite is the predominant mineral phase, as shown by the XRD pattern. The Raman spectrum (Figure 3) of the mineral sample powder shows vibrational bands associated with hematite and goethite mineral phases. The 222, 295, 406, 614, and 1350 cm^{-1} bands belong to hematite. Regarding goethite, the reported vibrational bands were 243, 299, 385, 479, 550, 685, and 993 cm^{-1} [25]. In the present spectrum, the 299, 385, and 479 cm^{-1} vibrational bands overlap with the hematite bands, whereas the peaks at 685 and 993 cm^{-1} were not present. The Raman spectrum confirms that the mineral sample is a mixture of goethite and hematite. Hematite is present as a minor constituent of the mineral phase.

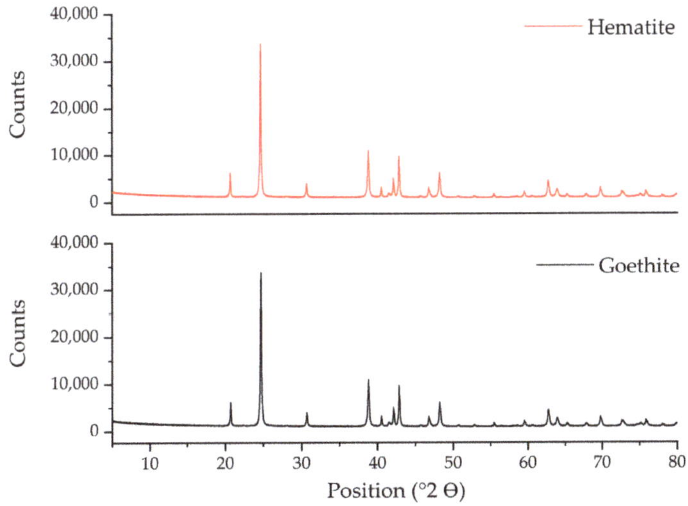

Figure 2. XRD pattern of collected natural goethite samples.

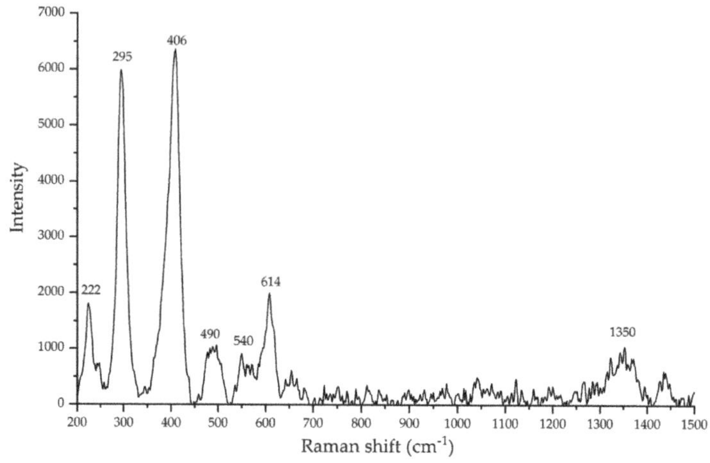

Figure 3. Raman spectrum of natural goethite mineral samples.

For DL-glyceraldehyde, the Raman spectrum shows that commercial (Sigma-Aldrich®) DL-glyceraldehyde isomers are prevalent when the molecule is in the crystalline state. Figure 4 shows the vibrational bands and their associated functional groups as reported in previous works [26]. The vibrational band associated with the carbonyl bond (1700 cm^{-1}) is not detected in the experimental samples. However, multiple peaks in the region of 200 to 700 cm^{-1} were found on the experimental samples. These vibration bands are not reported in the literature The last optical activity in the 200–700 cm^{-1} region is associated with the multiple deformations of an alkene (C=C) group [27]. This functional group is present on the isomer of DL-glyceraldehyde present in the sample, an enol.

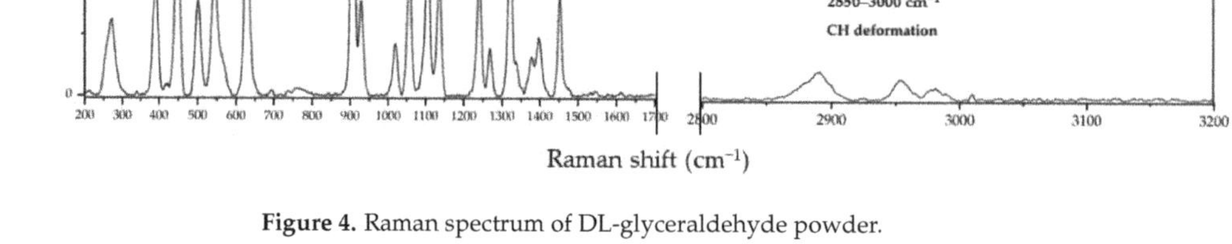

Figure 4. Raman spectrum of DL-glyceraldehyde powder.

3.2. *Iron(III) Oxide Hydroxide in a Simulated Archean Hydrothermal Area*

Iron(III) oxide hydroxide in its crystalline state was analyzed by Raman spectroscopy before and after sorption with aqueous DL-glyceraldehyde (Figure 5). Analysis of the region between 200 and 700 cm^{-1} reveals the overlapping of the 406 and 550 cm^{-1} vibrational bands of the goethite/DL-glyceraldehyde system with those corresponding to the skeletal deformations of alkenes (C=C) (Figure 5A). Using the Raman spectrum of unaltered crystalline goethite as a reference, the increase in the intensity of the 550 cm^{-1} band is notable. Additionally, the width of the 387 cm^{-1} band increase suggests that it is composed of at least two components, one of these being the 406 cm^{-1} vibrational band of DL-glyceraldehyde. An additional overlapping band is present in the region of 2500–3500 cm^{-1} (Figure 5B), associated with CH vibrations. As mentioned earlier, these vibrational bands in this region correspond to the enol form of glyceraldehyde.

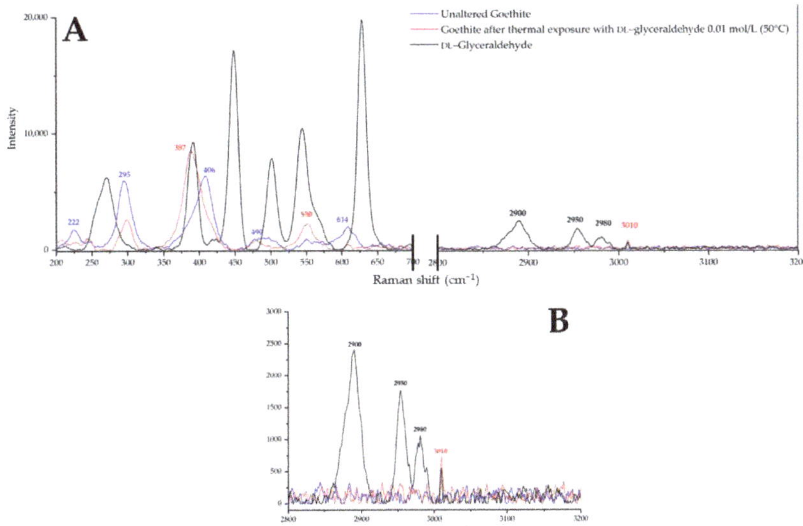

Figure 5. (**A**) Raman spectra of unaltered goethite (blue line) compared to altered goethite (red line) and DL-glyceraldehyde. Two vibrational bands of altered goethite overlap with bands associated with DL-glyceraldehyde in the 200–700 cm^{-1} region A sole band overlaps in the 2500–3500 cm^{-1} region. (**B**) 2800–3200 cm^{-1} region of the Raman spectra at a reduced-intensity scale.

To obtain additional information and confirmation regarding the adsorption and chemical forms of the organic constituents on the surface of goethite, the mineral was characterized by XPS spectroscopy before and after the sorption of aqueous DL-glyceraldehyde at defined time intervals (300 and 600 min). The Fe $2p_{3/2}$, C 1s, and O 1s peaks have been deconvoluted into multiple component curves, representing the best fit obtained for these regions. Figure 6 shows the deconvoluted core level spectra of Fe $2p_{3/2}$. The binding energies of the two main components of the Fe $2p_{3/2}$ peak are located at 711.9 and 710.3 eV, corresponding to FeO(OH) (goethite) and Fe_2O_3 (hematite), respectively. (Figure 6A–C). The remaining Fe component curves are associated with different iron species (such as metallic Fe and multiple oxides species).

As for the deconvolution of the core level spectra of C 1s, Figure 7 shows the binding energies of the three components of the C 1s peak—285.0, 286.6, and 288.5 eV, which correspond to C-H, C-C, and adventitious carbon, the second one to C-N, and the third one to carboxyl (C=O) compounds, respectively [28]. The 285.0 and 286.6 eV binding energies are associated with compounds formed by hydrocarbon and nitrogen impurities present in the original mineral sample (Figure 7C). The carboxyl compound component was also present in the unaltered mineral sample. After 300 min of thermal exposure to the organic component, the intensity of the component increases (Figure 7A); then, with increased thermal exposure time (6000 min), the intensity of the component decreases (Figure 7B).

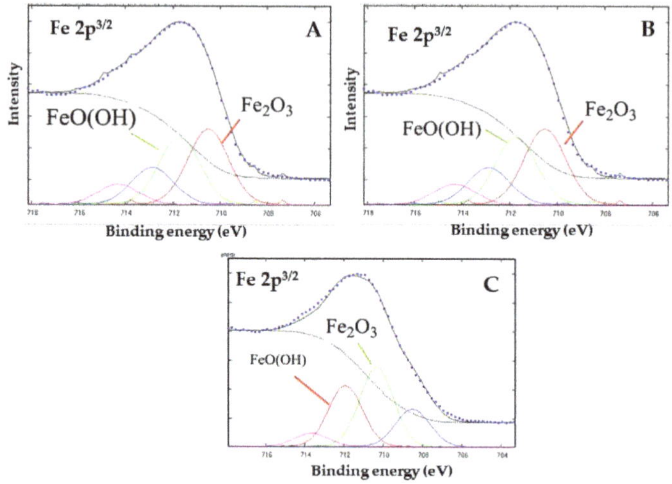

Figure 6. XPS core level peak of Fe $2p_{3/2}$ for goethite after absorption of 0.01 mol/L DL-glyceraldehyde solution, heated at 50 °C for 300 (**A**) and 600 (**B**) minutes. The spectrum of goethite samples before absorption (**C**) is shown as reference. The spectra consist of experimental (. . .) and fitted components (color lines).

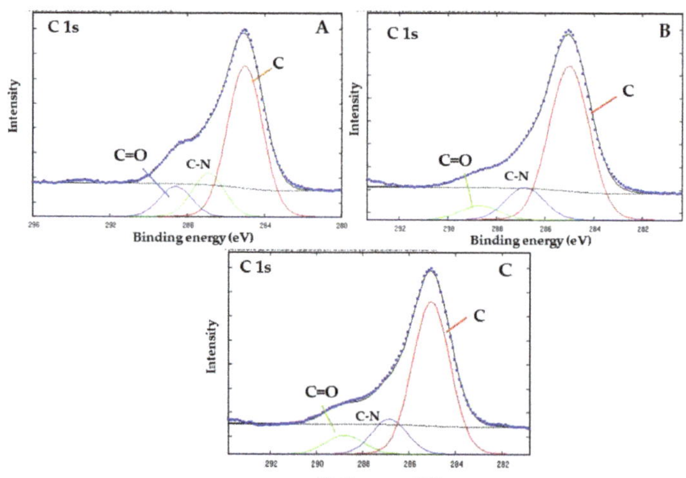

Figure 7. XPS core level peak of C 1s for goethite after adsorption of 0.01 mol/L DL-glyceraldehyde solution, heated at 50 °C for 300 (**A**) and 600 (**B**) minutes. The spectrum of goethite samples before absorption (**C**) is shown as reference. The spectra consist of experimental (. . .) and fitted components (color lines).

Figure 8A shows the deconvoluted core level spectra of O 1s. The binding energies of the three components of the O 1s peak were 530.4, 531.8, and 533.3 eV. The 530.4 eV component corresponds to the OH^- and oxides species. The component at 531.8 eV corresponds to the adventitious oxygen and O^{2-} surface atoms. This is coherent with the goethite structure (FeO(OH)). The peak at 533.3 eV corresponds to an organic compound with the functional group RC-OH [28]. The 533.3 eV binding energy intensity increases slightly when the thermal exposure time of the goethite/aqueous DL-glyceraldehyde mixture was increased to 600 min (Figure 8B). The 533.3 eV binding energy is present in the unaltered goethite mineral samples (Figure 8C).

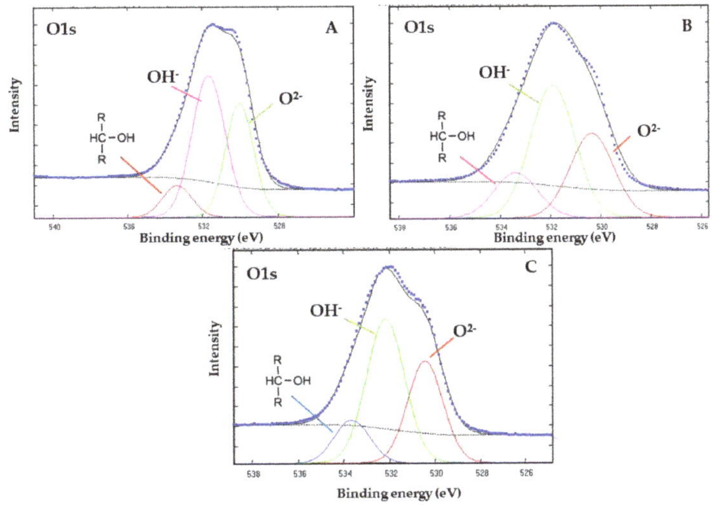

Figure 8. XPS core level peak of O 1s for goethite after adsorption of 0.01 mol/L DL-glyceraldehyde solution, heated at 50 °C for 300 (**A**) and 600 (**B**) minutes. The spectrum of goethite samples before absorption (**C**) is shown as reference. The spectra consist of experimental (. . .) and fitted components (color lines).

To determine whether the mineral phase or the temperature is responsible for the decomposition of DL-glyceraldehyde, the UV spectrum of the underivatized supernatant was analyzed by UV-Vis spectroscopy to monitor the absorption of the 285 nm peak after sorption at the maximum time of thermal exposure (Figure 9). The spectrum was compared to a control sample of the DL-glyceraldehyde solution heated at a constant temperature of 50 °C for 480 min. No notable change in the absorption spectra was observed between the DL-glyceraldehyde standard and the heated sample. However, an increase in overall absorption and a shift to higher wave numbers of the 285 nm band was discerned when the sample was heated in the presence of goethite.

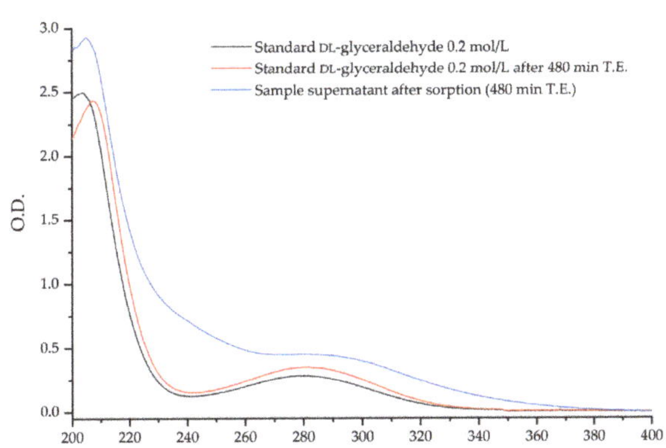

Figure 9. Comparison of the UV spectra of stock DL-glyceraldehyde solution (black line), stock DL-glyceraldehyde solution heated at 50 °C up to 480 min (red line), and underivatized supernatant analyzed after sorption (blue line).

3.3. Stability of DL-Glyceraldehyde in a Simulated Archean Hydrothermal Area

HPLC analysis shows that DL-glyceraldehyde is labile under the simulated hydrothermal conditions (Figure 10). The total percentage of remnant DL-glyceraldehyde in the solution decreases at higher thermal exposure times. The decomposition products are formaldehyde, acetaldehyde, glyoxal, and pyruvaldehyde. An unknown decomposition product, which could not be identified with the available carbonyl-DNPH standards, with an elution time of 5.9 min was also detected (Figure 11). The chromatogram of the DL-glyceraldehyde standard, derivatized with DNPH, shows the presence of multiple compounds (Figure 12) associated with the presence in solution of multiple isomers formed by keto-enol tautomerism of the sample. As for the unknown decomposition product, the HPLC mobility of the DNPH derivatives suggests that this compound could be glyceric acid, a previously identified decomposition product of DL-glyceraldehyde in an aqueous solution [11]. The decomposition products were stable for up to 480 min of thermal exposure (Figure 13).

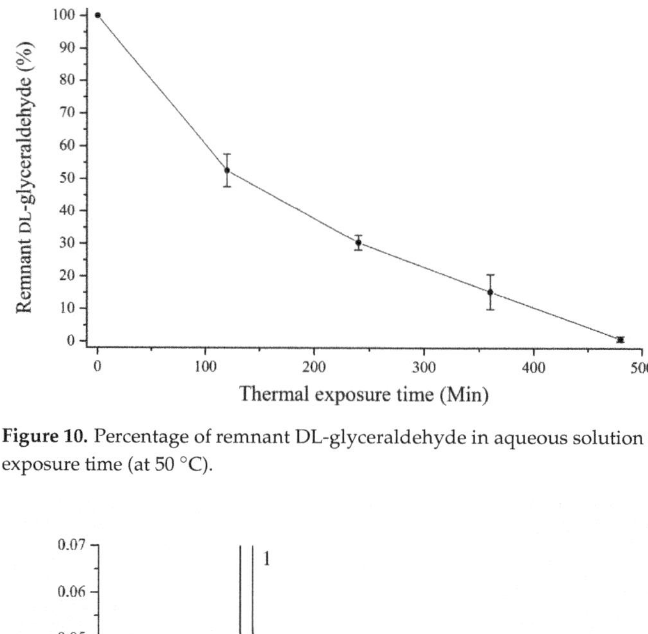

Figure 10. Percentage of remnant DL-glyceraldehyde in aqueous solution as a function of thermal exposure time (at 50 °C).

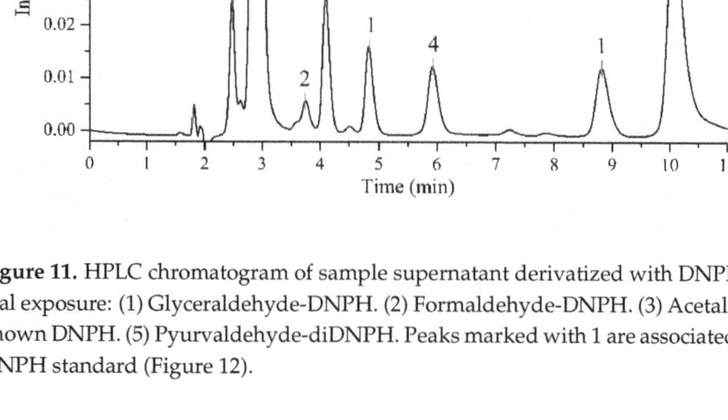

Figure 11. HPLC chromatogram of sample supernatant derivatized with DNPH after 120 min of thermal exposure: (1) Glyceraldehyde-DNPH. (2) Formaldehyde-DNPH. (3) Acetaldehyde-DNPH. (4) Unknown DNPH. (5) Pyurvaldehyde-diDNPH. Peaks marked with 1 are associated with Glyceraldehyde-DNPH standard (Figure 12).

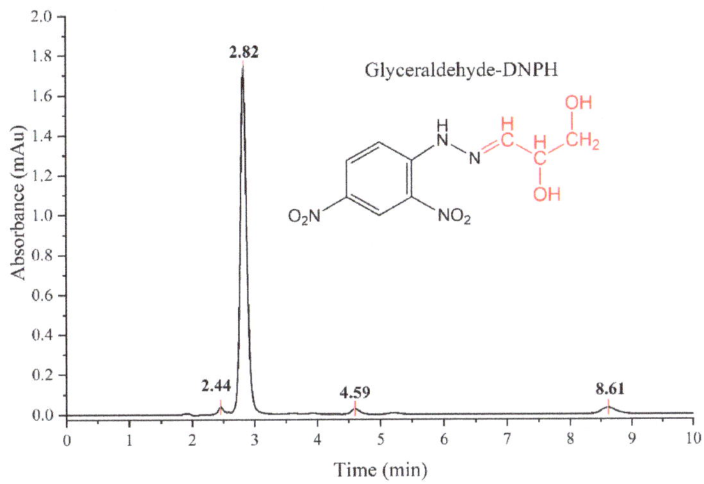

Figure 12. HPLC chromatogram of Glyceraldehyde-DNPH standard.

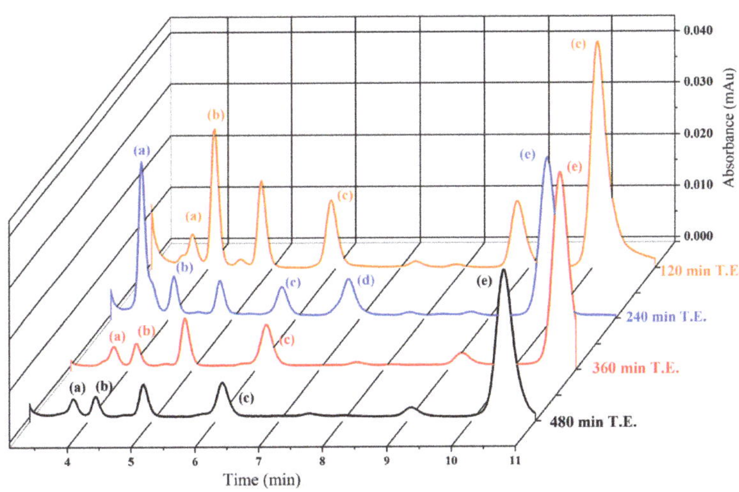

Figure 13. HPLC chromatogram of sample supernatants derivatized with DNPH at different times of thermal exposure (T.E). (a) Formaldehyde-DNPH. (b) Acetaldehyde-DNPH. (c) Unknown carbonyl-DNPH. (d) Glyoxal-diDNPH. (e) Pyurvaldehyde-diDNPH. Retention times of the samples exposed to 120 and 240 min of thermal exposure are slightly displaced when compared with the retention times of the samples exposed to 360 and 480 min of thermal exposure.

3.4. Mass Spectrometric Detection of Putative C5 and C6 Sugar-like Compounds

HPLC-ESIMS (Figure 14) was used for the detection of C5 and C6 (sugar-like) molecules in the sample supernatants, analyzed after the sorption with goethite. The applied technique was selected ion mode (SIM) HPLC-ESIMS monitoring. In negative ion mode, expected $[M\text{–}H]^-$ ions were registered; m/z 149 for pentoses and m/z 179 for hexoses. Five compounds ($C5'$and $C6^{1'}$, $C6^{2'}$, $C6^{3'}$, and $C6^{4'}$) were detected with a molecular weight of 150 ($C5'$) (Figure 14A) and 180 g/mol ($C6^{1\text{–}4'}$) (Figure 14B). These organic compounds were detected in all the sorption samples at all thermal exposure time intervals. Figure 14 shows the chromatogram of the supernatant exposed to 360 min of thermal exposure The retention times of these organic compounds were compared to the retention times of two carbohydrate standards (ribose and glucose). Although nominal m/z ratios of detected compounds coincide with those of ribose and glucose, the corresponding retention times are different. This suggests that the detected organic compounds are isomers of ribose and glucose, with molecular weights of 150 and 180 g/mol, respectively.

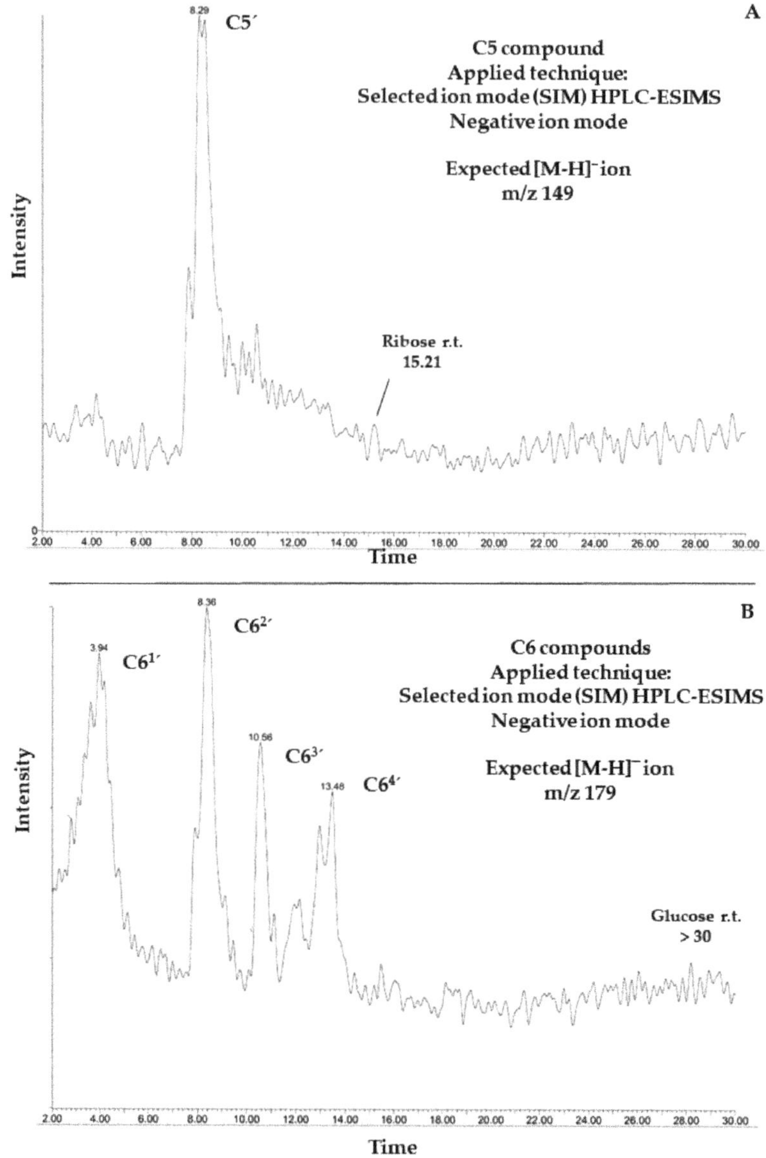

Figure 14. HPLC-ESIMS for C5 and C6 compounds detected in sample supernatants of DL-glyceraldehyde sorption in a simulated Archean hydrothermal area. (**A**): HPLC-ESIMS of the C5′, an organic compound with a molecular weight of 150 g/mol. (**B**): HPLC-ESIMS of four organic compounds (C6$^{1\prime}$, C6$^{2\prime}$, C6$^{3\prime}$, and C6$^{4\prime}$), each one of them with a molecular weight of 180 g/mol. Retention times of ribose and glucose are shown as reference.

4. Discussion

Investigations into the stability of DL-glyceraldehyde in aqueous solutions have seldom been conducted, and those that have focused on the role of this organic compound as a source of aldehyde monomers in autocatalytic reactions in sugar synthesis [5,6,9] and its stability under high-energy radiation fields in simulated prebiotic environments [10,11]. The novelty of the current work lies in the insight provided into the possible role of DL-glyceraldehyde as a source of aldehydes and C5/C6 compounds in simulated hydrothermal surface areas, where acidic conditions and variable temperatures are dominant. DL-glyceraldehyde in solution undergoes interconversion between different chemical species that exist in a complex equilibrium [24]. Under the experimental conditions, the dimeric cyclic hemiacetal forms of DL-glyceraldehyde are unstable; consequently, the keto and enol species are dominant (Figure 15). Intramolecular hydrogen bonds between the carbonyl carbon and the C-2 and C-3 hydroxyl hydrogens (1) stabilize the ketone form, shifting the equilibrium toward these species and ensuring that the enediol (1b) and hydrated forms (1a) are not dominant in solution. Therefore, it is highly probable that these molecules are not the dominant form of glyceraldehyde present in the solution. This chemical equilibrium is ubiquitous, regardless of whether the sample is in crystalline form or solution [24]. The Raman spectrum shows that crystalline DL-glyceraldehyde exists as the enol form (Figure 16, 1b), which contains the functional group C=C. The UV absorption peak of 285 nm (Figure 9) indicates the presence of this isomer in the solution. [29]. Derivatization with DNPH of DL-glyceraldehyde in solution shows that the carbonyl isomer is formed when the crystalline form is dissolved in water (Figure 12). Due to the complex keto enol tautomerism of the sample in solution, additional carbonyl compounds are present. However, the concentration of the impurities is low enough when compared to DL-glyceraldehyde-DNPH.

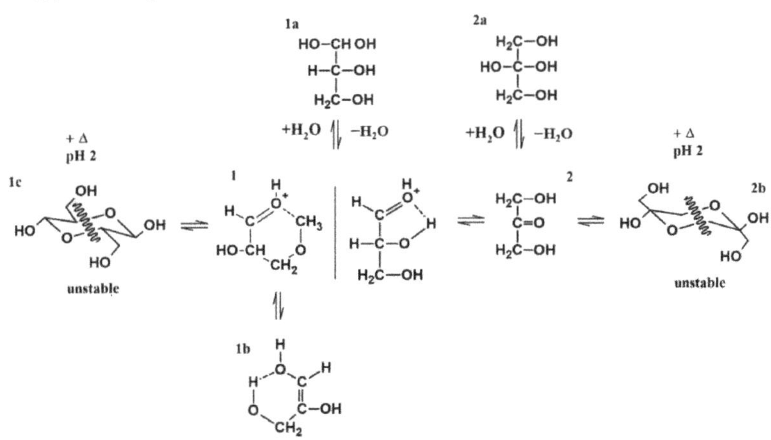

Figure 15. Proposed molecular interconversions of DL-glyceraldehyde in acid aqueous solution. (1) DL-glyceraldehyde (C-2 OH H bond), DL-glyceraldehyde (C-3 OH H bond). (**1a**) Hydrated glyceraldehyde. (**1b**) DL-glyceraldehyde enediol. (**1c**) Glyceraldehyde cyclic hemiacetal. (**2**) Dihydroxyacetone. (**2a**) Hydrated dihydroxyacetone. (**2b**) Dihydroxyacetone cyclic hemiketal.

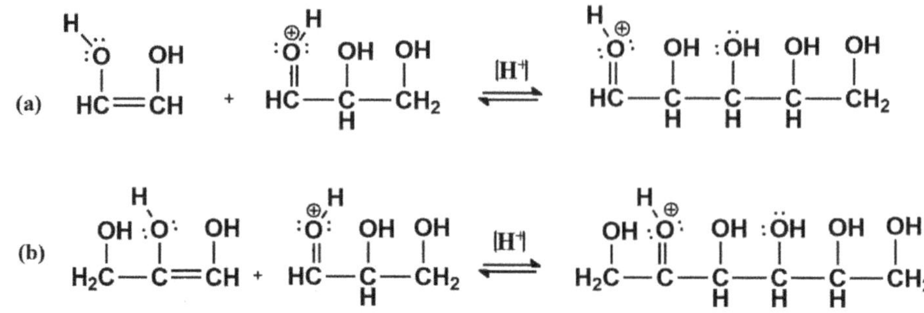

Figure 16. Proposed reaction mechanism for C5 and C6 compounds. Equation (**a**): Pentose isomer formation (aldol condensation of glyoxal and DL-glyceraldehyde). Equation (**b**): Hexose isomer formation (aldol condensation of DL-glyceraldehyde isomers).

The decomposition products of DL-glyceraldehyde, when under acidic thermal conditions, are formaldehyde (CH_2O), acetaldehyde (C_2H_4O), glyoxal ($C_2H_2O_2$), and pyruvaldehyde ($C_3H_4O_2$) (Figure 11). Additionally, one C5 and four C6 compounds, isomers of pentose and hexose with a molecular weight of 150 and 180 g/mol, respectively, were detected in the experimental samples (Figure 14). The formation of these isomers can be explained by the acid-catalyzed aldol condensation of the aldehydes with their respective enol forms. The protonated DL-glyceraldehyde reacts via the addition of 1,2-enediol (glyoxal isomer), leading to an aldol condensation product, a pentose isomer (Figure 16a). The addition of the keto and aldol forms of DL-glyceraldehyde can yield hexose isomers as condensation products (Figure 16b).

The instability of this molecule in the experimental conditions is due to the displacement of the hemiacetal equilibrium. In the presence of heat and aqueous acid, the formation of hemiacetals, which act as protective chemical groups of carbonyl compounds through the reduction of their reactivity, is hindered. This process increases the reactivity of the linear carbonyl compounds in solution, which, therefore, promotes the synthesis of different organic compounds. Nevertheless, the aldol condensation reaction is not selective to specific C5 and C6 compounds, and multiple products could form via crossed aldol reactions (i.e., condensation between two different aldehydes/ketones [30]). This is demonstrated by the detection of the C5′ and $C6^{1\text{–}4\prime}$ compounds (Figure 14) which are isomers of a pentose and a hexose. The lack of selectivity towards specific carbohydrates will yield a mixture of sugar-like compounds. The initial concentration of the stock solution hinders the formation of these compounds. C5′ and $C6^{1\text{–}4\prime}$ were detected only when the concentration of the stock DL-glyceraldehyde solution was increased to 0.2 mol/L, up from 0.01 mol/L. A high concentration of the initial aldehyde is required to obtain high yields of the C5′ and $C6^{1\text{–}4\prime}$ compounds. Therefore, the total yield of this compound must be low. In a hot, acidic environment, most of the DL-glyceraldehyde readily decomposes into low-molecular-weight aldehydes (<90 g/mol). Further analysis with higher-resolution techniques is required to reveal the structural features of the detected compounds.

Previous studies have investigated the role of goethite in carbohydrate synthesis. For example, reference [6] proposed the formation of C6 compounds via the catalytic action of goethite, which occurs through the aldol condensation of DL-glyceraldehyde with a mineral-stabilized glyceraldehyde enediolate on the surface of the iron(III) oxide hydroxide. Exhaustive analysis of the peak frequencies in goethite's Raman spectrum after exposure to solutions of DL-glyceraldehyde 0.01 mol/L (Figure 5) shows that two of the vibrational peaks (406 and 550 cm^{-1}) of goethite overlap with the vibrational peaks of DL-glyceraldehyde, which are associated with skeletal deformations of the C=C bond of the triose enol form. In particular, the 550 cm^{-1} band of goethite increases in intensity, while the 406 cm^{-1} band widens. These changes in the Raman spectrum suggest the adsorption of a DL-glyceraldehyde isomer (enol) on the surface of goethite. Regarding the XPS data, two binding energies are of special interest: 288.5 eV of core level C (1s) and 533.3 eV of core level O (1s). The increase in the intensity of the 288.5 eV component (associated with a carboxyl group) when the thermal exposure time is increased to 300 min can be explained by the adsorption of a carbonyl compound on the surface of goethite. The decrease exhibited when the thermal exposure time is further increased could be attributed to the decomposition of the same molecule. Due to the keto-enol equilibrium and decomposition of DL-glyceraldehyde (Figure 15), the adsorbed molecule could be either DL-glyceraldehyde, dihydroxyacetone, or one of the aldehydes formed by its decomposition (Figure 11). Regarding the 533.3 eV component, the increase in the intensity, proportional to the increase in thermal exposure time, can be explained by the gradual adsorption of RCOH-rich compounds on the mineral surface or by the decomposition compounds, which will continue to contribute to the increase in the same oxygen component (C-O, C-O-H). However, the exact chemical formulas of these compounds remain unknown. Considering the molecules formed in the solution due to the keto-enol tautomerism and decomposition of DL-glyceraldehyde, the adsorbed molecules could be either C5/C6 compounds, an enol isomer, an aldehyde, or hydrated glyceraldehyde/dihydroxyacetone.

According to the literature [6], the adsorption of DL-glyceraldehyde on the surface of goethite occurs due to the formation of an Fe^{3+}-O_2R bond between two OH groups of the glyceraldehyde enol isomer. However, neither the Raman nor XPS spectrum shows evidence of vibrational peaks associated with the formation of an organometallic bond. Therefore, the exact mechanism by which the organic compounds are adsorbed on the surface of the goethite samples remains elusive.

The existence of Fe^{3+} over the reduced Fe^{2+} form in an acidic aqueous environment can be explained by aqueous corrosion mechanisms, which can occur in both oxygenic and anoxic environments [31]. In the presence of oxygen, goethite (and Fe^{3+} oxyhydroxides) can form via the corrosion of Fe/Ni deposits or iron meteorites. Fe^0 can be oxidized by two oxidizing agents, O^2 and H^+, degrading the initial material into α- and β-FeO(OH) [32]. In acidic deaerated solutions, protons act exclusively as the oxidizing agent. Previous works provided evidence of the formation of hematite and goethite in an anaerobic acid environment [19,33], which confirms the ability of hot, low-pH environments to oxidize Fe^0 to Fe^{3+}. Goethite (and hematite) are considered the end members of many iron transformation routes due to their thermodynamic stability. Therefore, conversion to additional iron mineral phases is not expected [32].

Implications for Prebiotic Chemistry

Under the simulated hydrothermal conditions, formaldehyde, acetaldehyde, glyoxal, and pyruvaldehyde form via the thermal decomposition of DL-glyceraldehyde. Formaldehyde is considered the cornerstone organic molecule for carbohydrate synthesis via the formose reaction [3]. Strecker synthesis with acetaldehyde and ammonia can be used as a source for alanine [33] while glyoxal and pyruvaldehyde are usually categorized as intermediates in formose-like reaction pathways [4]. The results of the current work show that DL-glyceraldehyde could have been a prebiotic source of monomers in primitive hydrothermal environments, molecules that, under variable pH conditions, could act as energy sources for amino acid and carbohydrate synthesis [34]. Nevertheless, the aldol condensation of DL-glyceraldehyde provides an additional source of C5 and C6 compounds in hydrothermal environments. The possible pathway that can lead to the synthesis of these organic compounds is the acid-catalyzed aldol condensation between DL-glyceraldehyde and enol molecules. However, the yield of these acid-catalyzed aldol condensation reactions is low.

As mentioned earlier, researchers generally believe that the prebiotic synthesis of carbohydrates and carbohydrate precursors involved the condensation of low-molecular-weight aldehydes under basic conditions (e.g., pH > 11) in the presence of inorganic catalysts [1,13]. The synthesis of C5 and C6 compounds under acidic conditions seems notably more plausible than the reaction pathways commonly proposed, as surface hydrothermal systems were common during the early Archean [15,35]. Regarding goethite, prebiotic sources of this mineral could have arisen from the photooxidation of dissolved Fe(II), [6] and aqueous alterations of iron-rich rocks and minerals [19,22]. Therefore, it is highly likely that this mineral was common in the sediments of ancient hydrothermal regions [20,21].

5. Conclusions

Prebiotic carbohydrate synthesis is fundamental because of its role in metabolism and the abiotic formation of nucleotides and nucleic acid components. In this work, we investigated the stability of DL-glyceraldehyde in a simulated Archean hydrothermal environment and reached the following conclusions:

1. DL-glyceraldehyde readily decomposes in the presence of goethite under simulated hydrothermal conditions, and the decomposition products can react with remnant glyceraldehyde in an acid-catalyzed aldol condensation reaction to generate C5 and C6 compounds.
2. The thermal decomposition products of glyceraldehyde (formaldehyde, acetaldehyde, glyoxal, and pyruvaldehyde) are molecules of prebiotic significance; they are precursors of carbohydrates in formose-like reactions and other organic compounds.
3. Enol, aldehydes, and C5/C6 compounds, products of the decomposition of DL-glyceraldehyde in a simulated Archean hydrothermal area, successfully bind to the goethite surface.
4. C5 and C6 compounds could form through aldol condensation in an aqueous solution. These compounds are stable under acidic and thermal conditions for brief periods of time. The presence of the sugar-like compounds, $C5'$ and $C6^{1-4\prime}$, implies that a rather complex mechanism of sugar-like compound formation occurs in the experimental system. These reaction mechanisms are not selective towards specific pentose and hexoses, such as ribose or glucose.

Author Contributions: Conceptualization, C.A.F.-C. and A.N.-M.; methodology, C.A.F.-C.; validation, C.A.F.-C. and J.A.C.-C.; investigation, C.A.F.-C. and E.M.-M.; resources, A.N.-M. and E.M.-M.; writing—original draft preparation, C.A.F.-C.; writing—review and editing, J.A.C.-C., E.M.-M. and A.N.-M.; visualization, C.A.F.-C.; supervision, A.N.-M.; project administration, A.N.-M.; funding acquisition, A.N.-M. and E.M.-M. All authors have read and agreed to the published version of the manuscript.

Funding: This research was funded by DGAPA-PAPIIT grant No. IN114122, CONACYT grant No. 319818, and projects PID2019-104205GB-C21 from the Spanish Ministry of Science and Innovation.

Institutional Review Board Statement: Not applicable.

Informed Consent Statement: Not applicable.

Data Availability Statement: All experimental data in this study will be made available upon reasonable request from readers.

Acknowledgments: The authors acknowledge the CONACyT fellowship, Posgrado en Ciencias de la Tierra, and the Instituto de Ciencias Nucleares at the Universidad Nacional Autónoma de México (UNAM) for their support in this research. The authors express their gratitude to Claudia Camargo for her technical help in the Laboratorio de Evolución Química, ICN-UNAM; Karina Elizabeth Cervantes de la Cruz for her generous donation of the mineral samples; Teresa Pi Piug for his invaluable help in X-ray diffraction analysis, and José Rangel Gutierrez and Engineer Juan Eduardo Murrieta for their help in the maintenance of multiple laboratory equipment. The authors used the research facilities of the Centro de Astrobiología (CAB), CSIC-INTA, and were supported by the Instituto Nacional de Técnica Aeroespacial "Esteban Terradas" (INTA), by the projects PID2019-104205GB-C21 from the Spanish Ministry of Science and Innovation. Additionally, the authors are grateful to Santos Galvez Martinez for performing the XPS measurements. As a final note, the authors are grateful to the reviewers involved in the revision process, as the comments and suggestions made improved the overall quality of the work.

Conflicts of Interest: The authors declare no conflict of interest.

References

1. Kobayashi, K. Prebiotic synthesis of bioorganic compounds by simulation experiments. In *Astrobiology: From the Origins of Life to the Search for Extraterrestrial Intelligence*, 1st ed.; Yamagashi, A., Kakegawa, T., Usui, T., Eds.; Springer Nature: Singapore, 2019; pp. 43–61.
2. Weber, A.L. The triose model: Glyceraldehyde as a source of energy and monomers for prebiotic condensation reactions. *Orig. Life Evol. Biosph.* **1987**, *17*, 107–119. [CrossRef] [PubMed]
3. Miller, S.; Cleaves, H. Prebiotic chemistry on the primitive Earth. *Syst. Biol.* **2006**, *1*, 3–56.
4. Weber, A.L. The sugar model: Catalytic flow reactor dynamics of pyruvaldehyde synthesis from triose catalyzed by poly-l-lysine contained in a dialyzer. *Orig. Life Evol. Biosph.* **2001**, *31*, 231–240. [CrossRef] [PubMed]
5. Weber, A.L. Alanine synthesis from glyceraldehyde and ammonium ion in aqueous solution. *J. Mol. Evol.* **1985**, *21*, 351–355. [CrossRef]
6. Weber, A.L. Prebiotic sugar synthesis: Hexose and hydroxy acid synthesis from glyceraldehyde catalyzed by iron (III) hydroxide oxide. *J. Mol. Evol.* **1992**, *35*, 1–6. [CrossRef]
7. Kim, H.J.; Ricardo, A.; Illangkoon, H.I.; Kim, M.J.; Carrigan, M.A.; Frye, F.; Benner, S.A. Synthesis of carbohydrates in mineral-guided prebiotic cycles. *J. Am. Chem. Soc.* **2011**, *133*, 9457–9468. [CrossRef]
8. Shapiro, R. Prebiotic ribose synthesis: A critical analysis. *Orig. Life Evol. Biosph.* **1982**, *18*, 71–85. [CrossRef]
9. Weber, A.L. The sugar model: Autocatalytic activity of the triose–ammonia reaction. *Orig. Life Evol. Biosph.* **2007**, *37*, 105–111. [CrossRef]
10. Cruz-Castañeda, J.; Aguilar-Ovando, E.; Buhse, T.; Ramos-Bernal, S.; Meléndez-López, A.; Camargo-Raya, C.; Fuentes-Carreón, C.A.; Negrón-Mendoza, A. The importance of glyceraldehyde radiolysis in chemical evolution. *J. Radioanal. Nucl. Chem.* **2016**, *311*, 1135–1141. [CrossRef]
11. Aguilar-Ovando, E.; Cruz-Castañeda, J.; Buhse, T.; Fuentes-Carreón, C.; Ramos-Bernal, S.; Heredia, A.; Negrón-Mendoza, A. Irradiation of glyceraldehyde under simulated prebiotic conditions: Study in solid and aqueous state. *J. Radioanal. Nucl. Chem.* **2018**, *316*, 971–979. [CrossRef]
12. Gargaud, M.; Irvine, W.M.; Amils, R.; James, H.; Cleaves, H.J.; Pinti, D.L.; Quintanilla, J.C.; Rouan, D.; Spohn, T.; Tirard, S.; et al. (Eds.) *Encyclopedia of Astrobiology*, 2nd ed.; Springer: Berlin, Germany, 2011; pp. 357–358.
13. Dalai, P.; Kaddour, H.; Sahai, N. Incubating life: Prebiotic sources of organics for the origin of life. *Elements* **2016**, *12*, 401–406. [CrossRef]
14. Zahnle, K.; Arndt, N.; Cockell, C.; Halliday, A.; Nisbet, E.; Selsis, F.; Sleep, N.H. Emergence of a habitable planet. *Space Sci. Rev.* **2007**, *129*, 35–78. [CrossRef]
15. Zahnle, K.; Schaefer, L.; Fegley, B. Earth's earliest atmospheres. *Cold Spring Harb. Perspect. Biol.* **2010**, *2*, a004895. [CrossRef]
16. Pearce, B.K.D.; Tupper, A.S.; Pudritz, R.E.; Higgs, P.G. Constraining the Time Interval for the Origin of Life on Earth. *Astrobiology* **2018**, *18*, 343–364. [CrossRef]
17. Belmonte, L.; Mansy, S.S. Metal catalysts and the origin of life. *Elements* **2016**, *12*, 413–418. [CrossRef]
18. Olafsson, M.; Torfason, H.; Grönvold, K. Surface exploration and monitoring of geothermal activity in the Kverkfjöll geothermal area, central Iceland. In Proceedings of the World Geothermal Congress 2000, Kyoshu-Tohoku, Japan, 28 May–10 June 2000; pp. 1539–1545.
19. Bryant, D.E.; Greenfield, D.; Walshaw, R.D.; Johnson, B.R.G.; Herschy, B.; Smith, C.; Pasek, M.A.; Telford, R.; Scowen, I.; Munshi, T.; et al. Hydrothermal modification of the Sikhote-Alin iron meteorite under low pH geothermal environments. A plausibly prebiotic route to activated phosphorus on the early Earth. *Geochim. Cosmochim. Acta* **2013**, *109*, 90–112. [CrossRef]
20. Shanker, U.; Bhushan, B.; Bhattacharjee, G.; Kamaluddi. Oligomerization of glycine and alanine catalyzed by iron oxides: Implications for prebiotic chemistry. *Orig. Life Evol. Biosph.* **2012**, *42*, 31–45. [CrossRef]
21. Holm, N.G.; Ertem, G.; Ferris, J.P. The binding and reactions of nucleotides and polynucleotides on iron oxide hydroxide polymorphs. *Orig. Life Evol. Biosph.* **1993**, *23*, 195–215. [CrossRef]
22. Hazen, R.M.; Papineau, D.; Bleeker, W.; Downs, R.T.; Ferry, J.M.; McCoy, T.J.; Sverjensky, D.A.; Yang, H. Mineral evolution. *Am. Mineral.* **2008**, *93*, 1693–1720. [CrossRef]
23. Castillo-Rojas, S.; Landeros, J.C.; Negrón-Mendoza, A.; Navarro-González, R. Radiolysis of aqueous formaldehyde relevant to cometary environments. *Adv. Space Res.* **1992**, *12*, 57–62. [CrossRef]
24. Yaylayan, V.; Harty-Majors, S.; Ismail, A.A. Investigation of DL-glyceraldehyde-dihydroxyacetone interconversion by FTIR spectroscopy. *Carbohydr. Res.* **1999**, *318*, 20–25. [CrossRef]
25. De Faria, D.L.; Venâncio Silva, S.; De Oliveira, M.T. Raman microspectroscopy of some iron oxides and oxyhydroxides. *J. Raman Spectrosc.* **1997**, *28*, 873–878. [CrossRef]
26. Pecul, M.; Rizzo, A.; Leszczynski, J. Vibrational Raman and Raman optical activity spectra of D-lactic acid, D-lactate, and D-glyceraldehyde: Ab initio calculations. *J. Phys. Chem. A* **2002**, *106*, 11008–11016. [CrossRef]
27. Nyquist, R.A. Alkenes and other Compounds containing C=C double bonds. In *Interpreting Infrared, Raman, and Nuclear Magnetic Resonance Spectra*, 1st ed.; Academic Press: London, UK, 2001; pp. 55–92.
28. Moulder, J.F.; Stickle, W.F.; Sobol, P.E.; Bomben, K.D. *Handbook of X-ray Photoelectron Spectroscopy*, 1st ed.; Chastain, J., Ed.; Perkin-Elmer Corporation: Eden Prairei, MN, USA, 1992.

29. Simonov, A.N.; Pestunova, O.P.; Matvienko, L.G.; Snytnikov, V.N.; Snytnikova, O.A.; Tsentalovich, Y.P.; Parmon, V.N. Possible prebiotic synthesis of monosaccharides from formaldehyde in presence of phosphates. *Adv. Space Res.* **2007**, *40*, 1634–1640. [CrossRef]
30. McMurry, J. Carbonyl Chemistry. In *Organic Chemistry*, 2nd ed.; Lockwood, L., Kiselica, S., Woods, E., van Camp, S., Eds.; Cengage Learning: Belmont, CA, USA, 2012; pp. 712–944.
31. Cornell, R.M.; Schwertmann, U. Formation. In *The Iron Oxides. Structure, Properties, Reactions, Occurrences and Uses*, 2nd ed.; Wiley-VCH Verlag GambH & Co.: Weinhem, Germany, 2006; pp. 345–363.
32. Grokhovsky, V.I.; Oshtrakh, M.I.; Milder, O.B.; Semionkin, V.A. Mössbauer spectroscopy of iron meteorite Dronino and products of its corrosion. *Hyperfine Interact.* **2005**, *166*, 671–677. [CrossRef]
33. Shaw, A.M. Prebiotic Chemistry. In *Astrochemistry from Astronomy to Astrobiology*, 1st ed.; John Wiley & Sons: West Sussex, UK, 2007; pp. 225–257.
34. Weber, A.L. The sugar model: Catalysis by amines and amino acid products. *Orig. Life Evol. Biosph.* **2001**, *31*, 71–86. [CrossRef]
35. Zahnle, K.J. Earth's earliest atmosphere. *Elements* **2006**, *2*, 217–222. [CrossRef]

life

MDPI

Article

Komatiites as Complex Adsorption Surfaces for Amino Acids in Prebiotic Environments, a Prebiotic Chemistry Essay

Abigail E. Cruz-Hernández [1,*], María Colín-García [2,*], Fernando Ortega-Gutiérrez [2] and Eva Mateo-Martí [3]

1 Posgrado en Ciencias de la Tierra, Universidad Nacional Autónoma de México, Ciudad de México 04510, Mexico
2 Instituto de Geología, Universidad Nacional Autónoma de México, Ciudad de México 04510, Mexico
3 Centro de Astrobiología (CAB) CSIC-INTA, Carretera de Ajalvir km 4, 28850 Torrejón de Ardoz, Spain
* Correspondence: abigaeta@comunidad.unam.mx (A.E.C.-H.); mcolin@geologia.unam.mx (M.C.-G.)

Abstract: Komatiites represent the oldest known terrestrial rocks, and their composition has been cataloged as the closest to that of the first terrestrial crust after the cooling of the magma ocean. These rocks could have been present in multiple environments on the early Earth and served as concentrators of organic molecules. In this study, the adsorption of five amino acids (glycine, lysine, histidine, arginine, and aspartic acid) on a natural komatiite, a simulated komatiite, and the minerals olivine, pyroxene, and plagioclase were analyzed under three different pH values: acid pH (5.5), natural pH of the aqueous solution of each amino acid and alkaline pH (11). Adsorption experiments were performed in solid–liquid suspensions and organic molecules were analyzed by spectrophotometry. The main objective of this essay was to determine if the complex surfaces could have participated as concentrators of amino acids in scenarios of the primitive Earth and if the adsorption responds to the change of charge of the molecules. The results showed that komatiite is capable of adsorbing amino acids in different amounts depending on the experimental conditions. In total, 75 systems were analyzed that show different adsorptions, which implies that different interactions are involved, particularly in relation to the type of amino acid, the type of solid material and the conditions of the medium.

Keywords: komatiite; amino acids; adsorption; concentration mechanism; prebiotic chemistry

Citation: Cruz-Hernández, A.E.; Colín-García, M.; Ortega-Gutiérrez, F.; Mateo-Martí, E. Komatiites as Complex Adsorption Surfaces for Amino Acids in Prebiotic Environments, a Prebiotic Chemistry Essay. *Life* **2022**, *12*, 1788. https://doi.org/10.3390/life12111788

Academic Editors: Ranajay Saha and Alberto Vázquez-Salazar

Received: 8 October 2022
Accepted: 2 November 2022
Published: 4 November 2022

1. Introduction

Earth has followed a very complex and fascinating history since its formation. Very soon after accretion and differentiation, igneous activity dominated the geological processes of the early planet [1]. Regarding the primeval crust formed by crystallization of the surface magma ocean in the Hadean [2], this phase in planet evolution provided conditions for both mantle differentiation and volatiles distribution [3]. Further evolution depended on processes such as volcanism, degassing, fractional crystallization, and fluid-rock interactions, among others [4]. The study of the formation and composition of this crust is relevant not only for understanding the early ages of our planet, but also for elucidating the origin of life itself [5].

There is no record of the Hadean rocks, which shaped the primordial crust [2]. Although there is a debate about the composition of primitive crust, it has been proposed that the primordial oceanic crust had, most likely, an extremely similar composition to rocks called komatiites, named after the Komati river in South Africa [6,7]. The term komatiite refers to an ultrabasic volcanic rock, whose lava probably erupted at temperatures above 1600 °C. Due to secular variations in major, minor and trace elements, along with stratigraphic compositional and textural differences, there is no accurate definition for these igneous rocks [8]. However, these rocks are "primitive" lavas and likely play a key role in the understanding of early Earth evolution. Komatiites could represent a primitive environment where chemical evolution reactions may have occurred [6].

Chemical evolution may have occurred in different environments on our young planet [9], including in oceans, lakes, lagoons, tidal pools, submarine hydrothermal systems [10], and microenvironments [11]. In primitive water bodies, chemical compounds, necessary for the creation of the first living beings, could have been dissolved and accumulated [12]. Nonetheless, the estimated concentration of organics in primitive oceans would be very low; according to some authors, a 10^{-4} mol·L^{-1} solution would result from the entire dissolution of all the available organic compounds in oceans [12]. Thus, free waters represent an exceedingly harsh environment in which to accomplish prebiotic reactions. However, at the interface of oceanic waters and minerals, from oceanic crust, the presence of these solids may have enhanced the concentration processes. In smaller water bodies, the concentration was likely higher, and reactions could have occurred in shallow microenvironments in which there was a variety of rocks and minerals. Hence, surfaces could have increased the concentration of organics in all aqueous environments.

Minerals have been proposed as concentration sites for organic molecules in primitive environments [13–16]. Indeed, many minerals were present on early Earth when prebiotic chemistry was accomplished; more than 500 minerals already existed on Earth at 4 Ga [17]. However, laboratory models of primitive environments have not yet included complex surfaces, such as those present in primitive crusts, either oceanic or continental.

On the other hand, amino acids have been used commonly in adsorption experiments related to prebiotic chemistry (for a review, refer to) [17,18]. Amino acids are essential organic molecules for living beings, being the building blocks of proteins, and they were very likely available in primitive environments either by its synthesis under the conditions of the primitive Earth or by its delivery in the primitive oceans via external sources [19]. For example, amino acids could have originated from Strecker synthesis from precursors, compounds available in the primitive ocean and in the atmosphere, such as HCN, NH_3, aldehydes or ketones [20]. Also, it has been suggested that hydrothermal systems, with their temperatures and pH gradients, could have favored the synthesis of proteinogenic amino acids from hydrothermal fluids that contain H_2, NH_3 and H_2S; the latter has been confirmed experimentally under the simulation of an alkaline hydrothermal environment [21]. Likewise, another suggestion is that organic molecules in primitive environments could have entered through exogeneous sources, such as meteorites, comets or interplanetary dust particles (IDPs) [10,22–24].

The concentration of amino acids that was maintained in the primitive oceans must have been dependent on the balance between their input into the environment, the internal synthesis and the degradation of amino acids due to physical factors in primitive ocean [20]. Also, due to the lack of geological evidence on the composition of the early ocean and atmosphere, estimates of the possible concentration of amino acids have been modified over time. For example, Dose [25] estimated that amino acids, synthetized from several energy sources, were in an approximate concentration of 10^{-7} mol·L^{-1} in the primordial ocean, synthetized from several energy sources. Stribling and Miller [26] calculated a concentration of amino acids of 3×10^{-4} mol·L^{-1}, produced from a mixture of precursors in a simulated atmosphere and promoted mainly by sparks. Cleaves et al. [20] more recently consider that the concentration of these molecules should have been between 0.2 mol·L^{-1} (as the upper limit) and 3×10^{-4}–3×10^{-19} mol·L^{-1} (as the lower limit); their estimates depend on the amount of N in the Earth's crust, the oceanic reservoir, and on the production of amino acids in hydrothermal systems. Similarly, calculations have been made to estimate the contribution of amino acids from exogeneous sources. Sugahara and Mimura [23], based on comet impacts simulations, estimated the delivery of glycine to the early Earth to be 4×10^{-10} mol·L^{-1} and 1×10^{-8} mol·L^{-1} using the Lunar Cratering Model and the Nice Model, respectively. The general agreement among the scientific community has been that the concentration of amino acids in the early ocean must have been low. Thus, concentration mechanisms in primitive scenarios were extremely necessary. In these environments, other concentration mechanisms could have existed; however, adsorption by mineral surfaces is suggested to have been the most widespread mechanism on prebiotic Earth [19].

In this study, a first attempt to determine the ability of complex surfaces to adsorb amino acids was essayed. The solids employed as sorbents were komatiite, plagioclase, pyroxene, olivine, and a mixture including the most abundant minerals in those rocks (a compositional model for primitive oceanic crust).

As organics, proteinogenic amino acids (glycine, lysine, histidine, arginine and aspartic acid) were selected, whose abiotic synthesis (under simulated conditions of the primitive Earth, or the interstellar medium) or whose detection (in meteorites or interstellar medium) has been reported in the literature [19,27–30]. Histidine and arginine have been synthesized abiotically under different experimental conditions as mentioned above, but they have not been found on carbonaceous chondrites [31]. However, it has been observed that lysine, arginine, and histidine are important binding domains and contribute to the stabilization of catalytic conformations of ribozymes, which makes them important molecules in the origin of life [31]. In addition, it has been observed that they (lysine, histidine, and arginine) condense more extensively than non-proteinogenic amino acids (ornithine; 2,4-diaminobutyric acid; and 2,3-diaminopropionic acid); they are more selective in oligomerization because of the nature of their side chains, which represents an advantage for the formation of more complex molecules [32].

Due to their chemical structure, amino acids hold both a basic (-NH_2) and an acidic group (-CO_2H) in their structure. In aqueous solution, these groups ionize and, in general, the basic group has a p*K*a between 9 and 10, though acid p*K*a is close to 2; this implies that, at almost any pH value, amino acids exist as ions: cation (NH_3^+), anion (COO^-) or zwitterion form (NH_3^+, COO^-). This behavior is fundamental in the study of amino acid/mineral interactions. For this reason, a series of experiments was designed, including pH changes of the solution, as a determinant variable, for the sorption of amino acids. The main goal of this study was to determine whether complex surfaces could have actively participated in the adsorption of amino acids in different scenarios of primitive Earth. We focus on determining the best scenarios for amino acid adsorption onto those surfaces as a function of molecules charge change.

2. Materials and Methods

2.1. Materials

All experiments and procedures were realized with ultrapure water (Milli-Q, Millipore, Burlington, MA, USA). The olivine sample used in this work was the only one that was obtained commercially as peridote. The other minerals and the rock used were natural samples collected in the field and obtained through the donations of Professors Luca Ferrari (CGEO-UNAM), Fernando Ortega and Antoni Camprubí (IGl-UNAM). The chemical reagents used were purchased from Sigma-Aldrich Inc®., St. Louis, MO, USA.

2.2. Methods

2.2.1. Cleaning of Geological Material

Geological materials were cleaned individually. Samples were fragmented into small pieces no larger than one centimeter. All the fragments were weighed and deposited into a beaker. 10 mL of all the solutions for each gram of solid was used. A 3% KNO_3 solution was prepared and added to the geological material, and then the mixture was put into an ultrasonic bath for 30 min. After this process, the sample was decanted, and the material was rinsed with pure water. To eliminate all the rest of the solution, the material was stirred with enough pure water for 30 min, and then the water was poured out. A 3% HNO_3 solution was prepared and added to the material. An ultrasonic bath was conducted for 30 min. Subsequently, the solution was decanted, and the fragments were rinsed with pure water. Finally, enough water was added and stirred for 30 min. Once this step was concluded, the water was drained, and the geological material was maintained in the same beaker for drying at room temperature.

After the cleansing of geological materials, they were milled in an agate mortar and sifted. Particles <125 μm were used for all experiments.

2.2.2. Characterization of Komatiite Sample by X-ray Diffraction Analysis

Minerals phases in komatiite were determined by X-Ray Diffraction (XRD). Milled samples of rocks (particles < 75 μm) were placed in an aluminum sample holder. Measurement was made in an angular range 2θ (2 Theta) from 5° to 80° with a step of 0.003° and an integration time of 40 s per step. An Empyrean® Diffractometer with a nickel filter, fine focus copper tube, monochromator, and PIXcel3D detector was employed. This analysis was carried out at Laboratorio de Difracción de Rayos X, of the Laboratorio Nacional de Geoquímica y Mineralogía (LANGEM-IGl-UNAM).

2.2.3. Estimation of the Point of Zero Charge of Geological Materials

The point of zero charge (*pzc*) of each geological material was determined according to de Souza et al. [33] 100 mg of mineral or rock and 250 μL of either a KCl (1 mol·L^{-1}) solution or pure water were mixed. After 24 h, pH values were measured with a potentiometer Thermo Scientific Orion Versa Star Pro®. The process was conducted in triplicate for each geological material. The *pzc* values were calculated with the following equation:

$$pH_{pzc} = 2(pH_{KCl}) - pH_{H2O} \quad (1)$$

2.2.4. Quantification of Amino Acids

Glycine (Gly) (hydrophobic aliphatic), and the L-amino acids lysine (Lys), histidine (His), arginine (Arg) (the three hydrophilic basic), and aspartic acid (Asp) (hydrophilic acid) were selected; their principal chemical properties can be consulted in Table 1. These amino acids were quantified by UV-vis spectrophotometry through a colorimetric derivative, by the ninhydrin method [34]. According to this method, ninhydrin reacts with the amino group and gives a colored derived compound. The derivative is then measured by visible spectrophotometry at 570 nm. A Varian® Cary 100 Scan spectrophotometer was used to measure the absorbance. A calibration curve for each amino acid was made for quantification.

Table 1. The main chemical properties of the five amino acids used.

Amino Acid	Classification	Formula	pKa			pI
			pK_1	pK_2	pK_3	
Gly	Neutral	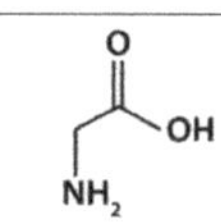	2.34	9.60	-	5.98
Lys	Basic	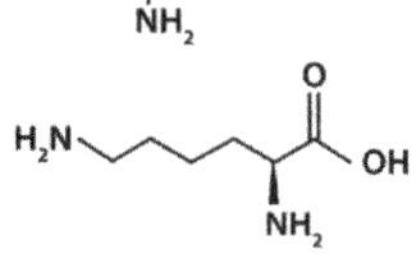	2.18	8.95	10.53	9.74
His	Basic	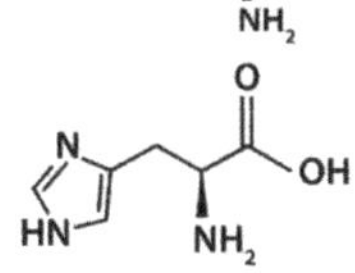	1.82	9.17	6.00	7.59

Table 1. *Cont.*

Amino Acid	Classification	Formula		pKa		pI
			pK_1	pK_2	pK_3	
Arg	Basic	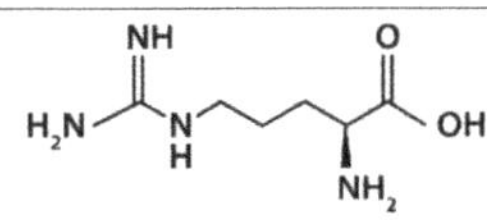	2.17	9.04	12.48	10.76
Asp	Acid	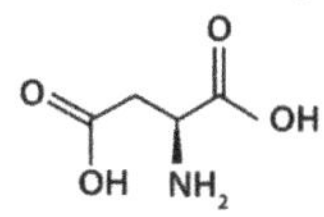	1.88	9.60	3.65	2.77

2.2.5. Effect of pH Variation

Three pH values were used in the experiments: (i) the natural pH value of each amino acid in an aqueous solution (which value is reported in Table 2); (ii) acidic pH (value 5.5) adjusted by the addition of an acetate buffer; and (iii) alkaline pH (value 11) modified by the addition of an NaOH solution (0.1 mol$\cdot$L^{-1}).

Table 2. Experimental pH values, determined for each amino acid in a solution without further modification.

Amino Acid	pH Value
Gly	5.8 ± 0.51
Lys	7.3 ± 0.26
His	6.7 ± 0.23
Arg	8.23 ± 0.6
Asp	4.3 ± 0.28

Note: The values presented here are the average of five independent measurements.

2.2.6. Experimental Systems

A komatiite sample and different samples of the main forming minerals of this rock (i.e., olivine, pyroxene, and plagioclase) were used. The purpose was to determine if any of these mineral phases had more of an effect on the adsorption processes. Since our komatiite sample presents alterations, we decided to use a mixture of the major minerals present in komatiites in approximately the same proportion that they would have occurred in an unaltered rock (olivine 49.29%, plagioclase 29.57%, and pyroxene 21.12%), as a compositional simulated komatiite, and hence a simulation of a primordial oceanic crust. Therefore, 75 different systems were studied, considering pH changes in a solution. All of them were treated with the same experimental adsorption procedure.

The experiments were carried out with the amino acids individually, and no mixtures of these were made. Under ideal conditions, the adsorption of amino acids is dependent on the value of the isoelectric point, when adsorbed by electrostatic interactions on a surface; thus, amino acids are expected to be selectively adsorbed. However, it is likely that when mixtures of amino acids are used, they could interact with each other, which scatters this ideality; therefore, the adsorption of certain amino acids in a mixture would depend on the presence of other amino acids, which modifies the behavior of adsorption [18]. Additionally, in mixtures, polymerization between them is also feasible, since intermolecular chemical forces drive the dynamics of these organic compounds towards self-recognition and assembly [35]. To avoid interactions between amino acids and to maintain the prebiotic context, a low concentration of amino acids was used. Also, when water is present, it has been observed that there is competition between the formation of H bonds that link amino acids (amino acid$_1$-amino acid$_2$) or amino acid and water molecules around it [35].

2.2.7. Adsorption Experiments

Solutions of each amino acid (1×10^{-4} mol·L^{-1}) were prepared in ultrapure water (Milli-Q, Millipore). 5 mL of each amino acid solution were used, mixed with 100 mg of each geological material [20:1 solid-solution ratio (mg/mL)]. These suspensions were content in polyallomer centrifuge tubes and shaken for 24 h onto a multi-tube platform Thermo ®. This duration was selected because in prebiotic chemistry, many experiments are performed considering 24 h as a suitable time to study sorption [36]. The suspensions were centrifuged using a Beckman Coulter ® Allegra 64R Centrifuge for 35 min at 36,000 rpm. The supernatant was recovered and analyzed by the ninhydrin method. With the calibration curves, the remaining concentration in the supernatant of each sample was determined. All experiments were conducted in triplicate, at room temperature (22 °C).

3. Results

3.1. Characterization of Geological Materials

3.1.1. Mineral Phases in Komatiite Sample

The results of XRD analysis can be observed in Table 3, and through this information, we determined that in the komatiite sample, just 10% of the total is fresh olivine, and a larger quantity is lizardite, a type of serpentine, which is evidence of an altered komatiite. In this rock, olivine is altered into hydrated minerals such as chlorite or serpentine, and those alterations are characteristics of the environment in which komatiites are generated [37]. Other mineral phases found in our komatiite sample were pyroxene, plagioclase, and other phyllosilicates.

Table 3. Mineral phases and the semiquantitative percentage of each one, found in the komatiite sample, all determined by X-ray diffraction analysis by the Reference Intensity Radio (RIR) method.

Mineral Phases	SemiQuant RIR (%)
Olivine (forsterite: Mg_2SiO_4)	10
Pyroxene (diopside: $CaMgSi_2O_8$)	14
Plagioclase (bytownite: $(NaCa)AlSi_3O_8$)	26
Serpentine (lizardite:$Mg_3Si_2O_5(OH)_4$)	29
Other phyllosilicates ~14 Å ppb type of chlorite	21
Total	100

3.1.2. Point of Zero Charge

The results of estimation of *pzc* of all geological materials, including the mix used as a simulation of an unaltered komatiite (komatiite-simulated), are shown in Table 4. The value of point of zero charge of olivine was found at 9.9, for pyroxene at 7.5, and in the case of plagioclase the value obtained was 8.9. The *pzc* value for the komatiite was 8.2, and it was observed that the simulated komatiite showed a higher value (10.01) compared to the natural rock.

Table 4. Estimated *pzc* values for geological materials used.

Geological Materials	*pzc*
Olivine	9.9
Pyroxene	7.5
Plagioclase	8.9
Komatiite	8.2
Komatiite-simulated *Olivine 49.29%* *Plagioclase 29.57%* *Pyroxene 21.12%*	10.05

Note: The komatiite-simulated sample is a solid mixture of the minerals named.

3.2. Adsorption Experiments

Results are shown as each system; each one is the combination of one mineral or geological material, and one amino acid at a specific pH value; these results can be observed in Figure 1. Adsorbed concentrations are presented in mmol of the adsorbed amino acid per milligram of solid used.

Figure 1. A graphic representation of the amount of amino acids adsorbed on different surfaces. The bar is proportional to the amount adsorbed. Three pH values are shown: acidic pH (5.5) in yellow; natural pH of the solution (see Table 2) in green, and alkaline pH (11) in red. The results are shown in mmol of amino acid adsorbed per mg of solid (see Tables 5–7).

Table 5. Number of amino acids adsorbed in each geological material; the pH of the solutions was not modified; see Table 2 for reference.

Geological Materials	Adsorption (mmol/mg)				
	Gly ± SD	Lys ± SD	His ± SD	Arg ± SD	Asp
Olivine	$3.18 \times 10^{-5} \pm 2.2 \times 10^{-6}$	$1.27 \times 10^{-4} \pm 4.9 \times 10^{-6}$	$7.17 \times 10^{-5} \pm 1.9 \times 10^{-6}$	$2.11 \times 10^{-4} \pm 5.1 \times 10^{-6}$	-
Pyroxene	$2.31 \times 10^{-4} \pm 7 \times 10^{-6}$	$1.12 \times 10^{-4} \pm 2 \times 10^{-6}$	-	-	-
Plagioclase	$1.55 \times 10^{-5} \pm 1.4 \times 10^{-7}$	$1.41 \times 10^{-4} \pm 9.8 \times 10^{-6}$	$2.15 \times 10^{-5} \pm 2.1 \times 10^{-8}$	$1.22 \times 10^{-4} \pm 1.8 \times 10^{-7}$	-
Komatiite	$6.14 \times 10^{-4} \pm 3.3 \times 10^{-6}$	$6.99 \times 10^{-4} \pm 4.3 \times 10^{-5}$	$2.64 \times 10^{-4} \pm 1.0 \times 10^{-6}$	$8.76 \times 10^{-4} \pm 1.4 \times 10^{-4}$	-
Komatiite-simulated	-	-	-	$5.14 \times 10^{-5} \pm 4.1 \times 10^{-7}$	-

Note: Amounts are presented as millimoles of amino acid adsorbed per milligram of solid used. The calculated standard deviation of the measurements is shown after the sorption values (± SD). For experiments in which adsorption was not observed, a dash (-) symbol is used.

Table 6. Adsorbed quantities of amino acids in each geological material, at acidic pH (5.5).

Geological Materials	Adsorption (mmol/mg)				
	Gly ± SD	Lys ± SD	His ± SD	Arg ± SD	Asp ± SD
Olivine	-	-	$4.52 \times 10^{-5} \pm 3.8 \times 10^{-6}$	$9.55 \times 10^{-5} \pm 1 \times 10^{-6}$	$1.52 \times 10^{-4} \pm 1.4 \times 10^{-6}$
Pyroxene	$4.74 \times 10^{-6} \pm 1.9 \times 10^{-8}$	-	$1.75 \times 10^{-4} \pm 4.9 \times 10^{-6}$	$8.94 \times 10^{-5} \pm 1.8 \times 10^{-7}$	$1.27 \times 10^{-4} \pm 2.1 \times 10^{-6}$
Plagioclase	-	-	$9.40 \times 10^{-4} \pm 1.2 \times 10^{-5}$	$8.93 \times 10^{-5} \pm 2.7 \times 10^{-7}$	$9.00 \times 10^{-5} \pm 5.4 \times 10^{-7}$
Komatiite	-	$1.69 \times 10^{-4} \pm 4.6 \times 10^{-6}$	$2.85 \times 10^{-4} \pm 4.2 \times 10^{-5}$	$3.96 \times 10^{-4} \pm 2.4 \times 10^{-6}$	$2.57 \times 10^{-4} \pm 1 \times 10^{-5}$
Komatiite-simulated	$1.65 \times 10^{-4} \pm 3.3 \times 10^{-6}$	-	$2.44 \times 10^{-4} \pm 2.5 \times 10^{-6}$	$1.72 \times 10^{-4} \pm 4.1 \times 10^{-6}$	$1.21 \times 10^{-4} \pm 4.5 \times 10^{-6}$

Note: Amounts are presented as millimoles of amino acid adsorbed per milligram of solid used. The calculated standard deviation of the measurements is shown after the sorption values (± SD). For experiments in which adsorption was not observed, a dash (-) is used.

Table 7. Adsorbed quantities of amino acids in each geological material, at alkaline pH (11).

Geological Materials	Adsorption (mmol/mg)				
	Gly ± SD	Lys ± SD	His ± SD	Arg ± SD	Asp ± SD
Olivine	$6.90 \times 10^{-5} \pm 2.6 \times 10^{-6}$	$1.08 \times 10^{-4} \pm 2.1 \times 10^{-6}$	$6.52 \times 10^{-5} \pm 2.7 \times 10^{-6}$	$3.40 \times 10^{-4} \pm 1.3 \times 10^{-5}$	$1.04 \times 10^{-5} \pm 3.5 \times 10^{-7}$
Pyroxene	$3.18 \times 10^{-5} \pm 3.4 \times 10^{-7}$	$1.35 \times 10^{-4} \pm 1.5 \times 10^{-6}$	-	$2.15 \times 10^{-4} \pm 1.6 \times 10^{-6}$	$1.34 \times 10^{-4} \pm 2.7 \times 10^{-6}$
Plagioclase	$4.68 \times 10^{-5} \pm 1.4 \times 10^{-6}$	$1.90 \times 10^{-4} \pm 1.2 \times 10^{-5}$	$1.72 \times 10^{-4} \pm 1.1 \times 10^{-5}$	$2.19 \times 10^{-4} \pm 9.3 \times 10^{-6}$	$5.29 \times 10^{-5} \pm 1.7 \times 10^{-6}$
Komatiite	$2.39 \times 10^{-5} \pm 2.9 \times 10^{-7}$	$4.76 \times 10^{-4} \pm 1.5 \times 10^{-5}$	$2.91 \times 10^{-5} \pm 3 \times 10^{-7}$	$3.82 \times 10^{-4} \pm 8.8 \times 10^{-5}$	$5.70 \times 10^{-7} \pm 2.7 \times 10^{-8}$
Komatiite-simulated	$2.27 \times 10^{-4} \pm 4.5 \times 10^{-6}$	$2.12 \times 10^{-4} \pm 3.6 \times 10^{-6}$	$6.45 \times 10^{-5} \pm 2.3 \times 10^{-6}$	$1.50 \times 10^{-4} \pm 3.3 \times 10^{-5}$	$5.57 \times 10^{-5} \pm 6.8 \times 10^{-7}$

Note: Amounts are presented as millimoles of amino acid adsorbed per milligram of solid used. The calculated standard deviation of the measurements is shown after the sorption values (± SD). For experiments in which adsorption was not observed, a dash (-) is used.

3.2.1. Adsorption Experiments at Natural pH Values

Natural pH refers to the pH obtained by dissolution of the amino acids in water, without further modifications. At this condition (natural pH of the amino acids) in 15 of the 25 systems, there was the adsorption of the organics onto minerals, the results of which are shown in Table 5. In the case of olivine, four of the amino acids were adsorbed; the amino acid with the highest adsorption in olivine was arginine with 2.11×10^{-4} mmol/mg, while the rest of amino acids present lower adsorption: lysine (1.27×10^{-4} mmol/mg), histidine (7.17×10^{-5} mmol/mg), and glycine (3.18×10^{-5} mmol/mg). For pyroxene, only two amino acids were adsorbed onto this mineral, glycine with 2.31×10^{-4} mmol/mg and lysine with 1.12×10^{-4} mmol/mg. In the case of plagioclase, lysine was the most adsorbed amino acid with 1.41×10^{-4} mmol/mg, and then arginine with 1.22×10^{-4}; the other two amino acids adsorbed onto plagioclase were histidine (2.15×10^{-5} mmol/mg) and glycine (1.55×10^{-5} mmol/mg), but to a lesser extent. The adsorption of lysine and arginine onto komatiite was observed, 6.99×10^{-4} mmol/mg and 8.76×10^{-4} mmol/mg, respectively; in addition, glycine (6.14×10^{-4} mmol/mg) and histidine (2.64×10^{-4} mmol/mg) were also adsorbed. In the case of the mixture of minerals, as a simulated-komatiite, only arginine was adsorbed (5.14×10^{-5} mmol/mg). Acid aspartic was not adsorbed in any of the systems at a natural pH of the solutions.

3.2.2. Adsorption Experiments at Acidic pH

This section may be divided by subheadings, which should provide a concise description of the experimental results, their interpretation, and the experimental conclusions that can be drawn. Table 6 shows all the quantities adsorbed in the experimental systems.

A vastly different behavior was witnessed in the sorption experiments at an acidic pH value, in comparison with the results mentioned above. Onto olivine, only three amino acids were adsorbed: aspartic acid (1.52×10^{-4} mmol/mg), arginine (9.55×10^{-5} mmol/mg), and histidine (4.52×10^{-5} mmol/mg), while in the pyroxene sample the amino acid with the highest adsorption value was histidine (1.75×10^{-4} mmol/mg). Other amino acids adsorbed were aspartic acid (1.27×10^{-4} mmol/mg), arginine (8.93×10^{-5} mmol/mg) and glycine (4.74×10^{-6} mmol/mg). In the case of plagioclase, the same three amino acids as for olivine were adsorbed (aspartic acid, arginine, and histidine); histidine was the one with the highest value of 9.40×10^{-4} mmol/mg. Finally, aspartic acid and arginine showed very similar adsorption amounts, 9×10^{-5} mmol/mg and 8.93×10^{-5} mmol/mg, respectively, on plagioclase.

In the case of adsorption onto komatiite, four of the amino acids were adsorbed: histidine (2.85×10^{-4} mmol/mg), aspartic acid (2.57×10^{-4} mmol/mg), and lysine (1.69×10^{-4} mmol/mg), while arginine was the molecule with the highest adsorption (3.96×10^{-4} mmol/mg). For the simulated-komatiite systems, four amino acids were also adsorbed: histidine (2.44×10^{-4} mmol/mg), arginine (1.72×10^{-4} mmol/mg), aspartic acid (1.21×10^{-4} mmol/mg), and glycine (1.65×10^{-4} mmol/mg), that, on this surface, was adsorbed at an acidic pH. In comparison, from the experiments at a natural pH, aspartic acid was adsorbed in all geological materials.

3.2.3. Adsorption Experiments at Alkaline pH

The results of the experiments at alkaline pH can be observed in Figure 1 and Table 7. There was a response for 23 of the 25 systems studied, in contrast with both pH values presented above; in fact, the alkaline pH was the one in which adsorption response was obtained for most of the systems. All the amino acids were adsorbed on olivine; arginine was the molecule with the highest value at 3.4×10^{-4} mmol/mg; the remaining amino acids were adsorbed in lower amounts—lysine (1.08×10^{-4} mmol/mg), glycine (6.90×10^{-5} mmol/mg), histidine (6.52×10^{-5} mmol/mg), and aspartic acid (1.04×10^{-5} mmol/mg). For the systems where pyroxene was used, the highest adsorption was for arginine with 2.15×10^{-4} mmol/mg, while, successive to this, lysine and aspartic acid with 1.35×10^{-4} mmol/mg and 1.34×10^{-4} mmol/mg, respectively, were found, and, finally, glycine (3.18×10^{-5} mmol/mg); histidine was not adsorbed in the pyroxene sample. In the case of plagioclase, all amino acids were adsorbed, and arginine was the amino acid with the highest sorption, 2.19×10^{-4} mmol/mg; the values of sorption for the rest of the amino acids were: lysine (1.9×10^{-4} mmol/mg), histidine (1.72×10^{-4} mmol/mg), aspartic acid (5.29×10^{-5} mmol/mg), and glycine (4.68×10^{-5} mmol/mg). In the case of komatiite, the molecules with the highest adsorption were lysine (4.76×10^{-4} mmol/mg) and arginine (3.82×10^{-4} mmol/mg); histidine and glycine were also adsorbed, with 2.91×10^{-5} mmol/mg and 2.39×10^{-4} mmol/mg, respectively. Compared with the natural rock, the komatiite-simulated sample mostly adsorbed glycine (2.27×10^{-4} mmol/mg), followed by lysine with 2.12×10^{-4} mmol/mg; the lowest values of adsorption were for aspartic acid (9.57×10^{-5} mmol/mg) and histidine (6.45×10^{-5} mmol/mg), and finally arginine was adsorbed in 1.5×10^{-4} mmol/mg.

3.2.4. Adsorption Trends under Different pH Values

As can be seen in Figure 2, if the adsorbed quantities of each amino acid in each geological material are compared at the natural pH of the solution and with the changes of pH to acid and alkaline, the pH value does indeed influence the adsorption.

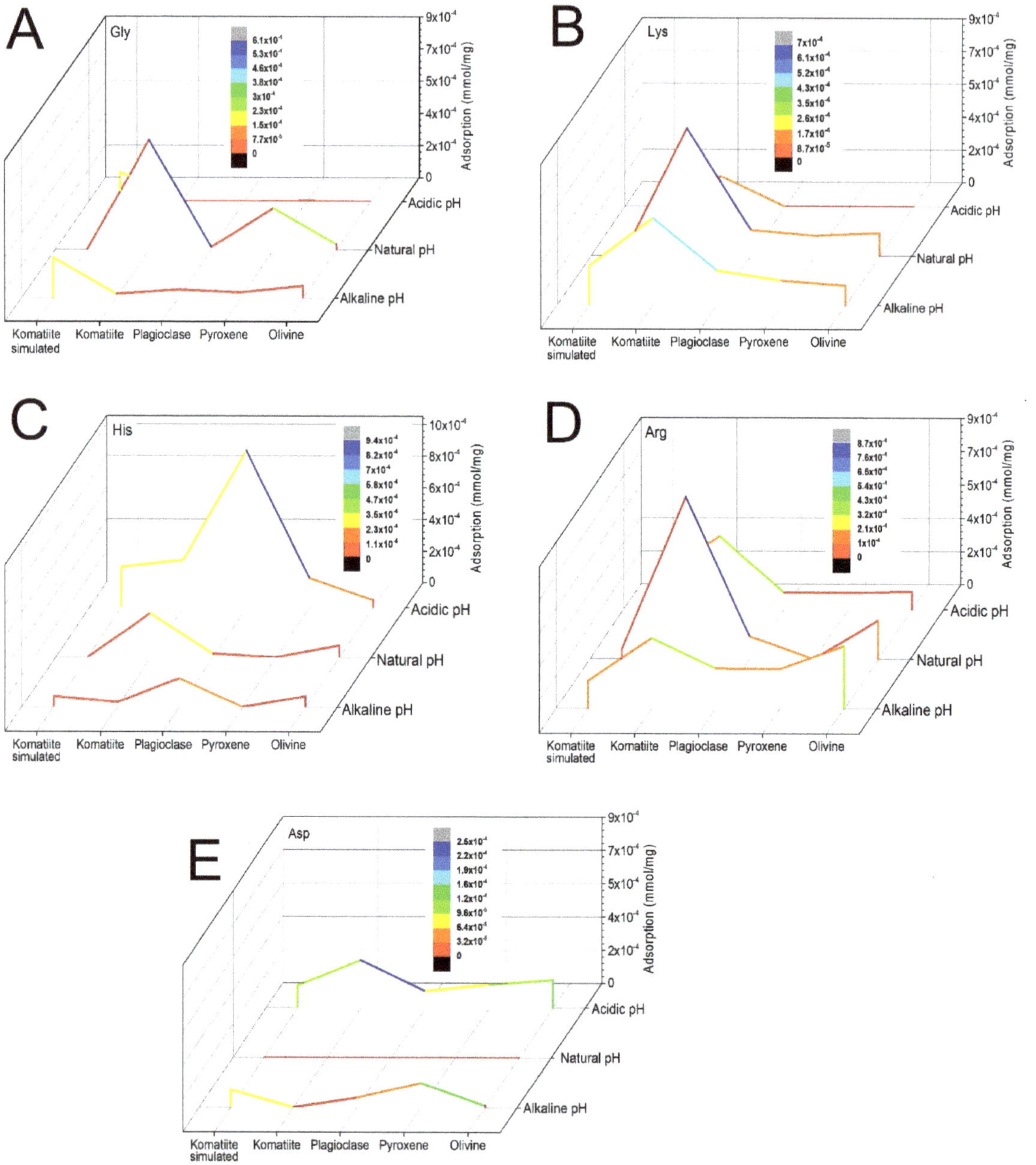

Figure 2. A comparison of the amounts adsorbed from each system at the three pH values used in the experiments: acid pH (5.5), natural pH of the solutions (see Table 2) and alkaline pH (11). (**A**) All glycine (Gly) systems; (**B**) all lysine (Lys) systems; (**C**) all histidine (His) systems; (**D**) all arginine (Arg) systems; (**E**) all aspartic acid (Asp) systems.

In general, the adsorption of glycine (Figure 2A) was higher using the natural pH of the solution. The acid pH did not favor the adsorption of this amino acid and when the alkaline pH was used, the highest amount adsorbed was in the system with the simulated komatiite.

Lysine (Figure 2B) is mostly adsorbed at a natural and alkaline pH in the komatiite system. It was observed that the acidic pH considerably favored histidine (Figure 2C) adsorption in all materials except for olivine, which maintained low amounts of adsorption in all cases. The arginine (Figure 2D) adsorption was favored using alkaline pH for most of the cases; however, it was observed that the adsorption was higher in the komatiite system at both natural and alkaline pH. The response of aspartic acid (Figure 2E) was very clear when using the natural pH; no adsorption was observed, and it showed the tendency to be adsorbed in greater amounts under acidic conditions.

In addition, it can be observed that for glycine, lysine, arginine and aspartic acid, komatiite was the system that had the highest adsorption at different pH values. In the case of histidine, the highest adsorption was recorded in plagioclase at acidic pH.

4. Discussion

4.1. Mineral Phases in Komatiite Sample

The mineralogy of an unaltered komatiite is mainly composed of Mg-rich olivine crystals due to the magma from which these rocks are formed; those olivine crystals constitute the skeleton of the characteristic spinifex texture of these rocks; the matrix in which olivine crystals are embedded is commonly composed of minerals such as pyroxenes, chromite, and plagioclase [37,38]. The komatiites that belong to the Gorgona Islands complex have been described as young komatiites (98.7 $\pm$ 7.7 to 64.4 $\pm$ 5 Ma) [39] whose spinifex texture is composed of spicules made of olivine forsterite-type and pyroxene with high calcium content and low aluminum values. In addition, andesite-type plagioclase and chloritized and vitrified glass have been found in the mineral composition of these komatiites from the Gorgona complex [40]. These mineral phases mentioned above are consistent with the mineralogical composition found by X-ray diffraction analysis performed in this study for our komatiite sample. Also, it has been described that komatiites suffer alterations in their original mineral phases due to the environment where they are found. In this process, olivine undergoes alteration to hydrated minerals such as serpentine [37]. Specifically for the komatiites of the Gorgona complex, it was observed that the rock-forming minerals are usually transformed into minerals such as chrysotile, lizardite, magnetite or chlorite [41]. The mineral phases that we found in the komatiite sample are consistent with the alterations reported in the literature for Gorgona komatiites. The low percentage of fresh olivine cores (10%) that was found is indicative that the sample has a high level of alteration, mainly due to the serpentinization of the olivine crystals.

4.2. pzc Estimations

One of the most important parameters in studies using solid-aqueous interfaces is the *pzc*, the pH value at which the surface charge of the solid equals zero. This value is important because in these interfaces the surface charge can be modified if the conditions of the medium change—the surficial charge—is pH-dependent, due to the presence of chemical species and ions that can exist in a solution [42]. The *pzc* value is widely used for identifying interactions through electrochemical forces between the surface and medium which includes the molecule in a solution [43].

It has been observed that *pzc* values published in the literature, for the same materials usually present variations, could be the result of the different methods used to estimate the *pzc* [44]. Moreover, these variations are governed by the nature of the materials, whether they are natural or synthetic samples, since properties such as the crystalline structure and the state of hydration of the material determine the adsorption capacity of ions on the surface, and these properties are different between synthetic and natural materials [45]. Nevertheless, what contributes the most to the variation in the *pzc* values, and the most important factor to consider, is the purity of the materials studied [46]. Hence, the fact that in this study only natural samples of all minerals and rocks have been worked with is probably the reason why the values obtained and presented in Table 4 show differences compared to other estimations.

The *pzc* values of olivine obtained in this study (9.9) were very close to those reported in the literature. Luce and Parks (1973) [43] reported the *pzc* value of forsterite-type olivine from a natural sample without alterations in 8.9, and Pokrovsky and Schott (2000) [47] estimated the pH value of *pzc* of a natural sample of forsterite by titration in 10. However, the *pzc* obtained for pyroxene (7.5) was discordant from those reported for this mineral. Lasaga (1984) [48] mentioned the *pzc* value for pyroxenes in a range of 2.6–3.8; therefore, the result we obtained was found to be above the reported range. No references were found to compare the point of zero charge obtained for the komatiite and the mixture of minerals that constitutes the simulated komatiite, since it is not common to make these measurements in rocks.

4.3. *Adsorption of Amino Acids in Geological Materials*

It has been observed that the adsorption of a molecule onto a surface is highly dependent on three fundamental aspects: the characteristics of the organic molecule, the properties of the surface, and the medium [49]. Here, the results obtained from the adsorptions for the systems used, based on the first two aspects mentioned, will be discussed. The medium relevance will be discussed later, in the next section.

4.3.1. Characteristics of the Amino Acids

In the first instance, the adsorption process can occur due to the differences in charges between the chemical functional groups of the organic molecules and the group that substrates expose on the surface. Theoretical studies of the adsorption of amino acids onto silicates can be found in the literature, and these describe an atomic level mechanism by which adsorption can be carried out. For example, Lomenech et al. [50] computationally analyzed the interaction of glycine on amorphous silica simulated by isolated silanol groups from the gas phase. Through the calculations carried out, they suggest that the amino acid preferentially interacts through the terminal -COOH group as an H acceptor, and with the silanol groups as an H donor, forming non-covalent interactions. In another approach, Escamilla-Roa and Moreno [51] studied the adsorption of glycine (in the gas phase) on forsterite via computational modeling and discovered different types of interactions: from H bonds to oxygen transfer between the amino acid and the top of the surface, depending on whether the glycine was in a neutral or zwitterionic form and whether the surface was dipolar or non-dipolar. These approximations are very useful in understanding the mechanisms of amino acid adsorption at the atomic level; nonetheless, it has been observed that the interactions between amino acids and mineral surfaces are different in the aqueous phase in respect to the gas phase [49]. At low organics concentrations, in an aqueous medium, the interactions through water molecules must be a relevant factor, since possible interactions between molecules of the same amino acid are avoided [52]. In an aqueous solution, amino acids change from a neutral to a zwitterionic form where they are solvated and present characteristics of divalent ions [53]. Water molecules in an aqueous medium are an important factor for the adsorption. When the solvation sphere is formed, zwitterionic amino acids are stabilized by water molecules, and this is a more stable structure (in the process of adsorption on a silicate surface) than one of the amino acids in neutral form. To continue with the stability of the solvation sphere of amino acids in the solid/liquid interface, the silanol groups on the top of the surface substitute one or more water molecules and establish physisorption-type interactions [54]. The silanol group (Si-O-H), present in all silicates, is commonly related to the interaction and retention of organic compounds, therefore acting as a surface adsorption site [55].

Due to the above information and to the fact that all the amino acids present a carboxylic acid (COOH) and an amino group (NH_2), both available to interact with the solid phases, we would expect that all the systems at the natural pH of the solution would have similar adsorption values, if the adsorption mechanism only depended on the amino and carboxyl groups. The results show that the systems presented high variations and differential adsorptions between each of them, suggesting that the adsorption must be largely governed by the possible interactions of the side chain, specific to each amino acid. On the one hand, the water molecules solvate the carboxyl and amino groups of the amino acids; the H atoms are oriented toward the COO^- and the O atoms are oriented toward the NH^{3+} group.

On the other hand, amino acids with long side chains including CH_2 groups, such as arginine and lysine in this case, can generate hydrophobic interactions with other charged groups and with water molecules. It has been observed that these interactions tend to increase with growing chain length of the amino acid side chain, and the presence of a polar terminal functional group increases the number and intensity of possible interactions with other molecules, groups, or charged ions, and with water [56]. This type of interaction in amino acids with a longer side chain can be considered an explanation for the fact that lysine and arginine were adsorbed more than the other amino acids in the adsorption experiments at the natural pH of the solutions.

As shown in Figure 3, histidine was the amino acid with the highest adsorption values. With this result, the unique structure of histidine can have a relevant weight. Histidine holds an imidazole group, as a side chain, that gives the amino acid (His) the ability to coordinate with metal cations and to be an acceptor or donor of H in H-bonds, even in different pH conditions [57]. Arginine and lysine were arranged into a subgroup due to their similar adsorption values; both are basic amino acids and have four carbons each in their side chains. In another subgroup, aspartic acid and glycine are found to be amino acids with similar adsorption values, and there the amino acids were adsorbed in smaller amounts. In general, a difference can be determined between amino acids with a 1:1 ratio of the carboxylic acid-amino group and the ones with different proportions. The presence of an extra amino group apparently increases adsorption, such as in the case of arginine and lysine.

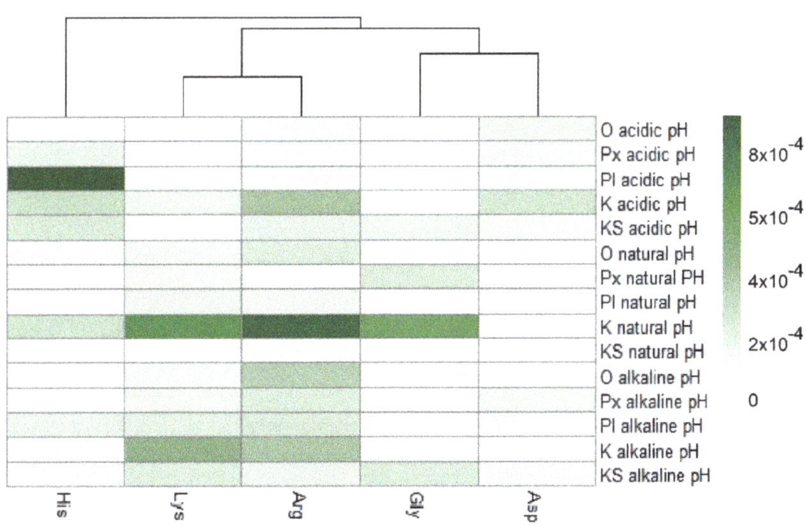

Figure 3. A heatmap showing the general ordering of the adsorptions presented by the different amino acids used. It is shown how aspartic acids and glycine are grouped by their similarity, and in another group arginine and lysine are clustered. In dark green, the highest values of adsorption are shown, and in light colors the lowest values are shown. Row names: O = olivine, Px = pyroxene, Pl = plagioclase, K = komatiite, and KS = komatiite-simulated.

In addition, it is known that functional groups, located on the side chains of many amino acids that shape proteins, are capable of forming complexes with metal ions; these groups are binding sites for metals in proteins, playing an important role in the catalytic activity of enzymes. For example, the imidazole group of histidine and the β-carboxyl group of aspartic acid are side chains capable of binding to metal centers [58]. In nature and especially in the Earth's crust, minerals containing metallic elements in their structure are very common. Silicates are the most abundant minerals in the crust; they are mainly composed of oxygen and silicon, but contain other elements (such as aluminum, iron, magnesium, sodium, potassium, and calcium) that are found in abundance and combined with silica to form 99% (wt) of the mass of the crust [59]. In the context of the early Earth, after magma cooling, the early Hadean crust would have been comprised of basalts and komatiites; these rocks are themselves comprised of olivine, pyroxene, magnetite, spinel, augite, and plagioclase minerals [60]. Therefore, the presence of metallic elements must have been very common on the surface of the Earth's early crust, and this sort of interaction between the lateral chain of amino acids and metals was possible. Nonetheless, in the case of aspartic acid, the adsorption at the natural pH of the solutions was null for all the geological materials, so the interactions between the β-carboxyl group of aspartic acid and the metallic elements of the surfaces are not fulfilled in our experiments.

4.3.2. Properties of the Geological Materials

The characteristics of the surface are another fundamental factor in sorption processes. In several works, the synthetic silica surface is widely used to simulate the surface of silicates. For example, Gao et al. [61] studied the adsorption of arginine, alanine, leucine, phenylalanine, and glutamic acid onto a mesoporous pure silica SBA-15. They observed that in this material the adsorption processes are governed by two types of interactions: (1) electrostatic forces between the ionic forms of amino acids in a solution and the positive ($\equiv SiOH_2^+$) and negative charges ($\equiv SiO^-$) present on the surface; and (2) the formation of hydrogen bonds between amino acids and water molecules; the possible formation of hydrogen bonds between the amino acids (in ionic form) and the $\equiv$SiOH groups are neglected. Undoubtedly, these works help us to understand the interactions between organic molecules and surfaces, but they present a homogeneity that natural geological materials do not show. The geological materials employed here were pulverized to increase the surface in which the amino acids can interact, but it was also sought to homogenize the natural samples, and they cannot be completely homogeneous, especially rock samples. Despite this, at the microscopic level, minerals have surface irregularities that generate variations in the adsorption of organics, such as amino acids. This effect was proposed by dos Santos et al. [52] after studying the adsorption of amino acids in 11 minerals of different nature, including forsterite-type olivine, two different pyroxene samples, and a basaltic lava. Corroboration of this proposal is the study of Galvez–Martinez et al. [62], who analyzed the adsorption of glycine on monocrystalline pyrite (100) by XPS and observed that the atomic ordering of the surface determines the form as the amino acid is adsorbed on the surface.

As can be seen in Figure 4, the surfaces adsorb amino acids in different ways. In the heatmap, olivine and pyroxene are grouped in a subgroup, since they present similar adsorption values. The other geological materials are not grouped. The hierarchical order of adsorption was: komatiite, plagioclase, simulated komatiite, olivine, and pyroxene from highest to lowest adsorption. The high adsorption of certain amino acids on komatiite suggests that complex surfaces (including a matrix and various mineral phases embedded in it and organized in a particular way) may favor sorption.

There is a high probability that other aspects, such as the superficial area of the materials and their chemical composition, influence the adsorption. However, these aspects were not studied in this work and their possible contributions go beyond what can be discussed in this manuscript; without a doubt, these aspects need to be considered for future work.

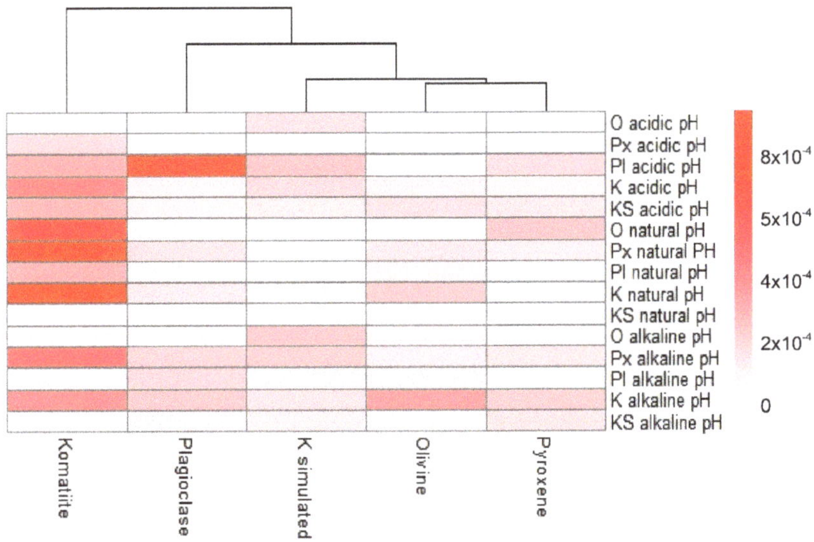

Figure 4. A heatmap showing the ordering of the surfaces by the amount of the adsorbed amino acids. In dark colors the systems that obtained the highest adsorption, in light colors those that presented the least adsorption. Komatiite-simulated is represented in the figure as K simulated.

4.4. Adsorption of Amino Acids in Geological Materials under Different pH Values

In the context of prebiotic chemistry, the role of pH has been highly discussed in the literature. The exact composition and values of parameters such as ionic strength or composition of primitive ocean are uncertain. Therefore, several pH values of those oceans have been proposed. Proposals are based on different processes and conditions, either by the ions-interchange present [63], by the interactions between the rocks of the primitive crust and the water [60], the primitive oceans seawater composition [64] or by more local environments, such as conditions prevailing on hydrothermal systems [65], to mention a few. In this work, three different pH values were used that could be representative of various environments of the early Earth.

As mentioned before, another fundamental aspect in the adsorption process is the medium in which this phenomenon occurs. In the first instance, the simplest form of adsorption occurs due to differences in charges between the surface and the organic molecule. The charges generated on the surfaces of solids depend highly on the pH of the medium. In this essay, we study the pH change in the suspensions as a variable that may or may not favor the adsorption of glycine, lysine, histidine, arginine, and aspartic acid on the five geological materials tested. The surface is positively charged when the pH value is below the *pzc* value and above this point, the surface is negatively charged [61]. Correspondingly, the chemical form of amino acids changes with the pH value due to their ionizable groups. When the pH value of the medium is equal to the isoelectric point, the molecule is electrically neutral [66]. Therefore, at that last pH value, interactions through electrostatic forces are not expected.

Gao and collaborators [61] reported the increase in the adsorption of arginine on pure silica as pH increases, for values above 2.5—results that agree with those obtained in this work. The increased adsorption of arginine at more alkaline pH can be explained by the increase in its charge density due to the side chain being positively ionized and adsorbed onto a surface of negative surface charge. This explanation is well suited to the experiments carried out at pH 11, as can be seen in Figure 5D. Arginine is positively charged, and the surfaces are negatively charged. In these systems, it can be thought that electrostatic forces govern the adsorption mechanism. In the case of glycine, as shown in Figure 5A, the amino acid is adsorbed despite having the same surface charge, so the adsorption mechanism must be associated with other processes.

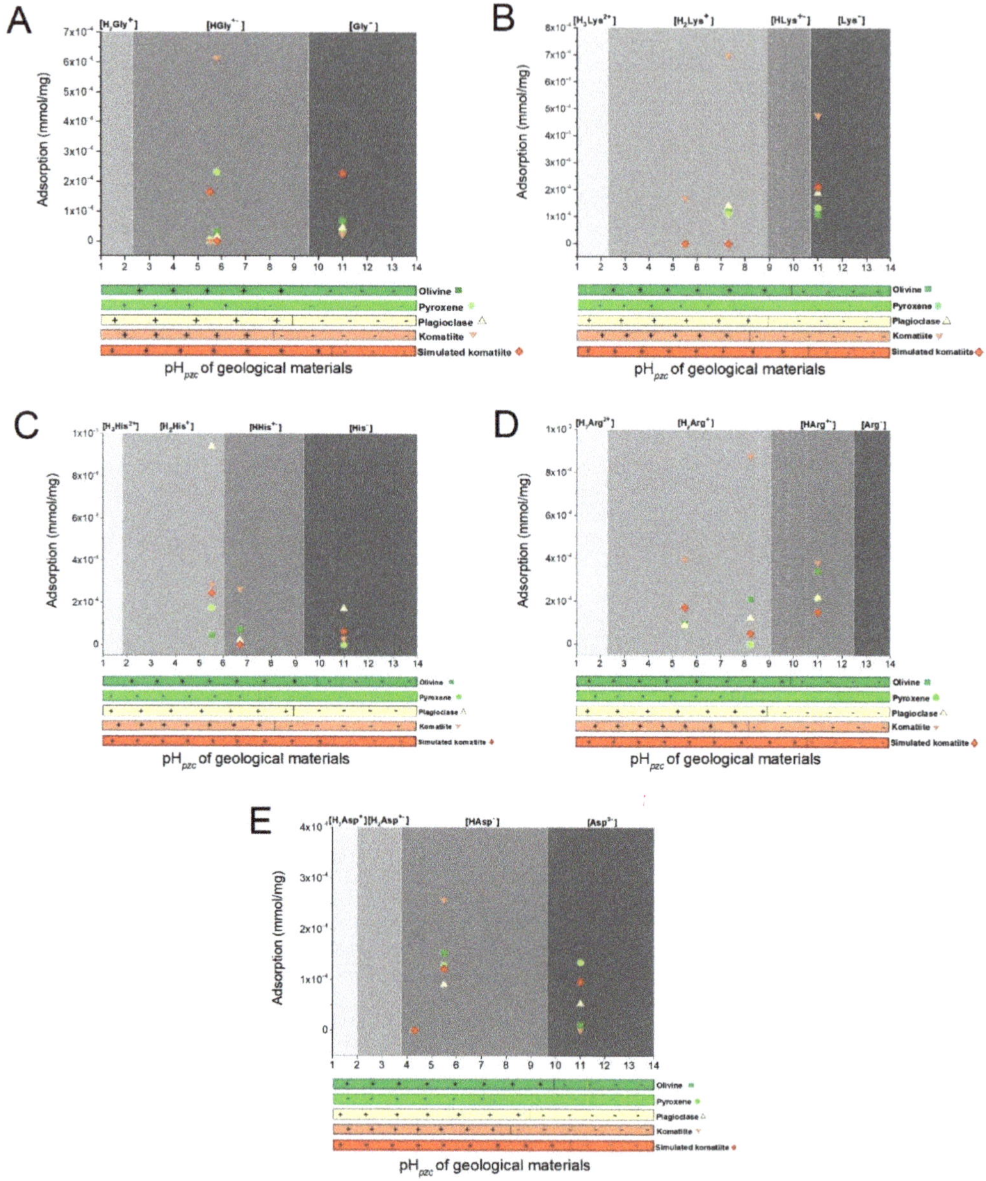

Figure 5. Amounts of each amino acid adsorbed in the geological materials at the three different pH values used. In addition, the distribution diagram of species of each amino acid can be observed and in the lower part the values obtained from the point of zero charge (*pzc*) of the geological materials. This figure shows the charge of the amino acid and the surface charge of the solids to observe if the adsorption is mainly directed by electrostatic interactions. (**A**) glycine; (**B**) lysine; (**C**) histidine; (**D**) arginine and (**E**) aspartic acid.

A greatly similar behavior can be observed for lysine and histidine (Figure 5B, Figure 5C, respectively). For aspartic acid, it is observed that when the acidic pH (5.5) is used, the amino acid is negatively charged and the surfaces are positively charged, whereby adsorption could also be due to opposite charge attraction. At the natural pH (Figure 5E), aspartic acid is also in the negative ion form, and it is not adsorbed in any of the surfaces. The differential adsorption of amino acids, even when the charge differences do not exist, may be explain in two different ways. First, it must be considered that the values of the points of zero charge are an average of processes occurring in all the faces of different crystals [67]. Therefore, at the physical scale in which the adsorption process occurs, the surfaces are heterogeneous. On the other hand, at a certain pH value, amino acids become charged (according to their species distribution diagram); however, these changes do not occur simultaneously in all molecules, and there may be some of them, in other ionic forms, that allow interaction with the surface to a lesser extent.

In addition, the constitution of the silicates must be weighed. The building units of silicates are silicon tetrahedrons (SiO_4), which have a net negative charge. This negative charge must be balanced, and for this silicon, tetrahedrons are associated with positively charged metal ions, such as $Fe^{2,3+}$, Mg^{2+}, or Na^+ [68]. It is with these metal atoms that organic molecules can associate, for example with carboxylic acids through ligands between the oxygen atoms and a metal cation [69]. It has been reported for forsterite (the same variety of olivine that was used in this work) that the Mg^{2+} ions, found on the surface, can interact with water molecules and form H-bonds with the H of the organic molecules and other functional groups thereof [68]. It should be noted that the two proposed scenarios are not mutually exclusive. Due to our results and what has been reviewed in the literature, we suggest that, in addition to electrostatic forces, the adsorption of amino acids may be directed by the formation of organometallic complexes. Specifically for histidine, it has been reported that it can bind to copper (II) metal cations; in these bonds, the metal atoms are able to coordinate up to 3 histidine binding sites, forming complexes of different types. The most determining factor to direct the site and the formation of union between copper and histidine is the pH, which defines the ionic state of the amino acid [70]. The results obtained here are quite interesting, and to corroborate the formation of these complexes it is necessary to carry out more exhaustive analyses.

In this essay, individual minerals were used to get an idea of the possible contribution of each of them in the sorption of organics, both in the komatiite sample and in the mineral mixture used as a komatiite simulation. However, the results must be treated with care, since the simulated komatiite model represents an approximation of the mineralogical composition of an unaltered komatiite, but not of its structure. Therefore, it does not reproduce important features, such as associations among minerals that compose it, and the interaction between the minerals and the matrix in which they are embedded.

Rocks are in fact heterogeneous adsorbents, real solids that exhibit surface disturbances such as polycrystalline and sometimes amorphous structures, cracks, fissures, dislocation of atoms (due to the presence of impurities), complex porous structure, etc. This surface heterogeneity affects the possible interactions of the solid with the organic molecules, and they vary at different points on the solid [71]. In contrast, all adsorption sites in the surfaces of pure or synthetic solids are considered equivalent, and along the surface they may have a constant and calculable response; in consequence, elementary theories of adsorption can be applied to describe the physical phenomenon occurring on them. However, for a heterogeneous solid, adsorption can be considered as the sum of the independent processes that take place in each phase, as a first approximation. Several attempts have been made to adjust the classical mathematical models of adsorption to describe adsorption in heterogeneous solids, but these solids are very complex systems, and it is also necessary for said mathematical description to have a real physical meaning, which makes it an important challenge [71].

4.5. Implications for Prebiotic Chemistry

On the primitive Earth, as today, the presence of isolated and pure minerals is practically nil. Mineral phases in nature are associated with rock constituents of the continental and oceanic crusts. Considering this, one of the constant criticisms of prebiotic chemistry models is that the simulations performed in the laboratory do not reflect a real scenario. The isolated, pure, or synthetic minerals that are commonly used did not exist on the primitive Earth's surface. In fact, the mineralogy of a terrestrial planet is considered a highly complex system [72]. Komatiites are the oldest rocks known to date and are believed to have the same composition as the rocks that formed the early oceanic crust. In addition, they can be associated with important primitive environments in prebiotic chemistry such as hydrothermal systems [73].

In these experiments, we wanted to contrast the capacity of adsorption of an altered natural komatiite sample versus a simulated unaltered komatiite. The results showed that the natural komatiite adsorbs more of a diversity of amino acids than the simulation, which is a positive result. The fact that surfaces of natural origin have many peculiarities that allow the adsorption of organic molecules has been much speculated. For example, the topology of the surface, which is a product of the crystallization of the mineral phases and depend on the conditions during the process, is not reproduced. In general, the surfaces of minerals and rocks are irregular or defective; these defects are probable sites where interactions between solids and organics occur [74,75]. In this case, all the characteristics of the komatiite favored the adsorption of organic molecules. The presence of serpentine in the altered komatiite can also be considered as one of the main causes of the differential adsorption between the natural sample and the komatiite-simulated. Said mineral is produced by the alteration of ultramafic rocks, and due to the high hydrothermal activity during the Hadean and Archean eons, serpentines must have been abundant and extremely common [76]. From our point of view, the presence of this alteration emulates more real conditions of the complex concentrator surfaces present in possible primitive scenarios. These results highlight the importance of further studies to analyze the concentrating role of serpentines. These complex surfaces are very difficult to study; despite this, it is important to understand their possible contribution to prebiotic processes. In addition to this, there is a particular interest in studying olivine since it is found not only on the surface of the Earth, but is also a mineral found in interstellar and interplanetary dust particles, as well as being present in meteorites and comets [68]. It can also be mentioned that forsterite is a relevant material for many environments, including Martian-type samples [77].

5. Conclusions

In this work, we studied the adsorption of five different amino acids (glycine, lysine, histidine, arginine and aspartic acid) on (1) a natural komatiite, (2) a simulated komatiite, which is a solid mixture of the three main constitutive minerals (olivine, pyroxene, and plagioclase), and (3) pure separates of each mineral. To do so, we explored the relevance of pH changes in the adsorption process.

There is a differential adsorption of the amino acids onto the solids. In general, amino acids are better adsorbed at alkaline conditions, regardless of their nature (either hydrophobic or hydrophilic) or the surface employed. This could be relevant in a prebiotic context, even when there is no consensus on the pH of primitive oceans, because a high pH would prevail in white smoker-like hydrothermal systems.

The amino acids histidine and arginine were adsorbed on most surfaces and under different pH conditions. The relevance of this finding relies on the fundamental role that these amino acids play in the stabilization of oligomers or as constituents of catalytic oligomers.

The adsorption mechanism of the organics is complex. The charge of the molecule (in the function of pH) turned out to be an important factor in the adsorption of amino acids since they could be adsorbed on surfaces by opposite charge attraction. However, electrostatic forces do not fully explain the observed adsorption in all systems. Other possible key factors in governing adsorption, associated with differences in adsorption, are the length and the functional groups of the side chain of the amino acids. Another relevant phenomenon for the matter is the formation of the sphere of solvation, which surrounds the amino acids in aqueous solution and contributes to the stabilization of the molecule. Further, the heterogeneity of surfaces and their chemical composition, in particular the number of cations on the surfaces, have a very important effect in sorption, as the occurrence cations is key in the formation of organometallic compounds.

Natural komatiite is the surface where the adsorption of most amino acids occurred, at different pH values. This is relevant since natural adsorption would have occurred on rocks, which are constituted by complex mineral interrelations. Individual minerals also adsorb amino acids, and the analyzed plagioclase is the mineral onto which more amino acids (at different pH) were adsorbed.

This work represents an effort to use komatiites as complex surfaces to study sorption processes. The results shown constitute a first approximation of further studies by using materials that recreate the likeliest substrates for prebiotic reactions. Our future experiments will include aqueous solutions with different concentrations of salts (simulating the composition of primitive oceans), and the use of mixtures of organics, in order to construct models that are closely related to prebiotic environments.

Author Contributions: Conceptualization, A.E.C.-H., M.C.-G. and F.O.-G.; Formal analysis, A.E.C.-H.; Funding acquisition, M.C.-G. and E.M.-M.; Investigation, A.E.C.-H. and M.C.-G.; Methodology, A.E.C.-H. and M.C.-G.; Project administration, M.C.-G. and E.M.-M.; Resources, M.C.-G., F.O.-G. and E.M.-M.; Supervision, M.C.-G.; Visualization, A.E.C.-H. and M.C.-G.; Writing—original draft, A.E.C.-H. and M.C.-G.; Writing—review & editing, F.O.-G. and E.M.-M. All authors have read and agreed to the published version of the manuscript.

Funding: This research was funded by CONACyT (A1-S-25341), DGAPA-PAPIIT (IN111720 and IN218323); and by the projects PID-2019-104205GB-C21 and PID-2019-107442RB-C32 by the Spanish Ministry of Science and Innovation/State Agency of Research MCIN/AEI. A.E.C.-H. acknowledges a CONACyT for a PhD grant (772180), and the Posgrado en Ciencias de la Tierra for the academic support.

Institutional Review Board Statement: Not applicable.

Informed Consent Statement: Not applicable.

Data Availability Statement: Not applicable.

Acknowledgments: Minerals used in the experiments were donated by Luca Ferrari (CGEO-UNAM), Fernando Ortega and Antoni Camprubí (IGl-UNAM). Teresa Pi i Puig (LANGEM) is acknowledged for the XRD analyses. The authors acknowledge the technical support of Alan U. Loredo-Jasso from the Laboratorio de Geoquímica Ambiental Molecular (LANGEM-IGL), Jessica Anaid Hernández Cano, and Francisco Martin Romero from the Laboratorio de Geoquímica Ambiental Aplicada (LANGEM-IGL). A.E.C.H. acknowledges the Laboratorio de Evolución Química (ICN-UNAM) for the use of its facilities.

Conflicts of Interest: The authors declare no conflict of interest.

References

1. Hazen, R.M.; Papineau, D.; Bleeker, W.; Downs, R.T.; Ferry, J.M.; McCoy, T.J.; Sverjensky, D.A.; Yang, H. Mineral Evolution. *Am. Mineral.* **2008**, *93*, 1693–1720. [CrossRef]
2. Ichikawa, H.; Gréaux, S.; Azuma, S. Subduction of the Primordial Crust into the Deep Mantle. *Geosci. Front.* **2017**, *8*, 347–354. [CrossRef]
3. Elkins-Tanton, L.T. Linked Magma Ocean Solidification and Atmospheric Growth for Earth and Mars. *Earth Planet. Sci. Lett.* **2008**, *271*, 181–191. [CrossRef]
4. Hazen, R.M.; Ferry, J.M. Mineral Evolution: Mineralogy in the Fourth Dimension. *Elements* **2010**, *6*, 9–12. [CrossRef]
5. Santosh, M.; Arai, T.; Maruyama, S. Hadean Earth and Primordial Continents: The Cradle of Prebiotic Life. *Geosci. Front.* **2017**, *8*, 309–327. [CrossRef]
6. Nna-Mvondo, D.; Martinez-Frias, J. Review Komatiites: From Earth's Geological Settings to Planetary and Astrobiological Contexts. *Earth Moon Planets* **2007**, *100*, 157–179. [CrossRef]

7. Martin, R. *Earth's Evolving Systems: The History of Planet Earth*; Jones & Bartlett Publishers: Burlington, MA, USA, 2013; ISBN 0-7637-8001-4.
8. Srivastava, R.K.; Ellam, R.M.; Gautam, G.C. Sr–Nd Isotope Geochemistry of the Early Precambrian Sub-Alkaline Mafic Igneous Rocks from the Southern Bastar Craton, Central India. *Mineral. Petrol.* **2009**, *96*, 71. [CrossRef]
9. Cleaves, H.J.; Lazcano, A. The Origin of Biomolecules. In *Chemical Evolution II: From the Origins of Life to Modern Society*; ACS Symposium Series; American Chemical Society: New York, NY, USA, 2009; Volume 1025, pp. 17–43. ISBN 978-0-8412-6980-4.
10. Kitadai, N.; Maruyama, S. Origins of Building Blocks of Life: A Review. *Geosci. Front.* **2018**, *9*, 1117–1153. [CrossRef]
11. Cockell, C.S. The Origin and Emergence of Life under Impact Bombardment. *Philos. Trans. R. Soc. B Biol. Sci.* **2006**, *361*, 1845–1856. [CrossRef]
12. Schwartz, A.W. Did Minerals Perform Prebiotic Combinatorial Chemistry? *Chem. Biol.* **1996**, *3*, 515–518. [CrossRef]
13. Ferris, J.P. Prebiotic Synthesis on Minerals: Bridging the Prebiotic and RNA Worlds. *Biol. Bull.* **1999**, *196*, 311–314. [CrossRef] [PubMed]
14. Goldschmidt, V. Geochemical Aspects of the Origin of Complex Organic Molecules on the Earth, as Precursors to Organic Life. *New Biol.* **1952**, *12*, 97–105.
15. Hartman, H.; Cairns-Smith, A.G. *Clay Minerals and the Origin of Life*; CUP Archive; Cambridge University Press: Cambridge, UK, 1986; ISBN 0-521-32408-4.
16. Schoonen, M.; Smirnov, A.; Cohn, C. A Perspective on the Role of Minerals in Prebiotic Synthesis. *Ambio* **2004**, *33*, 539–551. [CrossRef]
17. Zaia, D.A.M. A Review of Adsorption of Amino Acids on Minerals: Was It Important for Origin of Life? *Amino Acids* **2004**, *27*, 113–118. [CrossRef]
18. Lambert, J.-F. Adsorption and Polymerization of Amino Acids on Mineral Surfaces: A Review. *Orig. Life Evol. Biosph.* **2008**, *38*, 211–242. [CrossRef] [PubMed]
19. Zaia, D.A.M.; Zaia, C.T.B.V. A Few Experimental Suggestions Using Minerals to Obtain Peptides with a High Concentration of L-Amino Acids and Protein Amino Acids. *Symmetry* **2020**, *12*, 2046. [CrossRef]
20. Cleaves, H.J.; Aubrey, A.D.; Bada, J.L. An Evaluation of the Critical Parameters for Abiotic Peptide Synthesis in Submarine Hydrothermal Systems. *Orig. Life Evol. Biosph.* **2009**, *39*, 109–126. [CrossRef]
21. Kitadai, N. Energetics of Amino Acid Synthesis in Alkaline Hydrothermal Environments. *Orig. Life Evol. Biosph.* **2015**, *45*, 377–409. [CrossRef]
22. Pizzarello, S. Molecular Asymmetry in Prebiotic Chemistry: An Account from Meteorites. *Life* **2016**, *6*, 18. [CrossRef]
23. Sugahara, H.; Mimura, K. Peptide Synthesis Triggered by Comet Impacts: A Possible Method for Peptide Delivery to the Early Earth and Icy Satellites. *Icarus* **2015**, *257*, 103–112. [CrossRef]
24. Goldman, N.; Reed, E.J.; Fried, L.E.; William Kuo, I.-F.; Maiti, A. Synthesis of Glycine-Containing Complexes in Impacts of Comets on Early Earth. *Nat. Chem.* **2010**, *2*, 949–954. [CrossRef]
25. Dose, K. Peptides and Amino Acids in the Primordial Hydrosphere. In *The Origin of Life and Evolutionary Biochemistry*; Dose, K., Fox, S.W., Deborin, G.A., Pavlovskaya, T.E., Eds.; Springer: Boston, MA, USA, 1974; pp. 69–77. ISBN 978-1-4684-2115-6.
26. Stribling, R.; Miller, S.L. Energy Yields for Hydrogen Cyanide and Formaldehyde Syntheses: The Hcn and Amino Acid Concentrations in the Primitive Ocean. *Orig. Life Evol. Biosph.* **1987**, *17*, 261–273. [CrossRef] [PubMed]
27. Rode, B.M. Peptides and the Origin of Life1. *Peptides* **1999**, *20*, 773–786. [CrossRef]
28. Zaia, D.A.M.; Zaia, C.T.B.V.; De Santana, H. Which Amino Acids Should Be Used in Prebiotic Chemistry Studies? *Orig. Life Evol. Biosph.* **2008**, *38*, 469–488. [CrossRef]
29. Frenkel-Pinter, M.; Samanta, M.; Ashkenasy, G.; Leman, L.J. Prebiotic Peptides: Molecular Hubs in the Origin of Life. *Chem. Rev.* **2020**, *120*, 4707–4765. [CrossRef]
30. Colín-García, M.; Heredia, A.; Cordero, G.; Camprubí, A.; Negrón-Mendoza, A.; Ortega-Gutiérrez, F.; Beraldi, H.; Ramos-Bernal, S.; Colín-García, M.; Heredia, A.; et al. Hydrothermal Vents and Prebiotic Chemistry: A Review. *Boletín La Soc. Geológica Mex.* **2016**, *68*, 599–620. [CrossRef]
31. Raggi, L.; Bada, J.L.; Lazcano, A. On the Lack of Evolutionary Continuity between Prebiotic Peptides and Extant Enzymes. *Phys. Chem. Chem. Phys.* **2016**, *18*, 20028–20032. [CrossRef]
32. Frenkel-Pinter, M.; Haynes, J.W.; Petrov, A.S.; Burcar, B.T.; Krishnamurthy, R.; Hud, N.V.; Leman, L.J.; Williams, L.D. Selective Incorporation of Proteinaceous over Nonproteinaceous Cationic Amino Acids in Model Prebiotic Oligomerization Reactions. *Proc. Natl. Acad. Sci. USA* **2019**, *116*, 16338–16346. [CrossRef]
33. de Souza, C.M.D.; Carneiro, C.E.A.; Baú, J.P.T.; da Costa, A.C.S.; Ivashita, F.F.; Paesano, A.; di Mauro, E.; de Santana, H.; Holm, N.G.; Neubeck, A.; et al. Interaction of Forsterite-91 with Distilled Water and Artificial Seawater: A Prebiotic Chemistry Experiment. *Int. J. Astrobiol.* **2013**, *12*, 135–143. [CrossRef]
34. Starcher, B. A Ninhydrin-Based Assay to Quantitate the Total Protein Content of Tissue Samples. *Anal. Biochem.* **2001**, *292*, 125–129. [CrossRef]
35. Sacchi, M.; Jenkins, S.J. Co-Adsorption of Water and Glycine on Cu{110}. *Phys. Chem. Chem. Phys.* **2014**, *16*, 6101–6107. [CrossRef] [PubMed]
36. Samulewski, R.B.; Guimarães, R.T.D.; Zaia, D.A.M. Histidine Adsorption onto Modified Montmorillonite under Prebiotic Chemistry Conditions: A Thermodynamic and Kinetic Study. *Int. J. Astrobiol.* **2021**, *20*, 81–92. [CrossRef]

37. Arndt, N.; Lesher, M.C.; Barnes, S.J. *Komatiite;* Cambridge University Press: Cambridge, UK, 2008.
38. Aitken, B.G.; Echeverría, L.M. Petrology and Geochemistry of Komatiites and Tholeiites from Gorgona Island, Colombia. *Contrib. Mineral. Petrol.* **1984**, *86*, 94–105. [CrossRef]
39. Serrano, L.; Ferrari, L.; Martínez, M.L.; Petrone, C.M.; Jaramillo, C. An Integrative Geologic, Geochronologic and Geochemical Study of Gorgona Island, Colombia: Implications for the Formation of the Caribbean Large Igneous Province. *Earth Planet. Sci. Lett.* **2011**, *309*, 324–336. [CrossRef]
40. Gansser, A.; Dietrich, V.J.; Cameron, W.E. Palaeogene Komatiites from Gorgona Island. *Nature* **1979**, *278*, 545–546. [CrossRef]
41. Dietrich, V.J.; Gansser, A.; Sommerauer, J.; Cameron, W.E. Palaeogene Komatiites from Gorgona Island, East Pacific—A Primary Magma for Ocean Floor Basalts? *Geochem. J.* **1981**, *15*, 141–161. [CrossRef]
42. Kosmulski, M. IEP as a Parameter Characterizing the PH-Dependent Surface Charging of Materials Other than Metal Oxides. *Adv. Colloid Interface Sci.* **2012**, *171–172*, 77–86. [CrossRef]
43. Luce, R.W.; Parks, G.A. Point of Zero Charge of Weathered Forsterite. *Chem. Geol.* **1973**, *12*, 147–153. [CrossRef]
44. Cristiano, E.; Hu, Y.-J.; Siegfried, M.; Kaplan, D.; Nitsche, H. A Comparison of Point of Zero Charge Measurement Methodology. *Clays Clay Miner.* **2011**, *59*, 107–115. [CrossRef]
45. Bertus, L.M.; Carcel, R.A. Prediction of TiO_2 and WO_3 Nanopowders Surface Charge by Evaluation of Point of Zero Charge (PZC). *Environ. Eng. Manag. J. (EEMJ)* **2011**, *10*, 1021–1026. [CrossRef]
46. Kosmulski, M. The PH-Dependent Surface Charging and the Points of Zero Charge. *J. Colloid Interface Sci.* **2002**, *253*, 77–87. [CrossRef] [PubMed]
47. Pokrovsky, O.S.; Schott, J. Forsterite Surface Composition in Aqueous Solutions: A Combined Potentiometric, Electrokinetic, and Spectroscopic Approach. *Geochim. Cosmochim. Acta* **2000**, *64*, 3299–3312. [CrossRef]
48. Lasaga, A.C. Chemical Kinetics of Water-Rock Interactions. *J. Geophys. Res. Solid Earth* **1984**, *89*, 4009–4025. [CrossRef]
49. Lambert, J.-F.; Stievano, L.; Lopes, I.; Gharsallah, M.; Piao, L. The Fate of Amino Acids Adsorbed on Mineral Matter. *Planet. Space Sci.* **2009**, *57*, 460–467. [CrossRef]
50. Lomenech, C.; Bery, G.; Costa, D.; Stievano, L.; Lambert, J.F. Theoretical and Experimental Study of the Adsorption of Neutral Glycine on Silica from the Gas Phase. *ChemPhysChem* **2005**, *6*, 1061–1070. [CrossRef] [PubMed]
51. Escamilla-Roa, E.; Moreno, F. Adsorption of Glycine by Cometary Dust: Astrobiological Implications. *Planet. Space Sci.* **2012**, *70*, 1–9. [CrossRef]
52. dos Santos, R.; Patel, M.; Cuadros, J.; Martins, Z. Influence of Mineralogy on the Preservation of Amino Acids under Simulated Mars Conditions. *Icarus* **2016**, *277*, 342–353. [CrossRef]
53. Niehues, G.; Heyden, M.; Schmidt, D.A.; Havenith, M. Exploring Hydrophobicity by THz Absorption Spectroscopy of Solvated Amino Acids. *Faraday Discuss.* **2011**, *150*, 193–207. [CrossRef]
54. Costa, D.; Lomenech, C.; Meng, M.; Stievano, L.; Lambert, J.-F. Microsolvation of Glycine by Silanol Ligands: A DFT Study. *J. Mol. Struct. Theochem* **2007**, *806*, 253–259. [CrossRef]
55. Bilgiç, C. Investigation of the Factors Affecting Organic Cation Adsorption on Some Silicate Minerals. *J. Colloid Interface Sci.* **2005**, *281*, 33–38. [CrossRef]
56. Hossain, A.; Roy, S.; Dolui, B.K. Effects of Thermodynamics on the Solvation of Amino Acids in the Pure and Binary Mixtures of Solutions: A Review. *J. Mol. Liq.* **2017**, *232*, 332–350. [CrossRef]
57. Liao, S.-M.; Du, Q.-S.; Meng, J.-Z.; Pang, Z.-W.; Huang, R.-B. The Multiple Roles of Histidine in Protein Interactions. *Chem. Cent. J.* **2013**, *7*, 44. [CrossRef] [PubMed]
58. Shimazaki, Y.; Takani, M.; Yamauchi, O. Metal Complexes of Amino Acids and Amino Acid Side Chain Groups. Structures and Properties. *Dalton Trans.* **2009**, *38*, 7854–7869. [CrossRef] [PubMed]
59. Skinner, B.J. Earth Resources. *Proc. Natl. Acad. Sci. USA* **1979**, *76*, 4212–4217. [CrossRef] [PubMed]
60. Ueda, H.; Shibuya, T. Composition of the Primordial Ocean Just after Its Formation: Constraints from the Reactions between the Primitive Crust and a Strongly Acidic, CO_2-Rich Fluid at Elevated Temperatures and Pressures. *Minerals* **2021**, *11*, 389. [CrossRef]
61. Gao, Q.; Xu, W.; Xu, Y.; Wu, D.; Sun, Y.; Deng, F.; Shen, W. Amino Acid Adsorption on Mesoporous Materials: Influence of Types of Amino Acids, Modification of Mesoporous Materials, and Solution Conditions. *J. Phys. Chem. B* **2008**, *112*, 2261–2267. [CrossRef]
62. Galvez-Martinez, S.; Escamilla-Roa, E.; Zorzano, M.-P.; Mateo-Marti, E. Defects on a Pyrite(100) Surface Produce Chemical Evolution of Glycine under Inert Conditions: Experimental and Theoretical Approaches. *Phys. Chem. Chem. Phys.* **2019**, *21*, 24535–24542. [CrossRef]
63. Bada, J.L.; Miller, S.L. Ammonium Ion Concentration in the Primitive Ocean. *Science* **1968**, *159*, 423–425. [CrossRef]
64. Zaia, D.A.M. Adsorption of Amino Acids and Nucleic Acid Bases onto Minerals: A Few Suggestions for Prebiotic Chemistry Experiments. *Int. J. Astrobiol.* **2012**, *11*, 229–234. [CrossRef]
65. Villafañe-Barajas, S.A.; Colín-García, M.; Negrón-Mendoza, A.; Ruiz-Bermejo, M. An Experimental Study of the Thermolysis of Hydrogen Cyanide: The Role of Hydrothermal Systems in Chemical Evolution. *Int. J. Astrobiol.* **2020**, *19*, 369–378. [CrossRef]
66. Damodaran, S. Amino Acids, Peptides and Proteins. *Fennemas Food Chem.* **2008**, *4*, 217–329.
67. Churchill, H.; Teng, H.; Hazen, R.M. Correlation of PH-Dependent Surface Interaction Forces to Amino Acid Adsorption: Implications for the Origin of Life. *Am. Mineral.* **2004**, *89*, 1048–1055. [CrossRef]

68. Rimola, A.; Sodupe, M.; Ugliengo, P. Role of Mineral Surfaces in Prebiotic Chemical Evolution. In Silico Quantum Mechanical Studies. *Life* **2019**, *9*, 10. [CrossRef] [PubMed]
69. Meléndez-López, A.; Colín-García, M.; Ortega-Gutiérrez, F.; Cruz-Castañeda, J. Role of the Interchangeable Cations on the Sorption of Fumaric and Succinic Acids on Montmorillonite and Its Relevance in Prebiotic Chemistry. *Orig. Life Evol. Biosph.* **2021**, *51*, 87–116. [CrossRef] [PubMed]
70. Marti, E.M.; Methivier, C.; Dubot, P.; Pradier, C.M. Adsorption of (S)-Histidine on Cu(110) and Oxygen-Covered Cu(110), a Combined Fourier Transform Reflection Absorption Infrared Spectroscopy and Force Field Calculation Study. *J. Phys. Chem. B* **2003**, *107*, 10785–10792. [CrossRef]
71. Dąbrowski, A. Adsorption—From Theory to Practice. *Adv. Colloid Interface Sci.* **2001**, *93*, 135–224. [CrossRef]
72. Hazen, R.M.; Eldredge, N. Themes and Variations in Complex Systems. *Elements* **2010**, *6*, 43–46. [CrossRef]
73. Bost, N.; Loiselle, L.; Foucher, F.; Ramboz, C.; Westall, F.; Gaillard, F.; Auguste, J.-L. Synthesis of Gusev Crater Analogue Basalts, Mars: Interest for Astrobiology. In Proceedings of the 44th Lunar and Planetary Science Conference, The Woodlands, TX, USA, 18–22 March 2013; Lunar and Planetary Institute (LPI)/Universities Space Research Association (USRA). Abstract Number (14571). Available online: https://www.lpi.usra.edu/meetings/lpsc2013/pdf/1457.pdf (accessed on 1 November 2022).
74. Dass, A.V.; Hickman-Lewis, K.; Brack, A.; Kee, T.P.; Westall, F. Stochastic Prebiotic Chemistry within Realistic Geological Systems. *ChemistrySelect* **2016**, *1*, 4906–4926. [CrossRef]
75. Galvez-Martinez, S.; Escamilla-Roa, E.; Zorzano, M.-P.; Mateo-Marti, E. Ar+ Ion Bombardment Dictates Glycine Adsorption on Pyrite (100) Surface: X-Ray Photoemission Spectroscopy and DFT Approach. *Appl. Surf. Sci.* **2020**, *530*, 147182. [CrossRef]
76. Villafañe-Barajas, S.A.; Ruiz-Bermejo, M.; Rayo-Pizarroso, P.; Gálvez-Martínez, S.; Mateo-Martí, E.; Colín-García, M. A Lizardite-HCN Interaction Leading the Increasing of Molecular Complexity in an Alkaline Hydrothermal Scenario: Implications for Origin of Life Studies. *Life* **2021**, *11*, 661. [CrossRef]
77. Potenti, S.; Manini, P.; Fornaro, T.; Poggiali, G.; Crescenzi, O.; Napolitano, A.; Brucato, J.R.; Barone, V.; d'Ischia, M. Solid State Photochemistry of Hydroxylated Naphthalenes on Minerals: Probing Polycyclic Aromatic Hydrocarbon Transformation Pathways under Astrochemically-Relevant Conditions. *ACS Earth Space Chem.* **2018**, *2*, 977–1000. [CrossRef]

life

MDPI

Communication

Results of an Eight-Year Extraction of Phosphorus Minerals within the Seymchan Meteorite

Maheen Gull *, Tian Feng and Matthew A. Pasek

School of Geosciences, University of South Florida, Tampa, FL 33584, USA
* Correspondence: ambermaheen@yahoo.com

Abstract: In-fall of extraterrestrial material including meteorites and interstellar dust particles during the late heavy bombardment are known to have brought substantial amounts of reduced oxidation-state phosphorus to the early Earth in the form of siderophilic minerals, e.g., schreibersite ($(FeNi)_3P$). In this report, we present results on the reaction of meteoritic phosphide minerals in the Seymchan meteorite in ultrapure water for 8 years. The ions produced during schreibersite corrosion (phosphite, hypophosphate, pyrophosphate, and phosphate) are stable and persistent in aqueous solution over this timescale. These results were also compared with the short-term corrosion reactions of the meteoritic mineral schreibersite's synthetic analog Fe_3P in aqueous and non-aqueous solutions (ultrapure water and formamide). This finding suggests that the reduced-oxidation-state phosphorus (P) compounds including phosphite could be ubiquitous and stable on the early Earth over a long span of time and such compounds could be readily available on the early Earth.

Keywords: phosphorus; early Earth; origin of life; meteorites; phosphides; phosphite

Citation: Gull, M.; Feng, T.; Pasek, M.A. Results of an Eight-Year Extraction of Phosphorus Minerals within the Seymchan Meteorite. *Life* **2022**, *12*, 1591. https://doi.org/10.3390/life12101591

Academic Editors: Ranajay Saha and Alberto Vázquez-Salazar

Received: 20 September 2022
Accepted: 11 October 2022
Published: 12 October 2022

1. Introduction

Meteoritic impacts have been directly linked with the origin of life on the early Earth [1,2]. Since the identification of abundant organics in the Murchison meteorite, prebiotic chemists have assumed that meteorites supplied extraterrestrial compounds to the early Earth, which could have kick-started the events of prebiotic syntheses [3–13]. In addition to organic compounds, meteorites are also believed to have supplied a non-negligible portion of the phosphorus within the Earth's crust, possibly through a heavy bombardment [14,15]. Studies have shown that meteoritic mineral schreibersite [$(FeNi)_3P$]) is ubiquitous in iron meteorites and is common to many other meteorite classes [16,17]. This iron-nickel phosphide is known to be among the first inorganic P compounds to condense from the solar nebula as part of homogeneous accretion model and therefore, is considered to be one of the most ancient P minerals within our solar system [18].

Around 5–10% of all crustal P was at some stage delivered as meteoritic phosphide minerals [14,15,17]. Schreibersite is known to corrode in water by oxidation to release several inorganic P species in aqueous solution. These species include phosphite (HPO_3^{2-} with P (III)), orthophosphate (HPO_4^{2-} with P (V)), hypophosphate ($HP_2O_6^{3-}$ with P (IV)), and pyrophosphate ($HP_2O_7^{3-}$ with P (V)), respectively [11,15], with a concomitant release of H_2.

Schreibersite or its corrosion products can react with organic compounds to form C-O-P and C-P type compounds [18–22]. These reactions established the case for schreibersite as a prebiotically relevant P containing mineral. Since around 70% of the Earth's surface is covered with water, much of this meteoritic material would have hit oceans [23], and schreibersite would have corroded within water. The discovery of phosphite signatures in the ancient marine carbonates [20] supports the possibility that the corrosion products of schreibersite may have been present in Archean oceans. However, it is not known that the phosphite ions are persistent and can be stable for a long timespan.

In the present study, we describe the corrosion of schreibersite within the meteorite Seymchan (a pallasite that is compositionally similar to IIE irons [24], and which is relatively enriched in nickel at ~9 wt.%) that was left to corrode in deionized water for 8 years. Seymchan bears a few percent by volume of schreibersite [16]. This work specifically identifies whether the ions generated by schreibersite are ephemeral, or whether they can persist for years.

Furthermore, the corrosion reaction (P) products from the Seymchan meteorites were also compared with that of the synthetic analog Fe_3P. The corrosion reactions of Fe_3P were carried out at the room temperature (25 °C) for shorter timescales. Two types of solvents were employed; (1) ultrapure water as an aqueous medium, and (2) formamide as a non-aqueous medium. The latter option was chosen because of its availability and ubiquity in the interstellar regions and various habitable zones in our galaxy [25,26]. Furthermore, it is also considered to be a prebiotically relevant alternative solvent due to its remarkable solvation capabilities and ability to provide an anhydrous medium for phosphorylation and condensation reactions [27–31].

2. Materials and Methods

Both iron phosphide or Fe_3P (99.5% trace metals basis) and formamide (from BioUltra) for molecular biology, ≥99.5% were purchased from Sigma Aldrich (St. Louis, MO, USA). The doubly deionized water (DDI) was produced in house using a Barnstead (Dubuque, IA, USA) NANO pure® Diamond Analytical combined reverse osmosis-deionization system as reported previously [20–22]. It should be noted that the authentic meteoritic mineral schreibersite [$(Fe,Ni)_3P$] or its commercially available synthetic analogue (Fe_3P) upon corrosion release same P species as reported previously [15–22], but this study further investigates the use of authentic schreibersite in corrosion experiments.

The Seymchan meteorite sample was procured from meteorite dealers in the form of shavings covered with oil—done by the dealers during the preparation of Seymchan into slices to sell to collectors—which prevented oxidation and corrosion of these samples in the air. The oil was removed by placing the shavings into a clean flask containing acetone, sealing the container, and stirring the solution on a stir plate with the help of magnetic stirrer. The mixture was allowed to stir overnight to ensure dissolution of oil into the acetone. The shavings were then filtered and rinsed off with acetone. The process was repeated 2 times and after that was allowed to completely air dry and followed by heating in the oven at 50 °C for 5–6 h to ensure evaporation of the organic solvent. The mean surface area of the Seymchan was around 0.001 m^2/g based on the dimensions of the shavings (individually ~5 mm × ~1 mm × ~0.3 mm).

About 5 g Seymchan was added to 15 mL DDI (or ultrapure) water with no other additives and was sealed tightly in a glass vial under air and was set to corrode at room temperature for about 8 years. The glass vial was placed in a completely dark cabinet. No water evaporated from this vial over those 8 years, and the material was not stirred over this time. For comparing the corrosion reaction P products of meteoritic phosphide from Seymchan with that of Fe_3P, the following general procedure was followed. About 0.5 g Fe_3P (to each reaction sample) was added to a clean, small glass vial containing 5 mL DDI (or ultrapure) water (labeled as sample B) and to a glass vial containing 5 mL of pure formamide (labeled as sample F). A clean magnetic stirrer was added to each glass vial containing the reaction mixture. The glass vials of the solution mixtures were then tightly sealed and were allowed to stir at the room temperature for about 2 weeks.

The release of P from the Seymchan and the phosphide corrosion reactions in the ultrapure water and formamide were studied by ^{31}P-NMR (Varian, Palo Alto, CA, USA) as reported previously [32]. In case of Seymchan, the tightly sealed sample tube was opened after 8 years and to it 0.5 M Na_4EDTA solution was added until the pH reached ~12. This step was done to help with the extraction of P species from the meteorite sample as previously [15–22,32]. The mixture containing meteorite shavings in DI water and EDTA solution was transferred to a clean glass vial to let it stir at room temperature for 1 day, followed by filtration of the solution, and drying at the room temperature. Similarly, prior to ^{31}P-NMR analysis, both reaction samples B and F were also treated with 0.5 M Na_4EDTA solution until the pH reached to be around 11–12. The solution changed color to blackish brown. It was filtered and about 7 mL of the filtrate was transferred to a clean watch glass and was allowed to dry at the room temperature. The dried mixture was rehydrated with D_2O and was added to NMR tubes for the analysis. The dried mixture was rehydrated with 1 mL D_2O and was analyzed by the NMR. The molarity [M] of the solution was calculated by the formula (Equation (1)) as suggested by [19].

$$[\mathrm{M}] = 0.0075 \times \left(\frac{\left(\frac{\mathrm{S}}{\mathrm{N}}\right)}{\sqrt{\mathbf{Scans}}}\right)^2 + 0.0007 \times \left(\frac{\left(\frac{\mathrm{S}}{\mathrm{N}}\right)}{\sqrt{\mathbf{Scans}}}\right) + 0.0001 \quad (1)$$

where S/N is the signal to noise ratio, and Scans is the number of NMR scans taken. As reported in the previous studies [19]. this relationship was empirically determined and is accurate to about 10% over the range of 10^{-4} to 10^{-2} M based on the sample spectra obtained [19].

3. Results

The phosphide from the meteorite sample leached into the solution as various forms of P including orthophosphate, phosphite, pyrophosphate and hypophosphate. In addition, after being immersed in water the meteoritic shavings appeared to be somewhat rusty but all in all still retained the metallic luster of the meteoritic shavings (Figure 1). Interestingly, a few of the corroded shavings were brought closer to the magnet and were indeed pulled by the magnet, implying that even after 8 years the shavings were still not completely oxidized.

Figure 1. Seymchan meteorite (shavings) post 8 years corrosion reaction at room temperature.

The ^{31}P-NMR studies of the meteorite extract solutions (Figure 2, Table 1) identified four major inorganic P species including: orthophosphate, phosphite, pyrophosphate and hypophosphate. The concentrations of the P species in the Seymchan solution were determined by the integration of the ^{31}P-NMR spectra. It is important to mention here that NMR integration can be quantitative if the integrations are done over a narrow range of frequencies, e.g., less than 50 ppm [19]. In the proton-coupled mode of the NMR, the phosphite peak showed the typical split into a doublet which was identified by measuring the coupling constant (around 570 Hz) [33]. The measuring of the coupling constant distinguished phosphite from the orthophosphate peak which also located around the same region (~5.5 ppm). The pyrophosphate peak was located at −4.8 ppm and hypophosphate peak was identified to be around 13.5 ppm [19]. ^{31}P-NMR spectrum is shown in the H-coupled mode of ^{31}P-NMR.

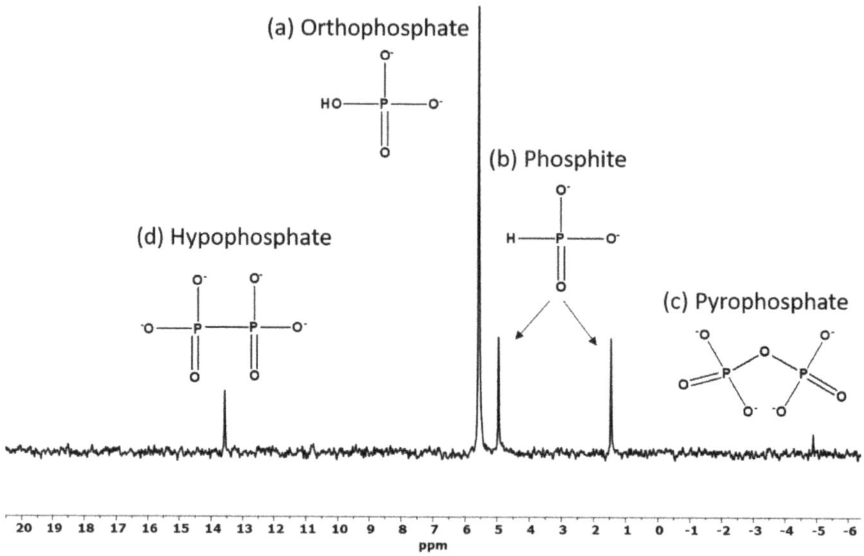

Figure 2. Corrosion reaction of Seymchan meteorite (sample SEY) stored at room temperature for 8 years. The *Y*-axis is in intensity and has arbitrary units.

Table 1. Yields [1] (%) and molarities of various P species released as a consequence of Fe_3P corrosion in the solution.

Sample	Yields (%)				Concentration of Various P Species in Solutions (mmolar)				[2] Total Molarity (mmolar)
	$(HPO_4)^{2-}$	$(HPO_3)^{2-}$	$(P_2O_7)^{4-}$	$(P_2O_6)^{4-}$	$[(HPO_4)^{2-}]$	$[(HPO_3)^{2-}]$	$[(P_2O_7)^{4-}]$	$[(P_2O_6)^{4-}]$	
SEY	71.12	23.40	0.20	5.27	1.20	0.40	0.05	0.10	1.75
F	29.24	54.39	16.37	ND	2.1	2.5	0.3	0.0	4.90
B	48.26	42.69	2.55	6.50	748	413.7	0.95	13.5	1176.1

[1] The yields of the corrosion P products of the corrosion reactions were calculated on the basis of the total phosphorus dissolved and by the peak integration method as previously reported [19]. [2] The total molarity of the solution here is the sum of the individual molarities of each of the inorganic P species including $(HPO_4)^{2-}$, $(HPO_3)^{2-}$, $(P_2O_7)^{4-}$, $(P_2O_6)^{4-}$. The molarities are represented in millimoles. In addition, the labeling of the samples and the solutions descriptions are mentioned in Table 1.

The ^{31}P-NMR studies of the corrosion reaction of Fe_3P in the formamide (sample F, Table 1) and in the ultrapure water (sample B, Table 1) revealed similar reaction products with quantitatively different ratios as observed in the meteorite Seymchan's solution extract and as also reported previously [19]. The major species in all samples included orthophosphate $(HPO_4)^{2-}$, phosphite $(HPO_3)^{2-}$, pyrophosphate $(P_2O_7)^{4-}$ and hypophosphate $(P_2O_6)^{4-}$, respectively. Interestingly, the highest yield of the pyrophosphate (condensed phosphates) was observed when formamide was used as a solvent (sample F) (Figure 3). This reaction sample also showed the highest amount of phosphite indicating that the oxidation of the P in non-aqueous media proceeds at rather slower rates as compared in the aqueous solutions. No hypophosphate was observed in sample F further indicating a relatively slower rate of corrosion and possibly the free radicals to form this P species were not generated as effectively in the formamide medium. The overall higher corrosion rate was observed in sample B (Figure 4) containing ultrapure water (or doubly de-ionized water) as a solvent thus indicating the ease of corrosion under aqueous conditions, also accompanied with facile oxidation. Furthermore, the highest concentration of P products was in sample B (Table 1).

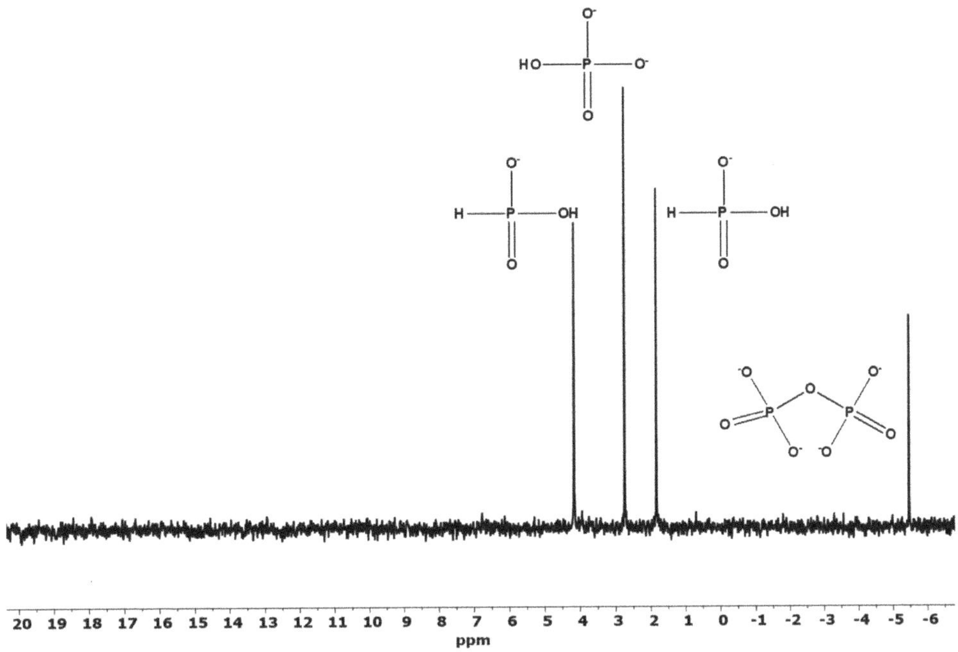

Figure 3. Corrosion reactions of Fe_3P in pure formamide (sample F). The *Y*-axis is in intensity and has arbitrary units.

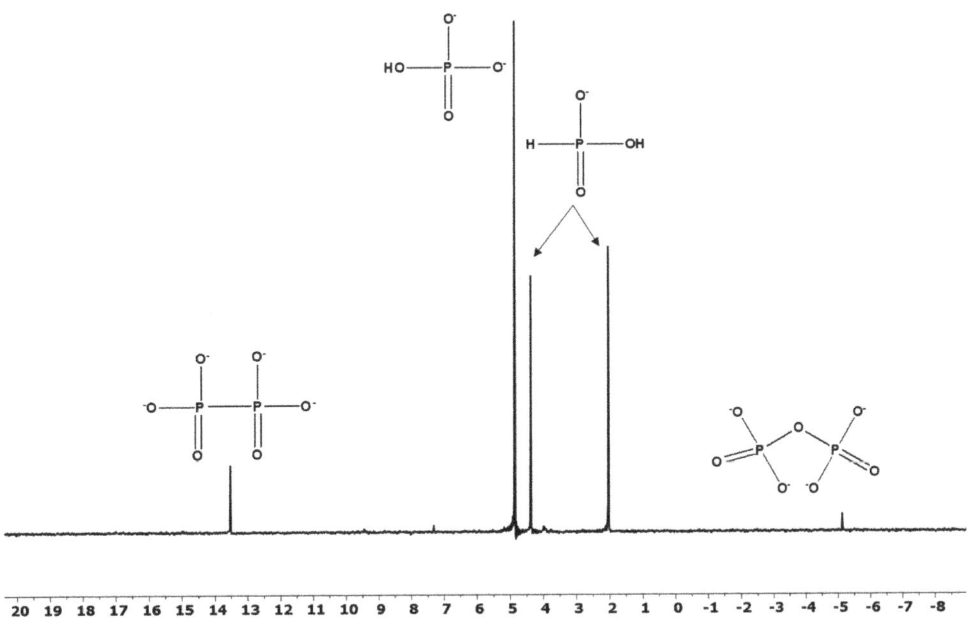

Figure 4. Corrosion reaction of Fe_3P in ultrapure water (sample B). The *Y*-axis is in intensity and has arbitrary units.

4. Discussion

After 8 years, the ^{31}P-NMR results demonstrated that the P compounds within the Seymchan–water solution were comparable to that of a typical ^{31}P-NMR spectrum of schreibersite's analog Fe_3P (samples B and F). Phosphite—typically considered to be highly reactive—was present even after 8 years of corrosion under air, suggesting the phosphite species is stable even though this compound is out of equilibrium and should have oxidized by reaction with water to form phosphate [19,34]. These results are also consistent with the results of the hydrothermal treatment of the Seymchan meteorite [24] that studied the schreibersite corrosion from Seymchan after ~2 weeks. The results showed that the treatment of Seymchan shavings afforded various P species including hypophosphate (4% of solution P), orthophosphate (34%), phosphite (49%) and pyrophosphate (13%), respectively [24]. These results are comparable to the present study (Table 1), though phosphite is lower after 8 years.

It is intriguing that room temperature reactions of schreibersite present in the meteorite can afford the formation of condensed phosphates including pyrophosphate, and the rather unstable phosphite is still present (Figure 2). Previous studies have shown that in the presence of Fe metal, the rate of conversion of phosphite to phosphate in one day is about 46% under air [19]. Such reactions are also possible for the Seymchan meteorite as it is known to be rich in Fe. Furthermore, the kinetic stability of orthophosphate, hypophosphate and pyrophosphate has been found to be more than phosphite in the presence of oxidizing radicals [19,35].

Considering that the ancient oceans were anoxic and Fe (II)-rich [36,37], phosphate is known to adsorb to iron oxyhydroxides, which may have depleted early oceans with respect to total phosphate. The effect of iron oxyhydroxides on reduced P compounds is unknown, but Pasek and colleagues have shown that the reduced-oxidation-state P compounds including phosphite and hypophosphite oxidize in the presence of Fe (II) and H_2O_2 under prebiotic conditions. This process gives orthophosphate, pyrophosphate and triphosphate. This process occurs under oxic as well as anoxic conditions [38].

Alternatively, it has also been suggested that Archean oceans may have been strongly P-limited due to the selective binding of phosphate to iron oxyhydroxide [39–41]. Routes around this issue may have included plausible pathway of the reduction of phosphate to phopshite by iron (II) at low diagenetic temperatures (160–200 °C) and under a dinitrogen atmosphere. This suggests a plausible geochemical pathway of solubilizing P in the Archean ocean and indicates that the reduction of phosphate to phosphite would have been widespread in the Archean [39].

Therefore, phosphite, and not necessarily phosphate, may have been a major P source in the early anoxic oceans and would definitely have played a key role in the origin of P biochemistry. Iron and phosphate (even as calcium phosphates) could have been ubiquitous in many Archean sediments, as these have already been co-located in hydrothermal plumes off the southern East Pacific Rise [42]. Therefore, the alteration of phosphate into phosphite in the presence of Fe (II) via diagenesis would have been possible in the ancient oceans [39].

Furthermore, considering that our leaching experiment was performed under neutral pH conditions (i.e., using doubly deionized water with pH = 7), similar to the increased pH values (~6.5 to 7.0) of the early Archean oceans as reported by Halevy and Bachan [43]. It should be noted that the increase in the pH could slightly decline the rates of the meteoritic phosphide corrosion; however, the P-speciation would remain the same [19].

Given the amounts of siderophilic P estimated to have impacted the early Earth during the late heavy bombardment [44], it is highly likely that highly water soluble and chemically reactive P species were readily available on the early Earth ready to kick-start the process of phosphorylation and phosphonylation of the organic compounds present on the early Earth.

Pasek and Lauretta [45] suggested that iron meteorites with schreibersite may have delivered that about 10^8 kg/yr of meteoritic P to the surface of the Earth, would have been exposed to aqueous modification [45]. Ritson et al. [46] similarly propose 3×10^7 kg/year delivered during the late accretion of meteorite material to the Earth. These models suggest that reduced P could have played a major role in delivering the reactive P to the early oceans and, even though these compounds are more reactive than phosphate, they can persist for longer timescales. These results also complement our previous finding of phosphite signatures in the marine carbonates [20], which would have been brought to the oceans possibly during late bombardment period as discussed above and would have played a major role in contributing to supplying the reduced form of P to the early Earth.

Furthermore, the successful corrosion of Fe_3P in pure formamide show remarkable promise for anhydrous chemistry and possible phosphorylation of the organics with better yields. For about 40 years, this solvent has been utilized both as a reactant as well as a solvent for the prebiotic syntheses of biomolecules [27–31,47]. One of the major advantages of considering formamide as a solvent in the prebiotic chemistry is due, in part, to the fact that it supports condensation reactions that are required for the phosphate ester formation [28].

Many successful experiments have shown its viability as a solvent for phosphorylation reactions [27–31]. Yet, using phosphates as phosphorylating agents still presents a challenge that can be overcome using schreibersite or Fe_3P. These results suggest that schreibersite readily corrodes in formamide and releases various P species as it would in water-based corrosion reactions. Future studies would help further solve the "problem of phosphorus chemistry" when schreibersite (or Fe_3P) is used as a phosphorylating agent and formamide is the solvent. This route can plausibly be beneficial in two ways; (1) to release more soluble forms of P and (2) favorable anhydrous conditions necessary for phosphorylation.

Furthermore, prior work has shown that the reduced-oxidation-state P compounds are quite capable of reacting with biomolecules to form organic-P compounds essential for the cellular membrane structures and DNA/RNA-forming units (simple nucleotide units) [20–22]. Our previous findings of detecting phosphite signatures in 3.5 billion-year-old marine carbonates [20] and the present findings thus strongly suggest that the reduced-oxidation-state P compounds were not only readily available but were also quite stable on the early Earth. These reduced-oxidation-state P compounds would have essentially taken part in kick-starting the origin of life chemistry on the early Earth.

Author Contributions: For Conceptualization, M.A.P. and M.G.; methodology, M.G.; formal analysis, M.G.; investigation, M.G. and M.A.P.; resources, M.A.P.; data curation, M.G. and T.F.; writing—original draft preparation, M.G. and T.F.; writing—review and editing, M.G., T.F. and M.A.P.; funding acquisition, M.A.P. All authors have read and agreed to the published version of the manuscript.

Funding: This work was supported by NASA Exobiology program No: 80NSSCC18K1288 and 80NSSC22K0509.

Data Availability Statement: Not Applicable.

Acknowledgments: This work has been supported in part by University of South Florida Interdisciplinary NMR Facility, The Department of Chemistry and the College of Arts and Sciences, Tampa, Florida. Authors thank the USF-NMR facility for the help with NMR. The authors also acknowledge Ryan Barkley for the help with the figures and the graphical abstract and to Romain Guilbaud for his valuable and helpful comments to improve this manuscript as well as to Professor Heather Abott-Lyon for the helpful discussions. Maheen Gull would dedicate this work to the loving memory of her wonderful grandfather Paul T. Barkley who has been so kind and loving and thanks to Luna and Nova Barkley for the wonderful source of inspiration and finally to Andrew Stella-Vega for the useful discussions about this manuscript.

Conflicts of Interest: Authors declare no conflicts of interest.

References

1. Cockell, C.S. The origin and emergence of life under impact bombardment. *Philos. Trans. R. Soc. Lond. B Biol. Sci.* **2006**, *361*, 1845–1856. [CrossRef]
2. Osinski, G.R.; Cockell, C.S.; Pontefract, A.; Sapers, H.M. The role of meteorite impacts in the origin of life. *Astrobiology* **2020**, *20*, 1121–1149. [CrossRef] [PubMed]

3. Mason, B. Organic matter from space. *Sci. Am.* **1963**, *208*, 43–49. [CrossRef]
4. Fegley, B.; Prinn, R.G.; Hartman, H.; Watkins, G.H. Chemical effects of large impacts on the Earth's primitive atmosphere. *Nature* **1986**, *319*, 305–308. [CrossRef] [PubMed]
5. Oró, J.; Mills, T. Chemical evolution of primitive solar system bodies. *Adv. Space Res.* **1989**, *9*, 105–120. [CrossRef]
6. Chyba, C.; Sagan, C. Endogenous production, exogenous delivery and impact-shock synthesis of organic molecules: An inventory for the origins of life. *Nature* **1992**, *355*, 125–132. [CrossRef]
7. Kobayashi, K.; Kasamatsu, T.; Kaneko, T.; Saito, T. Production of organic compounds in interstellar space. In *Exobiology: Matter, Energy, and Information in the Origin and Evolution of Life in the Universe*, 1st ed.; Chela-Flores, J., Raulin, F., Eds.; Springer: Dordrecht, The Netherlands, 1998; pp. 213–216.
8. Pohorille, A. From organic molecules in space to the origins of life and back. *Adv. Space Res.* **2002**, *30*, 1509–1520. [CrossRef]
9. Ehrenfreund, P.; Cami, J. Cosmic carbon chemistry: From the interstellar medium to the early Earth. *Cold Spring Harb. Perspect. Biol.* **2010**, *2*, a002097. [CrossRef]
10. Ehrenfreund, P.; Spaans, M.; Holm, N.G. The evolution of organic matter in space. *Philos. Trans. R. Soc. A Math. Phys. Eng. Sci.* **2011**, *369*, 538–554. [CrossRef]
11. Kwok, S. Complex organics in space from Solar System to distant galaxies. *Astron. Astrophys. Rev.* **2016**, *24*, 8. [CrossRef]
12. Nakano, H.; Hirakawa, N.; Matsubara, Y.; Yamashita, S.; Okuchi, T.; Asahina, K.; Tanaka, R.; Suzuki, N.; Naraoka, H.; Takano, Y.; et al. Precometary organic matter: A hidden reservoir of water inside the snow line. *Sci. Rep.* **2020**, *10*, 7755. [CrossRef] [PubMed]
13. Gull, M.; Pasek, M.A. The role of glycerol and its derivatives in the biochemistry of living organisms, and their prebiotic origin and significance in the evolution of life. *Catalysts* **2021**, *11*, 86. [CrossRef]
14. Macià, E.; Hernández, M.V.; Oró, J. Primary sources of phosphorus and phosphates in chemical evolution. *Orig. Life Evol. Biosph.* **1997**, *27*, 459–480. [CrossRef] [PubMed]
15. Pasek, M.A.; Lauretta, D.S. Aqueous corrosion of phosphide minerals from iron meteorites: A highly reactive source of prebiotic phosphorus on the surface of the early Earth. *Astrobiology* **2005**, *5*, 515–535. [CrossRef] [PubMed]
16. Pirim, C.; Pasek, M.A.; Sokolov, D.A.; Sidorov, A.N.; Gann, R.D.; Orlando, T.M. Investigation of schreibersite and intrinsic oxidation products from Sikhote-Alin, Seymchan, and Odessa meteorites and Fe_3P and Fe_2NiP synthetic surrogates. *Geochim. Cosmochim. Acta* **2014**, *140*, 259–274. [CrossRef]
17. Bryant, D.E.; Kee, T.P. Direct evidence for the availability of reactive, water soluble phosphorus on the early Earth. H-Phosphinic acid from the Nantan meteorite. *Chem. Commun.* **2006**, *22*, 2344–2346. [CrossRef]
18. Pasek, M.A. Rethinking early Earth phosphorus geochemistry. *Proc. Natl. Acad. Sci. USA* **2008**, *105*, 853–858. [CrossRef]
19. Pasek, M.A.; Dworkin, J.; Lauretta, D.S. A radical pathway for phosphorylation during schreibersite corrosion with implications for the origin of life. *Geochim. Cosmochim. Acta* **2007**, *71*, 1721–1736. [CrossRef]
20. Pasek, M.A.; Harnmeijer, J.P.; Buick, R.; Gull, M.; Atlas, Z. Evidence for reactive reduced phosphorus species in the early Archean Ocean. *Proc. Natl. Acad. Sci. USA* **2013**, *110*, 10089–10094. [CrossRef]
21. Gull, M.; Mojica, M.A.; Fernández, F.M.; Gaul, D.A.; Orlando, T.M.; Liotta, C.L.; Pasek, M.A. Nucleoside phosphorylation by the mineral schreibersite. *Sci. Rep.* **2015**, *5*, 17198. [CrossRef]
22. La Cruz, N.L.; Qasim, D.; Abbott-Lyon, H.; Pirim, C.; McKee, A.D.; Orlando, T.; Gull, M.; Lindsay, D.; Pasek, M.A. The evolution of the surface of the mineral schreibersite in prebiotic chemistry. *Phys. Chem. Chem. Phys.* **2016**, *18*, 20160–20167. [CrossRef] [PubMed]
23. Simonson, B.M.; Davies, D.; Wallace, M.; Reeves, S.; Hassler, S.W. Iridium anomaly but no shocked quartz from Late Archean microkrystite layer: Oceanic impact ejecta? *Geology* **1998**, *26*, 195–198. [CrossRef]
24. Bryant, D.E.; Greenfield, D.; Walshaw, R.D.; Evans, S.M.; Nimmo, A.E.; Smith, C.L.; Wang, L.; Pasek, M.A.; Kee, T.P. Electrochemical studies of iron meteorites: Phosphorus redox chemistry on the early Earth. *Int. J. Astrobiol.* **2009**, *8*, 27–36. [CrossRef]
25. Adande, G.R.; Woolf, N.J.; Ziurys, L.M. Observations of interstellar formamide: Availability of a prebiotic precursor in the galactic habitable zone. *Astrobiology* **2013**, *13*, 439–453. [CrossRef] [PubMed]
26. López-Sepulcre, A.; Balucani, N.; Ceccarelli, C.; Codella, C.; Dulieu, F.; Theulé, P. Interstellar Formamide (NH_2CHO), a Key Prebiotic Precursor. *ACS Earth Space Chem.* **2019**, *3*, 2122–2137. [CrossRef]
27. Gull, M.; Cafferty, B.J.; Hud, N.V.; Pasek, M.A. Silicate-promoted phosphorylation of glycerol in non-aqueous solvents: A prebiotically plausible route to organophosphates. *Life* **2017**, *7*, 29. [CrossRef]
28. Schoffstall, A.M. Prebiotic phosphorylation of nucleosides in formamide. *Orig. Life Evol. Biosph.* **1976**, *7*, 399–412. [CrossRef]
29. Furukawa, Y.; Kim, H.J.; Hutter, D.; Benner, S.A. Abiotic regioselective phosphorylation of adenosine with borate in formamide. *Astrobiology* **2015**, *15*, 259–267. [CrossRef]
30. Schoffstall, A.M.; Barto, R.J.; Ramos, D.L. Nucleoside and deoxynucleoside phosphorylation in formamide solutions. *Orig. Life Evol. Biosph.* **1982**, *12*, 143–151. [CrossRef]
31. Costanzo, G.; Saladino, R.; Crestini, C.; Ciciriello, F.; Di Mauro, E. Nucleoside phosphorylation by phosphate minerals. *J. Biol. Chem.* **2007**, *282*, 16729–16735. [CrossRef]
32. Pasek, M.A.; Omran, A.; Feng, T.; Gull, M.; Lang, C.; Abbatiello, J.; Garong, L.; Johnston, R.; Ryan, J.; Abbott-Lyon, H. Serpentinization as a route to liberating phosphorus on habitable worlds. *Geochim. Cosmochim. Acta* **2022**, *336*, 332–340. [CrossRef]
33. Pasek, M.A. *Phosphorus NMR of Natural Samples*, 1st ed.; Amazon: Seattle, WA, USA, 2018.
34. Gulick, A. Phosphorus as a factor in the origin of life. *Am. Sci.* **1955**, *43*, 479–489.

35. Schwartz, A.W.; Van der Veen, M. Synthesis of hypophosphate by ultraviolet irradiation of phosphite solutions. *Inorg. Nucl. Chem. Lett.* **1973**, *9*, 39–41. [CrossRef]
36. Poulton, S.W.; Canfield, D.E. Ferruginous Conditions: A Dominant Feature of the Ocean through Earth's History. *Elements* **2011**, *7*, 107–112. [CrossRef]
37. Guilbaud, R.; Poulton, S.; Butterfield, N.; Zhu, M.; Sheields-Zhou, G.A. A global transition to ferruginous conditions in the early Neoproterozoic oceans. *Nat. Geosci.* **2015**, *8*, 466–470. [CrossRef]
38. Pasek, M.A.; Kee, T.P.; Bryant, D.E.; Pavlov, A.A.; Lunine, J.I. Production of potentially prebiotic condensed phosphates by phosphorus redox chemistry. *Angew. Chem. Int. Engl.* **2008**, *47*, 7918–7920. [CrossRef]
39. Herschy, B.; Chang, S.J.; Blake, R.; Lepland, A.; Abbott-Lyon, H.; Sampson, J.; Atlas, Z.; Kee, T.P.; Pasek, M.A. Archean phosphorus liberation induced by iron redox geochemistry. *Nat. Commun.* **2018**, *9*, 1346. [CrossRef]
40. Ingalls, M.; Grotzinger, J.P.; Present, T.; Rasmussen, B.; Fischer, W.W. Carbonate-associated phosphate (CAP) indicates elevated phosphate availability in Neoarchean shallow marine environments. *Geophys. Res. Lett.* **2022**, *49*, e2022GL098100. [CrossRef]
41. Planavsky, N.J.; Rouxel, O.J.; Bekker, A.; Lalonde, S.V.; Konhauser, K.O.; Reinhard, C.T.; Lyons, T.W. The evolution of the marine phosphate reservoir. *Nature* **2010**, *467*, 1088–1090. [CrossRef] [PubMed]
42. Poulton, S.W.; Canfield, D.E. Co-diagenesis of iron and phosphorus in hydrothermal sediments from the southern east Pacific Rise: Implications for the evaluation of paleoseawater phosphate concentrations. *Geochim. Cosmochim. Acta* **2006**, *70*, 5883–5898. [CrossRef]
43. Halevy, I.; Bachan, A. The geologic history of seawater pH. *Science* **2017**, *355*, 1069–1071. [CrossRef]
44. Macia, E. The role of phosphorus in chemical evolution. *Chem. Soc. Rev.* **2005**, *34*, 691–701. [CrossRef]
45. Pasek, M.A.; Lauretta, D.S. Extraterrestrial flux of potentially prebiotic C, N, and P to the early Earth. *Orig. Life Evol. Biosph.* **2008**, *38*, 5–21. [CrossRef] [PubMed]
46. Ritson, D.J.; Mojzsis, S.J.; Sutherland, J. Supply of phosphate to early Earth by photogeochemistry after meteoritic weathering. *Nat. Geosci.* **2020**, *13*, 344–348. [CrossRef]
47. Gull, M. Prebiotic phosphorylation reactions on the early Earth. *Challenges* **2014**, *5*, 193–212. [CrossRef]

life

MDPI

Article

Novel Apparatuses for Incorporating Natural Selection Processes into Origins-of-Life Experiments to Produce Adaptively Evolving Chemical Ecosystems

Robert Root-Bernstein [1,*] and Adam W. Brown [2]

[1] Department of Physiology, Michigan State University, East Lansing, MI 48824, USA
[2] Department of Art, Art History and Design, Michigan State University, East Lansing, MI 48824, USA
* Correspondence: rootbern@msu.edu

Abstract: Origins-of-life chemical experiments usually aim to produce specific chemical end-products such as amino acids, nucleic acids or sugars. The resulting chemical systems do not evolve or adapt because they lack natural selection processes. We have modified Miller origins-of-life apparatuses to incorporate several natural, prebiotic physicochemical selection factors that can be tested individually or in tandem: freezing-thawing cycles; drying-wetting cycles; ultraviolet light-dark cycles; and catalytic surfaces such as clays or minerals. Each process is already known to drive important origins-of-life chemical reactions such as the production of peptides and synthesis of nucleic acid bases and each can also destroy various reactants and products, resulting selection within the chemical system. No previous apparatus has permitted all of these selection processes to work together. Continuous synthesis and selection of products can be carried out over many months because the apparatuses can be re-gassed. Thus, long-term chemical evolution of chemical ecosystems under various combinations of natural selection may be explored for the first time. We argue that it is time to begin experimenting with the long-term effects of such prebiotic natural selection processes because they may have aided biotic life to emerge by taming the combinatorial chemical explosion that results from unbounded chemical syntheses.

Keywords: prebiotic evolution; natural selection; selection; cycles; chemical ecosystems; chemical environments; chemical ecosystems; ultraviolet light; dark; heat; cold; freezing; drying; wetting

Citation: Root-Bernstein, R.; Brown, A.W. Novel Apparatuses for Incorporating Natural Selection Processes into Origins-of-Life Experiments to Produce Adaptively Evolving Chemical Ecosystems. *Life* **2022**, *12*, 1508. https://doi.org/10.3390/life12101508

Academic Editors: Ranajay Saha and Alberto Vázquez-Salazar

Received: 7 September 2022
Accepted: 24 September 2022
Published: 28 September 2022

1. Introduction

The iconic 1953 Miller experiment [1,2] producing amino acids from methane, hydrogen, ammonia and water vapor subjected to heat and electrical discharges opened the door to the modern era of prebiotic chemical experimentation. In the subsequent seventy years, many variations of the Miller apparatus have been constructed as well as novel designs better adapted to more specific prebiotic chemical syntheses (e.g., [3–8]). All previous apparatuses share the common characteristic of being meant mainly or solely to carry out chemical reactions, in most cases aimed at producing specific prebiotic chemical products such as amino acids, nucleic acids, sugars or lipids [9]. However, such chemical experiments fall short of permitting such chemical systems to evolve the spontaneous emergence of adaptive chemical ecosystems because the apparatuses used to carry them out do not incorporate two key processes required of evolutionary systems.

According to Darwinian evolutionary theory, four processes are at work in biologically evolving systems that presumably need to be present in analogous forms in evolving chemical ecosystems: (1) individuals within a species vary and species themselves differ from one another in the degree to which they are adapted (that is to say, survive relative to each other) in their environment; (2) individual and species variations are reproducible to the extent that some members of each generation resemble their parents more than other members; (3) natural processes act non-randomly to select among the reproducible variants, altering the rates at which variants survive and reproduce; (4) the most highly reproduced variants form the populations from which further variations can arise and be selected [10,11]. In this manner, biological entities continuously evolve to adapt to their changing environments. Of these four processes, prebiotic chemical experiments have satisfied only the first and second processes. The first process is satisfied by the ability to produce a wide range of chemical species that include biotically relevant ones. The reproducibility criterion can arguably be said to be met because the reactants in each of these chemical processes reliably generate these biotically relevant products so that there is a ready (often steady-state) supply within that chemical environment of the relevant species [12]. Such a continuous resupply of product can be equated with reproduction in biological systems, even though the mechanisms differ.

While the production of a range of prebiotic compounds is a necessary precursor to the evolution of adaptive chemical ecosystems, the greater the number of starting compounds and the longer reactions are run, the greater the number of types of compounds that result leading to what has been called the "prebiotic combinatorial chemical explosion" [13]. There are only three ways to control such combinatorial explosion and those are to selectively eliminate some of the compounds [13]); preferentially produce specific compounds through the emergence of stable synthetic or reproductive cycles [14,15]; or to stabilize and preserve selected compounds by the formation of molecular complexes [16,17]. We argue that it is time to begin experimenting with the long-term effects of the prebiotic natural selection processes that may have aided biotic life to emerge by taming the combinatorial chemical explosion. Little research, especially experimental, has been devoted to this problem [13,18–20]. Among the challenges has been the fact that most synthetic experiments have been limited to a week or less in duration [21] so that apparatuses that can explore longer-term experiments are needed such as that described, for example, by Asche et al. [22]. These new apparatuses need to make possible the evolution of chemical ecosystems through adaptation to selection pressures acting over long enough periods of time so that only some compounds "survive" and the selected chemical "species" can be subjected to further variation and adaptation.

A number of such natural selection processes would have been present during the origins of life, including light-dark cycles, freeze–thaw cycles, high heat-cool cycles, wet-dry cycles, and the presence of catalytic surfaces such as clays or minerals, sometimes working in various combinations. While each of these processes have been used previously to aid particular types of prebiotic chemical synthesis experiments (see below), few have been explored as means of prebiotic natural selection and, as far as we are aware, their combined effects have hardly been explored. In order to facilitate such selection experiments, we have prototyped two types of apparatuses that can be run continuously over long periods of time (months and perhaps years) and that incorporate all these selection processes and permit many of them to be run concurrently or serially.

The choice of prebiotic natural selection processes to incorporate into our apparatuses was made through consideration of physical factors present in prebiotic environments that could potentially act as selection agents on chemicals produced by Miller-like experiments. The following provides a brief review of these physical factors, which include cycles of wetness/dryness, heat/cold, light/dark, and presence/absence of potentially catalytic surfaces. We did not consider chemical factors (e.g., the types of gases or compounds, their purity, etc.) in the design of the apparatuses, although it is obvious that these would affect the outcomes of experiments because the choice of gases did not affect apparatus designs. We did, however, design the apparatuses to be re-gassed and the mineral content of the water supply replenished so that a constant resupply of chemical reactants was available to drive production of products ("species reproducibility"). A very brief, representative review of some ways in which each of potential selection factors has been used previously in prebiotic reaction processes follows, which is the result of a broad but not exhaustive literature search for studies using the four processes chosen above. This overview established the general nature of current uses of each process to drive prebiotic reactions, which was followed by a more intensive search on PubMed, and Scholar Google, using appropriate key words, to determine whether special apparatuses have been designed to carry out such processes and whether any are capable of long-term or automated use. Finally, a similar search was made for use of these processes for natural selection of prebiotic compounds. We make no claim of completeness and apologize if we have overlooked any obviously relevant sources.

1.1. Wet/Dry Cycles

Wet/dry cycles could have occurred in a number of ways under prebiotic conditions, ranging from daily or weekly cycles in tidal pools, ponds or puddles, to fluctuating hydrothermal pools, to seasonal cycles such as the monsoons that cause periodic flooding in some desert regions [23]. No studies using wet/dry cycles to select out prebiotic products or to evaluate destruction/disappearance of such products during cycling were located. Wet/dry cycles have been employed to drive a number of types of prebiotic reactions such as the polymerization of peptides from amino acids [24–27], ribonucleic acid (RNA) polymers from ribonucleotide precursors [23,28,29], and to drive the encapsulation of such polymers in lipid protocells [30–32]. Fox et al. [4] have created an automated, autonomous apparatus to create such wet-dry cycles called the 'wet-dry apparatus' (WDA), which permits variable-length cycles and temperatures during the cycle phases. It is capable of operating in an oxygen-free (pure nitrogen) atmosphere but not, apparently using the kind of ammonia-methane-hydrogen mixtures employed in Miller-type experiments. In fact, it appears that no one has subjected Miller-type product mixtures to such wet/dry cycles either to drive polymerization and lipid aggregation reactions or to study whether such cycles deplete some products while benefiting the survival of others.

1.2. Freeze/Thaw Cycles

Freeze/thaw cycles are a normal part of seasonal, and sometimes daily, temperature changes in much of the Northern and Southern hemispheres of the modern-day Earth, in mountainous regions even near the tropics, and occurs extraterrestrially on Mars, various planetary moons of the Solar System and may be of relevance to understanding prebiotic chemistry on comets. Because of the Earth's rotation and tilt, such cycles may have begun soon after the Earth cooled sufficiently to support prebiotic chemistry some 4 billion years ago or more [33].

Freeze/thaw cycles are often used in prebiotic chemistry experiments for promoting chemical syntheses. The basic premise of such experiments is that as a solution reaches its freezing point, solutes form pockets of highly concentrated eutectic solutions favoring some condensation and polymerization reactions. Lowering the temperature further may then protect the products thus formed. Using this approach, purine and pyrimidines have been synthesized from hydrogen cyanide or cyanoacetylene reactants [34]; ligation of nucleic acids into polymers [35]; RNA copying [36]; ribozyme assembly [37]; selection for optimized ribozyme function at freezing temperatures [38,39]; and the encapsulation of RNA within lipid membranes [40]. Notably absent from the literature are reports of the successful use of freeze/thaw cycles to polymerize peptides from amino acids or complex sugars from monomers, though whether this absence is due to the failure of this method for these compounds or lack of relevant experimentation is not clear.

Freeze/thaw cycles can also act as selection processes, although little emphasis has yet been put on these effect in prebiotic chemical experiments and most of what is known comes from studies from the food industry, analytical chemistry, biochemistry, and medicine. Cycles of freezing and thawing quickly degrade some amino acids (especially beta-alanine, cysteine, glutamic acid and aspartic acid) while others, such as valine and leucine, are not affected [41]. Some peptide hormones such as adrenocorticotrophin, are rapidly degraded as a result of repeated freeze–thaw cycles, but others, such as thyrotropin-releasing hormone, are not [42] and as little as 24 h at 40 °C causes some proteins to degrade into constituent amino acids and other metabolites, a phenomenon enhanced by repeated freeze–

thaw cycles [43]. On the other hand, the presence of some amino acids, such as glutamate and lysine, can protect peptides and proteins from freeze–thaw related degradation [44], so that small-molecule interactions need to be considered in complex prebiotic systems. Freeze–thaw cycles also promote cross-linking of proteins, especially those with oxidizable side chains such as cysteines [45]. Furthermore, freeze–thaw cycles preferentially degrade some RNA [46] and DNA [47] sequences in preference to others.

In sum, freeze/thaw cycles have clear relevance to prebiotic chemical conditions but have hardly been investigated in that context and no specialized apparatus to perform prebiotic freeze/thaw experiments appears to have been described.

1.3. Ultraviolet Light/Dark Cycles

Light-driven chemical reactions are a common aspect of life on Earth and light of various wavelengths, especially in the ultraviolet (UV) range (generally defined as wavelengths between 10 and 400 nm) has often been employed in prebiotic chemistry experiments since absorption of UV light can enhance the chemical reactivity of some classes of compounds. It is generally assumed that the amount of UV light reaching the prebiotic Earth was significantly greater than it is today because UV light is efficiently absorbed by ozone, which was essentially absent from the atmosphere until about 600 million years ago [48]. UV light is also absorbed below 204 nm by carbon dioxide (CO_2) and below 168 nm by water or water vapor [48]. The relative absence of UV absorbers on Mars and many other planets and in outer space also makes UV light a potential factor in driving prebiotic reactions in these environments. A more contentious question is whether bodies of water absorb sufficient UV light to prevent it from driving prebiotic chemical reactions or acting as a selection factor during the emergence of life. Experiments show that UV absorption is highly dependent on the solutes contained in any particular body of water, prebiotic freshwater sources being largely transparent to UV while salt waters specifically decrease shortwave ($\leq$220 nm) UV flux and iron-rich waters can be UV-opaque [49]. Thus, salt and iron-rich waters may protect prebiotic organic compounds from degradation by UV light [50]. Overall, the effects of UV light on prebiotic chemistry may vary by geography and hydrogeology and, in the cases in which UV light is combined with wet/dry and freeze/thaw cycles, by the stage of the cycle at which it is present or absent.

Notably, "UV light . . . may have both constructive and destructive effects for prebiotic syntheses" [49] and is thus another good candidate for acting as a selection factor (both positive and negative) on the products of prebiotic syntheses. Synthetically, UV light acting on simple mixtures of methanol-ammonia-water or acetone-ammonia-water yielded fundamental chemical precursors to many more complex molecules such as methylisourea and acetamide [51] and has driven a variety of prebiotic reactions involving hydrogen cyanide (HCN), sulfites, and sulfides [3]. UV light has also been used to drive the production of adenosine from its precursors and the synthesis of ATP from a combination of adenine and ribose [52,53]. UV-light-driven syntheses of nucleic acid bases, nucleosides, sugars, and ribonucleotides from reactants as simple as formamide and urea have been achieved with appropriate mineral catalysts such as titanium dioxide [54–56] and the UV light also has the benefit of destroying many of the side-products of the desired reaction [55].

UV light can also drive the synthesis of [2Fe-2S] and [4Fe-4S] clusters through the photooxidation of ferrous ions and the photolysis of organic thiols [57]. These iron-sulfur clusters can coordinate to and be stabilized by cysteine-containing peptides and mediate the assembly of iron-sulfur cluster-peptide complexes that can drive enzymatic reactions. Iron complexes have also been implicated in the production and breakdown of universal metabolic precursor compounds [58].

Less research seems to have gone into UV-driven synthesis of amino acids and peptides because many of these tend to be susceptible to UV degradation but Ponnamperuma and Peterson [59] reported UV-induced peptide formation from amino acids in the presence of cyanamide. More commonly, UV light tends to break down amino acids that can absorb it. In amino acid mixtures, UV light destroys tyrosine, tryptophan and cystine [60,61] whereas glycine and proline are quite stable except in the presence of salts, oxychlorines or high humidity [61]. Arginine is also stable to UV light but breaks down mainly into urea and ammonia in the presence of a combination of peroxides and UV light [62]. However, some of the breakdown products of amino acids are other amino acids: arginine exposed to UV light (254 nm) in the presence of hydrogen peroxide was partially converted to ornithine, norvaline, serine and aspartic acid [62] while UV light (200–400 nm) not only degraded phenylalanine, tyrosine and tryptophan into smaller fragments but converted a small proportion of each into the others [63–65]. Moreover, the breakdown products of these compounds tended to progressively protect the original molecule over time [65]. Thus, the effects of UV light on amino acids and their peptide and protein formation are complex and dependent on the presence or absence of oxidizing agents as well as other solutes.

Most UV light sources that are commercially available have limited spectral ranges but Rimmer et al. [3] have devised an apparatus specifically to simulate the range and intensity of actual solar UV radiation and they have explored the amount of UV exposure required to drive various prebiotic reactions.

1.4. Catalytic Surfaces

Catalytic surfaces including clays, minerals, rocks, glasses and even meteoritic dust, have also been explored as means of increasing rates of synthesis and polymerization of prebiotic molecules including amino acids, nucleic acids, peptides and polynucleic acids [12,54–56,66–73]. The literature concerning the use of such natural catalytic surfaces is so large that no attempt is made here to encompass its range. Notably, however, we found no references to the use of such materials as means of selecting among prebiotic molecules or their reaction pathways, although this would seem to be an area of possible importance. Additionally, no specific apparatuses appear to have been reported for the implementation of such catalytic surfaces, probably because these can easily be incorporated into existing Miller-style apparatuses.

1.5. Combining Selection Processes

In many prebiotic environments, multiple selection factors would have been at work on the chemical ecosystem, yet combinations of the four selection factors described above have rarely been explored. A few representative examples follow.

As noted above [54–56], titanium dioxide has been used as a catalyst to promote UV-light synthesis of nucleic aicds. Campbell et al. [74] have employed deliquescent minerals to modify and improve the effectiveness of dry-wet cycling in the production of prebiotic condensation reactions. Combinations of such deliquescent minerals are likely to be even more active because they exhibit increased water adsorption at lower relative humidities resulting in solid dissolution and an increase in chemical reactivity [75]. Shankar et al., [76] and Baú et al. [77] found that peptide polymerization was increased during wet-dry cycling when performed in the presence of goethite (Fe_2O_3). Ice has also been explored as a surface upon which UV light can act to synthesize amino acids, quinones and amphiphiles [78,79] and to promote the conversion of thymine into the other nucleobases [78]. Such conditions might have occurred both in space and in the polar regions of planets such as Earth and Mars. Finally, Lin et al. [80] have demonstrated that adsorption of peptides onto montmorillonite protects them from UV irradiation thus raising the possibility that combinations of these physicochemical processes act in competing ways that enhance the survival of some prebiotic species at the expense of others. Many other combinations are clearly possible but very few seem to have been tested.

Once again, no special apparatuses appear to have been devised for these combined process experiments.

1.6. Thermal Vents

One type of process that our apparatuses do not incorporate are those involved in thermal vent reactions. Alkaline hydrothermal systems containing catalytic minerals such as Fe(Ni) in the presence of sea water and vent gases such as S_2, H_2 and CO_2 might have driven organic syntheses capable of producing the range of prebiotic compounds necessary to drive the evolution of metabolic and genetic processes [81,82]. Several apparatuses for modeling such hydrothermal vent reactions already exist [83–85]. Because these reactors model processes that are largely isolated from the others described above and are difficult to incorporate into a single apparatus of the type we were designing, we felt that it was justifiable not to try to incorporate their features into our novel apparatuses.

1.7. Introduction Summary

As the preceding overview demonstrates, a range of physicochemical processes including wet/dry cycles, freeze/thaw cycles, UV light/dark cycles and the use of catalytic surfaces and materials, have been used to increase yields of prebiotic compounds. However, only rarely have such processes been used in prebiotic experiments as natural selection factors. A clear need therefore exists for apparatuses capable of modeling not only individual types of physicochemical selection process cycles but to combine them over extended periods of time so that their effects on complex chemical environments can be analyzed in terms of the evolution of the balances between both synthesis and selection acting simultaneously.

2. Materials and Methods

We have invented two apparatuses designed to combine the multiple physicochemical processes listed above either simultaneously or in alternating cycles over extended periods of time (weeks/months).

The first apparatus, called "ReBioGeneSys 1.0" deconstructed Miller-style apparatuses to create a series of connected mini-environments each capable of carrying out a single physicochemical process and connected in such a way that exposure to any particular environment can be carried out in any order desired. The second apparatus, called "ReBioGeneSys 2.0" reintegrates the various processes into a much simpler configuration that once again resembles a single environment exposed to multiple selection processes.

2.1. ReBioGeneSys 1.0

ReBioGeneSys 1.0 consists of six primary components along with an variety of hardware to connect and mount these components: (1) gas cylinders, a vacuum pump, and a manifold for controlling their connection to the rest of the apparatus; (2) five 5 L flasks with connections to the gas/vacuum manifold and with connections between the flasks for moving liquid between them; (3) an array of peristaltic pumps for moving liquid between the five flasks; (4) a bespoke, programmable electronic control system for controlling the peristaltic pumps; (5) a specially designed shelf for holding the flasks and their connectors; (6) a Marx generator for reliably producing high voltage sparks at frequent (5 to 10 s) intervals (Figure 1 and Table 1).

The gas/vacuum manifold (Figure 2) essentially consists of a series of Swagelok valves connected in such a way as to permit all of the gas cylinders to access all five of the flasks via a single line so that the flasks share a common atmosphere at a common pressure. The vacuum pump is also connected through the manifold via a valve that shuts off all of the gases permitting all of the flasks to be evacuated simultaneously and any liquids in them to be degassed. The manifold also permits the apparatus to be regularly regassed so that the synthetic conditions are maintained at a relatively constant concentration, thereby avoiding the need to seed new containers at each step of a repeated synthetic process as is done in some current experiments (e.g., [12,13]).

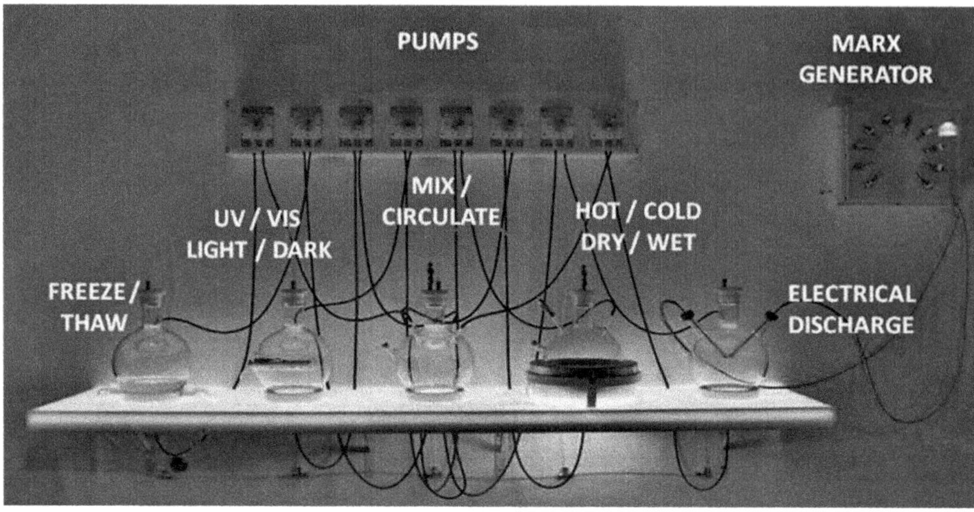

Figure 1. Overview of the main elements of ReBioGeneSys showing the peristaltic pump array at the top that circulates the fluid from one flask to another in any order desired and, from left to right, the flask through which the chiller fluid circulates antifreeze fluid (left), which permits freezing/thawing depending on the temperature setting of the chiller; a flask modified to permit insertion of an ultraviolet light or visible light source; a flask (center) through which all the other flasks can circulate their fluid to enable passage to any other flask; a flask sitting in a heating mantle permitting boiling or evaporation of the fluid; and (right) a flask fitted with electrodes connected to the Marx generator that creates electrical discharges mimicking lightning.

Table 1. List of parts for ReBioGeneSys 1.0 keyed to Figures below.

Item Number	Part Description	Quantity
1	High pressure regulator for hydrogen	1
2	High pressure regulator for methane	1
3	Anhydrous ammonia regulator	1
4	Custom aluminum and stainless steel gassing manifold	1
5	Aluminum standoff mounts for the gassing manifold	4
6	Adixen (Pfeiffer) Chemical Pascal 2005C1 3.5CFM vacuum	1
7	Stainless steel KF-25 to $\frac{1}{4}''$ compression fitting, centering ring and clamp	3
8	Stainless steel braided low pressure vacuum hose	2
9	Visitrap water and ammonia vacuum trap	1
10	Custom switching controller for peristaltic pump array	1
11	Custom peristaltic pump array made of Delrin	1
12	Ultraviolet light power transformer	1
13	Cole Parmer recirculating heater/chiller	1
14	$\frac{1}{2}''$ OD Tygon tubing	2
15	$\frac{1}{4}''$ stainless steel Swagelok Ultra Torr to compression fitting: SS-4-UT-6-400	1

Table 1. *Cont.*

Item Number	Part Description	Quantity
16	50 mm PFA threaded plug	5
17	Analytic Jena UV(AB) light	1
18	Eating mantle for 5 L round bottom flask	1
19	3/8″ PFA tubing	4
20	3/8″ stainless steel Swagelok Ultra Torr to compression fitting: SS-6-UT-6-600	8
21	Custom stainless steel electrodes with round ball caps	2
22	Heating mantle controller	1
23	Custom Marx generator, HV electronics, stainless steel and Delrin	1
24	12 VDC variable power transformer for Marx generator	1
25	Power for illuminated shelf	1
26	Custom made Corian shelf with LED illumination	1
27	Corian support shelf made to match main Corian shelf	1
28	5L custom borosilicate glass round bottom flask, modified to accept input and output tubing, threaded plugs, UV light, chiller condenser, heater, or electrodes	5
29	Support ring	4
30	Masterflex FDA Viton tubing L/S #96412-25	16
31	Heavy insulated high voltage cable	2

Each five liter flask has a unique function and was modified specifically for it. All flasks are connected to the gas/vacuum line via Swagelock connectors into threaded plastic (PFA) plugs (Figure 3). The central flask has four inlets in its sides and four outlets in the bottom. The four inlets permit fluids to be pumped in from the outlet ports of the other four flasks via Tygon tubing while the four outlets permit the fluid in the central flask to be pumped into the inlets via Tygon tubing in any of the other four flasks. In addition to a single inlet and outlet port, each of the four other flasks has its own function. One has been modified to incorporate a condenser coil through which antifreeze solution is pumped by a chiller/circulator with variable temperature control (Figure 3). The condenser can be used to freeze and thaw the liquid in the flask by varying the temperature appropriately. It can also act as a heat sink for the system if the heating mantle is used to increase the temperature in the next flask, which can be varied from room temperature to boiling thus permitting this flask to either be evaporated via the tubing into the other flasks or to act as a model of a hot spring. Another flask has an insert into which a light source such as a UV lamp can be placed to irradiate the liquid in the flask (Figure 3), modeling light/dark cycles. Furthermore, the final flask has been modified to accept a pair of electrodes that are attached to the Marx generator, permitting the atmosphere to be subjected to rapid, periodic, very high voltage electrical discharges for extended periods of time.

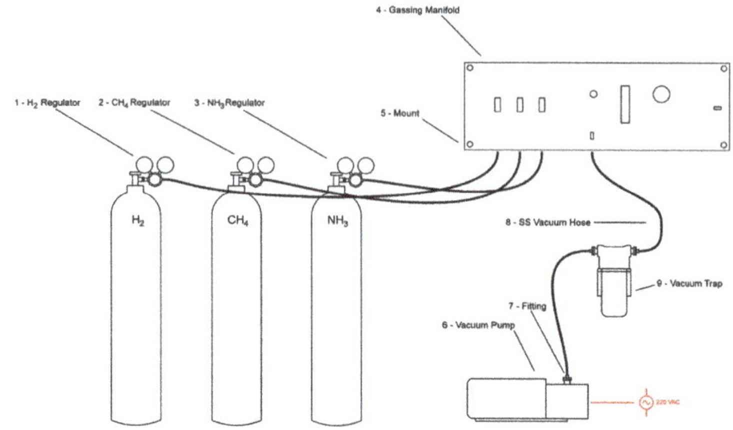

Figure 2. Diagram illustrating how the gas/vacuum manifold integrates gas flow from the gas cylinders into the manifold and into the rest of the apparatus, each controlled by a Swagelok valve, and exiting the manifold through a single Tygon tube to the right (not shown). The valve to the vacuum pump (lower right) is closed during gassing of the apparatus. If the three valves to the gasses are all closed, the valve to the vacuum pump can be opened and the apparatus evacuated. The vacuum pump attaches to the manifold by means of a pair of stainless steel low-pressure vacuum hoses that integrate a Visitrap water and ammonia trap to protect the pump.

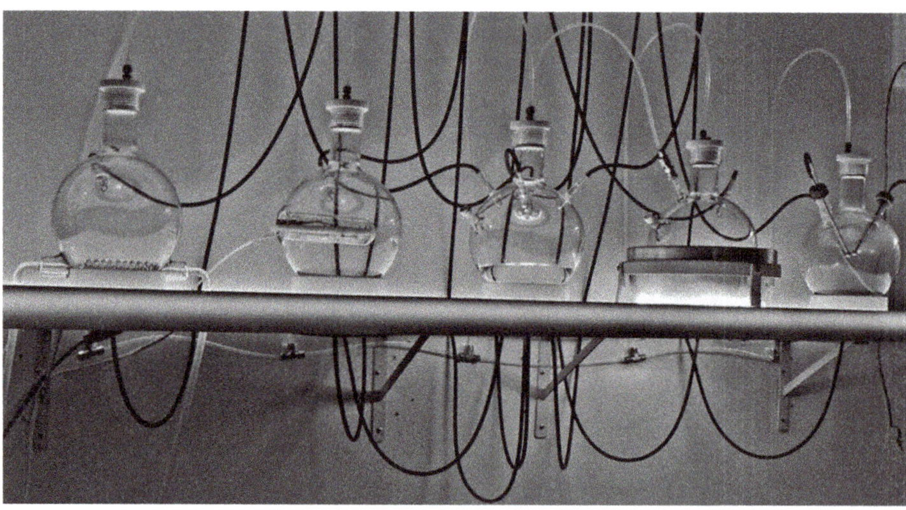

Figure 3. Close up of the flasks showing the two ways in which they are interconnected. The black Tygon tubing connects all of the flasks through the center one via the programmable peristaltic pumping apparatus (not shown). Fluid enters at the upper sides of the flasks and is drained through connectors at the bottom (not visible). The white PFA tubing that connects through the white PFA plugs at the top of each flask connect all of the flasks to the manifold controlling gas entry to the system. All of the flasks are share the same gas environment as there are no valves or pumps in the gas tubing connecting the flasks. The freezing flask has a coiled borosilicate tube inserted through which antifreeze is circulated by a chiller (not shown), which permits the temperature of the flask to be maintained at whatever temperature is desires, including below the freezing point. The UV light flask has an insert into which a light source can be placed. A better version is illustrated in the integrated version, ReBioGeneSys 2.0, illustrated below.

Liquid flow between the flasks is controlled by a set of eight peristaltic pumps (four to move liquid into the central flask from any of the others and four to move liquid from the central flask to any one of the others) via Viton tubing. The pumps are mounted in a Delrin case (Figure 4) and controlled by a programmable electronic switch box (Figure 5). The organization of the pump/tubing connections to the flasks is illustrated in Figure 5.

Figure 4. Peristaltic pump array permitting every flask to be connected to the central receiving flask from which every other flask can be reached. The Viton tubing is shown connected only to the left-most pump and the wiring connecting the pump array to the custom programmable electronic control box is visible at the extreme left.

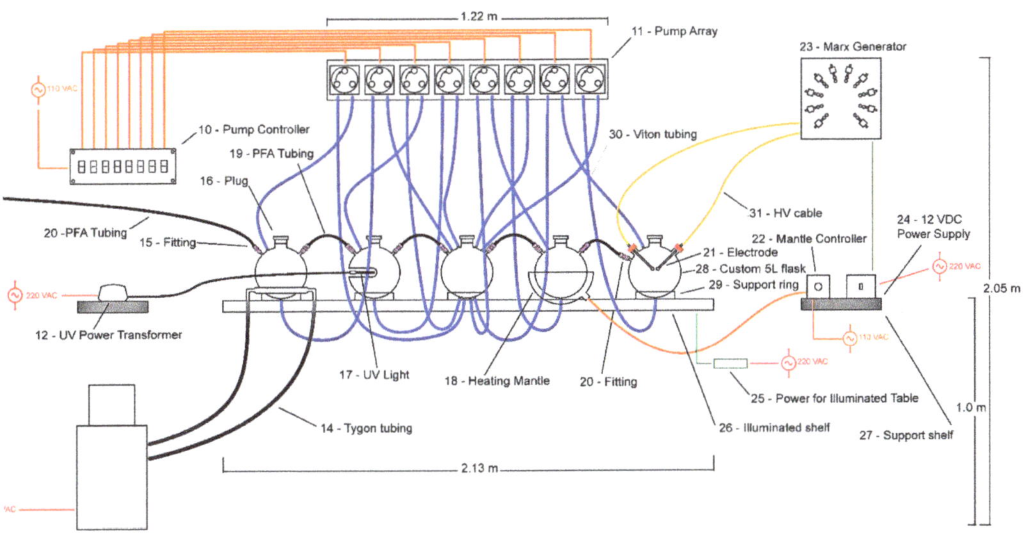

Figure 5. Schematic diagram showing the arrangement of the pump apparatus and its tube connections to the flasks. Each flask has an inlet near its top to allow liquid to be pumped in and an outlet at the bottom to allow liquid to be pumped out. Inlets and outlets are controlled by separate peristaltic pumps (top) controlled by a series of programmable switches (upper left). When a pump is inactive, no liquid can flow in or out of a flask.

The Marx generator supplying electrical discharges to the apparatus (far right flask in Figures 1 and 5) is based on well-known principles and schematics and consists essentially of a series of ten capacitors that each accumulate a charge permitting the voltage to be increased from about 9 volts direct current to about 250,000 volts every five to ten seconds (Figures 6 and 7).

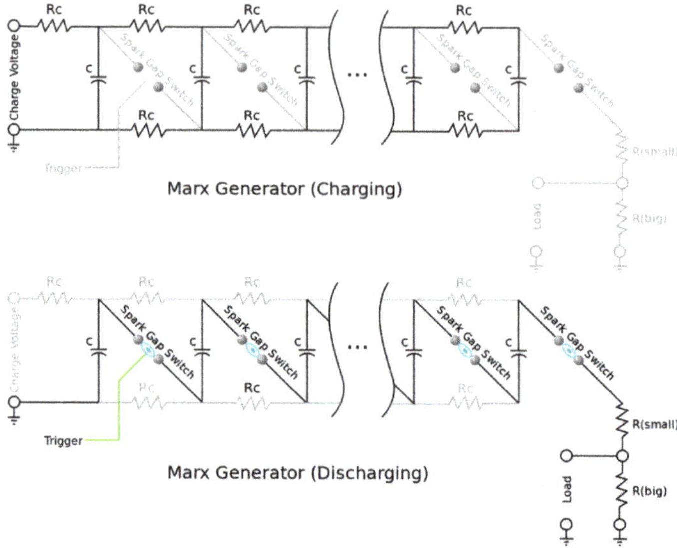

Figure 6. Schematic diagram of a Marx generator, a high voltage circuit used in insulation testing and scientific research. It generates a pulse of high voltage by charging multiple capacitors in parallel and then suddenly connecting them together in series by spark gaps. The capacitors are charged by the resistor network. Although the left capacitor has the greatest charge rate, the generator is typically allowed to charge for a long period of time, and all capacitors eventually reach the same charge voltage. (Wikimedia Commons, https://en.wikipedia.org/wiki/File:Marx_Generator.svg Accessed 28 August 2022).

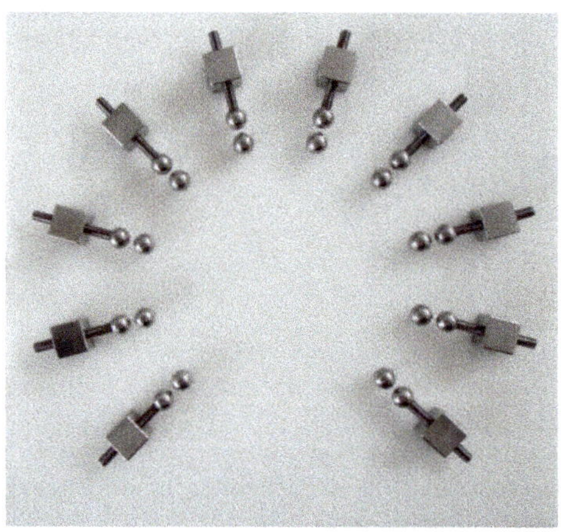

Figure 7. Arrangement of the custom Marx generator showing the spark gap electrodes, which permit the spark gap distance to be varied by means of a threaded screw arrangement.

2.2. *ReBioGeneSys 2.0*

While ReBioGeneSys 1.0 permits automation and a great amount of experimental flexibility in terms of combining selection processes in varying orders for varying amounts of time, we recognized that a simpler and less expensive apparatus might also have benefits. ReBioGeneSys 2.0. integrates the five flasks of ReBioGeneSys 1.0 into a single apparatus with three glass-connected flasks performing the same range of functions (Figure 8). This configuration brings the apparatus back to something resembling the original Urey-Miller apparatus but with additional functions integrated into the modified design.

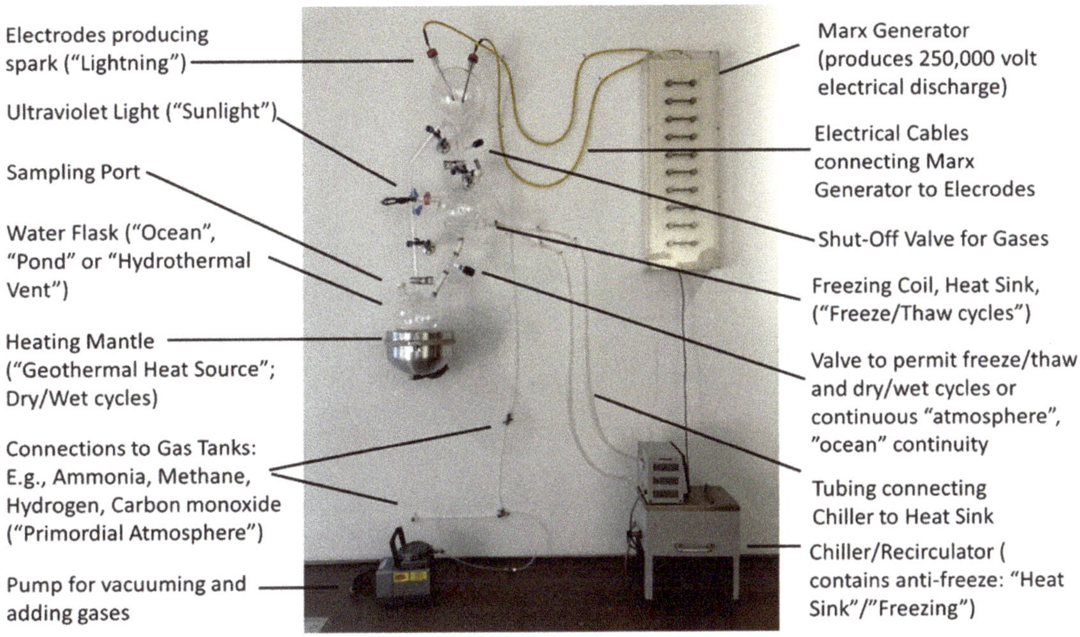

Figure 8. Overview of ReBioGeneSys version 2.0, in which the elements from ReBioGeneSys 1.0 have been reintegrated into a simpler form.

The essence of the redesign is begin with the modified Urey-Miller apparatus that we hope to report elsewhere and to integrate an additional flask between the one housing the electrodes and the one that is heated. This additional flask incorporates an opening into which a UV light can be inserted, sealed and clamped in place, as well as a condenser coil that can act as a heat sink or means of freezing and thawing the liquid condensing in the flask. In order to accumulate this liquid, a valve has been inserted below this flask attached to it by means of Masterflex ™ I/P ™ Precision C-Flex tubing. This is special, very thick tubing that resists collapsing during the high vacuum conditions used to evacuate the apparatus and to de-gas liquids added to it, as well as being highly resistant to any ammonia that may be in the experimental atmosphere used during experimentation. Additional details concerning unique components of the apparatus are provided in Figure 9 but, in general, the components do not differ significantly from those already in use in many other Urey-Miller type apparatuses or those described above for ReBioGeneSys 1.0.

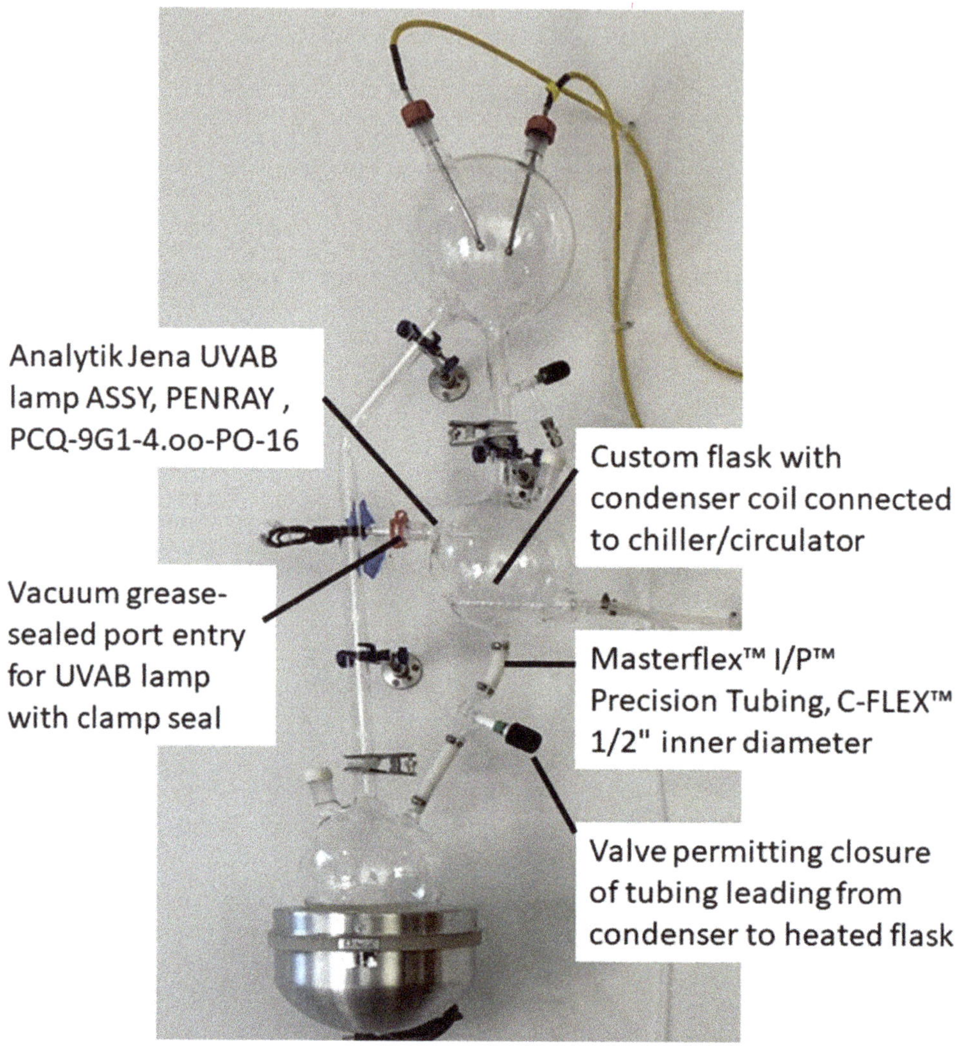

Figure 9. Close-up view of ReBioGeneSys 2.0 highlighting some of the key innovations. A single flask (center) now performs both freeze/thaw and UV light/dark functions, either simultaneously of separately. To use the flask for freezing, water from the heating flask (lower left) can be evaporated into the freezing flask by closing the valve below the freezing flask (the flow of anti-freeze at a temperature below 0 °C acts as a heat sink to balance the heat source evaporating the water in the heating flask). In order to accommodate freezing of the water in the tube connecting the condenser flask to the valve, and to accommodate the need to resist vacuum prior to gassing and re-gassing, and to resist the ammonia in the experimental atmosphere, special Masterflex tubing is required (as noted in the image). Unlike ReBioGeneSys 1.0, the UVAB lamp is inserted directly into the flask via a clamped, vacuum grease-sealed port rather than being set into an indentation in the flask.

3. Results

3.1. Tests of the Apparatuses

Unlike apparatuses designed to mimic one particular environment, which can be tested by determining whether the apparatus functions to the specifications of that chosen environment by producing specific results under tightly determined conditions (e.g., [1,2,82–85], the apparatuses described here have no specific function but are designed to be extremely flexible and have a large number of possible experimental permutations. Tests of the apparatuses therefore require comparisons between the results obtained using one selection pressure (say wet/dry) versus another (freeze/thaw) under otherwise strictly comparable conditions. We have not yet performed such experiments and cannot, therefore, validate the utility of the apparatuses in comparison with any existing set of data.

However, we have performed long-term fundamental tests of robustness: both apparatuses are capable of performing simple Miller-type experiments in which each component is activated serially. Both apparatuses can hold a vacuum for extended periods of time (days); liquids move through the apparatuses as designed; all the individual elements (heating, freezing, electrical discharging, etc.) function over extended periods of time; the electrical control system for ReBioGeneSys 1.0 performs as designed, moving the liquids from one flask to another in any programmed order; the freezing elements are robust enough to perform for many weeks through freezing and thawing cycles without cracking; the Marx generator design is robust enough to perform for many months or years (which we know because we have used the same Marx generators in previous Miller-like designs (unpublished data).

3.2. Limitations of the Apparatuses

One important limitation which we have found is that the Marx generator needs to be shielded from other electrical elements of the apparatus such as the controller for the pumps and the chiller, using a Faraday cage-like wire mesh. Failure to turn off an unshielded Marx generator while attempting to initiate pumping in ReBioGeneSys 1.0 can short out the electronic controller while failure to turn it off an unshielded Marx generator in both apparatuses can result in interruption of the chiller/circulator function.

The apparatuses have several other experimental limitations as well. They cannot accurately reproduce the conditions that would be found at temperature extremes such as well below the freezing point of water or above its boiling point. They cannot accurately reproduce phenomena such as undersea vents. Their efficiency and effectiveness are very difficult to test completely because they permit a very large number of process permutations, some of which may turn out to interfere with each other or to cause apparatus failures that were not foreseen. If ammonia is employed as one of the gases, the rubber sampling ports degrade of a period of a couple of months and need to be replaced, which may interrupt long term experiments. Furthermore, the borosilicate glass begins to degrade after several years of use (unpublished data) eluting boron compounds into the reaction vessel. Thus, the glass components may need to be replaced regularly as well.

An additional limitation of the current apparatuses is that they are not robotized nor connected directly to chemical detection equipment such as gas chromatography-mass spectrometry as Asche et al. [22] have done with their long-term prebiotic apparatus. Thus, additional innovations that improve performance are undoubtedly possible and hopefully these apparatuses will stimulate such innovations.

4. Discussion

The development of apparatuses that can run for extended periods of time and that incorporate physicochemical processes capable of naturally selecting among the diverse products of prebiotic chemical experiments is a necessity if origins of life research is to move beyond chemical syntheses to prebiotic evolutionary systems. As Brunk and Marshall [86] have argued, the next phase of origins-of-life experimentation needs explore more complex chemical ecosystems over longer periods of time under non-equilibrium conditions involving "containment, steady energy and material flows, and structured spatial heterogeneity from the outset". The apparatuses described here are capable of performing these necessary functions and ReBioGeneSys 2.0 in particular is based on a simpler, Urey-Miller-like apparatus that has been extensively tested over several years to run multi-week and multi-month prebiotic syntheses (unpublished data). The novelty of these new apparatuses is to incorporate for the first time wet/dry, freeze/thaw, and UV light/dark cycles while also permitting simultaneous or separate use of catalytic materials such as clays or minerals. These options open a wide range of novel possibilities for new types of long-term experiments in the evolution of prebiotic chemical ecosystems under varying selection pressures. Such experiments would help to fulfill the suggestion by Stüeken et al. [87] that future prebiotic evolution experiments try to mimic the interaction of multiple environments that may have been present in the Hadean Earth and explore the interactions of their various products.

Some aspects of the apparatuses could undoubtedly be improved in order to optimize their adoption by scientists experimenting with prebiotic systems. One of the major limitations of the two systems built thus far is the large scale and the size makes them difficult to install, expensive to build, and potentially dangerous because of the large amounts of gases employed. While ReBioGeneSys 2.0 attempts to remedy some of these problems, it continues to fall short of one of the intended goals and that is to be an open source platform to enable many researchers to explore many new types of experiments. To remedy this the authors have been steadily working on a miniaturized, microfluidic version of ReBioGeneSys. The intent of the new Mini-ReBioGeneSys is to producible it in multiples and small enough to easily fit in a chemical hood or even in one's hand.

One option is miniaturization of the current designs, which is clearly possible without great difficulties for both types of apparatus. However, some limitations are inherent in the materials, such as the fragility of glass as it becomes smaller and thinner, the need for the glass to support Swagelok connectors, tubing and compression fittings to control gassing of the apparatus, and the lack of availability of these below 2-to-3 mm in diameter (https://www.swagelok.com accessed on 17 September 2022). There are also increasing challenges making bespoke glass apparatuses of ever smaller scales that can incorporate sampling ports, electrodes, chiller coils, etc. Sleeve stopper septa, which are used for sampling ReBioGeneSys 2.0 do not appear to be available below 2.4 mm inner diameter (https://www.sigmaaldrich.com/US/en/product/aldrich/z565695 accessed on 17 September 2022). Furthermore, there appears to be a lack of heating mantles smaller than 250 mL capacity (https://chemglass.com/mantles-heating-tops-glas-col accessed on 17 September 2022), miniaturized power sources for these, or very small chiller/circulators. Thus, miniaturization of the apparatuses to a desk-top size is feasible but unlikely to decrease the costs of fabrication or operation significantly.

An alternative is to explore hand-sized microfluidic types of apparatuses printed in glass or other chemical and heat resistant materials. Such devices would be reproducible in multiple copies at significantly lower costs than the current apparatuses and therefore easily contained in a chemical hood for safety when using gases such as ammonia, hydrogen sulfide, carbon monoxide, etc. Another advantage would be that microfluidic devices can be designed to operate at very high gas pressures, such as might be found on gas giant planets, and at very low temperatures such as are usually found there. Such apparatuses might integrate all of the selection factors of ReBioGeneSys 2.0 or separate devices could be designed, each to carry out one of the selection processes, as in ReBioGeneSys 1.0. Separate devices could then be ganged together in whatever order desired. Yet, another possibility would be to have a single, very minimalistic design that could be exposed to each selection factor independently, e.g., to UV light/dark cycles delivered by a separate UV light source and then to freeze/thaw cycles implemented by literally placing the apparatus in a freezer and taking it out. One additional advantage of such microfluidic apparatuses would be the ease with which they might be directly integrated into HPLC or GC/Mass Spec analytical equipment for continuous monitoring of products while obviating risk of contamination that might occur from human sampling of the apparatus. The use of multiple, miniaturized apparatuses would also make it much easier to run side-by-side experiments comparing the effects of different selection factors on a common set of starting reactants, or the effects of different combinations of such selection factors or their order.

These apparatuses, whether large or miniaturized, may also permit experimental exploration of strategies by which nature controlled the combinatorial chemical explosion problem that must accompany experiments utilizing complex mixtures of compounds (so-called "dirty experiments") over long periods of times. While it is logical to assume that the longer a chemical experiment is run, the worse the combinatorial chemical explosion problem will become [22], we predict that the addition of selection processes will mitigate the problem. Indeed, Cronin et al. [13] have found that it is possible to tame the combinatorial explosion of the formose reaction when run over extended periods of time by "seeding" the product mixture into a fresh version of the reaction and by using various mineral surfaces. They observed that the overall number of products decreased as the number of cycles increased, suggesting that as more complex molecules evolve, they begin to compete for available chemical precursors, thereby restraining the combinatorial explosion. Further experiments of this type augmented by cycles of selection may be similarly revealing.

Another possible outcome of implementing selection pressures may result in another outcome that can control the combinatorial chemical explosion problem and that is to evolve metabolic replication cycles. Baum and his colleagues have demonstrated that chemical ecosystem selection can be performed by repeatedly seeding a synthetic "prebiotic soup" onto pyrite grains to yield mutually reinforcing sets of catalyzed reactions [12]. Evolving such autocatalytic sets of reactions from mutually reinforcing sets or reactions are a clear desiderata for prebiotic chemical studies, since all known biological systems implement such cycles [88].

Finally, implementing selection pressures may also yield the third of limiting a combinatorial chemical explosion, which is through the evolution of functional modules through molecular complementarity [89]. Molecular complementarity may be observable in small-molecule interactions that protect the constituents of the complex against degradation processes thereby increasing the concentrations of the participating compounds compared with other, non-complementary molecules; it may foster the emergence of peptide or ribonucleotide catalysts; and polymerization reactions driven by freeze–thaw and wet/dry reactions (see Sections 1.1 and 1.2) may yield self-organizing molecules (phospholipids, polysaccharides, polypeptide aggregates, or polyribonucleosides) as well as self-replicating polymers. Regarding these possibilities, it is important to note that some amino acids, have been demonstrated to catalyze many types of chemical reactions (e.g., [90,91], many of relevance to prebiotic chemical reactions (e.g., synthesis of sugars [92]); that some peptides catalyze specific chemical reactions [93]; while other peptides, such as poly-serines, are capable of self-replication reactions [94]. Thus, future studies should look not only for the emergence of polyribonucleosides and their autocatalytic sets [94–96], but also for the emergence small-molecule catalyzed reactions more broadly and synergies between nucleic acids, amino acids, lipids, etc. [97].

In sum, new apparatuses have been designed to explore the roles of selection processes that may have been at work in various prebiotic chemical ecosystems, either alone or in combination, and to make it possible to determine whether prebiotic natural selection leads to increased synthesis of polymerized products, emergence of self-organization and replication of selected sets of products, and control of the combinatorial chemical explosion that would be expected in the absence of selection. These apparatuses are, hopefully, just the first of many, improved versions that foster novel experimentation in prebiotic chemical ecosystems.

Author Contributions: Conceptualization, R.R.-B. and A.W.B.; methodology, R.R.-B.; validation, R.R.-B., A.W.B.; resources, R.R.-B. and A.W.B.; writing—original draft preparation, R.R.-B.; writing—review and editing, R.R.-B. and A.W.B.; visualization, R.R.-B. and A.W.B.; project administration, R.R.-B.; funding acquisition, R.R.-B. and A.W.B.; design of apparatuses: R.R.-B. and A.W.B.; construction of apparatuses: A.W.B. All authors have read and agreed to the published version of the manuscript and have contributed substantially to the work reported.

Funding: National Science Foundation EAGER Grant Program, (Biology, Chemistry and Informal Science Directorates) NSF DRL-1212365 "Exploring Public Engagement with Real-Time Experimentation in Different Public Venues" Robert Root-Bernstein, Primary Investigator; Adam Brown, co-investigator, September 2012–August 2014. Additional funding was provided by a Humanities and Arts Research Projects Production grant from Michigan State University: HARP-P-I-Root-Bernstein-FY20-21, 10032668-PHYSIOLOGY NATURAL SCIENCE, 2021–2022.

Data Availability Statement: There is no data associated with this paper.

Acknowledgments: We thank Scott Bankroff (The Glass Lab, MSU Chemistry Department) for glasswork; Barry Tignor (Electronics Facility, MSU Physics and Astronomy) for electronics; the staff of the Machine Shop (Physics and Astronomy, MSU Physics and Astronomy).

Conflicts of Interest: The authors declare no conflict of interest.

References

1. Miller, S.L. A Production of Amino Acids Under Possible Primitive Earth Conditions. *Science* **1953**, *117*, 528–529. [CrossRef]
2. Bada, J.L. New insights into prebiotic chemistry from Stanley Miller's spark discharge experiments. *Chem. Soc. Rev.* **2013**, *42*, 2186–2196. [CrossRef] [PubMed]
3. Rimmer, P.B.; Thompson, S.J.; Xu, J.; Russell, D.A.; Green, N.J.; Ritson, D.J.; Sutherland, J.D.; Queloz, D.P. Timescales for Prebiotic Photochemistry Under Realistic Surface Ultraviolet Conditions. *Astrobiology* **2021**, *21*, 1099–1120. [CrossRef]
4. Fox, S.; Pleyer, H.L.; Strasdeit, H. An automated apparatus for the simulation of prebiotic wet-dry cycles under strictly anaerobic conditions. *Int. J. Astrobiol.* **2019**, *18*, 60–72. [CrossRef]
5. Parker, E.T.; Cleaves, J.H.; Burton, A.S.; Glavin, D.P.; Dworkin, J.P.; Zhou, M.; Bada, J.L.; Fernández, F.M. Conducting Miller-Urey Experiments. *J. Vis. Exp.* **2014**, *83*, e51039. [CrossRef]
6. Ruiz-Bermejo, M.; Menor-Salván, C.; Osuna-Esteban, S.; Veintemillas-Verdaguer, S. Prebiotic Microreactors: A Synthesis of Purines and Dihydroxy Compounds in Aqueous Aerosol. *Orig. Life Evol. Biosph.* **2007**, *37*, 123–142. [CrossRef]
7. Dobson, C.M.; Ellison, G.B.; Tuck, A.F.; Vaida, V. Atmospheric aerosols as prebiotic chemical reactors. *Proc. Natl. Acad. Sci. USA* **2000**, *97*, 11864–11868. [CrossRef]
8. Imai, E.-I.; Honda, H.; Hatori, K.; Brack, A.; Matsuno, K. Elongation of Oligopeptides in a Simulated Submarine Hydrothermal System. *Science* **1999**, *283*, 831–833. [CrossRef]
9. Baum, D.A.; Vetsigian, K. An Experimental Framework for Generating Evolvable Chemical Systems in the Laboratory. *Orig. Life Evol. Biosph.* **2017**, *47*, 481–497. [CrossRef] [PubMed]
10. Godfrey-Smith, P. Conditions for Evolution by Natural Selection. *J. Philos.* **2007**, *104*, 489–516. Available online: https://www.petergodfreysmith.com/ConditionsNS-07-JP-web.pdf (accessed on 18 September 2022). [CrossRef]
11. Dennett, D.; Sterelny, K.; Queller, D.; Godfrey-Smith, P. Darwinian Populations and Natural Selection. No Date. Available online: https://www.google.com/url?sa=t&rct=j&q=&esrc=s&source=web&cd=&cad=rja&uact=8&ved=2ahUKEwiF5NDUzKH6AhV4kGoFHSJBATMQFnoECA8QAQ&url=https%3A%2F%2Fpetergodfreysmith.com%2FAgents_Acacias_PGS.pdf&usg=AOvVaw0m-0y1HUwpB-PQwNxYNZ4k (accessed on 18 September 2022).
12. Vincent, L.; Berg, M.; Krismer, M.; Saghafi, S.T.; Cosby, J.; Sankari, T.; Vetsigian, K.; Cleaves, H.J., II; Baum, D.A. Chemical Ecosystem Selection on Mineral Surfaces Reveals Long-Term Dynamics Consistent with the Spontaneous Emergence of Mutual Catalysis. *Life* **2019**, *9*, 80. [CrossRef]
13. Colón-Santos, S.; Cooper, G.J.T.; Cronin, L.; Cronin, L. Taming the Combinatorial Explosion of the Formose Reaction via Recursion within Mineral Environments. *ChemSystemsChem* **2019**, *1*, e1900014. [CrossRef]
14. Eigen, M.; Schuster, P. The hypercycle. A principle of natural self-organization. Part A, Emergence of the hypercycle. *Die Nat.* **1977**, *64*, 541–565. [CrossRef]
15. Hordijk, W.; Steel, M.; Kauffman, S.A. Molecular Diversity Required for the Formation of Autocatalytic Sets. *Life* **2019**, *9*, 23. [CrossRef] [PubMed]
16. Root-Bernstein, R.S.; Dillon, P.F. Molecular Complementarity I: The Complementarity Theory of the Origin and Evolution of Life. *J. Theor. Biol.* **1997**, *188*, 447–479. [CrossRef] [PubMed]
17. Hunding, A.; Kepes, F.; Lancet, D.; Minsky, A.; Norris, V.; Raine, D.; Sriram, K.; Root-Bernstein, R. Compositional complementarity and prebiotic ecology in the origin of life. *BioEssays* **2006**, *28*, 399–412. [CrossRef]
18. Charlat, S.; Ariew, A.; Bourrat, P.; Ruiz, M.F.; Heams, T.; Huneman, P.; Krishna, S.; Lachmann, M.; Lartillot, N.; D'Hendecourt, L.L.S.; et al. Natural Selection beyond Life? A Workshop Report. *Life* **2021**, *11*, 1051. [CrossRef]
19. Ruiz-Mirazo, K.; Briones, C.; de la Escosura, A. Chemical roots of biological evolution: The origins of life as a process of development of autonomous functional systems. *Open Biol.* **2017**, *7*, 170050. [CrossRef]
20. Jia, T.; Caudan, M.; Mamajanov, I. Origin of Species before Origin of Life: The Role of Speciation in Chemical Evolution. *Life* **2021**, *11*, 154. [CrossRef] [PubMed]
21. Keserű, G.M.; Soós, T.; Kappe, C.O. Anthropogenic reaction parameters—The missing link between chemical intuition and the available chemical space. *Chem. Soc. Rev.* **2014**, *43*, 5387–5399. [CrossRef] [PubMed]

22. Asche, S.; Cooper, G.J.T.; Keenan, G.; Mathis, C.; Cronin, L. A robotic prebiotic chemist probes long term reactions of complexifying mixtures. *Nat. Commun.* **2021**, *12*, 3547. [CrossRef] [PubMed]
23. Damer, B.; Deamer, D.; Damer, B.; Deamer, D. Coupled Phases and Combinatorial Selection in Fluctuating Hydrothermal Pools: A Scenario to Guide Experimental Approaches to the Origin of Cellular Life. *Life* **2015**, *5*, 872–887. [CrossRef] [PubMed]
24. Hawker, J.R.; Oró, J. Cyanamide mediated syntheses of peptides containing histidine and hydrophobic amino acids. *J. Mol. Evol.* **1981**, *17*, 285–294. [CrossRef]
25. Forsythe, J.G.; Yu, S.-S.; Mamajanov, I.; Grover, M.A.; Krishnamurthy, R.; Fernández, F.M.; Hud, N.V. Ester-Mediated Amide Bond Formation Driven by Wet-Dry Cycles: A Possible Path to Polypeptides on the Prebiotic Earth. *Angew. Chem. Int. Ed.* **2015**, *54*, 9871–9875. [CrossRef] [PubMed]
26. Higgs, P.G. The effect of limited diffusion and wet–dry cycling on reversible polymerization reactions, implications for prebiotic synthesis of nucleic acids. *Life* **2016**, *6*, 24. [CrossRef]
27. Frenkel-Pinter, M.; Haynes, J.W.; C, M.; Petrov, A.S.; Burcar, B.T.; Krishnamurthy, R.; Hud, N.V.; Leman, L.J.; Williams, L.D. Selective incorporation of proteinaceous over nonproteinaceous cationic amino acids in model prebiotic oligomerization reactions. *Proc. Natl. Acad. Sci. USA* **2019**, *116*, 16338–16346. [CrossRef]
28. Becker, S.; Schneider, C.; Okamura, H.; Crisp, A.; Amatov, T.; Dejmek, M.; Carell, T. Wet-dry cycles enable the parallel origin of canonical and non-canonical nucleosides by continuous synthesis. *Nat. Commun.* **2018**, *9*, 163. [CrossRef]
29. Hassenkam, T.; Deamer, D. Visualizing RNA polymers produced by hot wet-dry cycling. *Sci. Rep.* **2022**, *12*, 10098. [CrossRef]
30. Damer, B.; Deamer, D. The Hot Spring Hypothesis for an Origin of Life. *Astrobiology* **2020**, *20*, 429–452. [CrossRef] [PubMed]
31. Joshi, M.P.; Sawant, A.A.; Rajamani, S. Spontaneous emergence of membrane-forming protoamphiphiles from a lipid–amino acid mixture under wet–dry cycles. *Chem. Sci.* **2021**, *12*, 2970–2978. [CrossRef]
32. Steller, L.H.; Van Kranendonk, M.J.; Wang, A. Dehydration Enhances Prebiotic Lipid Remodeling and Vesicle Formation in Acidic Environments. *ACS Central Sci.* **2022**, *8*, 132–139. [CrossRef] [PubMed]
33. Valley, J.; Cavosie, A.; Ushikubo, T.; Reinhard, D.A.; Lawrence, D.F.; Larson, D.J.; Clifton, P.H.; Kelly, T.F.; Wilde, S.; Moser, D.E.; et al. Hadean age for a post-magma-ocean zircon confirmed by atom-probe tomography. *Nat. Geosci.* **2014**, *7*, 219–223. [CrossRef]
34. Menor-Salván, C.; Marín-Yaseli, M.R. Prebiotic chemistry in eutectic solutions at the water–ice matrix. *Chem. Soc. Rev.* **2012**, *41*, 5404–5415. [CrossRef] [PubMed]
35. Fraccia, T.P.; Zanchetta, G.; Rimoldi, V.; Clark, N.A.; Bellini, T. Evidence of Liquid Crystal–Assisted Abiotic Ligation of Nucleic Acids. *Orig. Life Evol. Biosph.* **2015**, *45*, 51–68. [CrossRef]
36. Zhang, S.J.; Duzdevich, D.; Ding, D.; Szostak, J.W. Freeze-thaw cycles enable a prebiotically plausible and continuous pathway from nucleotide activation to nonenzymatic RNA copying. *Proc. Natl. Acad. Sci. USA* **2022**, *119*, e2116429119. [CrossRef]
37. Mutschler, H.; Wochner, A.; Holliger, P. Freeze–thaw cycles as drivers of complex ribozyme assembly. *Nat. Chem.* **2015**, *7*, 502–508. [CrossRef]
38. Attwater, J.; Wochner, A.; Holliger, P. In-ice evolution of RNA polymerase ribozyme activity. *Nat. Chem.* **2013**, *5*, 1011–1018. [CrossRef]
39. Vlassov, A.V.; Johnston, B.H.; Landweber, L.F.; Kazakov, S.A. Ligation activity of fragmented ribozymes in frozen solution: Implications for the RNA world. *Nucleic Acids Res.* **2004**, *32*, 2966–2974. [CrossRef]
40. Qiao, H.; Hu, N.; Bai, J.; Ren, L.; Liu, Q.; Fang, L.; Wang, Z. Encapsulation of Nucleic Acids into Giant Unilamellar Vesicles by Freeze-Thaw: A Way Protocells May Form. *Orig. Life Evol. Biosph.* **2016**, *47*, 499–510. [CrossRef]
41. An, Z.; Shi, C.; Li, P.; Liu, L. Stability of amino acids and related amines in human serum under different preprocessing and pre-storage conditions based on iTRAQ®-LC-MS/MS. *Biol. Open* **2021**, *10*, bio055020. [CrossRef]
42. Hu, K.; Stewart, A.J.; Yuen, K.Y.; Hinrichsen, S.; Dryburgh, E.L.; Bertin, F. The effect of freeze-thaw cycles on determination of immunoreactive plasma adrenocorticotrophic hormone concentrations in horses. *J. Veter- Intern. Med.* **2020**, *34*, 1350–1356. [CrossRef] [PubMed]
43. Wandro, S.; Carmody, L.; Gallagher, T.; LiPuma, J.J.; Whiteson, K. Making It Last: Storage Time and Temperature Have Differential Impacts on Metabolite Profiles of Airway Samples from Cystic Fibrosis Patients. *mSystems* **2017**, *2*, e00100-17. [CrossRef] [PubMed]
44. Heinz, K.A.; Glofcheski, D.J.; Lepock, J.R.; Kruuv, J. Mechanism of freeze—Thaw damage to liver alcohol dehydrogenase and protection by cryoprotectants and amino acids. *Cryobiology* **1990**, *27*, 521–538. [CrossRef]
45. Cui, X.; Xu, S.; Su, W.; Sun, Z.; Yi, Z.; Ma, X.; Chen, G.; Chen, X.; Guo, B.; Li, X. Freeze–thaw cycles for biocompatible, mechanically robust scaffolds of human hair keratins. *J. Biomed. Mater. Res. Part B Appl. Biomater.* **2019**, *107*, 1452–1461. [CrossRef] [PubMed]
46. Kellman, B.P.; Baghdassarian, H.; Pramparo, T.; Shamie, I.; Gazestani, V.; Begzati, A.; Li, S.; Nalabolu, S.; Murray, S.; Lopez, L.; et al. Multiple freeze-thaw cycles lead to a loss of consistency in poly(A)-enriched RNA sequencing. *BMC Genom.* **2021**, *22*, 69. [CrossRef] [PubMed]
47. Shao, W.; Khin, S.; Kopp, W.C. Characterization of Effect of Repeated Freeze and Thaw Cycles on Stability of Genomic DNA Using Pulsed Field Gel Electrophoresis. *Biopreservation Biobanking* **2012**, *10*, 4–11. [CrossRef]
48. Ranjan, S.; Kufner, C.L.; Lozano, G.G.; Todd, Z.R.; Haseki, A.; Sasselov, D.D. UV Transmission in Natural Waters on Prebiotic Earth. *Astrobiology* **2022**, *22*, 242–262. [CrossRef]
49. Ranjan, S.; Sasselov, D.D. Influence of the UV Environment on the Synthesis of Prebiotic Molecules. *Astrobiology* **2016**, *16*, 68–88. [CrossRef]

50. Cleaves, H.J.; Miller, S.L. Oceanic protection of prebiotic organic compounds from UV radiation. *Proc. Natl. Acad. Sci. USA* **1998**, *95*, 7260–7263. [CrossRef]
51. Dondi, D.; Merli, D.; Pretali, L.; Fagnoni, M.; Albini, A.; Serpone, N. Prebiotic chemistry: Chemical evolution of organics on the primitive Earth under simulated prebiotic conditions. *Photochem. Photobiol. Sci.* **2007**, *6*, 1210–1217. [CrossRef]
52. Ponnamperuma, C.; Mariner, R.; Sagan, C.; Ponnamperuma, R.M.C. Formation of Adenosine by Ultra-violet Irradiation of a Solution of Adenine and Ribose. *Nature* **1963**, *198*, 1199–1200. [CrossRef] [PubMed]
53. Ponnamperuma, C.; Sagan, C.; Mariner, R. Synthesis of Adenosine Triphosphate Under Possible Primitive Earth Conditions. *Nature* **1963**, *199*, 222–226. [CrossRef] [PubMed]
54. Saladino, R.; Ciambecchini, U.; Crestini, C.; Costanzo, G.; Negri, R.; Di Mauro, E. One-Pot TiO_2-Catalyzed Synthesis of Nucleic Bases and Acyclonucleosides from Formamide: Implications for the Origin of Life. *ChemBioChem* **2003**, *4*, 514–521. [CrossRef]
55. Powner, M.; Gerland, B.; Sutherland, J.D. Synthesis of activated pyrimidine ribonucleotides in prebiotically plausible conditions. *Nature* **2009**, *459*, 239–242. [CrossRef] [PubMed]
56. Ritson, D.; Sutherland, J.D. Prebiotic synthesis of simple sugars by photoredox systems chemistry. *Nat. Chem.* **2012**, *4*, 895–899. [CrossRef] [PubMed]
57. Bonfio, C.; Valer, L.; Scintilla, S.; Shah, S.; Evans, D.J.; Jin, L.; Szostak, J.W.; Sasselov, D.D.; Sutherland, J.D.; Mansy, S.S. UV-light-driven prebiotic synthesis of iron–sulfur clusters. *Nat. Chem.* **2017**, *9*, 1229–1234. [CrossRef] [PubMed]
58. Muchowska, K.; Varma, S.J.; Moran, J. Synthesis and breakdown of universal metabolic precursors promoted by iron. *Nature* **2019**, *569*, 104–107. [CrossRef]
59. Ponnamperuma, C.; Peterson, E. Peptide Synthesis from Amino Acids in Aqueous Solution. *Science* **1965**, *147*, 1572–1574. [CrossRef]
60. Burke, M.; Augenstein, L. A comparison of the effects of ultraviolet and ionizing radiations on trypsin activity and on its constituent amino acids. *Biochem. J.* **1969**, *114*, 535–545. [CrossRef] [PubMed]
61. Liu, D.; Kounaves, S.P. Degradation of Amino Acids on Mars by UV Irradiation in the Presence of Chloride and Oxychlorine Salts. *Astrobiology* **2021**, *21*, 793–801. [CrossRef]
62. Ansari, A.S.; Khan, I.A.; Ali, R. UV degradation of arginine in the presence of hydrogen peroxide: Involvement of hydroxyl radical in the photolytic process. *J. Radiat. Res.* **1985**, *26*, 321–329. [CrossRef]
63. Scappini, F.; Casadei, F.; Zamboni, R.; Monti, S.; Giorgianni, P.; Capobianco, M. Laboratory simulation of UV irradiation from the Sun on amino acids. I: Irradiation of tyrosine. *Int. J. Astrobiol.* **2007**, *6*, 123–129. [CrossRef]
64. Scappini, F.; Capobianco, M.; Casadei, F.; Zamboni, R.; Giorgianni, P. Laboratory simulation of UV irradiation from the Sun on amino acids. II. Irradiation of phenylalanine and tryptophan. *Int. J. Astrobiol.* **2007**, *6*, 281–289. [CrossRef]
65. Scappini, F.; Capobianco, M.; Casadei, F.; Zamboni, R. Laboratory simulation of ultraviolet irradiation from the Sun on amino acids. III. irradiation of glycine-tyrosine. *Int. J. Astrobiol.* **2009**, *8*, 63–68. [CrossRef]
66. Paecht-Horowitz, M.; Berger, J.; Katchalsky, A. Prebiotic Synthesis of Polypeptides by Heterogeneous Polycondensation of Amino-acid Adenylates. *Nature* **1970**, *228*, 636–639. [CrossRef] [PubMed]
67. Lawless, J.G.; Levi, N. The role of metal ions in chemical evolution: Polymerization of alanine and glycine in a cation-exchanged clay environment. *J. Mol. Evol.* **1979**, *13*, 281–286. [CrossRef]
68. Hansma, H.G. Potassium at the Origins of Life: Did Biology Emerge from Biotite in Micaceous Clay? *Life* **2022**, *12*, 301. [CrossRef] [PubMed]
69. Hayatsu, R.; Studier, M.H.; Oda, A.; Fuse, K.; Anders, E. Origin of organic matter in early solar system–II. Nitrogen compounds. *Geochim. Cosmochim. Acta* **1968**, *32*, 175–190. [CrossRef]
70. Liu, R.; Orgel, L.E. Polymerization on the rocks, beta-amino acids and arginine. *Orig. Life Evol. Biosph.* **1998**, *28*, 245–257. [CrossRef] [PubMed]
71. Hill, A.R., Jr.; Böhler, C.; Orgel, L.E. Polymerization on the rocks, negatively-charged alpha-amino acids. *Orig. Life Evol. Biosph.* **1998**, *28*, 235–243. [CrossRef] [PubMed]
72. Jerome, C.A.; Kim, H.-J.; Mojzsis, S.J.; Benner, S.A.; Biondi, E. Catalytic Synthesis of Polyribonucleic Acid on Prebiotic Rock Glasses. *Astrobiology* **2022**, *22*, 629–636. [CrossRef] [PubMed]
73. Mizuuchi, R.; Blokhuis, A.; Vincent, L.; Nghe, P.; Lehman, N.; Baum, D. Mineral surfaces select for longer RNA molecules. *Chem. Commun.* **2019**, *55*, 2090–2093. [CrossRef]
74. Campbell, T.D.; Febrian, R.; McCarthy, J.T.; Kleinschmidt, H.E.; Forsythe, J.G.; Bracher, P.J. Prebiotic condensation through wet–dry cycling regulated by deliquescence. *Nat. Commun.* **2019**, *10*, 4508. [CrossRef] [PubMed]
75. Salameh, A.K.; Taylor, L.S. Role of Deliquescence Lowering in Enhancing Chemical Reactivity in Physical Mixtures. *J. Phys. Chem. B* **2006**, *110*, 10190–10196. [CrossRef]
76. Shanker, U.; Bhushan, B.; Bhattacharjee, G. Kamaluddin Oligomerization of Glycine and Alanine Catalyzed by Iron Oxides: Implications for Prebiotic Chemistry. *Orig. Life Evol. Biosph.* **2012**, *42*, 31–45. [CrossRef] [PubMed]
77. Baú, J.P.T.; Carneiro, C.E.A.; da Costa, A.C.S.; Valezi, D.F.; di Mauro, E.; Pilau, E.; Zaia, D.A.M. The Effect of Goethites on the Polymerization of Glycine and Alanine Under Prebiotic Chemistry Conditions. *Orig. Life Evol. Biosph.* **2021**, *51*, 299–320. [CrossRef]
78. Sandford, S.A.; Bera, P.P.; Lee, T.J.; Materese, C.K.; Nuevo, M. Photosynthesis and Photo-Stability of Nucleic Acids in Prebiotic Extraterrestrial Environments. *Top. Curr. Chem.* **2015**, *356*, 123–164. [CrossRef]

79. Lee, C.-W.; Kim, J.-K.; Moon, E.-S.; Minh, Y.C.; Kang, H. Formation of glycine on ultraviolet-irradiated interstellar ice-analog films and implications for interstellar amino acids. *Astrophys. J. Lett.* **2009**, *697*, 428–435. [CrossRef]
80. Lin, R.; Wang, Y.; Li, X.; Liu, Y.; Zhao, Y. pH-Dependent Adsorption of Peptides on Montmorillonite for Resisting UV Irradiation. *Life* **2020**, *10*, 45. [CrossRef] [PubMed]
81. Altair, T.; Borges, L.; Galante, D.; Varela, H. Experimental Approaches for Testing the Hypothesis of the Emergence of Life at Submarine Alkaline Vents. *Life* **2021**, *11*, 777. [CrossRef] [PubMed]
82. Sojo, V.; Herschy, B.; Whicher, A.; Camprubí, E.; Lane, N. The Origin of Life in Alkaline Hydrothermal Vents. *Astrobiology* **2016**, *16*, 181–197. [CrossRef]
83. Mielke, R.E.; Russell, M.J.; Wilson, P.R.; McGlynn, S.E.; Coleman, M.; Kidd, R.; Kanik, I. Design, Fabrication, and Test of a Hydrothermal Reactor for Origin-of-Life Experiments. *Astrobiology* **2010**, *10*, 799–810. [CrossRef]
84. Herschy, B.; Whicher, A.; Camprubi, E.; Watson, C.; Dartnell, L.; Ward, J.; Evans, J.R.G.; Lane, N. An Origin-of-Life Reactor to Simulate Alkaline Hydrothermal Vents. *J. Mol. Evol.* **2014**, *79*, 213–227. [CrossRef]
85. White, L.M.; Shibuya, T.; Vance, S.D.; Christensen, L.E.; Bhartia, R.; Kidd, R.; Hoffmann, A.; Stucky, G.D.; Kanik, I.; Russell, M.J. Simulating Serpentinization as It Could Apply to the Emergence of Life Using the JPL Hydrothermal Reactor. *Astrobiology* **2020**, *20*, 307–326. [CrossRef]
86. Brunk, C.; Marshall, C. 'Whole Organism', Systems Biology, and Top-Down Criteria for Evaluating Scenarios for the Origin of Life. *Life* **2021**, *11*, 690. [CrossRef]
87. Stüeken, E.E.; Anderson, R.; Bowman, J.; Brazelton, W.J.; Colangelo, J.; Goldman, A.D.; Som, S.M.; Baross, J.A. Did life originate from a global chemical reactor? *Geobiology* **2013**, *11*, 101–126. [CrossRef]
88. Peng, Z.; Linderoth, J.; Baum, D.A. The hierarchical organization of autocatalytic reaction networks and its relevance to the origin of life. *PLoS Comput. Biol.* **2022**, *18*, e1010498. [CrossRef]
89. Root-Bernstein, R. A Modular Hierarchy-Based Theory of the Chemical Origins of Life Based on Molecular Complementarity. *Accounts Chem. Res.* **2012**, *45*, 2169–2177. [CrossRef]
90. List, B.; Pojarliev, P.; Castello, C. Proline-Catalyzed Asymmetric Aldol Reactions between Ketones and α-Unsubstituted Aldehydes. *Org. Lett.* **2001**, *3*, 573–575. [CrossRef]
91. Howard, T.S.; Cohen, R.D.; Nwajiobi, O.; Muneeswaran, Z.P.; Sim, Y.E.; Lahankar, N.N.; Yeh, J.T.-H.; Raj, M. Amino-Acid-Catalyzed Direct Aldol Bioconjugation. *Org. Lett.* **2018**, *20*, 5344–5347. [CrossRef]
92. Breslow, R.; Cheng, Z.-L. L-amino acids catalyze the formation of an excess of D-glyceraldehyde, and thus of other D sugars, under credible prebiotic conditions. *Proc. Natl. Acad. Sci. USA* **2010**, *107*, 5723–5725. [CrossRef]
93. Duschmalé, J.; Kohrt, S.; Wennemers, H. Peptide catalysis in aqueous emulsions. *Chem. Commun.* **2014**, *50*, 8109–8112. [CrossRef]
94. Kosikova, T.; Philp, D. Exploring the emergence of complexity using synthetic replicators. *Chem. Soc. Rev.* **2017**, *46*, 7274–7305. [CrossRef]
95. Arsène, S.; Ameta, S.; Lehman, N.; Griffiths, A.D.; Nghe, P. Coupled catabolism and anabolism in autocatalytic RNA sets. *Nucleic Acids Res.* **2018**, *46*, 9660–9666. [CrossRef]
96. Hayden, E.J.; von Kiedrowski, G.; Lehman, N. Systems Chemistry on Ribozyme Self-Construction: Evidence for Anabolic Autocatalysis in a Recombination Network. *Angew. Chem. Int. Ed.* **2008**, *47*, 8424–8428. [CrossRef]
97. Liu, B.; Pappas, C.G.; Ottelé, J.; Schaeffer, G.; Jurissek, C.; Pieters, P.F.; Altay, M.; Marić, I.; Stuart, M.C.A.; Otto, S. Spontaneous Emergence of Self-Replicating Molecules Containing Nucleobases and Amino Acids. *J. Am. Chem. Soc.* **2020**, *142*, 4184–4192. [CrossRef]

life

Review

Some Factors from Theory, Simulation, Experiment and Proteomes in the Current Biosphere Supporting Deep Oceans as the Location of the Origin of Terrestrial Life

J. W. Halley

School of Physics and Astronomy, University of Minnesota-Twin Cities, Minneapolis, MN 55455, USA; halle001@umn.edu

Abstract: Some standard arguments are reviewed supporting deep ocean trenches as a likely location for the origin of terrestrial life. An analysis of proteomes of contemporary prokaryotes carried out by this group is cited as supporting evidence, indicating that the original proteins were formed by quenching from temperatures close to the boiling point of water. Coarse-grained simulations of the network formation process which agree quite well with experiments of such quenches both in drying and rapid fluid emission from a hot to a cold fluid are also described and cited as support for such a scenario. We suggest further experiments, observations and theoretical and simulation work to explore this hypothesis.

Keywords: origin of life; complex chemical systems; prebiotic chemistry; ocean trenches; prions

Citation: Halley, J.W. Some Factors from Theory, Simulation, Experiment and Proteomes in the Current Biosphere Supporting Deep Oceans as the Location of the Origin of Terrestrial Life. *Life* **2022**, *12*, 1330. https://doi.org/10.3390/life12091330

Academic Editors: Ranajay Saha and Alberto Vázquez-Salazar

Received: 25 June 2022
Accepted: 16 August 2022
Published: 28 August 2022

1. Introduction

It is well known that the chemistry of terrestrial life is essentially universal throughout the biosphere. That appears to imply that the successful initiation of life on earth only occurred once. That is very well known, but an implication which appears to follow quite unambiguously is less often discussed in the literature: the absence of any hint of 'exotic' life depending on different molecular systems appears to imply that the biochemical events leading from nonlife to life were rare events, in the sense that the probability per year of their occurence was less than something of the order of 10^{-9}. That, in turn strongly suggests that a huge number of unsuccessful natural 'experiments' had to be performed before natural processes stumbled on a successful combination permitting Darwinian evolution to start. An ultimately successful model accounting for life's origin will, it seems, have to take into account the rare nature of the initiating events and will therefore have to involve a process that occurred rapidly and repetitively over hundreds of millions of years.

In such a model, one will need to understand how a very large number of natural experiments could have been proceeding on early earth in order to sort through enough chemical combinations to encounter a successful one. The rare nature of several later key developments in the history of terrestrial life, such as the appearance of eukaryotes and of multicellularity, also needs to be acknowledged and taken into account to explain those events as well, but I will focus here on the earlier steps.

Taking the rare nature of the initiating event into account helps one to understand why laboratory experiments have had so little success in initiating much of any aspect of cellular life from abiogenic material despite enormous effort. It suggests that to understand the natural originating event, one needs to conceive of trial and error on a scale that may need to be hundreds of orders of magnitude larger than is feasible in any human laboratory. A very simple example of the sort of numbers involved is the well-known estimate [1,2] of the time required to form an initial genome by a random assembly of nucleotides, which results in times which are 1000 s of orders of magnitude longer than the present age of the universe. This is sometimes called 'Eigen's paradox'. Setting the issue of forming an initial genome aside, others suggest that the formation of a complex of interacting proteins with lifelike properties, possibly somewhat like modern prions and amyloids, is more likely. A primitive estimate illustrates that such a scenario greatly reduces the expected times to initiation of lifelike chemistry, although the times remain extremely long:

Suppose that one protein such as the one [3–7] in prions must be formed to start the process. I choose the prion protein (there is only one) here as an example to illustrate the orders of magnitude involved in the time required for a protein first model to generate a lifelike start. (Our models, described in Section 2, do not explicitly model that protein nor do they assume that prions initiated terrestrial life.) The PrP protein in prions is reported [4] to contain 209 amino acids. (I am not concerned here with the secondary structure, although it is central to functioning of the prion form PrP^{Sc} of the protein.) It appears that the total number of types of amino acids chemically possible is not known. A total of at least 52 from meteorites is reported [8], and more than 1739 are listed in the norine database [9]. The task of a natural process to generate the starting protein (assuming only one will work) for initiating prion-like reproduction is then to sort through all the possible polypeptides of the needed length for the one that 'works' to start a Darwinian evolutionary process. The order of magnitude of the result only depends logarithmically on the number of available amino acids, and I will use the number 500 for the estimates in this paragraph. Then, there are roughly $500^{209} \approx 10^{564}$ possibilities. The corresponding numbers if the number of available amino acids is 100 or 1000 are 10^{418} and 10^{627}, respectively. These numbers are smaller than the number of possible initial genomes in a genome-first model ($4^{10,000} \approx 10^{6020}$), but they are still very formidable.

Are the fluxes from ocean trenches large enough to make such a natural random search likely to be successful? The global water flux from high-temperature 'smoker'-type emissions from trenches is roughly estimated at 10^{13} kg/year [10], and the concentrations of amino acids in high-temperature smokers have been estimated to be in the range 10^{-5} molar [11]. With a water density of 1 gm/cm^3, this gives about 10^{40} amino acids flowing through the rift system per year. If the required polymer is 200 units long, it will the take, on average in the case of 500 available amino acid types, $200 \times 10^{564}/10^{40} \approx 10^{526}$ years for the needed polymer to appear assuming, very optimistically, that all the amino acids flowing through form polypeptides about 200 units long. Thus, even a rift mechanism leading to an abiogenic appearance of a protein-based initial life form seems likely to be an extremely rare event.

One way to further reduce the estimate is to suppose that there are a great many initial polymers, here assumed so far to be just one polypeptide, which will work. In the context of an amyloid model [12] for the origin of life, there might be some indication that many polypeptides would work because a substantial number (in the order of 10 to 100) of polypeptides have been found to form replicating, metastable amyloid fibrils, many of which cause various neurological diseases. However, to make the estimated average time for natural processes to form one of the needed polymers comparable to the present estimated age of the universe (or, for these purposes, the age of the earth, which is nearly exactly one-third of the age of the universe) would require that the fraction f of the total number of available polymers that would work be $f \approx 10^{-50}$. The fraction is small, but it means that a huge number $\approx 10^{514}$ of forms of 'exotic' lifelike systems would be possible. Because the fraction is so small, one may argue that if that is the resolution of Eigen's paradox, then one would not expect to find any 'exotic' life on earth, and also that, in the event that lifelike systems are found on exoplanets, they will be extremely likely to be 'exotic', sharing very few of the specific biochemical features found on earth. That would make the task of identifying lifelike chemistry on exoplanets both qualitatively different and much more difficult than it is sometimes conceived to be.

I am referring to Eigen's, not Fermi's, paradox here: The latter refers to the fact that if one assumes that life initiation was NOT a rare event, then the failure, so far, to observe extraterrestrial life is paradoxical. I have argued here that the evidence suggests that the event WAS rare and therefore that there is no Fermi paradox. In the preceding paragraph, I entertained the possibility that the number of possible chemical starting points is very large. Then, one might expect some kind of lifelike chemistry on every 'habitable' exoplanet. However, in that case, although both Eigen's and Fermi's paradox would be, in a sense, resolved, the lifelike chemistry on other planets would be expected to be 'exotic', that is, different, and quite possibly extremely different, from that on earth. (There is a vast speculative literature on the Fermi paradox, which is not reviewed here. The present author's perspective on these issues as of 2012 is described in [13].)

Here and elsewhere in this paper, I am avoiding any attempt at a precise general definition of 'life'. Despite much discussion [14–16], there is no consensus on such a definition, and for our purposes here, we do not need one: For the terrestrial biosphere, it is clear what is meant because we have a vast trove of detailed data on the properties of 'living' things on earth. For extraterrestrial systems, we have extremely limited data on the chemistry happening near solid planetary and lunar surfaces, none of which has been claimed to correspond to 'lifelike' chemistry. For extraterrestrial systems, I will refer to our goal as the identification of 'lifelike' processes occurring on or near those surfaces. (I do not mean just mineral surfaces, although they may play a role in some prebiotic chemistry [17–19].) By 'lifelike', I will mean one or more generic processes or properties which have been deemed essential in one or more of the many proposed generic definitions of 'life'. In practice, most of our work to date has focused on the property of sustained dynamic disequilibrium. The quantitative measures of the property we use are described in the next section. Disequilibrium is qualitatively similar to the property of 'homeostasis' as described in [16]. Sustained dynamic disequilibrium has sometimes been dismissed as 'trivial', but in fact, because of the second law of thermodynamics, it is far from trivial and is not easy to realize in a dense fluid. However, although that property is 'lifelike', I do not mean to imply that it is sufficient to establish that a system is 'alive'. We have also studied [20] one variant of our model in which a property related to the property of 'replication' described in [16] is found, but that will not be discussed further here. If complex systems with the property of dynamic disquilibrium and a few other of the proposed generic features are found extraterrestrially, they are, on the basis of the preceding argument, very likely not to look very much like our terrestrial biosphere at all. If and/or when we have data on a few such systems, a more meaningful discussion concerning an appropriate general generic definition of 'life' will be possible.

I infer from these considerations that the determination of the initial chemical steps in the origin of life is a less critical question than the question of how those steps could have taken place, given that random processes which failed to produce lifelike chemistry would have been overwhelmingly more likely. I therefore suggest that a high priority is to find mechanisms for extremely rapid natural experimentation taking place in parallel over hundreds of millions of years. The general idea that many natural experiments are required is not new, but much work in this field emphasizes how slow the process must have been, whereas here, I emphasize that given the enormous natural search task, the processes had to be extremely fast at the molecular level. In the next sections, I will review some recent simulation and experimental work which suggest that the rapid quenching of hot fluids containing large amounts of biomonomers after a suitable incubation time may be the most likely process for achieving the needed massive parallel processing, such as in well known models of abiogenic evolution in the fluids emitted from ocean rifts [21]. Several ways that such quenching could have happened on early earth have been proposed, for example in hot springs or impact craters [22] as well as in ocean rifts. In view of the centrality of the need for massive processing, however, ocean rifts, which have continuously quenched very large volumes of hot fluid over hundreds of millions of year, seem at least in that respect the most favorable. Ocean rifts have other favorable features, some of which are discussed in [21]. For example, the reactions near the ocean floors would be relatively protected from ultraviolet light and high-energy cosmic rays that could interrupt the low energy chemical development required, whereas it seems less obvious that such protection would be likely to be present to the same degree in hot springs.

However, here, we emphasize another advantage of the quenches of hot fluids emerging from within the earth into cold ocean water: namely, the large volumetric flow of fluid which increases the chance of a successful initiation of lifelike chemistry. We envision the importance of a quench (which also occurs in hot springs) as follows: Proceeding at a temperature near ambient, covalent bonds such as peptide bonds have very long lifetimes as discussed further below, and the chemical sorting that would produce many combinations of monomers (amino acids in the protein case) would be extremely slow. On the other hand, at high enough temperatures, the covalent bonds will break and form, on average, much more quickly, and more polypeptides will be sampled in a 'dynamical chemical network' [23]. However, the instability produced by the frequent thermal breaking of covalent bonds would not be favorable to the development of lower energy processes of prebiotic evolution that use those polymers as building blocks. Thus, one needs a 'sampling step' in which each combination of prospective biopolymers produced at high temperatures is stabilized and allowed to evolve at low temperatures. That is quite precisely what happens in a quench when the thermal bond breaking is frozen out as the system goes quite suddenly from a hot stage to a cold one, as it emerges from the ocean rift into the cold ocean water. To freeze the high-temperature distribution of polymer lengths so that it is retained in the low-temperature, quenched, phase, the quench must occur faster than the (adiabatic) rate, which would allow the covalent bonding to completely equilibrate to the low-temperature equilibrium distribution during cooling.

In this paper, I review the analytical and computational efforts of my group over the last ten years to obtain insight concerning how nature might have produced the rare event leading to life on earth and possibly other rare events leading to lifelike systems extraterrestrially. It is not a review of the entire literature on the ocean trench model for the origin of life or of other computational models intending to provide insight into the origin of terrestrial life.

2. Results

I hope I have convincingly reiterated the argument that the process leading to life's origin was extremely unlikely over the timescale set by the age of the earth, but we would like to know more quantitatively how (un)likely it was. For that, one needs some more specific ideas concerning where and by what detailed chemical processes the initiation took place. If a model for those processes is too chemically specific, it may miss some essential generic features which are promising because the chemical details embedded in the model obscure the promise. On the other hand, if the model is too coarse grained, it is likely to be uninformative. In polymer science, scientists navigate between those two dangers by coarse graining of the atomic-scale description [24]. Formally, the coarse graining can be mathematically carried out by convoluting the variables of the original atomic description with a smoothing function, such as a Gaussian. However, more commonly, some physically plausible assumptions about the essential features are made in order to specify a coarse-grained model. For example, early on, polymer chemists used a 'united atom' model of hydrocarbon polymers by treating the carbon atoms and the hydrogen entities bonded to them as an entity located at one point in three-dimensional space. Similarly, biochemists use models of biopolymers in aqueous solution which treat the water as a continuum ('implicit water') and the motions and positions of the monomers in the biopolymers with atomic specificity.

In addition to the practical computational advantages, coarse graining can reveal features which may be obscured by the mass of noisy detail in a more fine-grained description. If features appear in the coarse-grained description of a system which were obscured in a more detailed description, statistical physicists speak of 'emergence' [25,26]. One may regard terrestrial life itself as an emergent phenomenon in that sense, since it is difficult to discern the essential features of life by regarding, for example, a molecular dynamics simulation of protein motions on angstrom and picosecond scales. (However, the problems in attempting to make such a statement quantitative are outlined in [26].) The distinction between fine-grained and coarse-grained descriptions is qualitatively similar (but not isomorphic) to that between biological descriptions in terms of genotype (an atomic-scale description) and phenotype(usually a more coarse-grained description of structure or behavior), respectively. There is another advantage to coarse graining if one is interested in the likelihood of extraterrestrial lifelike systems, possibly on exoplanets or moons: Given the enormous combinatorial space available, it is much more likely than not that such lifelike exobiological systems, if ever found, will be exotic in the sense that the basic biochemistry will be very different from the terrestrial one. Such possiblities will be missed in models which are too chemically specific. Thus, coarse graining avoids some aspects of terracentricity.

The coarse-grained descriptions which we have been using in my group in efforts to gain insight into the origin of life were initially inspired by the work of Stuart Kauffman and coworkers [27–29], who introduced a coarse-grained model for the origin of life in the 1980s. In it, there were 'monomers' of an integer number b of different types which were allowed to link together to form 'polymers' through a process termed 'ligation' and to become unlinked through the reverse process of 'scission'. A parameter p in the range $[0,1]$ was introduced which specified the likelihood that any given reaction among all the possible ones would actually occur. There was no spatial component: one could say that the model described a 'well-mixed' reactor. There was also no temperature and no energy. By starting with a finite set of 'monomers' and 'dimers', Kauffman et al. built a set of reaction networks and then, by randomly assigning rates to the reactions included, followed the dynamics stochastically by numerical simulation for each one. They found that for essentially all values of p, the populations of polymers grew without limit as long as the population of monomers and dimers was maintained (one could say by 'feeding'). The results were regarded as suggesting that lifelike properties could be expected to appear quite readily on earth and elsewhere, which is in contrast to the discussion in the introduction to this paper.

In our first extension of the Kauffman work [30], we noted that, without extending the model, one could compute the combinatorial entropy of the states resulting from the dynamic simulations and compare it to its maximum possible value which, since the model was essentially working at infinite temperature, would be the value expected in thermodynamic equilibrium at that temperature. When we counted the fraction of times that the dynamics led to an equilibrium state by that criterion, we found that the dynamics led to equilibrium most of the time. Only a small fraction, which we interpreted as having the lifelike property of disequilibrium, was produced by the dynamics. The others were growing in population, but they were just adding molecules to a system in equilibrium, and we interpreted them as not lifelike. One could ask, in that context, why all the final states were not in equilibrium since it is predicted by the second law of thermodynamics that dynamic processes always increase the entropy. However, the states we found that were not in equilibrium were kinetically trapped by the sparsity of the network in a kind of cul de sac of the chemical space. They were taking a longer time to escape such 'kinetic traps' than the length of the computer run and were thus in a metastable disequilibrium state, as living things are.

Although we found that interesting, the model left out a great deal. In particular, there was no energy, no temperature and no spatial dependence (as well as no detailed chemical description of the monomers, but we regarded the latter as an advantage for the reasons cited above). There were several stages [20,31–36] in the effort to take account of these features, but I will skip the intervening steps and describe our most recent efforts in which, as a kind of bonus, we found a way to interpret the somewhat mysterious Kauffman parameter p in terms of more directly accessible experimental quantities. We retain a description of the available reactions as the ligation and scission of polymers of monomers of b species. It should be noted that this assumption is open to serious question if one wants, as we do, to use our results to model what might happen during the quenches taking place as fluids are emitted from ocean rifts. That is because although the ligation reactions leading to polypeptides and nucleic acids are endergonic and high temperatures can therefore enhance peptide bond formation for example, the formation of the monomers themselves, amino acids in the case of peptides, is believed to be exergonic so that the monomers themselves would not survive [21]. On the other hand, we noted above [11] that amino acids are detected observatonally in the fluids emitted, and studies of the effects of hydrogen in stabilizing amino acids in hydrothermal environments [37] suggest that amino acids survive in black smokers.

To take account of energetics, we introduced a binding energy Δ which is the energy difference between two unbonded monomers and the same two monomers when bonded. If $\Delta < 0$, it means that the bonded system is at higher energy than the unbonded one, as is the case for both peptide bonds in proteins and for nucleic acids in RNA and DNA in the absence of enzymes. In our most recent work, we therefore consider only that $\Delta < 0$ case and also introduce an activation energy Δ_a which controls the rate of scission. Unlike Δ, Δ_a takes different values for each reaction, as selected from a Gaussian distribution with a mean $\overline{\Delta_a}$ fixed by experiment. In particular, in our application to polypeptides, we fix $\overline{\Delta_a}$ to be the measured average activation energy for the hydrolysis of the glycine–glycine peptide bond in water [38]. The two energy parameters are used to fix the rates in the dynamics simulation. The 'forward' (scission) rate for a reaction $1 \rightarrow 2+3$ is given $vk_d n_1$ and the reverse rate is vn_2n_3/k_d, where $k_d^2 = \overline{n_2 n_3}/\overline{n_1}$ the line over the $n's$ in the definition of k_d^2 indicates the equilibrium distribution of the species. The factors v take the form $v = e^{-\Delta_a/k_BT}$ and the Δ_a values are drawn during network formation from a Gaussian distribution centered on the average activation energy as experimentally determined. The physical rates are interpreted to be $f_a v$, where f_a is the measurable prefactor, sometimes termed the 'attempt rate', in the rate constant for scission, or in the case of the polypeptides, the hydrolysis of the peptide bond. With that parameterization, the model as written mathematically and coded is expressing time in units of $1/f_a$, and we can convert that to physically measured time in the laboratory or observation by multiplying computed rates by f_a and times by $1/f_a$. The factors k_d guarantee that the system is being driven toward equilibrium in accordance with the second law, although we are interested in cases in which it stalls in a metastable lifelike state before it gets there. The width of the Gaussian is another parameter. We think of the Gaussian distribution as arising from fluctuations in the activation energy arising from a heterogeneous environment, and in the case of peptides, we found very little indication of an appropriate value for its width σ in the literature. We obtain an order of magnitude estimate from the data in [38] on glycine–glycine hydrolysis where the activation energy is reported to vary substantially for glycine–glycine bonds in a few different polypetides.

In this model, we have taken account of some missing features, but now every possible reaction is allowed, so we need to understand how kinetic trapping can occur. That is because the experiments, natural or made by humans, take a finite time say of order t_{expt}. Even at the average activation energy, the mean time for hydrolysis to occur in peptides can be as much as a century, and the Gaussian distribution allows the possibility of much longer times. Thus, those reactions for which the rate $f_a v < 1/t_{expt}$ will not occur during the experiment and can be left out of the networks. Therefore, we can predict, on the basis of parameters which are experimentally known (within some estimatable uncertainties), the value of the fraction p_{eff} of the reactions which should be included in the network, just as they were in the original Kauffman model. Working out the details [36], the parameter p_{eff} can be expressed in terms of combinations of the experimental quantities $\xi = \overline{\Delta_a}/\sigma$ and $T_c = \overline{\Delta_a}/\ln(t_{expt} f_a)$ as described in detail in Section 2.2 below.

It turns out that the model predicts that a swift quench from above the temperature T_c to a temperaturre below T_c will take the system from a state in which nearly all possible reactions involving the breaking and forming of covalent bonds are occuring to one in which they almost all stop abruptly. ('Swift' means fast compared to the rate at which the system equilbrates to the temperature of its external bath.) This is one of several indications that suggest that the original peptides were formed in a high-temperature environment and then quenched. Before describing that model result in more detail, in the next section, I review more model-independent direct experimental evidence, which we found [32] in data on the proteomes of prokaryotes indicating that quenches may have led to the formation of proteins.

2.1. Evidence from Biodata for an Origin in a Quench

The first indication of a quench origin for terrestrial life which we noted came from a study which we carried out [32] of length distributions in proteomes of prokaryotes, where the length of each protein was the number of amino acids in the chains of proteins. We analyzed data on the proteomes of all the prokaryotes listed in the KEGG database [39] in 2018. The motivation for the study was not initially to establish quenching as a mechanism for the origin of life but to test our hypothesis that a measure that we had been using to detect lifelike systems in our simulations did indeed distinguish real living systems from nonliving ones. To clarify the meaning which we attribute to one of the results of [32] in the context of the present paper, the definition of the measures used there is reviewed here. It should be emphasized that the results of reference [32] were obtained without use of any of the details of the network and dynamics models which we used for simulations in references [20,30,31,35] as described in the next section, although we used the same measure to analyze the simulation results in some of those papers that we used in [32] to analyze experimental results from the proteome study. We assumed in [32] that the bond energies (parameterized by a parameter $\Delta < 0$ as discussed in the previous section) of all the possible amino acid pairings in the proteomes were the same and that no other energies were relevant in establishing the equilibrium state of the proteome system with regard to its polymer length distribution. That is a kind of coarse graining in the energy description as further discussed in [35]. Using that parameterization, the data provide the relevant numbers $\{N_L\}/V$ and $e = E/V$ where V is the volume, N_L/V is the number of proteins per unit volume having (amino acid) length L and E is the total bonding energy. The energy of a protein of length L is $-(L-1)\Delta$ in such a description, and the total energy E per amino acid of a state of the proteome characterized by the distribution $\{N_L\}$ is $-\sum_L (L-1)\Delta/N$ where $N = \sum_L N_L$.

Now, we suppose that the system is in contact with a thermal bath at temperature T. We can define two different equilibrium distributions for the system, one in which the system is in thermal equilibrium with the bath and a second in which the system is self-equilibrated but has not equilibrated to the bath. This is a textbook distinction in physics [40], but it is less familiar in the context of origin of life studies, so I will briefly dwell somewhat further on it. The first form of equilibrium will be attained in a nonliving system (and even in a living system that has died) after enough time has passed for the system to equilibrate to an external bath. Because the bath might be outside the system and only accessing it through a two-dimensional surface in space, that equilibration can take a long time. When it does, one has the familiar canonical distribution reviewed below. The second, self-equilibration can occur faster, although it only does so in so-called ergodic systems. The length distributions in that second equilibrium state have the same form as they do in the first kind of equilibrium, but the effective temperature is determined differently and depends on E but not on the external T, as seen in the equations reviewed below. One can say in textbook language that the second distribution is microcanonical because the energy and not the outside temperature is fixed, but it is not described by a δ function describing the energy condition. INstead, it is described by a distribution of the canonical form but with an effective temperature determined from the total energy. The description of this isolated second type of equilibrium state in terms of a delta function and the description in terms of a canonical distribution with a temperature determined from the total energy E turn out to be identical in the thermodynamic limit of large volumes and large particle numbers [40]. That thermodynamic limit will be very nearly reached in most macroscopically observed systems of interest here. The effective temperature of the isolated state can be different from the bath temperature if the system has not been in contact with the bath long enough. If it has been in contact with the bath for a very long time, then the two forms of equilibrium are expected to be the same. The relevance to a quench is that a system of amino acids in contact with a high-temperature bath may attain equilibration of the first type because the processes are fast at high temperature but then take a very long time to equilibrate to the cold bath into which it is plunged by the quench. Then, the distribution of polymer lengths in the system will reflect the temperature of the hot bath and not of the bath with which it is currently in contact. In that way, we may be able to read something about the thermal history of the system from its length distribution. That is how we proposed to interpret the results of reference [32].

I provide a few details of the determination of an equilbrium length distribution from [32] to indicate the meaning of the measures of disequilibrium which we used: We denote the total number of polymers N in a sample by $N = \sum_{L=1}^{l_{max}} N_L$. However, in contrast to the situation in the dynamic simulations described in [31], the input data for the calculation of equilibrium distributions are not N and E but the volumetric polymer concentration $\rho = N/V$ where V is the solution volume and the volumetric energy density $e = E/V$. To take entropic account of the dilution of the experimental sample, we introduced a microscopic length $l_p L^\nu$ where l_p is the (microscopic) polymer persistence length and ν is a dimensionless index, which would be 1/2 for a random walk. For denatured proteins, ν has been determined experimentally [41] and is close to the value expected for a self-avoiding walk. We then modified the expression for the entropy used in [31] to take account of the number of ways to distribute N polymers in a volume V giving entropy $S/k_B = \ln W$ with

$$W = \sum_L \frac{(N_L + G_L - 1)!}{N_L!(G_L - 1)!}$$

and $G_L = b^L V/v_L$ and $v_L = l_p^3 L^{3\nu}$. b is the number of monomers available for inclusion in the polymers in the system. The expression is identical to the one used in [31] except for the factor V/v_L in the degeneracy G_L.

Proceeding in the standard way to maximize the entropy under these constraints, we have when both energy density $e = E/V$ and polymer number density $\rho = N/V$ are fixed that the values $\overline{N_L}$ of the populations that maximize this entropy are

$$\overline{N_L} = \frac{G_L - 1}{\exp(-\beta(e,\rho)\mu(e,\rho) - \beta(e,\rho)\Delta(L-1)) - 1} \tag{1}$$

Here, the parameters $\beta(e,\rho)$ and $\mu(e,\rho)$ are determined from the total energy density e and polymer number density $\rho = N/V$ by the implicit equations (with (4))

$$e = -(1/V)\sum_{L=1}^{l_{max}} (L-1)\overline{N_L}\Delta \tag{2}$$

and

$$\rho = (1/V)\sum_{L=1}^{l_{max}} \overline{N_L} \tag{3}$$

We use the definition of G_L and define $v_p = l_p^3$ to write these relations as

$$\overline{N_L(v_p/V)} = \frac{b^L/L^{3\nu} - (v_p/V)}{\exp(-\beta(e,\rho)\mu(e,\rho) - \beta(e,\rho)\Delta(L-1)) - 1} \tag{4}$$

and

$$ev_p = -\sum_{L=1}^{l_{max}} (L-1)\overline{N_L(v_p/V)}\Delta \tag{5}$$

and

$$\rho v_p = \sum_{L=1}^{l_{max}} \overline{N_L(v_p/V)} \tag{6}$$

These are in dimensionless form, which is convenient for solving for $\beta(e,\rho)\mu(e,\rho)$ and $\beta(e,\rho)\Delta$ numerically because they do not involve macroscopically large numbers. $1/\beta((e,\rho)$ is Boltzmann's constant times the temperature fixed by self-equilibration and the given energy and particle density. $\mu(e,\rho)$ is the corresponding self-equilibrated chemical potential. ρv_p is the estimated (fractional) number of polymers in a cube of volume v_p and is a small number. The term v_p/V on the right-hand side of (4) is in all cases much less than $b^L/L^{3\nu}$ and is dropped in the numerical analysis. This is the distribution arising from the second kind of equilibrium discussed above in which the system is either isolated from any bath or, more often, has not had time to equilbrate to an external bath. We have sometimes [31] referred to this equlibrium as 'isolated'. There is no reference to the temperature of an external bath.

To obtain the first kind of equilibrium distribution, we solved (6) for μ where we replaced $\beta(e,\rho)$ and $\mu(e,\rho)$ by $1/k_BT$ and $\mu(T,\rho)$, respectively, obtaining $\mu(T,\rho)$ with a fixed value of T. We used values of the ambient temperature in determining that first kind of equilibrium distribution and made no use of (5). In a quench, the ambient temperature is the high temperature of the fluid before the quench and the low temperature of the fluid after the quench. That first kind of equilibrium distribution then describes the expected distribution when the system has had time to equilibrate to an external bath. In each case, we use (4) to evaluate the polymer length density distributions expected in those two equilibrium states. After a quench, if it is fast, the first kind of equilibrium is often not achieved by the system for a long time. Note that operationally 'fast' here means fast compared to the rate at which the system equilibrates to the lower external temperature in the quenched state.

Because the systems of interest are not necessarily (and in fact are often found not to be) in either kind of chemical equilibrium, we used the experimentally observed values of the polymer length densities $N_L(v_p/V)$ to evaluate Euclidean distances in the space of values of sets $\{N_Lv_p/V\}$ between the actual population set $\{N_Lv_p/V\}$ and the ones corresponding to the two kinds of equilibria given by (4) with $\beta(e,\rho)$, $\mu(e,\rho)$ in the isolated case and with a fixed β and $\mu(T,\rho)$ in the case in which the system is equilibrated to an external bath. Thus, we define two Euclidean distances R_L and R_T in the l_{max}-dimensional space of sets $\{N_L\}$ which characterize how far the system of interest is from the two kinds of equilbrium:

$$R_L = \sqrt{\sum_L (v_p/V)^2(N_L - \overline{N_L(\beta(e,\rho),\mu(e,\rho))})^2/(\sqrt{2}v_p\rho)} \tag{7}$$

for distance from the locally equilibrated state and

$$R_T = \sqrt{\sum_L (v_p/V)^2(N_L - \overline{N_L(\beta,\mu(T,\rho))})^2/(\sqrt{2}v_p\rho)} \tag{8}$$

for distance from the equilibrium state with the external bath. R_L and R_T are, respectively, measures of the distances from the two kinds of equilibrium discussed above. A similar measure in the context of origin of life studies was suggested in [33].

In [32], we reported on the use of those measures in several ways, but here, I focus on some of the results for the proteomes. With a given polymer length distribution, we could plot R_T as a function of the temperature assumed in the equation for R_T. (R_L does not depend on the external temperature.) Most of the prokaryotes considered currently live in an environment with a temperature close to ambient, but we explored how the R_T value varied with different assumed external bath temperatures. We show a typical result for one of the proteomes and also the average R_T for all 4555 of them in Figure 1. To our surprise, there was a very sharp drop in the calculated value of R_T as a function of T at a temperature quite close to the boiling point of water. At the minimum, the R_T value is essentially identical to the R_L value, and the effective temperature of the isolated equilibrium was also found to be near the same temperature at which the dip in R_T was observed.

We interpreted this to suggest that the original proteins in the prokaryotes were formed at the high temperature around the boiling point of water and were then quenched to a cooler temperature which froze out most of the reactions which broke or formed peptide bonds so that the higher temperature length distribution was frozen into the subsequent evolution of the system. The data leading to Figure 1 thus suggest that all 4555 proteomes arose from a collection of proteins which formed at the same temperature. Many events which we are not attempting to model here occurred after that initial collection of proteins was formed (including, for example, the introduction of a genetic code, the appearance of the ribosome, and the adaptation of various prokaryotes to different environments including high-temperature ones).

There are issues associated with time scales in this interpretation. Peptide bonds in water can have reaction constants for hydrolysis, leading to average times for bond breaking at ambient temperatures of up to centuries. However, the suggestion requires that the length distributions in the prokaryotes survive not centuries but rather more than 3 billion years. To avoid a contradiction, we suggest that once a sufficiently promising system was quenched by chance, the transition to a lifelike system evolving and growing by natural selection might occur much more rapidly than equiibration to the external bath. For example, in the case of polypetides, that would require that the initiation through Darwinian evolution of processes such as autocatalysis to replace damaged polymers and protect the nonequilibrium state would need to occur in a time less than the order of a century. The resulting nonequilibrium steady state could then be self-sustained by those evolved processes, and the associated polymer length distributions also could be preserved for much longer times, possibly up to billions of years. If the starter protein was like PrP^{Sc}, for example, that might be conceivable, since prion reproduction [34] does not appear to make much use of the elaborate reproductive machinery of modern cells, and only a few steps might be required.

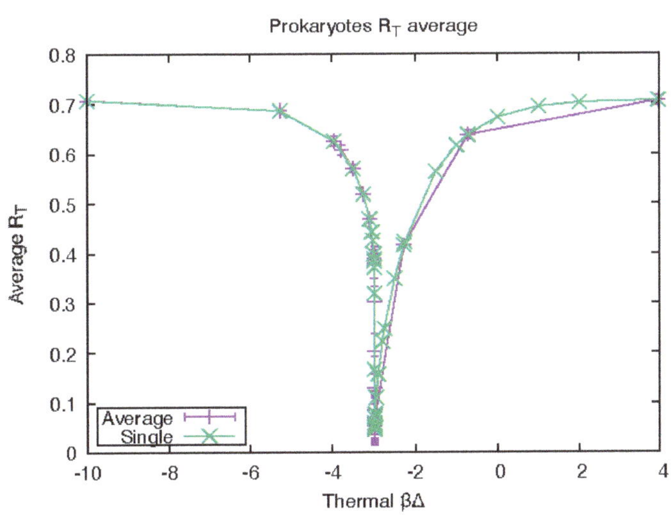

Figure 1. R_T as a function of $\beta\Delta$ for one prokaryote (KEGG code ebw, green crosses) and the average values over the entire KEGG database of 4555 prokaryotes (purple plusses). The error bars (purple) indicate the standard deviation over the full data set. The variable $\beta\Delta$ is the bond energy Δ divided by $k_B T$ where T is the assumed ambient temperature. The graph shows that the proteome has a length distribution at a temperature close to equilibrium at about 400,000, although most of the prokaryotes commonly live at much lower temperatures.

2.2. Evidence from Theory and Simulation for an Origin in a Quench

Turning to theory and simulation, we show the form of the effective fraction p_{eff} of the possible scission and ligation reactions briefly described at the end of the Section 1 as predicted by the model of reference [35] in Figure 2. The model provides an analytical expression for p_{eff} which depends only on the combinations of experimentally determinable quantites $\xi = \overline{\Delta_a}/\sigma$ and $T_c = \overline{\Delta_a}/\ln(t_{expt}f_a)$. T_c is a temperature, and the figure shows that for parameters suitable for the description of peptides, p_{eff} undergoes a very sharp transition from nearly one to nearly zero when T passes from above T_c to below it. (The analytical form or p_{eff} in terms of ξ and T_c is determind by evaluation of the integral

$$p_{eff} = \int_{f_a t_{expt}}^{\infty} (dP/dv)dv \tag{9}$$

where dP/dv is the probability distribution of the values of v determined from the assumptions cited above. The result is

$$p_{eff} = \frac{erf(\xi) - erf((1 - T/T_c)\xi)}{erf(\xi) + 1} \tag{10}$$

where erf is the error function [42]. (Please see [36] for details.) T_c is about 370,000 in the peptides and is independent of the poorly known parameter σ.

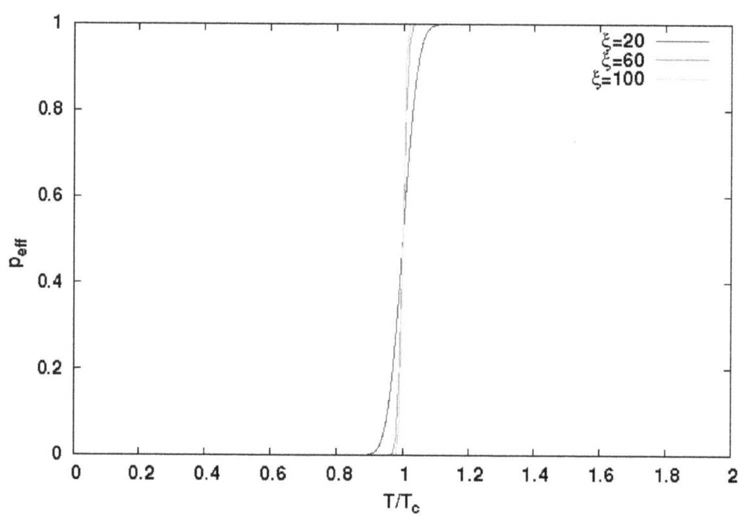

Figure 2. p_{eff} as a function of temperature for various values of ξ using Equation (10).

The sharp change from 1 to 0 in p_{eff} illustrated in Figure 2 and predicted by the detailed model of reference [36] thus occurs at a temperature quite similar to the one at which there is a relatively model independent dip in R_T, which we found in [32] to occur in the proteomes. We suggest that the model thus provides additional support for an origin of life scenario such as the one described in Section 1 in which a quench took the system from a p_{eff} of nearly 1 at a temperature above T_c to a value of T below T_c with a p_{eff} of nearly zero, at which point spontaneous hydrolysis and peptide bond formation almost stopped. The temperature of the dip in R_T arose from the proteome data, but the sharp shift in p_{eff} depends on just a few parameters: The average value $\overline{\Delta_a}$ of the activation energy for peptide bond hydrolysis, the 'experimental' time t_{expt} which the system spends at high temperatures before quench and the prefactor f_a, interpreted as a frequency, in the Arrenhius relation for the hydrolysis rate. The detailed relation is

$$T_c = \frac{\overline{\Delta_a}/k_B}{\ln(t_{expt} f_a)} \tag{11}$$

The most poorly known quantity here is t_{expt}. (T_c is independent of the parameter σ.) We have some knowledge of t_{expt} in laboratory experiments briefly described below. In the case of smokers from ocean trenches, the relevant time is presumably the average time which the fluids spend in the high-temperature part of the circulation pattern, which takes low-temperature water through the porous crust of the ocean floor near a ridge to a horizontal path near the hot magma of the interior, after which it rises through the rift to emerge in the smokers [10,43]. These events, while believed to occur quite generally, are reported to be extremely heterogeneous in their time scales. However, time scales of the order of 1 to 10 years are typically suggested, even though some processes are much faster. Solving Equation (11) gives $t_{expt} = f_a^{-1} exp(\overline{\Delta_a}/k_B T_c)$, and using $\overline{\Delta_a}/k_B \approx 1.2 \times 10^4$ K and $f_a \approx 1.3 \times 10^7\ \mathrm{s}^{-1}$ as extracted from the data in [38] for peptide hydrolysis rates, we find $t_{expt} \approx 0.3$ years using $T_c = 370$ K. Thus, this scenario appears to be roughly consistent with what is currently known. However, the values of the activation energies reported in [38] vary by as much as a factor of 2 depending on what pair of bonded amino acids one is hydrolyzing, so this conclusion is tentative.

The measurements of amino acid concentrations in the emissions from smokers in ocean rifts reported in [11] and cited earlier provide another piece of possibly relevant information: The concentrations of peptides were measured in a way that permitted the concentration of all polypeptide molecules including monomers to be measured as well as the concentration of amino acid monomers alone. The former was much larger than the latter, suggesting that the amino acids were mainly in bonded polypeptides, although a detailed length distribution was not found. That might suggest support for an origin of life scenario, since long polypeptides would be needed, but it might be in contradiction with laboratory experiments discussed below where many more monomers than polypeptides were found. One possible explanation is that the high-temperature phase of the quench in the laboratory experiments was much lower than the average temperature of the fluids in the relevant natural ocean rift experiments, and the higher temperature would lead to the formation of more long polymers, which would be retained upon quenching. Some of the fluid temperatures in the high-temperature phase of the quenches occurring in ocean rifts are reported to be much higher (up to nearly 600,000) than those in the reported laboratory experiments (usually around $373,000$).

2.3. Comparison of Model Predictions with Laboratory Results of Quenches

Laboratory experiments to determine whether quench-like processes could result in enhanced polypeptide synthesis from monomeric solutions of amino acids have been reported by groups led by Matsuno [44–46] and somewhat less directly by John Yin [47–49]. Both groups observed polypeptide formation, although the numbers of polypeptides were small. The Matsuno et al. experiments directly modelled aspects of the proposed scenario in ocean ridges: Hot solution was forced through an orifice into a cold bath and then recirculated to be heated again. The experiment differed from the envisioned oceanic one because the dwell times were much shorter—of the order of a minute compared to dwell times in the hot part of the cycle in an ocean trench estimated, as reviewed above to be of the order of a year or more. As seen in Equation (11), the critical temperature above which we expect rapid polymer formation depends inversely on the logarithm of the dwell time t_{expt}, suggesting a significantly higher value of T_c in the Matsuno experiments than in the ocean trenches. Thus, it is possible that the hot phase in the Matsuno cycle was at or below T_c where polypetide formation would not be significantly enhanced. Furthermore, because the ocean ridge dwell times are believed to be longer, the predicted T_c from

Equation (11) is lower, increasing the likelihood that ocean ridge quenches would have hot phase temperatures above T_c.

I show some preliminary attempts to fit our model to the data of the Yin and Matsuno groups in Figures 3 and 4. Qualitative features are reproduced with reasonable parameters, but this is hardly a proof that the model exactly describes the processes taking place in these experiments. However, it is clear that the numbers of polypeptides produced is small compared to the monomers, which is consistent with the predictions of the model and in contrast to the limited information from ocean trenches as discussed above. It is of some interest that the model reproduces the even–odd oscillations with respect to L, which are observed in both sets of experiments, although more markedly in the Yin et al. results. We have also formulated a simplified version of the model for the case in which there are a lot more monomers than polymers, which reproduces very similar even–odd oscillations.

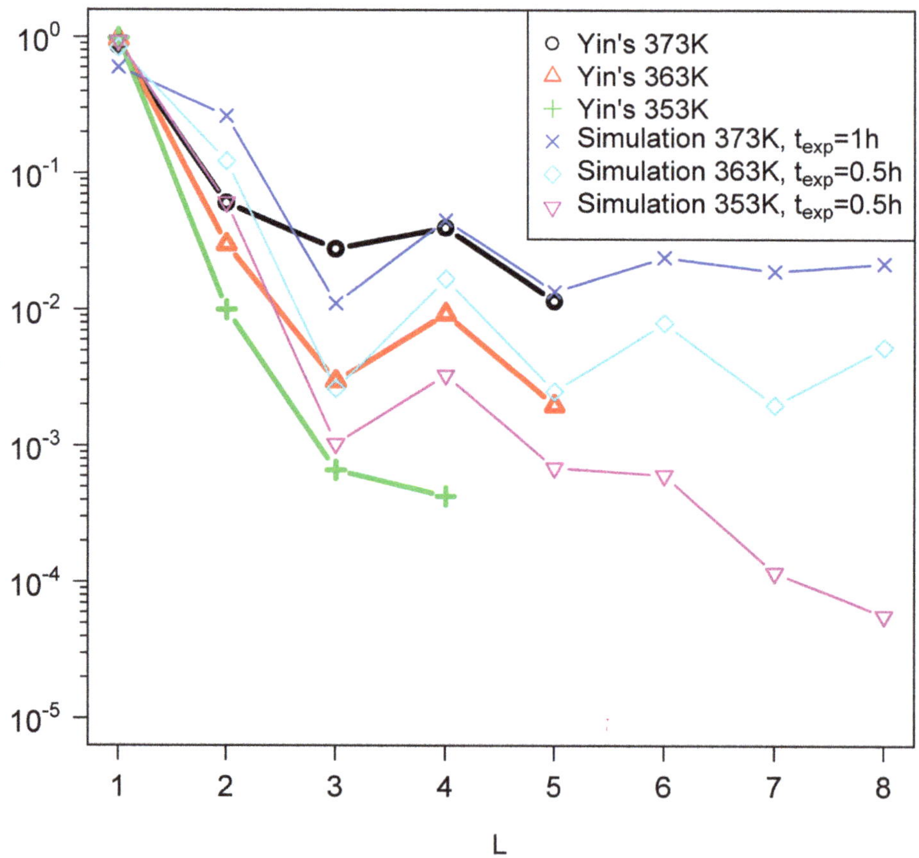

Figure 3. Comparison of the length distributions reported from reference [48] on drying experiments with analine–glycine mixtures with results from simulated quenches. The value of σ was assumed to be the same for all the experiments but was fitted. Parameters were $\Delta_a = 99.7$ kcal/mol, $f_a = 5.96 \times 10^{-7}\,\mathrm{s}^{-1}, \sigma = 0.13\Delta_a)$.

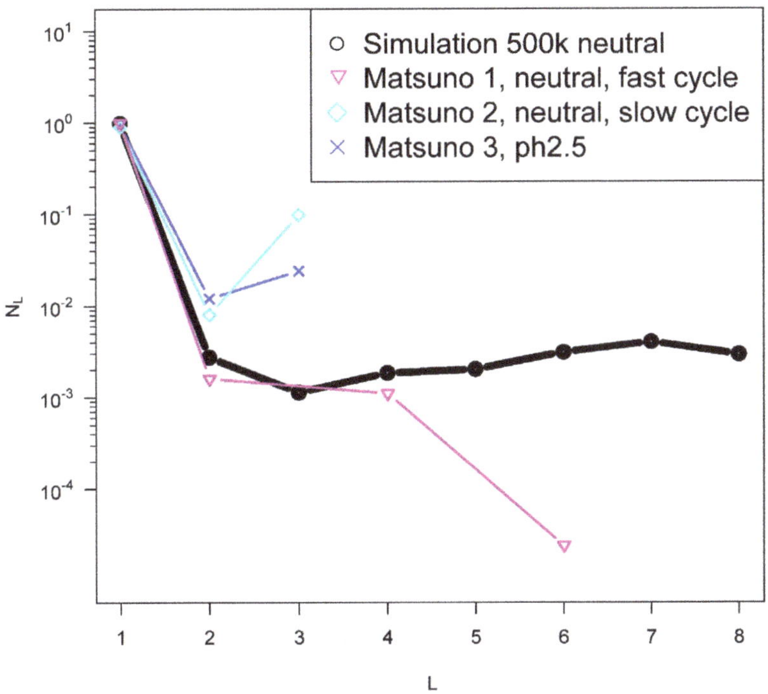

Figure 4. Similar to Figure 3 for data reported from quenches from hot to cold water in [44]. The black simulation was obtained with $\Delta_a = 99.7$ kcal/mol, $\sigma = 0.13\Delta_a$ and $p_{eff} = 0.5$. We can infer from (10) that the high-temperature phase in these experiments was at temperature close to 373 K.

3. Discussion and Conclusions

As reviewed in [21], the idea that life may have originated on earth in the emission of hot fluids from ocean ridges is not at all new. Here, I have reviewed aspects of our work that explore the issues associated with that hypothesis from a somewhat novel perspective, which we believe contributes some additional support to the idea. In particular, we are not aware of other work that has used the polymer length distributions in existing organisms to infer the temperature during the time when the initial synthesis of biopolymers took place, as reviewed briefly here and in detail in [32]. Our models have extended the work of Kauffman to permit an interpretation of the network parameter p in terms of measurable quantities including the time the fluid spends at high temperature and the distribution of activation energies for bond formation. We find evidence in the models of a sharp nonequilibrium transition from long polymers and a dense reaction network to a sparse reaction network populated by a nonequilibrium population of long polymers in which the long polymers retain a kind of memory of their thermal history at low temperatures after a quench.

Some of these results might be further tested experimentally without extreme difficulty. For example, our analysis of the experiments of reference and [44–46,48] suggests that the high temperature in those quenches was very close to the transition point T_c and possibly below it. The model suggests that quenching from a higher temperature would result in many more long polypeptides as possibly consistent with the fragmentary results from ocean trench measurements. The practical problem for a laboratory experiment is that to go to higher initial temperatures requires working at high pressures, so that the aqueous fluid does not boil. That brings more trouble and expense but is certainly doable in principle.

An alert reader will note that we have suggested two origins for a kind of critical temperature around the boiling point of water in this work. In this paper (11) and in [36], we have introduced a T_c depending on kinetic factors including the distribution of activation energies for bond dissolution and the experimental time which the fluid spends at high temperature. In references [32,35] on the basis of the equilibrium distributions of polymer lengths at high and low temperatures, we suggested that a temperature near the boiling point of water at atmospheric pressure might arise because the equilibrium distribution of lengths would be nearly uniform at that temperature, leading to a wide distribution of lengths required by lifelike processes after quench. The latter temperature was estimated to be $-\Delta/\ln b$, where b is the number of types of available monomers and Δ is the equilibrium bond strength as defined above. It appears to be a coincidence that the two temperatures are close to one onother when $b = 20$. However, it will not be hard to design experiments in which they are different, as indeed they are for example when $b < 20$, and thereby to clarify the role of each.

I have emphasized protein synthesis leading to reproducing prion-like entities as a possible origin of life scenario because of its simplicity, although our models could be used with a mere change in parameters to study networks of nucleic acids. However, the parameters needed for such a simulation do not really favor a nucleic acid origin within our framework. In particular, the hydrolysis of nucleotide bonds is much more rapid in a prebiotic environment devoid of supporting proteins [50]. It is sometimes argued that the prion-like model cannot be right because there is no information-carrying molecule. However, I think that is to misunderstand the nature of information, which is essentially the negative of entropy up to an additive constant. The prion protein PrP^{Sc}, for example , is quite obviously carrying information in its secondary structure as evidenced by the fact that other varieties of the PrP protein with the same amino acid sequence but different secondary structures behave entirely differently and do not form reproducing prions. (Some work [51] has reported the use of data on secondary structures in modern biomolecules to construct a tree giving the history of terrestrial biological life. The results are interpreted as suggesting that proteins preceded nuclei acids in the history by several hundred million years).

If life originated in quenches of hot fluids from ocean rifts, then a much justified concern is that the fluids emitted would be rapidly dispersed from their origin by hydrodynamic processes, which would dilute them so much that any kind of lifelike process might be impossible. Examples of suggestions for trapping mechanisms which might prevent this are found, for example, in [52,53]. However, it appears that one might look for signs of prebiotic chemistry in contexts in which the bath of water into which the fluids were emitted was less turbulent and of smaller volume than around most ocean ridges. Existing ocean ridges approximating such conditions have been studied for example in the Bay of California [54] where the rift in the Guaymas Basin, unlike the Pacific ridges, is surrounded by detritus of biological origin on the ocean floor that is so deep that drilling projects have not yet been able to penetrate it to the bedrock beneath. The situation in the Red Sea [55] is reported to be similar. The age of the ocean floor around such slowly spreading rifts is believed to be too young to be the site of prebiotic chemistry. However, recent work [56] reports the discovery of more ancient material, up to nearly 3 billion years old, in and around such rifts. Whether its origin is from ancient continents or from recycling through the mantle after earlier subduction appears to be currently unknown. In any case, an ocean rift origin model suggests that the geological remanents of prebiotic evolution will be found in material near ancient slowly spreading rifts in ocean floors of several billion years ago, if such material can be identified.

Funding: Work was supported in part by NASA grant NNX14AQ05G and by the Minnesota Supercomputing Institute and by the Open Science Grid.

Data Availability Statement: Data associated with the publications in references [20,30–32,35,36] has been or will shortly be publihed in the Data Repository of the University of Minnesota (DRUM) accessible at https://conservancy.umn.edu/handle/11299/166578.

Acknowledgments: This paper reviews and interprets work by people associated with the Halley group over a decade or more. They include Ivan Federov, Aaron Wynveen, Simon Schneider, Noah Rogers, Ben Frederick Intoy, and Qianyi Sheng. I particularly thank B, Intoy and Q. Sheng for help with this manuscript preparation. Patrick Schelling of the University of Central Florida read a preliminary draft of this paper and made many helpful suggestions. I thank John Yin for fruitful discussions and David Baum for discussions and for introducing me to people working on the origin of life problem at the University of Wisconsin. William Seyfried, Dionysis Foustoukos and John Yin participated in proposals to carry out quenches from high temperatures as discussed here. William Seyfried helpfully provided guidance to the oceanography literature. The constructive criticisms of many anonymous referees to our papers in Physical Review E as well as reviewers of the first submission of this manuscript have contributed substantially to improvement and development of the work.

Conflicts of Interest: The author declares no conflict of interest.

References

1. Eigen, M. Selforganization of Matter and the Evolution of Biological Macromolecules. *Naturwissenschaften* **1971**, *58*, 465–523. [CrossRef] [PubMed]
2. Hart, M. *Atmosphereic Evolution, the Drake Equation and DNA: Sparse Life in an Infinite Universe, Chapter 22 in Extraterrestrials, Where Are They?* 2nd ed.; Zuckerman, B., Hart, M.H., Eds.; Cambridge University Press: Cambridge, UK, 1995.
3. Maury, C.P.J. Self-Propagating β-Sheet Polypeptide Structures as Prebiotic Informational Molecular Entities: The Amyloid World. *Orig. Life Evol. Biosph.* **2009**, *39*, 141–150. [CrossRef] [PubMed]
4. Riesner, D. Biochemistry and structure of PrP^C and PrP^{Sc}. *Br. Med. Bull.* **2003**, *66*, 21–33. [CrossRef] [PubMed]
5. Baskakov, I.V. Autocatalytic Conversion of Recombinant Prion Proteins Displays a Species Barrier. *J. Biol. Chem.* **2004**, *279*, 7671–7677. [CrossRef]
6. Portillo, A.; Hashemi, M.; Zhang, Y.; Breydo, L.; Uversky, V.N.; Lyubchenko, Y.L. Role of monomer arrangement in the amyloid self-assembly *Biochim. Biophys. Acta* **2015**, *1854*, 218–228.
7. Hegde, R.S.; Mastrianni, J.A.; Scott, M.R.; DeFea, K.A.; Tremblay, P.; Torchia, M.; DeArmond, S.J.; Prusiner, S.B.; Lingappa, V.R. A Transmembrane Form of the Prion Protein in Neurodegenerative Disease. *Science* **1998**, *279*, 827–834. [CrossRef]
8. Cronin, J.R.; Pizzarello, S. Amino Acids in Meteorites. *Space Res.* **1983**, *3*, 5–18. [CrossRef]
9. Flissi, A.; Ricart, E.; Campart, C.; Chevalier, M.; Dufresne, Y.; Michalik, J.; Jacques, P.; Flahaut, C.; Lisacek, F.; Leclère, V.; et al. Norine: Update of the nonribosomal peptide resource. *Nucleic Acids Res.* **2020**, *48*, D465–D469. [CrossRef]
10. German, C.; Seyfriedm, W.E., Jr. Hydrothermal Processes. In *Treatise on Geochemistry*, 2nd ed.; Holland, H.D., Turekian, K.K., Eds.; Oxford Elsevier: Oxford, UK, 2014; pp. 191–233.
11. Fuchida, S.; Mizuno, Y.; Masuda, H.; Toki, T.; Makita, H. Concentrations and distributions of amino acids in black and white smoker fluids at temperatures over 200 °C. *Org. Geochem.* **2014**, *66*, 98–106. [CrossRef]
12. Iadanza, M.G.; Jackson, M.P.; Hewitt, E.W.; Ranson, N.A.; Radford, S.E. A new era for understanding amyloid structures and disease. *Nat. Rev. Mol. Cell Biol.* **2018**, *19*, 755–773. [CrossRef]
13. Halley, J.W., *How Likely Is Extraterrestrial Life*; Springer Briefs in Astronomy series; Hillebrandt, W., Ratcliffe, M., Eds.; Springer: Berlin/Heidelberg, Germany; Dordrecht, The Netherlands; London, UK; New York, NY, USA, 2012.
14. Popa, R. *Between Necessity and Probability: Searching for the Definition and Origin of Life*; Springer: Berlin, Germany, 2004.
15. Pross, A. Toward a general theory of evolution: Extending Darwinian theory to inanimate matter. *J. Syst. Chem.* **2011**, *2*, 1. [CrossRef]
16. Dyson, F. *Origins of Life*; Cambridge University Press: Cambridge, UK, 1999.
17. Cairns-Smith, A.G. The origin of life and the nature of the primitive gene. *J. Theor. Biol.* **1966**, *10*, 53–88. [CrossRef]
18. Zaia, D.A.; Zaia, C.T.B. A few experimental suggestions using minerals to obtain peptides with a high concentration of L-amino acids and protein amino acids. *Symmetry* **2020**, *12*, 2046. [CrossRef]
19. Kitadai, N.; Nishiuchi, K.; Takahagi, W. Thermodynamic Impact of Mineral Surfaces on Amino Acid Polymerization: Aspartate Dimerization on Two-Line Ferrihydrite, Anatase, and γ-Alumina. *Minerals* **2021**, *11*, 234. [CrossRef]
20. Intoy, B.F.; Wynveen, A.; Halley, J.W. Effects of spatial diffusion on nonequilibrium steady states in a model for prebiotic evolution. *Phys. Rev. E* **2016**, *94*, 042424. [CrossRef]
21. Matsuno, K. Hydrothermal Vent Origin of Life Models. In *Encyclopedia of Astrobiology*; Gargaud, M., Irvine, W., Amils, R., Cleaves, H.J., II, Pinti, D.L., Quintanilla, J.C., Rouan, D., Spohn, T., Tirard, S., Viso, M., Eds.; Springer: Berlin/Heidelberg, Germany, 2020; pp. 1162–1166. [CrossRef]
22. Nejdl, L.; Petera, L.; Šponer, J.; Zemánková, K.; Pavelicová, K.; Knížek, A.; Adam, V.; Vaculovičová, M.; Ivanek, O. Ferus, M. Quantum Dots in Peroxidase-like Chemistry and Formamide-Based Hot Spring Synthesis of Nucleobases. *Astrobiology* **2022**, *5*, 541–551. [CrossRef] [PubMed]
23. Taran, O.; Chen, C.; Omosun, T.O.; Hsieh, M.C.; Rha, A.; Goodwin, J.T.; Mehta, A.K.; Grover, M.A.; Lynn, D.G. Expanding the informational chemistries of life: Peptide/RNA networks. *Phys. Trans. R. Soc. A* **2017**, *375* 20160356. [CrossRef] [PubMed]

24. Müller-Plathe, F. Coarse-graining in polymer simulation: From the atomistic to the mesoscopic scale and back. *Chem. Phys. Chem.* **2002**, *3*, 754–769. [CrossRef]
25. Anderson, P.W. More is Different. *Science* **1972**, *177*, 393. [CrossRef]
26. Kivelson, S.; Kivelson, S. Defining emergence in physics. *Quantum Mater.* **2016** *1*, 16024. [CrossRef]
27. Kauffman, S. A. *The Origins of Order*; Oxford University Press: New York, NY, USA, 1993; Chapter 7.
28. Farmer, J.D.; Kauffman, S.A.; Packard, N.H. Autocatalytic eplication of polymers. *Physica* **1986**, *220*, 50–67.
29. Bagley, R.; Farmer, J.D. Spontaneous Emergence of a Metabolism. In *Artificial Life II*; Langton G.D., Taylor C., Farmer J.D., Rasmussen, S., Eds.; Addison Wesley: Redwood City, CA, USA, 1991; p. 93.
30. Wynveen, A.; Fedorov, I.; Halley, J.W. Nonequilibrium steady states in a model for prebiotic evolution. *Phys. Rev. E* **2014**, *89*, 022725. [CrossRef] [PubMed]
31. Intoy, B.F.; Halley, J.W. Energetics in a model of prebiotic evolution. *Phys. Rev. E* **2017**, *96*, 062402. [CrossRef] [PubMed]
32. Intoy, B.F.; Halley, J.W. Some generic measures of the extent of chemical disequilibrium applied to living and abiotic systems. *Phys. Rev. E* **2019**, *99*, 062419. [CrossRef] [PubMed]
33. Baum, D. The origin and early evolution of life in chemical complexity space. *J. Theor. Biol.* **2018** , *456*, 295–304. [CrossRef]
34. Marín-Moreno, A.; Fernández-Borges, N.; Espinosa, J.C.; Andréoletti, O.; Torres, J.M. Transmission and Replication of Prions, *Prog. Mol. Biol. Transl. Sci.* **2017**, *150*, 181–201.
35. Sheng, Q.; Intoy, B.F.; Halley, J.W. Quenching to fix metastable states in models of Prebiotic Chemistry. *Phys. Rev. E* **2020**, *102*, 062412. [CrossRef]
36. Sheng, Q.; Intoy, B.F.; Halley, J. W. Effects of activation barriers on quenching to stabilize prebiotic chemical systems. *Phys. Rev. E* 2022, *in press*.
37. Lee, N.; Foustoukos, D.I.; Sverjensky, D.A.; Hazen, R.M.; Cody, G.D Hydrogen enhances the stability of glutamic acid in hydrothermal environments. *Chem. Geol.* **2014**, *386*, 184–189. [CrossRef]
38. Radzicka, A.; Wolfenden, R. Rates of uncatalyzed peptide bond hydrolysis in neutral solution and the transition state affinities of proteases. *J. Am. Chem. Soc.* **1996**, *118*, 6105–6109. [CrossRef]
39. Kanehisa, M.; Goto, S. KEGG: Kyoto Encyclope- dia of Genes and Genomes. *Nucleic Acids Res.* **2000**, *28*, 27. [CrossRef] [PubMed]
40. Tolman, R.C. *The Principles of Statistical Mechanics*; Oxford University Press: Oxford, UK, 1938; pp. 58–59.
41. Kohn, J.E.; Millett, I.S.; Jacob, J.; Zagrovic, B.; Dillon, T.M.; Cingel, N.; Dothager, R.S.; Seifert, S.; Thiyagarajan, P.; Sosnick, T.R.; Hasan, M.Z. Random-coil Behavior and the Dimensions of Chemically Unfolded Proteins. *Proc. Natl. Acad. Sci. USA* **2004**, *101*, 12491–12496. Erratum in *Proc. Natl. Acad. Sci. USA* **2005**, *102*, 14475. [CrossRef] [PubMed]
42. Gautschi, W. Error Function and Fresnel Integrals. In *Handbook of Mathematical Functions*; Abramowitz, M., Stegun, I., Eds.; Dover Publications: New York, NY, USA, 1965; pp. 295–296.
43. Liu, L.; Zhai, S. Basic mathematical model for the normal black smoker system and the hydrothermal megaplume formation. *Acta Oceanol. Sin.* **2007**, *26*, 30–40.
44. Imai, E.I.; Honda, H.; Hatori, K.; Brack, A.; Matsuno, K. Elongation of Oligopeptides in a Simulated Submarine Hydrothermal System. *Science* **1999**, *283*, 831. [CrossRef] [PubMed]
45. Ogata, Y.; Imai, E.I.; Honda, H.; Hatori, K.; Matsuno, K., Hydrothermla Circulation of Seawater through hot vents and contribution of Interface Chemistry to Prebiotic Synthesis. *Orig. Life Evol. Biosph.* **2000**, *30*, 527–537. [CrossRef]
46. Tsukahara, H. Prebiotic oligomerization on or inside lipid vesicles in hydrothermal environments. *Orig. Life Evol. Biosph.* **2002**, *32*, 13–21. [CrossRef]
47. Sibilska, I.K.; Chen, B.; Li, L.; Yin, J. Effects of trimetaphosphate on abiotic formation and hydrolysis of peptides. *Life* **2017**, *7*, 50. [CrossRef]
48. Sibilska, I.; Feng, Y.; Li, L.; Yin, J. Trimetaphosphate Activates Prebiotic Peptide Synthesis across a Wide Range of Temperature and pH. *Orig. Life Evol. Biosph.* **2018**, *48*, 277–287. [CrossRef]
49. Sibilska-Kaminski, I.K.; Yin, J. Toward Molecular Cooperation by De Novo Peptides. *Orig. Life Evol. Biosph.* **2021**, *51*, 71–82. doi: 10.1007/s11084-021-09603-6. [CrossRef]
50. Larralde, R.; Robertson, M.P.; Miller, S.L. Rates of decomposition of ribose and other sugars: Implications for chemical evolution *Proc. Natl. Acad. Sci. USA* **1995**, *92*, 8158–8160. [CrossRef]
51. Caetano-Anollés, G.; Caetano-Anollés, D. Computing the origin and evolution of the ribosome from its structure—Uncovering processes of macromolecular accretion benefiting synthetic biology. *Comput. Struct. Biotechnol. J.* **2015**, *13*, 427–447. [CrossRef] [PubMed]
52. Baaske, P.; Weinert, F.M.; Duhr, S.; Lemke, K.H.; Russell, M.J.; Braun, D. Extreme accumulation of nucleotides in simulated hydrothermal pore systems. *Proc. Natl. Acad. Sci. USA* **2007**, *104*, 9346–9351. [CrossRef] [PubMed]
53. Navrotsky, A.; Hervig, R.; Lyons, J.; Seo, D.-K.; Shock, E. ; Voskanyan,A. Cooperative formation of porous silica and peptides on the prebiotic Earth. *Proc. Natl. Acad. Sci. USA* **2021**, *118*, e2021117118. [CrossRef]
54. Damm, V. Chemistry of submarine hydrothermal solutions at Guaymas Basin, Gulf of California. *Gechemica Cosmochim. AGU* **1985**, *49*, 2221–2237. [CrossRef]

55. Potter, R.J.; Barnes, H.L. *The Kuroko and Related Volcanogenic Massive Sulfide Deposits*; Ohmoto, H., Skinner, B.J., Eds.; Society of Economic Geologists: Littleton, CO, USA, 1983; pp. 198–223.
56. Liu, C.Z.; Dick, H.J.; Mitchell, R.N.; Wei, W.; Zhang, Z.Y.; Hofmann, A.W.; Yang, J.F.; Li, Y. Archean cratonic mantle recycled at a mid-ocean ridge. *Sci. Adv.* **2022**, *8*, eabn6749. [CrossRef]

MDPI
St. Alban-Anlage 66
4052 Basel
Switzerland
www.mdpi.com

Life Editorial Office
E-mail: life@mdpi.com
www.mdpi.com/journal/life

www.ingramcontent.com/pod-product-compliance
Lightning Source LLC
Chambersburg PA
CBHW061830210926
51850CB00006B/847